Grundlehren der mathematischen Wissenschaften 316

A Series of Comprehensive Studies in Mathematics

Springer
Berlin
Heidelberg
New York
Barcelona
Budapest
Hong Kong
London
Milan
Paris
Santa Clara
Singapore
Tokyo

Edward B. Saff · Vilmos Totik

Logarithmic Potentials with External Fields

With 18 Figures

Springer

Edward B. Saff

University of South Florida
Institute for Constructive
Mathematics
Department of Mathematics
4202 East Fowler Avenue, PHY 114
Tampa, FL 33620-5700, USA

esaff@math.usf.edu

Vilmos Totik

Jozsef Attila University
Bolyai Institute
Aradi v. tere 1
Szeged, 6720 Hungary

totik@inf.u-szeged.hu

and

University of South Florida
Department of Mathematics
4202 East Fowler Avenue, PHY 114
Tampa, FL 33620-5700, USA

totik@math.usf.edu

Library of Congress Cataloging-in-Publication Data

Saff, E. B., 1944-
Logarithmic potentials with external fields / Edward B. Saff, Vilmos Totik.
p. cm. - (Grundlehren der mathematischen Wissenschaften,
ISSN 0072-7830; 316)
Includes bibliographical references and index.
ISBN 3-540-57078-0 (alk. paper)
1. Potential theory (Mathematics) I. Totik, V. II. Title. III. Series.
QA404.7.S24 1997 515'.9-dc21 97-28048 CIP

Mathematics Subject Classification (1991):
31A15, 30C85, 41A17, 42A50, 33C25

ISSN 0072-7830

ISBN 3-540-57078-0 Springer-Verlag Berlin Heidelberg New York

Printed in Germany

Cover design: MetaDesign plus GmbH, Berlin
Typesetting: Authors' input files edited and reformatted by Kurt Mattes, Heidelberg, using a Springer TEX macro-package
SPIN: 10124084 41/3143-5 4 3 2 1 0 Printed on acid-free paper

To Loretta and Veronika

Preface

In recent years approximation theory and the theory of orthogonal polynomials have witnessed a dramatic increase in the number of solutions of difficult and previously untouchable problems. This is due to the interaction of approximation theoretical techniques with classical potential theory (more precisely, the theory of logarithmic potentials, which is directly related to polynomials and to problems in the plane or on the real line). Most of the applications are based on an extension of classical logarithmic potential theory to the case when there is a weight (external field) present. The list of recent developments is quite impressive and includes: creation of the theory of non-classical orthogonal polynomials with respect to exponential weights; the theory of orthogonal polynomials with respect to general measures with compact support; the theory of incomplete polynomials and their widespread generalizations, and the theory of multipoint Padé approximation. The new approach has produced long sought solutions for many problems; most notably, the Freud problems on the asymptotics of orthogonal polynomials with respect to weights of the form $\exp(-|x|^\alpha)$; the "1/9-th" conjecture on rational approximation of $\exp(x)$; and the problem of the exact asymptotic constant in the rational approximation of $|x|$.

One aim of the present book is to provide a self-contained introduction to the aforementioned "weighted" potential theory as well as to its numerous applications. As a side-product we shall also fully develop the classical theory of logarithmic potentials.

Perhaps the easiest way to describe the main aspects of this work is to use the electrostatic interpretation of the underlying basic extremal problem. We assure the mathematically oriented reader that in what follows we do not use any deep concepts from physics, and do not appeal to anything in our "physical" interpretation that is not intuitively absorbable.

The fundamental electrostatics problem concerns the equilibrium distribution of a unit charge on a conductor. If the conductor is regarded as a compact set E in the complex plane $\mathbf{C}$ and charges repel each other according to an inverse distance law, then in the absence of an external field, equilibrium will be reached when the total energy

$$I(\mu) = \int\int \log \frac{1}{|z-t|} d\mu(z) d\mu(t)$$

is minimal among all possible charge distributions (measures) μ on E having total charge one. There is a unique distribution μ_E supported on E for which this minimal energy is attained; this *equilibrium distribution* μ_E is actually supported on the outer boundary of E, and its logarithmic potential

$$U^{\mu_E}(z) = \int \log \frac{1}{|z-t|} d\mu_E(t)$$

is essentially constant on E. The facts that the support set of μ_E is known and that the potential $U^{\mu_E}(z)$ is essentially constant on this set enable the use of Stieltjes-type inversion formulas to readily determine μ_E.

The distribution μ_E arises in a variety of problems encountered in constructive analysis. For example, it describes the limiting behavior (as $n \to \infty$) of n points on E, the product of whose mutual distances is maximal. These so-called *Fekete points* provide nearly optimal choices for points of polynomial interpolation. In the study of orthogonal polynomials with respect to a large class of (regular) measures on a compact set $E \subset \mathbf{R}$, the equilibrium measure μ_E gives the limiting distribution of the zeros.

The introduction of an external field $Q(z)$ in the electrostatics problem creates some significant differences in the fundamental theory, but opens much wider doors to applications. The problem now becomes that of minimizing the *weighted energy*

$$I_w(\mu) = \int\int \log \frac{1}{|z-t|w(z)w(t)} d\mu(z)d\mu(t) = I(\mu) + 2\int Q d\mu,$$

where the weight $w = e^{-Q}$, and the minimum is again taken over all unit charges μ supported on E.

The external field problem has its origins in the work of C. F. Gauss, and is sometimes referred to as the *Gauss variation problem*. O. Frostman investigated the problem and the Polish school headed by F. Leja made important contributions during the period 1935–1960 that have greatly influenced the present work. A rebirth of interest in the Gauss variational problem occurred in the 1980's when E. A. Rakhmanov and, independently, collaborators H. N. Mhaskar and E. B. Saff used potentials with external fields to study orthogonal polynomials with respect to exponential weights on the real line.

The external field problem is often treated in the literature as an addendum to the classical theory—a generalization for which the similarities with the unweighted case ($Q \equiv 0$) are the main emphasis. On the other hand, this energy problem can be viewed as a special case of the potential theory developed for energy integrals having symmetric, lower-semicontinuous kernels in locally compact spaces. But in this generality many of the unique features of the external field problem, as well as its concrete applications to constructive analysis, remain hidden.

Our goal in writing this book has been to present a self-contained and fairly comprehensive treatment of the Gauss variation problem in the plane, beginning with a review in Chapter 0 on harmonic functions. This is followed by a detailed treatment of Frostman type for the existence and uniqueness of the *extremal measure* μ_w satisfying

$$I_w(\mu_w) = \min\{I_w(\mu) | \operatorname{supp}(\mu) \subset E, \quad ||\mu|| = 1\}.$$

Our analysis applies even for unbounded closed sets E, under suitable assumptions on the weight w (or, equivalently, on the external field Q). In this early stage of the development we encounter one of the most glaring differences with the classical (unweighted) electrostatics problem; namely, the support S_w of the extremal measure μ_w need not coincide with the outer boundary of E and, in fact, can be quite an arbitrary subset of E (depending on w), possibly with positive area. Determining the support set S_w and its properties are two of the main themes of this work that distinguish it from standard treatments in the literature.

There are several important aspects of the external field problem (and its extension to signed measures) that justify its special attention. The most striking is that it provides a unified approach to several (seemingly different) problems in constructive analysis. These include, among others, the following:

(a) The asymptotic analysis of polynomials orthogonal with respect to a weight function on an *unbounded* interval (e.g., exponential weights of the form $\exp(-|x|^\alpha)$, $\alpha > 0$, on $\mathbf{R}$).
(b) The asymptotic behavior (as $n \to \infty$) of *weighted Fekete points* that maximize the product

$$\prod_{1 \le i < j \le n} |z_i - z_j| w(z_i) w(z_j)$$

among all n-tuples of points $(z_1, \ldots, z_n)$ lying in a closed set E.
(c) The existence and construction of *fast decreasing polynomials*; that is, polynomials $p_n(x)$ of degree n that satisfy for a prescribed nonnegative function $\varphi(x)$ on $[-1, 1]$ the restrictive growth estimates

$$p_n(0) = 1, \quad |p_n(x)| \le \exp(-n\varphi(x)) \text{ for } x \in [-1, 1].$$

(d) The study of *incomplete polynomials* of the form $\Sigma_{k=s}^{n} a_k x^k$ with $s \ge \theta n$ $(\theta > 0)$.
(e) The *numerical conformal mapping* of simply and doubly connected domains onto a disk and annulus, respectively.
(f) A generalization of the *Weierstrass approximation theorem* wherein, for a given weight function w on a closed set E, one seeks to characterize those continuous functions f on E that are uniform limits of weighted polynomials of the form $w^n p_n$, where the power n of the weight is the same as the degree of the polynomial p_n.
(g) The asymptotic behavior of "ray sequences" of *Padé approximants* (interpolating rational functions) to Markov and Stieltjes functions.
(h) The determination of *rates of convergence* of best approximating rational functions to certain classes of functions f (for example, $f(x) = e^{-x}$ on $[0, +\infty)$).

(i) The mathematical modelling of elasticity problems where the shape of the elastic medium is distorted by the insertion of an object under pressure.

In addition, the external field problem provides a rather natural setting for several important concepts in potential theory itself. These include:

(a) Solving simultaneous *Dirichlet's problems*, which arises from the fact that the equilibrium potential $U^{\mu_w}(z)$, with $w = e^{-Q}$, solves this problem (up to a constant) for boundary data $-Q(z)$ on each bounded component of the complement of the support set S_w.
(b) The *balayage* (sweeping) of a measure ν to a compact set E, which is simply given by the extremal measure for the external field $Q(z) = -U^{\nu}(z)$ on E.
(c) The problem of finding the *best Green potential approximation* to a given superharmonic function with respect to an energy norm, which is given by the solution to a Gauss variational problem.
(d) Solving *constrained minimal energy problems* for which one seeks a unit measure λ that minimizes the (unweighted) energy integral for unit measures on E subject to the constraint $\lambda \leq \sigma$, where σ is a given positive measure with $\operatorname{supp}(\sigma) = E$ and $||\sigma|| > 1$.

In developing the theory for potentials in the presence of an external field (Chapters I and II), we provide motivations and detailed proofs for many of the basic results from potential theory, such as generalized maximum principles, the Riesz decomposition theorem, the principle of domination, Evans' theorem, etc. These results are presented as they are needed and, as an aid for the reader, we provide a listing of them in the Appendix along with their locations in the text. Wiener's theorem and the Dirichlet problem are also treated in the Appendix.

At the end of each of the main chapters we have included a section entitled "Notes and Historical References," that includes discussion of related results along with citations for many of the theorems presented in the text. There are, however, many new results and proofs that appear here for the first time, such are the ones that are not referenced in the Notes sections.

While our analysis of the weighted energy problem proceeds along the lines of classical potential theory, alternative approaches are being developed, most notably by L. A. Pastur and his collaborators who use random matrix techniques (see Section IV.9). Furthermore, inverse spectral methods have recently been employed by P. Deift, T. Kriecherbauer and K. T-R. McLaughlin to derive more detailed information about the equilibrium distributions for certain smooth fields Q (see the Notes section for Chapter IV).

The theory of weighted potentials in $\mathbf{C}^N$, $N \geq 2$, is still in its infancy relative to the single variable case. To introduce the reader to this vital subject we have included an appendix written by Thomas Bloom that contains generalizations of several theorems in the text to the multidimensional case. This presentation emphasizes the role that the Monge-Ampère operator plays in extending the external field problem to the pluripotential setting.

We are indebted to a large and distinguished cast of students and colleagues who have provided us with valuable feedback on this project. To T. Bloom, A. B. J. Kuijlaars, N. Levenberg, A. L. Levin, D. S. Lubinsky, H. Mhaskar, V. Prokhorov, and H. Stahl we extend our sincere appreciation for their input and encouragement. For their careful reading of the manuscript we especially wish to acknowledge S. Damelin, P. Dragnev, I. Ivanov, P. Simeonov, and Y. Zhou.

Our research activities pertinent to the writing of this book have spanned several years during which the U.S. National Science Foundation and the Hungarian National Science Foundation for Research have provided essential support for which we are truly appreciative.

An important phase of this project was conducted at the Mathematics Research Institute in Oberwolfach, Germany, under the auspices of the Research in Pairs program. To Director Matthias Kreck and the Oberwolfach staff, we are sincerely grateful for their hospitality and for providing us with such a stimulating and enjoyable environment. Finally, we want to thank the staff at Springer-Verlag, and particularly Catriona Byrne, for their dedicated assistance in bringing this work to fruition.

E. B. Saff · V. Totik

Table of Contents

Chapter 0. Preliminaries

An understanding of harmonic and subharmonic functions in the complex plane provides the foundation for the study of logarithmic potentials. In this chapter we review the definitions and basic properties of such functions. In so doing we assume that the reader is acquainted with the fundamentals of analytic function theory.

0.1 Weak* Topology and Lower Semi-continuity

In this book we shall work either on the complex plane $\mathbf{C}$ or on its completion $\overline{\mathbf{C}} := \mathbf{C} \cup \{\infty\}$, which is called the Riemann sphere. We can identify $\overline{\mathbf{C}}$ with a real sphere by the usual stereographic projection, and thereby the Euclidean metric on the sphere induces the topology on $\overline{\mathbf{C}}$ (recall that for this topology a basis of neighborhoods for the point infinity is the set of exteriors of closed disks centered at the origin).

Let $K \subset \overline{\mathbf{C}}$ and $f : K \to \mathbf{R} \cup \{+\infty\}$ an extended real-valued function. Then f is said to be *lower semi-continuous* (l.s.c.) if for every $x \in K$ and $\alpha < f(x)$ there is a neighborhood O of x such that for every $y \in O \cap K$ we have $f(y) > \alpha$. If $-f$ is l.s.c., then f is called *upper semi-continuous*. It is clear that the continuity of f is the same as its lower and upper semi-continuity. It is a standard exercise that a l.s.c. function on a compact set attains its minimum, and it is uniformly l.s.c.

Theorem 1.1. *f is lower semi-continuous on the compact set K if and only if it is the pointwise limit of an increasing sequence of continuous functions.*

Proof. It is easy to verify that the limit function of an increasing sequence of continuous functions is l.s.c., so we only consider the necessity part of the theorem.

Assume f is l.s.c. If f is identically $+\infty$ on K, then the statement is obvious. Otherwise, the functions

$$f_n(x) := \inf_{y \in K} \{f(y) + n|x - y|\}$$

are easily seen to be continuous (when the point infinity belongs to K, then $|x - y|$ is understood to be the distance on the Riemann sphere). It is clear that $f_n \leq f_{n+1}$

for every n; furthermore, by setting $y = x$ in the expression after the infimum we see that $f_n(x) \le f(x)$ for all n and x. Thus, it remains to prove that

$$\liminf_{n\to\infty} f_n(x) \ge f(x)$$

for every $x \in K$. This, however, is a consequence of the lower semi-continuity of f. In fact, let $-M$ be a lower bound for f on K. If $\alpha < f(x)$ and δ is such that for $|y - x| \le \delta$, $y \in K$, we have $f(y) > \alpha$, then for $n > (M + \alpha)/\delta$ we get $f_n(x) \ge \alpha$ because $f(y) + n|x - y| > \alpha$ for all $y \in K$. □

We remark that the result is also true for other, perhaps noncompact sets; in particular, for the case when $K = \mathbf{C}$. In fact, repeat the preceding proof with

$$f_n(x) := \inf_{y\in K} \{f(y) + nH(|x - y|)\},$$

where H is a positive continuous function on $[0, \infty)$ with the properties that $H(0) = 0$ and H increases sufficiently fast. Then the infimum defining f_n is attained, and the remainder of the proof is the same.

Concerning the continuity properties of lower semicontinuous functions, the following is true.

Theorem 1.2. *If f is a lower semi-continuous function on a closed set K, then f is continuous on a dense G_δ subset of K (with continuity meant in the topology of $\mathbf{R} \cup \{\infty\}$ at points where f is $+\infty$).*

Recall that a set is called a G_δ set if it is the intersection of countably many open sets.

Proof of Theorem 1.2. In fact, by applying the arctangent function to f we can assume that f is bounded. Now if the oscillation of f at x is defined as

$$o_f(x) := \limsup_{y\to x} f(y) - \liminf_{z\to x} f(z),$$

then, for every $m = 1, 2, \ldots$, the set

$$V_m := \{x \mid o_f(x) \ge \frac{1}{m}\}$$

is a closed and nowhere dense set in K (the density of V_m in any relatively open subset of K together with the lower semi-continuity of f would imply the existence of a sequence of points $x_1, x_2, \ldots$ with $f(x_{n+1}) - f(x_n) \ge 1/2m$, which is impossible by the boundedness of f). But then f is continuous on the complement of $\cup_m V_m$, and by the Baire category theorem [195] this complement is a dense G_δ subset of K. □

We shall frequently use the so-called *weak* topology*. Let K be a compact subset of $\overline{\mathbf{C}}$, and consider the set $C(K)$ of continuous complex (or real)-valued functions on K (if $\mathbf{R}$ is the scalar field, then in what follows, "complex-valued measure" must be replaced by "signed measure"). By the Riesz representation theorem [195, Theorem 2.14] every bounded linear functional L on $C(K)$ has the form

$$Lf = \int f\,d\mu$$

for some complex-valued Borel measure μ. Furthermore, the norm of L coincides with the total mass of the measure $|\mu|$, which is called the *total variation* of μ and is defined as

$$|\mu|(E) := \sup\left\{\sum_{j=1}^{\infty} |\mu(E_j)| \;\middle|\; E = \bigcup_{j=1}^{\infty} E_j,\ \ E_i \cap E_j = \emptyset \text{ if } i \neq j\right\}.$$

In the terminology of functional analysis this amounts to saying that the dual space of $C(K)$ is the space $\mathcal{M}_c(K)$ of all complex-valued Borel measures on the set K. Now the weak* topology on $\mathcal{M}_c(K)$ is the topology describing pointwise convergence of linear functionals. In other words, we say that $\mu_n \to \mu$ in the weak* topology, symbolically $\mu_n \xrightarrow{*} \mu$, if

$$\int f\,d\mu_n \to \int f\,d\mu$$

for every $f \in C(K)$. It is not hard to see that there is such a topology; in fact, it can even be given explicitly by the metric

$$d(\mu, \nu) := \sum_{j=1}^{\infty} \frac{1}{2^j \|f_j\|_K} \left| \int f_j\,d\mu - \int f_j\,d\nu \right|,$$

where $\{f_j\}_{j=1}^{\infty}$ is a countable dense set in $C(K)$ and $\|f\|_K$ denotes the supremum norm (of $|f|$) on K. Since norm convergence implies pointwise convergence, we can deduce that if $|\mu - \mu_n|(K) \to 0$, then $\mu_n \xrightarrow{*} \mu$. Furthermore, it is easy to prove that if all the measures μ_n have support in a compact set $K_1 \subset K$ and $\mu_n \xrightarrow{*} \mu$, then μ is also supported on K_1. Recall that the *support* of a positive measure μ, denoted $\operatorname{supp}(\mu)$, consists of all points z such that $\mu(D_r(z)) > 0$ for every open disk $D_r(z)$ of radius $r > 0$ and with center at z.

We adopt the following convention. In writing $\mu_n \xrightarrow{*} \mu$, we mean convergence with respect to the dual space of $C(K)$, where K is any compact subset of $\overline{\mathbf{C}}$ containing the supports of all of the measures μ_n. We can, of course, always take $K = \overline{\mathbf{C}}$; however, selecting K as the smallest compact set containing the supports of the μ_n serves to remind us that the limit measure μ is likewise supported on such K.

We shall frequently use the following selection theorem.

Theorem 1.3 (Helly's Selection Theorem). *If $\{\mu_n\}$ is a sequence of complex-valued measures on a compact set K with bounded total mass $|\mu_n|(K)$, then we can select from $\{\mu_n\}$ a weak* convergent subsequence.*

The assumption of the theorem is nothing else than the boundedness of the measures in question in the dual of $C(K)$, and it is a general feature of Banach spaces that in the dual space bounded sets are compact in the weak* topology (which is an immediate consequence of the fact that the product of compact spaces is compact). However, we present here an elementary proof of Helly's theorem.

Proof of Theorem 1.3. Let $\{f_j\}_{j=1}^{\infty}$ be a countable dense set in $C(K)$. Since we have assumed that $|\mu_n|(K) \le M$ for some M and all n, it follows that the sequence of integrals

$$S_j = \left\{ \int f_j \, d\mu_n \right\}_{n=1}^{\infty}$$

is bounded for each j. Therefore, we can select a convergent subsequence, and by the standard diagonal argument we can even select a subsequence $\{n_k\}_{k=1}^{\infty}$ of the natural numbers such that each of the corresponding subsequences of the S_j's has a limit, i.e. the limits

$$Lf_j := \lim_{k\to\infty} \int f_j \, d\mu_{n_k}$$

exist for each j. If $f \in C(K)$ is arbitrary, then for every $\varepsilon > 0$ there is a j such that $\|f - f_j\|_K \le \varepsilon$, which implies that independently of k

$$\left| \int f \, d\mu_{n_k} - \int f_j \, d\mu_{n_k} \right| \le M\varepsilon.$$

Therefore

$$\limsup_{k,l\to\infty} \left| \int f \, d\mu_{n_k} - \int f \, d\mu_{n_l} \right| \le 2M\varepsilon,$$

which yields that

$$Lf := \lim_{k\to\infty} \int f \, d\mu_{n_k}$$

exists. It is clear that L is a linear functional on $C(K)$; furthermore since $|Lf| \le M\|f\|_K$, it is bounded. Hence, by the Riesz representation theorem, L can be represented as an integral against a complex-valued measure μ. But then the very definition of weak* convergence gives that $\mu_{n_k} \xrightarrow{*} \mu$ as $k \to \infty$. □

Theorem 1.4. *Let μ_n, $n = 1, 2, \ldots$, be positive measures with support in a compact set $K \subseteq \overline{\mathbf{C}}$, and assume that $\mu_n \xrightarrow{*} \mu$. Then for every lower semi-continuous function Q on K we have*

$$\int Q \, d\mu \le \liminf_{n\to\infty} \int Q \, d\mu_n.$$

In particular, if O is an open subset of $\overline{\mathbf{C}}$, then

$$\mu(O) \le \liminf_{n\to\infty} \mu_n(O). \tag{1.1}$$

By taking negatives we can conclude that if q is upper semi-continuous, then

$$\int q\,d\mu \geq \limsup_{n\to\infty} \int q\,d\mu_n, \tag{1.2}$$

and furthermore, for every closed subset F of the complex plane we have

$$\mu(F) \geq \limsup_{n\to\infty} \mu_n(F). \tag{1.3}$$

Proof of Theorem 1.4. Since Q is lower semi-continuous on K, it is the pointwise limit of an increasing sequence $\{h_m\}$ of continuous functions. Therefore, since $\mu_n \to \mu$ in the weak* topology, we have by the monotone convergence theorem,

$$\begin{aligned} \liminf_{n\to\infty} \int Q\,d\mu_n &\geq \lim_{m\to\infty}\lim_{n\to\infty} \int h_m\,d\mu_n \\ &= \lim_{m\to\infty} \int h_m\,d\mu = \int Q\,d\mu. \end{aligned} \tag{1.4}$$

□

The two relations (1.1) and (1.3) imply that if H is any Borel set with the property $\mu(\overline{H} \setminus \mathrm{Int}(H)) = 0$, then

$$\mu(H) = \lim_{n\to\infty} \mu_n(H). \tag{1.5}$$

There is a wealth of sets H with this property: if $\{H_\tau\}$ is any family of sets with disjoint boundaries $\overline{H_\tau} \setminus \mathrm{Int}(H_\tau)$, then all but at most a countable of the H_τ's satisfy the condition $\mu(\overline{H}_\tau \setminus \mathrm{Int}(H_\tau)) = 0$. Apply this to a family of concentric circles with center at a rational point of the plane, and it immediately follows that the set of those open H's for which $\mu(\overline{H}\setminus H) = 0$, forms a basis for the Euclidean topology, and of course, for every such H we have (1.5) provided $\mu_n \stackrel{*}{\to} \mu$. Now it is an easy task to check that the converse of this fact is also true (use that every continuous function can be uniformly approximated by linear combinations of characteristic functions of members of a basis). Therefore we have the following result.

Theorem 1.5. *Let μ, μ_n, $n = 1, 2, \ldots$, be positive measures with support in a compact set $K \subseteq \overline{\mathbf{C}}$, and let $\mathcal{H}$ be a basis for the topology of $\overline{\mathbf{C}}$ such that for each member H of $\mathcal{H}$ we have $\mu(\overline{H} \setminus H) = 0$. Then $\mu_n \stackrel{*}{\to} \mu$ if and only if*

$$\mu(H) = \lim_{n\to\infty} \mu_n(H) \tag{1.6}$$

for every $H \in \mathcal{H}$.

0.2 Fundamentals of Harmonic Functions

A real-valued function $u(z) = u(x, y)$ defined in a domain D of the complex plane is said to be *harmonic* in D if u and its first and second order partial derivatives are continuous in D and satisfy Laplace's equation:

$$u_{xx}(z) + u_{yy}(z) = 0, \quad z \in D. \tag{2.1}$$

By a *domain* we mean an open connected set. A function u is said to be *harmonic at a point* z_0 if it is harmonic in some neighborhood centered at z_0. Hence u is harmonic in D if and only if it is harmonic at each point of D.

If $f = u + iv$ is analytic in D, then it follows from the Cauchy-Riemann equations

$$u_x = v_y, \quad v_x = -u_y$$

that both $u = \operatorname{Re} f$ and $v = \operatorname{Im} f$ are harmonic in D. In the converse direction we have

Theorem 2.1. *A function u is harmonic in a simply connected domain D if and only if there exists a single-valued function f analytic in D such that* $u(z) = \operatorname{Re} f(z)$ *in D.*

The assumption that D is simply connected (D has no "holes") is essential as the example $D = \{z \mid 0 < |z| < 1\}$, $u(z) = \log |z|$ shows. In the proof of Theorem 2.1 we shall make use of the fact that every $f = u + iv$ with continuous partial derivatives satisfying the Cauchy-Riemann equations in a domain is analytic there. In particular, we require the following lemma.

Lemma 2.2. *If u is harmonic in a domain D, then* $g(z) := u_x(z) - iu_y(z)$ *is analytic in D.*

Proof. The Cauchy-Riemann equations for the given function are

$$u_{xx} = -u_{yy} \quad \text{and} \quad -u_{xy} = -u_{yx},$$

which are clearly satisfied for any harmonic function u. □

Note that Lemma 2.2 is valid for *any* domain D.

Proof of Theorem 2.1. As previously remarked, if $u = \operatorname{Re} f$ and f is analytic in D, then u is harmonic in D.

So assume now that u is harmonic in D, and set

$$g(z) := u_x(z) - iu_y(z).$$

Since D is simply connected and, by Lemma 2.2, g is analytic in D, the function

$$F(z) := \int_{z_0}^{z} g(t)\, dt$$

is single-valued and analytic in D, where $z_0 \in D$ is fixed, and the integration is along any path in D joining z_0 to z. Observe that

$$\begin{aligned} F(z) &= \int_{z_0}^{z} (u_x - iu_y)(dx + idy) \\ &= \int_{z_0}^{z} (u_x dx + u_y dy) + i \int_{z_0}^{z} (u_x dy - u_y dx) \\ &= u(z) - u(z_0) + i \int_{z_0}^{z} (u_x dy - u_y dx). \end{aligned}$$

Thus, $u(z)$ is the real part of the analytic function $f(z) := F(z) + u(z_0)$ in D. □

If u is harmonic in D and $f = u + iv$ is analytic in D, then v is called a *harmonic conjugate* of u and f is an *analytic completion* of u. The functions v and f (if they exist) are unique up to an additive constant.

Remark 2.3. As the proof of Theorem 2.1 shows, if u is harmonic in a simply connected domain D, then

$$v(z) = \int_{z_0}^{z} (-u_y dx + u_x dy)$$

is a harmonic conjugate of u in D.

If f is analytic on a closed disk $\{z \mid |z - a| \le r\}$, then the Cauchy integral formula asserts that

$$f(a) = \frac{1}{2\pi i} \int_{|t-a|=r} \frac{f(t)}{t-a}\, dt = \frac{1}{2\pi} \int_0^{2\pi} f(a + re^{i\theta})\, d\theta. \tag{2.2}$$

In other words, the value of f at the center of a disk is the average of its values over the circumference. The same is true for harmonic functions as can be seen from Theorem 2.1. More precisely, we have

Theorem 2.4 (Mean-Value Property). *If u is harmonic in the open disk $|z-a| < r$ and continuous on its closure, then*

$$u(a) = \frac{1}{2\pi} \int_0^{2\pi} u(a + re^{i\theta})\, d\theta \tag{2.3}$$

and

$$u(a) = \frac{1}{\pi r^2} \iint_{|z-a| \le r} u(x, y)\, dx dy, \quad z = x + iy. \tag{2.4}$$

Proof. If u is harmonic on the closed disk $\{z \mid |z-a| \leq r\}$, then Theorem 2.1 applies in some domain containing this disk and, by taking the real part in (2.2), we get (2.3). For the case when u is merely assumed to be continuous on $\{z \mid |z-a| = r\}$, property (2.3) follows by considering $u(a+\rho(z-a))$ for $\rho < 1$ and letting $\rho \nearrow 1$.

The area mean-value property (2.4) follows immediately from (2.3) on expressing the integral in terms of polar coordinates. □

Remark 2.5. As we shall show in Theorem 4.6, the mean-value property (2.3) actually characterizes harmonic functions in the sense that if u is continuous in D and (2.3) holds for each $a \in D$ and each r sufficiently small ($0 < r < r_a$), then u is harmonic in D. This easily follows from the maximum principle for functions with the mean value property and from the possibility of solving Dirichlet's problem, both of which are to be discussed below.

As an important consequence of the mean-value property we get the following.

Theorem 2.6 (Maximum Principle). *If u is harmonic in a domain D and attains its maximum or minimum value in D, then u is constant in D.*

Proof. It is enough to consider the case when u attains its maximum at some $z_0 \in D$. First we show that the set

$$E := \{z \in D \mid u(z) = u(z_0)\}$$

is open. Indeed, suppose $z_1 \in E$ and choose $r > 0$ so small that D contains the disk $D_r(z_1)$ with center z_1 and radius r. Then $u(z) \leq u(z_1) = u(z_0)$ for all $z \in D_r(z_1)$ and, by property (2.3), for each $\rho < r$ we can write

$$u(z_0) = u(z_1) = \frac{1}{2\pi} \int_0^{2\pi} u(z_1 + \rho e^{i\theta})\, d\theta. \tag{2.5}$$

Suppose that $u(z_1 + \rho_0 e^{i\theta_0}) < u(z_0)$ for some $\rho_0 < r$ and some θ_0. Then, by continuity, there exists an interval about θ_0 on which $u(z_1 + \rho_0 e^{i\theta}) < u(z_0)$. But then

$$\frac{1}{2\pi} \int_0^{2\pi} u(z_1 + \rho_0 e^{i\theta})\, d\theta < u(z_0),$$

which contradicts (2.5). Thus $u(z) = u(z_0)$ for all $z \in D_r(z_1)$; that is, E is open.

On the other hand, the set $F := D \setminus E$ is also open by the continuity of u. Since $D = E \cup F$ is the union of two disjoint open sets and D is connected, one of these sets must be empty. By assumption, $z_0 \in E$ and so $F = \emptyset$; that is, $u(z) = u(z_0)$ for all $z \in D$. □

Corollary 2.7. *If u is harmonic in the interior of a compact set K and continuous on K, then u attains its maximum and minimum values on the boundary of K.*

0.3 Series Representations of Harmonic Functions

Coupled with Theorem 2.1, the familiar Taylor representation $\sum_0^\infty a_n(z-z_0)^n$ for a function analytic in a neighborhood $D_R(z_0)$ with center z_0 and radius R yields the following.

Theorem 3.1. *If $u(z)$ is harmonic in the disk $D_R(z_0)$, then there are unique constants $\{\alpha_n\}_0^\infty$, $\{\beta_n\}_1^\infty$ such that for $z = z_0 + re^{i\theta}$, $0 \le r < R$,*

$$u(z) = \frac{\alpha_0}{2} + \sum_{n=1}^{\infty}(\alpha_n \cos n\theta - \beta_n \sin n\theta)r^n. \tag{3.1}$$

Furthermore, if $v(z)$ is a harmonic conjugate to $u(z)$ in $D_R(z_0)$, then for some constant β_0,

$$v(z) = \frac{\beta_0}{2} + \sum_{n=1}^{\infty}(\beta_n \cos n\theta + \alpha_n \sin n\theta)r^n. \tag{3.2}$$

Both series (3.1) and (3.2) converge uniformly on every closed subdisk $|z - z_0| \le \rho < R$.

The constants α_n, β_n are given by

$$\alpha_n = \frac{1}{\pi\rho^n}\int_0^{2\pi} u(z_0 + \rho e^{i\theta}) \cos n\theta \, d\theta, \quad n = 0, 1, 2, \ldots, \tag{3.3}$$

$$\beta_n = -\frac{1}{\pi\rho^n}\int_0^{2\pi} u(z_0 + \rho e^{i\theta}) \sin n\theta \, d\theta, \quad n = 1, 2, \ldots, \tag{3.4}$$

for any $0 < \rho < R$.

Conversely, if real sequences $\{\alpha_n\}_0^\infty$, $\{\beta_n\}_0^\infty$ are given, then the right-hand sides of (3.1) and (3.2) define conjugate harmonic functions. More precisely, we have

Theorem 3.2. *If*

$$\frac{1}{R} := \limsup_{n\to\infty} |\alpha_n + i\beta_n|^{1/n} < \infty, \tag{3.5}$$

then the series

$$u(z) := \frac{\alpha_0}{2} + \sum_{n=1}^{\infty}(\alpha_n \cos n\theta - \beta_n \sin n\theta)r^n,$$

$$v(z) := \frac{\beta_0}{2} + \sum_{n=1}^{\infty}(\beta_n \cos n\theta + \alpha_n \sin n\theta)r^n,$$

where $z = z_0 + re^{i\theta}$, converge uniformly on compact subsets of $D_R(z_0)$. Moreover, $u(z)$ and $v(z)$ are conjugate harmonic functions in $D_R(z_0)$.

Proof. From (3.5), it follows that the function $f(z) := \sum_0^\infty a_n(z - z_0)^n$ with $a_n := \alpha_n + i\beta_n$ is analytic in the open disk $D_R(z_0)$. We get the stated result by taking real and imaginary parts of this series. □

To derive the series expansion for a function $u(z)$ harmonic in an annulus, say $A := \{z \mid R_0 < |z| < R_1\}$, we proceed as follows. Consider the (simply connected) cut annulus

$$B := A \setminus \{z \mid R_0 < z < R_1\}$$

and let F be an analytic completion of u in B. Next, recall from Lemma 2.2 that $g(z) := u_x(z) - iu_y(z)$ is analytic in A and so, by Laurent's theorem,

$$g(z) = \frac{a_{-1}}{z} + \sum_{n=-\infty,\, n\neq -1}^{\infty} a_n z^n, \quad z \in A, \tag{3.6}$$

for suitable coefficients a_n. Since $F' = g$ in B, it follows that for some constant K

$$F(z) = K + a_{-1}\log z + \sum_{n=-\infty,\, n\neq -1}^{\infty} \frac{a_n}{n+1} z^{n+1} =: K + a_{-1}\log z + h(z) \tag{3.7}$$

for $z \in B$, where $\log z$ denotes a branch of the logarithm that is analytic in B. On taking real parts in (3.7) we find that

$$u(z) = \operatorname{Re} K + (\operatorname{Re} a_{-1})\log|z| - (\operatorname{Im} a_{-1})\arg z + \operatorname{Re} h(z) \tag{3.8}$$

holds in B. But this implies $\operatorname{Im} a_{-1} = 0$ in view of the continuity in A of u, $\log|z|$, and $\operatorname{Re} h(z)$, and the discontinuity of any branch of $\arg z$ in A. Thus from (3.7) and (3.8) we get that for all $z = re^{i\theta} \in A$

$$u(z) = (\operatorname{Re} a_{-1})\log r + \operatorname{Re} K$$
$$+ \sum_{n=-\infty,\, n\neq -1}^{\infty} \left(\operatorname{Re}\frac{a_n}{n+1}\cos(n+1)\theta - \operatorname{Im}\frac{a_n}{n+1}\sin(n+1)\theta\right) r^{n+1}.$$

More generally the following result holds.

Theorem 3.3. *If $u(z)$ is harmonic in the annulus*

$$A := \{z \mid R_0 < |z - z_0| < R_1\},$$

then there are unique constants c, $\{\alpha_n\}_{-\infty}^{\infty}$, and $\{\beta_n\}_{n=-\infty,\, n\neq 0}^{\infty}$ such that for $z = z_0 + re^{i\theta}$, $R_0 < r < R_1$,

$$u(z) = c\log r + \frac{\alpha_0}{2} + \sum_{n=-\infty,\, n\neq 0}^{\infty} (\alpha_n \cos n\theta - \beta_n \sin n\theta) r^n. \tag{3.9}$$

The series (3.9) converges uniformly on any compact subset of A.

From the representation (3.9) we can deduce the qualitative behavior of a harmonic function near an isolated singularity.

Theorem 3.4. *Let u be harmonic in the punctured disk*

$$D'_R(z_0) := \{z \mid 0 < |z - z_0| < R\}$$

and (3.9) *be its expansion there.*

(a) *If $\alpha_n \neq 0$ or $\beta_n \neq 0$ for some $n \leq -1$, then in every punctured neighborhood of z_0, the function u assumes all real values.*
(b) *If $\alpha_n = \beta_n = 0$ for all $n \leq -1$ and $c \neq 0$, then*

$$\lim_{z \to z_0} u(z) = \begin{cases} \infty & \text{if } c < 0 \\ -\infty & \text{if } c > 0. \end{cases}$$

(c) *If neither* (a) *nor* (b) *holds, then u can be defined at z_0 so that u is harmonic at z_0.*

Proof. To establish (a) it suffices, by continuity, to show that in every punctured neighborhood of z_0, $u(z)$ is not bounded above or below. Assume to the contrary that $u(z) < M$ in the punctured neighborhood $D'_\rho(z_0)$. Suppose also that $\alpha_{-n_0} \neq 0$, where $n_0 \geq 1$ (the case $\beta_{-n_0} \neq 0$ is similar). Then clearly

$$I_{r,n_0} := \int_0^{2\pi} u(z_0 + re^{i\theta})(1 \pm \cos n_0\theta)\, d\theta \leq 4\pi M \quad \text{for} \quad 0 < r < \rho. \tag{3.10}$$

On the other hand, from the representation (3.9), we get

$$I_{r,n_0} = 2\pi c \log r + \pi\alpha_0 \pm \alpha_{-n_0}\pi r^{-n_0} \pm \alpha_{n_0}\pi r^{n_0}.$$

Thus if we choose the $\pm$ sign to match the sign of α_{-n_0} we get

$$I_{r,n_0} = 2\pi c \log r + \pi|\alpha_{-n_0}|r^{-n_0} + A(r),$$

where $A(r)$ is bounded near $r = 0$. Consequently $I_{r,n_0} \to +\infty$ as $r \to 0+$, which contradicts (3.10). This proves that u cannot be bounded from above in $D'_\rho(z_0)$. The proof that u is not bounded from below follows by similar reasoning. Thus (a) is established.

Now suppose that $\alpha_n = \beta_n = 0$ for all $n \leq -1$. Then $u(z) = c \ln r + \operatorname{Re} H(z)$, where H is analytic at z_0. Thus, as $z \to z_0$ ($r \to 0+$), we have $u(z) \to +\infty$ if $c < 0$ and $u(z) \to -\infty$ if $c > 0$.

If neither (a) nor (b) holds, then $\alpha_n = \beta_n = 0 = c$ for all $n \leq -1$, and so $u(z) = \operatorname{Re} H(z)$, where H is analytic at z_0. Consequently, u is harmonic in a neighborhood of z_0. □

Corollary 3.5. *If u is harmonic and bounded in some punctured disk about z_0, then u can be defined (or redefined) at z_0 so that u is harmonic at z_0.*

Corollary 3.6. *If u is harmonic and bounded from one side in some punctured disk about z_0, then u is of the form*

$$u(z) = c \log |z - z_0| + v(z),$$

where v is harmonic at z_0.

By making use of the transformation $z \to 1/z$ we can easily transform the preceding definitions and results to the case $z_0 = \infty$, i.e. to speak about harmonicity on a subdomain of the Riemann sphere $\overline{\mathbb{C}}$. For example, Corollary 3.6 takes the following form.

Corollary 3.7. *If u is harmonic and bounded from one side in a neighborhood of $z_0 = \infty$, then u is of the form*

$$u(z) = c \log |z| + v(z),$$

where v is harmonic at ∞.

0.4 Poisson's Formula and Applications

Cauchy's integral formula leads to the following integral representation for functions u harmonic in a disk.

Theorem 4.1. *If $u(z)$ is harmonic in the open disk $D_R(0) := \{z \mid |z| < R\}$ and continuous on its closure $\overline{D_R(0)}$, then*

$$u(z) = \frac{1}{2\pi} \int_0^{2\pi} \frac{R^2 - |z|^2}{|Re^{i\phi} - z|^2} u(Re^{i\phi})\, d\phi, \quad z \in D_R(0). \tag{4.1}$$

Proof. Assume at first that u is harmonic on $\overline{D_R(0)}$ and let f be an analytic completion of u in a domain containing $\overline{D_R(0)}$. By Cauchy's formula,

$$f(z) = \frac{1}{2\pi i} \int_{|t|=R} \frac{f(t)}{t - z}\, dt = \frac{1}{2\pi} \int_0^{2\pi} \frac{f(t)t}{t - z}\, d\phi, \quad t = Re^{i\phi}, \quad z \in D_R(0). \tag{4.2}$$

Furthermore, for any w with $|w| > R$, Cauchy's theorem asserts that

$$\frac{1}{2\pi} \int_0^{2\pi} \frac{f(t)t}{t - w}\, d\phi = 0. \tag{4.3}$$

On selecting $w = R^2/\bar{z}$, we get from (4.2) and (4.3)

$$f(z) = \frac{1}{2\pi} \int_0^{2\pi} f(t) \left[\frac{t}{t - z} - \frac{t}{t - R^2/\bar{z}} \right] d\phi = \frac{1}{2\pi} \int_0^{2\pi} f(t) \frac{R^2 - |z|^2}{|Re^{i\phi} - z|^2}\, d\phi, \tag{4.4}$$

where $t = Re^{i\phi}$ and $z \in D_R(0)$. By taking the real parts in (4.4) we obtain (4.1).

If u is not harmonic on $|z| = R$, then the first part of the proof can be applied to $u_r(z) := u(rz)$ for $r < 1$. Appealing to the uniform continuity of u in $\overline{D_R(0)}$ and letting $r \to 1$ we obtain (4.1). □

The function

$$P(t, z) := \frac{|t|^2 - |z|^2}{|t - z|^2}, \quad t = Re^{i\phi}, \tag{4.5}$$

appearing in (4.1) is called the *Poisson kernel* and we note that

$$P(t, z) = \mathrm{Re}\left(\frac{t+z}{t-z}\right) = \frac{R^2 - r^2}{R^2 + r^2 - 2Rr\cos(\phi - \theta)} \tag{4.6}$$

for $z = re^{i\theta}$, $t = Re^{i\phi}$. Given a function $U(t)$ that is integrable in the Lebesgue sense on the circle $|t| = R$, observe that the *Poisson integral*

$$u(z) := \frac{1}{2\pi}\int_0^{2\pi} P(t, z)U(t)\,d\phi \tag{4.7}$$

$$= \mathrm{Re}\left(\frac{1}{2\pi}\int_0^{2\pi}\frac{t+z}{t-z}U(t)\,d\phi\right), \quad t = Re^{i\phi},$$

is harmonic in $D_R(0)$ since it is the real part of an analytic function. The important question that arises is the relationship between the values of $U(t)$ and the limiting values of $u(z)$ as $|z| \to R$. This issue is resolved in the following result known as *Schwarz's theorem.*

Theorem 4.2. *Suppose that $U(t)$ is integrable on $|t| = R$ and continuous at $t = Re^{i\alpha}$. Then the function $u(z)$ in (4.7) is harmonic in $D_R(0)$ and satisfies*

$$\lim_{z \to Re^{i\alpha},\ |z|<R} u(z) = U(Re^{i\alpha}).$$

In the proof of this theorem we appeal to the following simple lemma.

Lemma 4.3. *The Poisson kernel $P(t, z)$ satisfies*

(a) $P(t, z) > 0$ *for* $|z| < |t|$;

(b) $\displaystyle\frac{1}{2\pi}\int_0^{2\pi} P(t, z)\,d\phi = 1, \quad |z| < R, \quad t = Re^{i\phi}$;

(c) *if $U(t)$ is integrable on $|t| = R$ and $\beta \in [0, 2\pi]$, then for each $1 > \varepsilon > 0$*

$$\lim_{z \to Re^{i\beta},\ |z|<R}\left\{\int_0^{\beta-\varepsilon} + \int_{\beta+\varepsilon}^{2\pi}\right\} P(t, z)U(t)\,d\phi = 0, \quad t = Re^{i\phi}.$$

Naturally, in case $\beta - \varepsilon < 0$ or $\beta + \varepsilon > 2\pi$, the range of integration is understood mod 2π, e.g. if $\beta - \varepsilon < 0$, then it is $\int_{\beta+\varepsilon}^{2\pi+\beta-\varepsilon}$.

Proof. Part (a) is trivial, and part (b) is immediate from Theorem 4.1 with $u \equiv 1$.

To establish part (c), choose $z = re^{i\theta}$ sufficiently close to $Re^{i\beta}$ so that $|\beta - \theta| < \varepsilon/2$. Then

$$\cos(\phi - \theta) \le \cos(\varepsilon/2) \quad \text{for} \quad 0 \le \phi \le \beta - \varepsilon \quad \text{and} \quad \beta + \varepsilon \le \phi \le 2\pi.$$

Hence, from the representation (4.6), we have

$$0 < P(t, z) \le \frac{R^2 - r^2}{R^2 + r^2 - 2Rr\cos(\varepsilon/2)},$$

and so

$$\left|\left\{\int_0^{\beta-\varepsilon} + \int_{\beta+\varepsilon}^{2\pi}\right\} P(t, z)U(t)\,d\phi\right| \le \frac{R^2 - r^2}{R^2 + r^2 - 2Rr\cos(\varepsilon/2)} \int_0^{2\pi} |U(t)|\,d\phi.$$

As $z \to Re^{i\beta}$, we have $r \to R$, and so the right-hand side of the last inequality tends to zero. □

We can now give the

Proof of Theorem 4.2. Let $\varepsilon > 0$. Then by parts (a) and (c) of Lemma 4.3 we have

$$\begin{aligned}
\limsup_{z\to Re^{i\alpha},\,|z|<R} u(z) &= \limsup_{z\to Re^{i\alpha},\,|z|<R} \frac{1}{2\pi}\left\{\int_0^{\alpha-\varepsilon} + \int_{\alpha-\varepsilon}^{\alpha+\varepsilon} + \int_{\alpha+\varepsilon}^{2\pi}\right\} P(t, z)U(t)\,d\phi \\
&= \limsup_{z\to Re^{i\alpha},\,|z|<R} \int_{\alpha-\varepsilon}^{\alpha+\varepsilon} P(t, z)U(t)\,d\phi \\
&\le \frac{1}{2\pi}\left[\sup_{\phi\in[\alpha-\varepsilon,\alpha+\varepsilon]} U(Re^{i\phi})\right] \cdot \limsup_{z\to Re^{i\alpha},\,|z|<R} \int_{\alpha-\varepsilon}^{\alpha+\varepsilon} P(t, z)\,d\phi.
\end{aligned}$$

Also note from parts (b) and (c) of Lemma 4.3 that

$$\limsup_{z\to Re^{i\alpha},\,|z|<R} \frac{1}{2\pi}\int_{\alpha-\varepsilon}^{\alpha+\varepsilon} P(t, z)\,d\phi = 1.$$

Hence

$$\limsup_{z\to Re^{i\alpha},\,|z|<R} u(z) \le \sup_{\phi\in[\alpha-\varepsilon,\alpha+\varepsilon]} U(Re^{i\phi}) =: A_\varepsilon. \tag{4.8}$$

Similarly, we obtain

$$\liminf_{z\to Re^{i\alpha},\,|z|<R} u(z) \ge \inf_{\phi\in[\alpha-\varepsilon,\alpha+\varepsilon]} U(Re^{i\phi}) =: B_\varepsilon. \tag{4.9}$$

Now since U is continuous at $t = Re^{i\alpha}$, both A_ε and B_ε tend to $U(Re^{i\alpha})$ as $\varepsilon \to 0$. Thus, from (4.8) and (4.9) it follows that $\lim_{z\to Re^{i\alpha}} u(z)$ exists, and equals $U(Re^{i\alpha})$. □

As an immediate consequence of Theorem 4.2 we get

Corollary 4.4. *Suppose that $U(z)$ is continuous on the circle $C_R(a) := \{z \mid |z-a| = R\}$. Then there exists a unique function $u(z)$ harmonic in $D_R(a)$, continuous in $\overline{D_R(a)}$, such that $u(z) = U(z)$ on $C_R(a)$. Moreover,*

$$u(z) = \frac{1}{2\pi} \int_0^{2\pi} P(t-a, z-a) U(t)\, d\phi, \quad t = a + Re^{i\phi}.$$

The function u is called the solution of the Dirichlet problem with boundary function U. For more on Dirichlet problems see Section I.2 and Appendix A.2.

As a simple application of Schwarz's theorem we show that the assumption of continuity of the first and second order partial derivatives in the definition of a harmonic function can be deleted.

Theorem 4.5. *Suppose that the function u is continuous in a domain D and that u_{xx}, u_{yy} exist at each point of D and satisfy $u_{xx} + u_{yy} = 0$. Then u is harmonic in D.*

Proof. Let $a \in D$ and select $r > 0$ so that the closed neighborhood $\overline{D_r(a)}$ is part of D. Since u is continuous on the circle $C_r(a)$, it follows from Corollary 4.4 that the Poisson integral around this circle

$$v(z) := \frac{1}{2\pi} \int_0^{2\pi} P(t-a, z-a) u(t)\, d\theta, \quad t = a + re^{i\theta},$$

is harmonic in $D_r(a)$, continuous on $\overline{D_r(a)}$ and satisfies $v(z) = u(z)$ on $C_r(a)$. We shall show that $v = u$ in $D_r(a)$.

Suppose to the contrary that, say, $u(\xi) - v(\xi) > 0$ for some $\xi \in D_r(a)$. Choose $\eta > 0$ so that

$$\max_{|z-a|=r} \eta (\operatorname{Re} z)^2 < u(\xi) - v(\xi), \tag{4.10}$$

and consider the function

$$h(z) := \eta (\operatorname{Re} z)^2 + u(z) - v(z), \quad z \in \overline{D_r(a)}.$$

Clearly h is continuous on $\overline{D_r(a)}$. Moreover, h attains its maximum at some point c in the *open* disk $D_r(a)$ since on $C_r(a)$ we have $h(z) = \eta(\operatorname{Re} z)^2$, while (4.10) implies that

$$h(\xi) > \eta(\operatorname{Re} \xi)^2 + \max_{|z-a|=r} \eta(\operatorname{Re} z)^2 \geq \max_{|z-a|=r} h(z).$$

At the maximum point c we must therefore have $h_{xx}(c) \leq 0$ and $h_{yy}(c) \leq 0$. But then, since $u - v$ satisfies Laplace's equation, we get

$$0 \geq h_{xx}(c) + h_{yy}(c) = 2\eta + 0 = 2\eta,$$

which contradicts the positivity of η.

Thus $u(z) - v(z) \leq 0$ in $D_r(a)$ and a similar argument establishes the reverse inequality. Hence $u \equiv v$ in $D_r(a)$, which proves that u is harmonic $D_r(a)$. Since $a \in D$ is arbitrary, u is harmonic in D. □

We next show that the mean-value property established in Theorem 2.4 does indeed characterize harmonic functions.

Theorem 4.6. *Suppose that u is continuous in a domain D and that for each point $a \in D$ there exists a constant $r_a > 0$ such that*

$$u(a) = \frac{1}{2\pi}\int_0^{2\pi} u(a + re^{i\theta})d\theta \quad \text{if } 0 < r < r_a.$$

Then u is harmonic in D.

Proof. The crucial observation is that the proof of the maximum principle (Theorem 2.6) only requires a *local* mean-value property. Hence Theorem 2.6 and Corollary 2.7 hold for the given function u. Now suppose $\overline{D_\rho(z_0)} \subset D$, with $\rho < r_{z_0}$, and let v denote the Poisson integral of u around the circumference $|z - z_0| = \rho$. Then by Corollary 4.4 and Theorem 2.4, the function v and hence also the function $u - v$ have the mean-value property in $D_\rho(z_0)$. But then Corollary 2.7 applies and we get that $u - v$ attains its maximum and minimum on the circumference where $u - v = 0$. Thus $u \equiv v$ in $D_\rho(z_0)$, so that u is harmonic at z_0. □

Since the mean-value property is clearly preserved under uniform convergence, we have the following consequence of Theorem 4.6.

Corollary 4.7. *Let $\{u_n\}$ be a sequence of harmonic functions in a domain D that converges uniformly on every compact subset of D to the function u. Then u is harmonic in D.*

To describe the behavior of monotone sequences of harmonic functions, the following inequality will be useful.

Lemma 4.8 (Harnack's Inequality). *Suppose that $u(z)$ is nonnegative and harmonic in the disk $D_R(0)$. Then for $|z| \le r < R$,*

$$u(0)\frac{R-r}{R+r} \le u(z) \le u(0)\frac{R+r}{R-r}. \tag{4.11}$$

Proof. By considering $u(\rho z)$ with $\rho < 1$ and then letting $\rho \to 1$ we may assume that u is continuous on $\overline{D_R(0)}$. Then by Theorem 4.1 we can write

$$u(z) = \frac{1}{2\pi}\int_0^{2\pi} P(t, z)u(t)\,d\theta, \quad t = Re^{i\theta}. \tag{4.12}$$

For $|z| = r (< R)$ we see from (4.5) that

$$\frac{R-r}{R+r} = \frac{R^2 - r^2}{(R+r)^2} \le P(t, z) \le \frac{R^2 - r^2}{(R-r)^2} = \frac{R+r}{R-r}.$$

On multiplying this string of inequalities by $u(Re^{i\theta})$ and integrating with respect to $d\theta$ we obtain from (4.12) that (4.11) holds for $|z| = r$. But then, by the maximum principle, it must also hold for $|z| \le r$. □

For arbitrary domains D, Harnack's inequality yields the following result.

Lemma 4.9 (Harnack's Lemma). *Suppose that $u(z)$ is nonnegative and harmonic in a domain D that contains the compact set K and the point ζ. Then there exists a constant $C = C(K, \zeta, D)$ independent of u such that*

$$\max_{z \in K} u(z) \le Cu(\zeta). \tag{4.13}$$

Proof. Since K is compact, it can be covered by a finite number of closed disks, say $\{\overline{D_{r_i}(z_i)}\}_{i=1}^m$, such that $\overline{D_{2r_i}(z_i)} \subset D$ for $i = 1, \ldots, m$. Then, from (4.11) we have

$$\max_{z \in K} u(z) \le \max_{1 \le i \le m} \left\{ u(z_i) \frac{2r_i + r_i}{2r_i - r_i} \right\} = 3 \max_{1 \le i \le m} \{u(z_i)\}. \tag{4.14}$$

On the other hand, ζ can be joined to each z_i by a polygonal line Γ_i that lies in D. Let ∂D denote the boundary of D and choose $\rho > 0$ such that $\operatorname{dist}(\Gamma_i, \partial D) > 4\rho$ for $i = 1, \ldots, m$. On covering each Γ_i by a finite number, say k_i, of closed disks with radius ρ and centers on Γ_i we deduce, on successively applying the lower estimate in (4.11), that

$$\frac{u(z_i)}{3^{k_i}} \le u(\zeta), \quad i = 1, \ldots, m.$$

Combining this with (4.14) we obtain (4.13). □

We can now prove

Theorem 4.10 (Harnack's Principle). *Let $\{u_n(z)\}$ be a sequence of functions harmonic in a domain D that satisfy $u_n(z) \le u_{n+1}(z)$ for all $z \in D$ and n sufficiently large. Then either $\{u_n(z)\}$ converges uniformly on every compact subset of D to a function harmonic in D or $\{u_n(z)\}$ tends to $+\infty$ at every $z \in D$.*

Proof. Set $u(z) := \lim_{n \to \infty} u_n(z)$, which exists as an extended real number for every $z \in D$, and fix N so that $\{u_n(z)\}$ is monotone for $n \ge N$. If $u(z) \equiv +\infty$, we are done. So suppose that there is a point $\zeta \in D$ such that $u(\zeta) < \infty$. Then, from Lemma 4.9, for each compact subset $K \subset D$, there exists a constant $C = C(K, \zeta, D)$ such that, for $j \ge k \ge N$,

$$0 \le \max_{z \in K} \left\{ u_j(z) - u_k(z) \right\} \le C \left[u_j(\zeta) - u_k(\zeta) \right]. \tag{4.15}$$

Since $\{u_n(\zeta)\}$ is a Cauchy sequence of real numbers, it follows from (4.15) that $\{u_n(z)\}$ is a uniform Cauchy sequence of functions on K. Hence $\{u_n(z)\}$ converges uniformly on every compact subset of D. Corollary 4.7 then asserts that u is harmonic in D. □

Another simple consequence of Harnack's inequality is the following theorem of Picard.

Theorem 4.11. *If u is harmonic and bounded from one side in the whole plane $\mathbf{C}$, then it is constant.*

Proof. It suffices to assume that u is nonnegative on $\mathbf{C}$. Then on letting $R \to \infty$ in (4.11) we deduce that $u(z) \equiv u(0)$. □

0.5 Superharmonic Functions

Definition 5.1. An extended real-valued function f on a domain $D \subseteq \mathbf{C}$ is called *superharmonic* on D if it is not identically $+\infty$ and satisfies the following three conditions:

a) $f(z) > -\infty$ for all $z \in D$,
b) f is lower semi-continuous on D,
c) $f(z) \geq \dfrac{1}{2\pi}\displaystyle\int_{-\pi}^{\pi} f(z + re^{i\theta})\, d\theta$ for all $z \in D$ and $r > 0$ such that the disk $\overline{D_r(z)} := \{z' \,|\, |z' - z| \leq r\}$ is contained in D.

We remind the reader that the property b) of lower semi-continuity on D is equivalent to each of the following (cf. Section 0.1):

b') For each $z_0 \in D$, we have $f(z_0) \leq \liminf_{z \to z_0} f(z)$;
b'') For each $\alpha \in \mathbf{R}$, the set $\{z \in D \,|\, f(z) > \alpha\}$ is open;
b''') For each compact set $E \subset D$, there exists a monotone increasing sequence of continuous functions on E with limit f.

We further observe that properties a) and b) ensure that f is lower bounded on compact subsets of D; hence the integral in c) is well defined.

A function f is called *subharmonic* on D if $-f$ is superharmonic on D. We note from Theorems 2.4 and 4.6 that a function is harmonic in D if and only if it is both superharmonic and subharmonic in D. It is customary to work with subharmonic functions, for they are directly tied to holomorphic functions and to their applications. However, in this book we have potentials in mind, and so it is more convenient to state results for superharmonic functions rather than for subharmonic ones.

The notion "superharmonic" is explained by the following property that can replace c) above.

c') If D_1 is a bounded domain that is contained in D together with its boundary ∂D_1, u is a function harmonic in D_1 and continuous on the closure $\overline{D_1} = D_1 \cup \partial D_1$ of D_1, and if $f(x) \geq u(x)$ for $x \in \partial D_1$, then $f(z) \geq u(z)$ for all $z \in D_1$.

The equivalence of c) and c') can be seen as follows. First let us assume property c') and let us prove that c) is true for the disks indicated. By property b''') there are continuous functions $\{u_n\}$ defined on $\partial D_r(z)$ such that $u_n(z') \nearrow f(z')$ for $z' \in \partial D_r(z)$. We can extend this u_n to $D_r(z)$ as a harmonic function via Corollary 4.4. On applying property c') to f and u_n we can conclude from the mean-value property

$$u_n(z) = \frac{1}{2\pi} \int_{-\pi}^{\pi} u_n\left(z + re^{i\theta}\right) d\theta$$

that

$$f(z) \geq \frac{1}{2\pi} \int_{-\pi}^{\pi} u_n\left(z + re^{i\theta}\right) d\theta,$$

from which property c) follows if we let $n \to \infty$ and apply the monotone convergence theorem.

To prove that c) implies c'), consider the function

$$g(z) := f(z) - u(z).$$

This is superharmonic in D_1 and, by property b'),

$$\liminf_{z \to z', z \in D_1} g(z) \geq g(z') \geq 0$$

for every $z' \in \partial D_1$. Hence, the desired conclusion follows from the following principle.

Theorem 5.2 (Minimum Principle). *Let D be a bounded domain and g a superharmonic function on D such that*

$$\liminf_{z \to z', z \in D} g(z) \geq m \tag{5.1}$$

for every $z' \in \partial D$. Then $g(z) > m$ for $z \in D$ unless g is constant.

A more precise minimum principle will be proved in Section I.2.

Proof of Theorem 5.2. The argument is essentially the same as that used in the proof of Theorem 2.6. We assume as we may that $m = 0$. First we show that $g(z) \geq 0$ for every $z \in D$. Indeed, if this is false, then $g(z) = -\varepsilon < 0$ for some $z \in D$. Since g is lower semi-continuous, the assumption implies that the set

$$E := \{z \in D \mid g(z) \leq -\varepsilon\}$$

is a nonempty compact set. A lower semi-continuous function attains its minimum on a compact set; hence there is a $z_0 \in E$ such that

$$g(z_0) = m_g := \inf_{z \in D} g(z).$$

Notice that m_g is finite by property a). On applying property c) for g we get

$$m_g = g(z_0) \geq \frac{1}{2\pi} \int_{-\pi}^{\pi} g(z_0 + re^{i\theta}) d\theta$$

provided the closed disk with center at z_0 and of radius r lies in D. But then the inequality $g(z_0 + re^{i\theta}) \geq m_g$ implies that $g(z_0 + re^{i\theta}) = m_g$ for almost every θ in $[-\pi, \pi]$, and the lower semi-continuity of g yields the same relation for every θ. Since this is true for every small r, we can conclude that g is constant $(= m_g)$ in a neighborhood of z_0.

By a chain of overlapping disks we can reach any point in D from z_0 and so the reasoning above gives that g is constant and is equal to $m_g \leq -\varepsilon < 0$ throughout D. But this contradicts (5.1); hence $g(z) \geq 0$ is satisfied for all z.

Finally, if $g(z) = 0$ at a point $z \in D$, then g attains its minimum at some point of D, and so the preceding argument shows that g is constant. □

Remark 5.3. Notice that in the proof of Theorem 5.2 we used only that the mean-value inequality holds in some small neighborhood of each $z \in D$. Thus property c) is also equivalent to

c") $f(z) \geq \frac{1}{2\pi}\int_{-\pi}^{\pi} f(z+re^{i\theta})d\theta$ for all $z \in D$ and $0 < r < r(z)$.

From this fact we observe: *f is superharmonic on D if and only if it is locally superharmonic in D.*

A rich supply of subharmonic functions is provided by the following example.

Example 5.4. If F is analytic in a domain D and $p > 0$, then $|F(z)|^p$ is subharmonic in D; furthermore, $\log|F(z)|$ is subharmonic in D provided F is not identically zero. Indeed, suppose that $\overline{D_r(z_0)} \subset D$. If $F(z_0) = 0$, then clearly

$$|F(z_0)|^p \leq \frac{1}{2\pi}\int_{-\pi}^{\pi} \left|F(z_0+re^{i\theta})\right|^p d\theta. \tag{5.2}$$

If $F(z_0) \neq 0$, then there exists a single-valued branch of $F(z)^p$ that is analytic on some closed disk $\overline{D_\rho(z_0)}$. But then $F(z)^p$ has the mean-value property in $\overline{D_\rho(z_0)}$ (recall (2.2)), which implies that inequality (5.2) holds for $r < \rho$; that is, $-|F(z)|^p$ satisfies the local condition c"). A similar (even simpler) argument can be used for $\log|F(z)|$. □

For smooth functions f, we can determine if f is superharmonic by examining the sign of its Laplacian:

Theorem 5.5. *Let f and its first and second order partial derivatives be continuous in a domain D. Then f is superharmonic in D if and only if*

$$\Delta f(z) := f_{xx}(z) + f_{yy}(z) \leq 0, \quad z \in D. \tag{5.3}$$

Proof. Suppose first that (5.3) holds in D. Let $\overline{D_r(z_0)} \subset D$ and let $v_\varepsilon(z)$ be the Poisson integral of the function $f_\varepsilon(z) := f(z) - \varepsilon|z|^2$, $\varepsilon > 0$, around the circumference $C_r(z_0) := \{z \mid |z - z_0| = r\}$. By Corollary 4.4 we know that v_ε is continuous on $\overline{D_r(z_0)}$, harmonic in $D_r(z_0)$ and $v_\varepsilon(z) = f_\varepsilon(z)$ for $z \in C_r(z_0)$. Now the assumption of (5.3) implies that $f_\varepsilon - v_\varepsilon$ cannot attain its minimum over $\overline{D_r(z_0)}$ at a point of the open disk $D_r(z_0)$; indeed, $\Delta(f_\varepsilon - v_\varepsilon) \leq -4\varepsilon$ in $D_r(z_0)$ (compare with the proof of Theorem 4.5). Thus $f_\varepsilon - v_\varepsilon$ attains its minimum on $C_r(z_0)$, and so $f_\varepsilon \geq v_\varepsilon$ on $\overline{D_r(z_0)}$. Consequently, since v_ε satisfies the mean-value property, we have

$$f_\varepsilon(z_0) \geq v_\varepsilon(z_0) = \frac{1}{2\pi}\int_{-\pi}^{\pi} v_\varepsilon(z_0 + re^{i\theta})\,d\theta = \frac{1}{2\pi}\int_{-\pi}^{\pi} f_\varepsilon(z_0 + re^{i\theta})\,d\theta.$$

On letting $\varepsilon \to 0+$, we find that f has the mean-value inequality property c) and hence is superharmonic in D.

Now suppose that f is superharmonic in D but that $\Delta f(a) > 0$ at some point $a \in D$. By continuity, $\Delta f(z) > 0$ for z in some open disk $D_r(a) \subset D$. But then, by the first part of the proof, f is also *sub*harmonic in $D_r(a)$. Hence f is harmonic in $D_r(a)$, which contradicts the fact that $\Delta f > 0$ there. □

We leave as simple exercises the verifications of the facts that if $f_1, \dots, f_n$ are each superharmonic in a domain D, then so are the functions $\inf_{1\leq k\leq n} f_k(z)$ and $\sum_{1\leq k\leq n} c_k f_k(z)$, $c_k \geq 0$, $k = 1, \dots, n$.

Since harmonicity is conformally invariant, we get from properties a), b), and c') that the same is true of superharmonicity. This allows us to define superharmonicity around ∞, i.e. on a domain of the Riemann sphere $\overline{\mathbf{C}} := \mathbf{C} \cup \{\infty\}$.

Superharmonic functions are closely related to logarithmic potentials of positive measures. Let μ be a finite positive Borel measure of compact support. Its *logarithmic potential* is defined by

$$U^\mu(z) := \int \log\frac{1}{|z-t|}\,d\mu(t).$$

This integral is well-defined (and may equal $+\infty$), and we now show that it is superharmonic.

Theorem 5.6. *The potential U^μ is superharmonic in* $\mathbf{C}$ *and harmonic at each point z not in the support of μ.*

Proof. To verify the superharmonicity of U^μ we have to prove a) – c) with $f = U^\mu$. Indeed, a) is obvious and b) follows from the representation

$$U^\mu(z) = \lim_{M\to\infty}\int \min\left(M, \log\frac{1}{|z-t|}\right) d\mu(t),$$

where the functions on the right-hand side are continuous and increasing with M (recall that the limit of an increasing sequence of continuous functions is lower semi-continuous).

To establish property c) we first note that the kernel $\log(1/|z-t|)$ is superharmonic in $\mathbf{C}$ for each fixed t (cf. Example 5.4). Thus

$$\frac{1}{2\pi}\int_{-\pi}^{\pi} \log\frac{1}{|z+re^{i\theta}-t|}\,d\theta \leq \log\frac{1}{|z-t|}, \quad z, t \in \mathbf{C}. \tag{5.4}$$

Turning now to the potential of μ, it follows from the Fubini–Tonelli theorem and (5.4) that

$$\begin{aligned}\frac{1}{2\pi}\int_{-\pi}^{\pi} U^{\mu}\left(z+re^{i\theta}\right)d\theta &= \int \frac{1}{2\pi}\int_{-\pi}^{\pi}\log\frac{1}{\left|z+re^{i\theta}-t\right|}\,d\theta\, d\mu(t) \\ &\leq \int \log\frac{1}{|z-t|}\,d\mu(t) = U^{\mu}(z),\end{aligned}$$

which proves property c).

Finally, we observe that for each fixed t, the function $\log(1/|z-t|)$ is harmonic in $\mathbf{C}\setminus\{t\}$. Hence for $z \notin \operatorname{supp}(\mu)$ the Laplacian of U^{μ} satisfies

$$\Delta U^{\mu}(z) = \int \Delta \log\frac{1}{|z-t|}\,d\mu(t) = 0,$$

because interchanging the order of integration and differentiation is permitted in this case. Thus U^{μ} is harmonic in $\mathbf{C}\setminus\operatorname{supp}(\mu)$. □

A converse of Theorem 5.6, called the Riesz decomposition theorem, will be proved in Theorem II.3.1 of Chapter II.

In the proof of Theorem 5.6 and in several other instances throughout the book the potential corresponding to arc measure over a circle arises. This potential can easily be evaluated, as we now show.

Example 5.7. For each $r > 0$,

$$\frac{1}{2\pi}\int_{-\pi}^{\pi}\log\frac{1}{|z-re^{i\theta}|}\,d\theta = \begin{cases}\log 1/r & \text{if } |z|\leq r\\ \log 1/|z| & \text{if } |z|>r.\end{cases} \tag{5.5}$$

This formula is immediate for $|z| > r$ since $\log(1/|z-\zeta|)$ is a harmonic function of ζ for $|\zeta| \leq r$ and we have only to apply the mean-value property at $\zeta = 0$. The argument is similar when $|z| < r$ if we first factor out $e^{i\theta}$ from the absolute value:

$$\begin{aligned}\frac{1}{2\pi}\int_{-\pi}^{\pi}\log\frac{1}{|z-re^{i\theta}|}\,d\theta &= \frac{1}{2\pi}\int_{-\pi}^{\pi}\log\frac{1}{|ze^{-i\theta}-r|}\,d\theta \\ &= \frac{1}{2\pi}\int_{-\pi}^{\pi}\log\frac{1}{|\overline{z}e^{i\theta}-r|}\,d\theta = \log\frac{1}{r}.\end{aligned}$$

Finally, if $|z| = r$, then for $0 < \rho < r$ we have from the dominated convergence theorem and the above evaluations that

$$\frac{1}{2\pi}\int_{-\pi}^{\pi}\log\frac{1}{|z-re^{i\theta}|}\,d\theta = \lim_{\rho\to r^-}\frac{1}{2\pi}\int_{-\pi}^{\pi}\log\frac{1}{|z-\rho e^{i\theta}|}\,d\theta = \log\frac{1}{r}.$$

□

Chapter I. Weighted Potentials

In this chapter we discuss a minimal energy (or equilibrium) problem with logarithmic kernel in the presence of a weight (external field). The results form the basis for all later developments and applications.

Our main concern will be to solve the energy problem which consists of minimizing the energy expression

$$I_w(\mu) = \int\int \log \frac{1}{|z-t|}\, d\mu(z)\, d\mu(t) + 2\int Q\, d\mu,$$

where $Q = \log(1/w)$ is a given function (which we call the external field) and where the infimum is taken for all unit charges μ supported on some closed set Σ. This is a variational problem that goes back to Gauss. The emphasis is on the effect that the external field Q has on the equilibrium distribution; therefore the theory of potentials with general kernels is too vague for the concrete applications we shall discuss in later chapters of the book. Nevertheless, very often the results run parallel with the theory for general kernels, and many of the proofs also follow standard arguments.

The classical theory corresponds to the case when the conductor Σ is compact and the external field is zero. Therefore, if one removes the external field, then the classical case remains. However, our generalized treatment is more flexible and has many advantages over the classical setting, as will become clear from later chapters. In fact, besides numerous applications we shall see that there are several extremal problems that lead to the aforementioned minimal energy problem under suitable reformulation; furthermore, such concepts as balayage, Green function, Dirichlet solution, etc. arise as natural occurrences of the weighted case. Of most importance will be the case of an unbounded conductor when the external field keeps the charge within a bounded region.

In the presentation of the results we shall always keep later applications in mind. Therefore, we shall adhere to the logarithmic case which is intimately connected with the complex plane.

First we establish that, under mild conditions on the external field Q, the equilibrium problem has a unique solution μ_w which is a measure of compact support. The compactness is enforced by the assumption that Q increases sufficiently fast around infinity. In the connected components of the complement of the support of μ_w the associated logarithmic potential

$$U^{\mu_w}(z) := \int \log \frac{1}{|z-t|} d\mu_w(t)$$

turns out to be the solution of the Dirichlet problem (modulo an additive constant) with boundary function $-Q$. The extremal/equilibrium measures μ_w have some features that are missing in the classical theory (indeed, almost all measures can appear as equilibrium measures with respect to an appropriate field); but the μ_w can be characterized via inequalities for their potentials that resemble the classical inequalities of Frostman. Even though the extremal measures can be very different from classical equilibrium distributions, the continuity properties of the equilibrium potentials are very similar to the continuity properties in the unweighted case. We shall develop the theory of fine topology to the extent that allows us to establish these basic continuity properties. Although of less importance, for the sake of completeness, we include the analogue of the notion of capacity for the weighted case.

In developing the subject we shall encounter in this chapter many results of classical potential theory, such as a basic unicity theorem, the principle of descent, the lower envelope theorem, Wiener's criterion for regular boundary points, etc. Since these are scattered throughout the text (appearing where we need them or where they naturally fit in the discussion), we have provided a list of theorems in classical potential theory at the end of the monograph to serve as a guide for the reader.

I.1 The Energy Problem

There are several possible starting points for the development of the classical theory of logarithmic potentials. The most commonly accepted one is the concept of energy of measures and sets, and the theory is first developed for compact sets. Thus, let $\Sigma \subset \mathbf{C}$ be a compact subset of the complex plane and $\mathcal{M}(\Sigma)$ the collection of all positive unit Borel measures with support in Σ. The *logarithmic energy* $I(\mu)$ of a $\mu \in \mathcal{M}(\Sigma)$ is defined as

$$I(\mu) := \iint \log \frac{1}{|z-t|} d\mu(z) d\mu(t), \tag{1.1}$$

and the energy V of Σ by

$$V := \inf\{I(\mu) \mid \mu \in \mathcal{M}(\Sigma)\}. \tag{1.2}$$

Then V turns out to be finite or $+\infty$, and in the finite case there is a unique measure $\mu = \mu_\Sigma \in \mathcal{M}(\Sigma)$ for which the infimum defining V in (1.2) is attained. This μ_Σ is called the *equilibrium distribution* or *measure* of the compact set Σ, and for its *logarithmic potential*

$$U^{\mu_\Sigma}(z) := \int \log \frac{1}{|z-t|} d\mu_\Sigma(t) \tag{1.3}$$

we have

$$U^{\mu_\Sigma}(z) \le V \tag{1.4}$$

for all $z \in \mathbf{C}$. The quantity

$$\operatorname{cap}(\Sigma) := e^{-V} \tag{1.5}$$

is called the *logarithmic capacity* of Σ. For example, if Σ is a circle C or disk D of radius r, then $\operatorname{cap}(\Sigma) = r$, and the equilibrium measure is the normalized arc measure on $C = \partial D$ (cf. Example 3.4 of Section I.3). In this case (see (0.5.5))

$$U^{\mu_\Sigma}(z) = \begin{cases} \log 1/r & \text{if} \quad |z| \le r \\ \log 1/|z| & \text{if} \quad |z| > r. \end{cases} \tag{1.6}$$

Further, when Σ is a segment of length l, say $\Sigma = [-l/2, l/2]$, then $\operatorname{cap}(\Sigma) = l/4$, and the equilibrium distribution is the arcsine distribution:

$$d\mu_\Sigma(x) = \frac{1}{\pi\sqrt{l^2/4 - x^2}}dx, \quad x \in [-l/2, l/2]. \tag{1.7}$$

The corresponding equilibrium potential is easily seen to be

$$U^{\mu_\Sigma}(z) = \log\frac{4}{l} - \log\left|\frac{2z}{l} + \sqrt{\left(\frac{2z}{l}\right)^2 - 1}\right| \tag{1.8}$$

(cf. Example 3.5).

The capacity of an arbitrary Borel set E is defined as

$$\operatorname{cap}(E) := \sup\{\operatorname{cap}(K) \mid K \subseteq E, \quad K \text{ compact}\},$$

and every set (not necessarily a Borel set) that is contained in a Borel set of zero capacity is considered to have capacity zero. Since a Borel set is of positive capacity if and only if it supports a positive measure of finite logarithmic energy, it is then easy to show that the union of countably many sets of zero capacity is again of zero capacity. A property is said to hold *quasi-everywhere* (q.e.) on a set E if the set of exceptional points is of capacity zero.

With this notion it is known, in addition to (1.4), that

$$U^{\mu_\Sigma}(z) = V \tag{1.9}$$

for quasi-every $z \in \Sigma$. For the proofs of all these results see the presentation of the more general "weighted" theory below (for (1.9) see Theorem 1.3(f) and Corollary 4.5, while (1.4) is stated in Corollary II.3.4 in Section II.3), or see [222, Chapter III].

These classical concepts have the following electrostatic interpretation: suppose that Σ is a conductor and a positive unit charge is placed on Σ. If the force between two charged particles is proportional to the reciprocal of their distance, then μ_Σ will be the equilibrium distribution of the charge, i.e. μ_Σ describes the state where the charge attains its minimal energy. Now how is all this changed if, in addition, there is an external electrostatic field present? This will obviously influence the minimal energy state and to get a mathematical model we need a "weighted" version of the classical theory sketched above. In this weighted version we no longer need to restrict Σ to be compact, for a sufficiently strong weight (or external field) will not permit positive mass (or charge) around infinity.

Thus, let $\Sigma \subseteq \mathbf{C}$ be a closed set and $w : \Sigma \to [0, \infty)$. We call such a function a *weight function* on Σ.

Definition 1.1. A weight function w on Σ is said to be *admissible* if it satisfies the following three conditions:

$$\begin{array}{ll} \text{(i)} & w \text{ is upper semi-continuous;} \\ \text{(ii)} & \Sigma_0 := \{z \in \Sigma \mid w(z) > 0\} \text{ has positive capacity}^{\dagger}; \\ \text{(iii)} & \text{if } \Sigma \text{ is unbounded, then } |z|w(z) \to 0 \text{ as } |z| \to \infty,\ z \in \Sigma. \end{array} \tag{1.10}$$

We define $Q = Q_w$ by

$$w(z) =: \exp(-Q(z)). \tag{1.11}$$

Then $Q : \Sigma \to (-\infty, \infty]$ is lower semi-continuous, $Q(z) < \infty$ on a set of positive capacity and if Σ is unbounded, then

$$\lim_{|z|\to\infty,\, z\in\Sigma} \{Q(z) - \log|z|\} = \infty.$$

Let $\mathcal{M}(\Sigma)$ be the collection of all positive unit Borel measures μ with $\operatorname{supp}(\mu) \subseteq \Sigma$ and define the weighted energy integral

$$\begin{aligned} I_w(\mu) &:= \iint \log[|z-t|w(z)w(t)]^{-1} d\mu(z)d\mu(t) \qquad (1.12) \\ &= \iint \log\frac{1}{|z-t|}\, d\mu(z)d\mu(t) + 2\int Q\, d\mu, \end{aligned}$$

where the last representation is valid whenever both integrals exist and are finite. It follows from property (1.10)(iii) that the first integral is well defined. The classical case corresponds to choosing Σ to be compact and $w \equiv 1$ on Σ.

Remark 1.2. There is a certain redundancy in the definitions because if we define

$$\tilde{w}(z) := \begin{cases} w(z) & \text{if } z \in \Sigma \\ 0 & \text{if } z \notin \Sigma, \end{cases}$$

then $\tilde{w}$ will be an admissible weight function on $\mathbf{C}$ and the energy problem for the pair (Σ, w) is equivalent to the one for the pair $(\mathbf{C}, \tilde{w})$, so we could have assumed Σ to be equal to $\mathbf{C}$. However, the above presentation is closer to the classical case and usually Σ will be the essential support of w (e.g. if one considers a problem on the real line, then it is natural to set $\Sigma = \mathbf{R}$).

† Since $\Sigma_0 \subseteq \Sigma$, condition (ii) implies that Σ has positive capacity. Thus the statement "w is an admissible weight on Σ" tacitly assumes that $\operatorname{cap}(\Sigma) > 0$.

In this section we will be primarily interested in the measure that minimizes the weighted energy integral. Every subsequent section will be based on the considerations here. Our basic theorem is

Theorem 1.3. *Let w be an admissible weight on the closed set Σ and let*

$$V_w := \inf\{I_w(\mu) \mid \mu \in \mathcal{M}(\Sigma)\}. \tag{1.13}$$

Then the following properties hold.

(a) *V_w is finite.*

(b) *There exists a unique element $\mu_w \in \mathcal{M}(\Sigma)$ such that*

$$I_w(\mu_w) = V_w.$$

Moreover, μ_w has finite logarithmic energy, i.e.,

$$-\infty < \iint \log \frac{1}{|z-t|} \, d\mu_w(z) d\mu_w(t) < \infty.$$

(c) *$S_w := \operatorname{supp}(\mu_w)$ is compact, is contained in Σ_0 (cf. property* (ii) *above), and has positive capacity.*

(d) *Setting*

$$F_w := V_w - \int Q \, d\mu_w, \tag{1.14}$$

the inequality

$$U^{\mu_w}(z) + Q(z) \geq F_w$$

holds quasi-everywhere on Σ.

(e) *The inequality*

$$U^{\mu_w}(z) + Q(z) \leq F_w$$

holds for all $z \in S_w$.

(f) *In particular, for quasi-every $z \in S_w$,*

$$U^{\mu_w}(z) + Q(z) = F_w.$$

The measure μ_w is called the *equilibrium* or *extremal measure* associated with w. The important constant F_w in (1.14) is called the *modified Robin constant* for w.

Remark 1.4. The proof of Theorem 1.3 actually shows that (cf. (c)) the support S_w of the extremal measure μ_w is contained in

$$\Sigma_\varepsilon := \{z \mid w(z) \geq \varepsilon\}$$

for some $\varepsilon > 0$. Hence Q is bounded on S_w.

Remark 1.5. In Section I.3 we shall see that properties (d) and (f) uniquely characterize the extremal measure μ_w in the sense that if $\mu \in \mathcal{M}(\Sigma)$ has compact support and finite logarithmic energy and satisfies

$$\int \log \frac{1}{|z-t|} d\mu(t) + Q(z) = C \tag{1.15}$$

for q.e. $z \in \text{supp}(\mu)$ and

$$\int \log \frac{1}{|z-t|} d\mu(t) + Q(z) \geq C$$

for q.e. $z \in \Sigma$, then $\mu = \mu_w$ and $C = F_w$. We also mention that there are many other μ's satisfying (1.15) alone.

Remark 1.6. The (first) maximum principle for logarithmic potentials asserts that if μ has compact support and

$$U^\mu(z) \leq M \quad \text{for} \quad z \in \text{supp}(\mu), \tag{1.16}$$

then (1.16) is true for all $z \in \mathbf{C}$ (see Corollary II.3.3 in the next chapter). Using this principle and property (e) for μ_w we can conclude

$$U^{\mu_w}(z) \leq \max_{t \in S_w}\{-Q(t)\} + F_w \tag{1.17}$$

for all $z \in \mathbf{C}$.

In the classical case: Σ compact, $\text{cap}(\Sigma) > 0$, $w \equiv 1$ on Σ, Theorem 1.3 reduces to a theorem of O. Frostman. In fact, in this case, we have $Q \equiv 0$ and hence $F_w = V$. Thus, in this case (f) reduces to (1.9) while (1.17) reduces to (1.4).

Remark 1.7. It easily follows from the inner regularity of Borel measures (see e.g. [195, Theorem 2.18]) and the definition of capacity above that if E is a Borel set of zero capacity and μ has finite logarithmic energy, then $\mu(E) = 0$. In particular, the inequality in (d) and the equality in (f) hold μ_w-almost everywhere.

Proof of Theorem 1.3(a). First we note that since w is finite-valued, upper semi-continuous, and (if Σ is unbounded) $w(z)|z| \to 0$ as $|z| \to \infty$, it follows that w is bounded from above on Σ and the function

$$\log\,[|z-t|w(z)w(t)]^{-1}$$

is bounded from below on $\Sigma \times \Sigma$. Hence the integral defining $I_w(\mu)$ is well-defined for every $\mu \in \mathcal{M}(\Sigma)$ and $V_w > -\infty$.

For $\varepsilon > 0$, let

$$\Sigma_\varepsilon := \{z \mid w(z) \geq \varepsilon\}. \tag{1.18}$$

Then Σ_ε is compact and

$$\Sigma_0 = \bigcup_{n=1}^{\infty} \Sigma_{1/n}.$$

Since w is admissible, Σ_0 has positive capacity, and so there is an n such that $\mathrm{cap}(\Sigma_{1/n}) > 0$. This means that there is a probability measure μ_n supported on $\Sigma_{1/n}$ such that

$$\int\int \log \frac{1}{|z-t|} d\mu_n(z) d\mu_n(t) < \infty.$$

On the support of μ_n (which is contained in $\Sigma_{1/n}$) w is bounded by $1/n$ from below; hence

$$\int\int \log[w(z)w(t)]^{-1} d\mu_n(z) d\mu_n(t)$$

is also finite. Consequently $I_w(\mu_n) < \infty$, and it follows that $V_w < \infty$. □

To prove the unicity in (b) we need

Lemma 1.8. *Let $\mu = \mu_1 - \mu_2$ be a signed Borel measure with compact support and total mass $\mu(\mathbf{C}) = 0$. Suppose further, that each of the positive measures μ_1 and μ_2 has finite logarithmic energy. Then the logarithmic energy of μ is nonnegative:*

$$I(\mu) := \int\int \log \frac{1}{|z-t|} d\mu(z) d\mu(t) \geq 0,$$

and it is zero if and only if $\mu = 0$.

The proof will be given after we have completed the proof of Theorem 1.3; here let us only mention that the finiteness of the logarithmic energies of μ_1 and μ_2 implies that $I(\mu)$ is well-defined. Having Lemma 1.8 at our disposal we can continue with the proof of assertion (b) in the main theorem.

Proof of Theorem 1.3(b). We claim that, for sufficiently small $\varepsilon > 0$,

$$V_w = \inf\{I_w(\mu) \mid \mu \in \mathcal{M}(\Sigma_\varepsilon)\}, \tag{1.19}$$

where Σ_ε is defined in (1.18). Since Σ_ε is compact, Theorem 1.3(b) can then be proved using standard arguments.

To prove the claim (1.19) we shall first show that, for sufficiently small $\varepsilon > 0$,

$$\log[|z-t|w(z)w(t)]^{-1} > V_w + 1 \quad \text{if} \quad (z,t) \notin \Sigma_\varepsilon \times \Sigma_\varepsilon. \tag{1.20}$$

For this purpose it is enough to prove that if $\{(z_n, t_n)\}_{n=1}^{\infty}$ is a sequence with

$$\lim_{n\to\infty} \min(w(z_n), w(t_n)) = 0, \tag{1.21}$$

then

$$\lim_{n\to\infty} \log[|z_n - t_n| w(z_n) w(t_n)]^{-1} = \infty. \tag{1.22}$$

Without loss of generality we may assume that $z_n \to z$, $t_n \to t$ as $n \to \infty$, where z or t or both may be infinity. If z and t are both finite, then (1.22) is obvious from (1.21). If, say, $z = \infty$ but t is finite, then from the admissibility property (iii) of w it follows that

$$|z_n - t_n| w(z_n) \to 0 \quad \text{as} \quad n \to \infty,$$

and we again get (1.22). Finally, if both z and t are infinite, then again from (iii) we have

$$|z_n - t_n| w(z_n) w(t_n) \to 0 \quad \text{as} \quad n \to \infty,$$

and this proves (1.22). Consequently (1.20) holds.

Consider the $\varepsilon > 0$ of (1.20). We shall show that for every $\mu \in \mathcal{M}(\Sigma)$ with $\operatorname{supp}(\mu) \cap (\mathbf{C} \setminus \Sigma_\varepsilon) \neq \emptyset$ and $I_w(\mu) < V_w + 1$, there is a $\tilde{\mu} \in \mathcal{M}(\Sigma_\varepsilon)$ such that $I_w(\tilde{\mu}) < I_w(\mu)$. This clearly implies (1.19) and, moreover, that $I_w(\mu) = V_w$ is possible only for measures with support in Σ_ε. For the μ above we define $\tilde{\mu}$ as $(\mu|_{\Sigma_\varepsilon})/\mu(\Sigma_\varepsilon)$ and notice that because of (1.20) and $I_w(\mu) < V_w + 1$ we must have $\mu(\Sigma_\varepsilon) > 0$. Now (1.20) yields

$$\begin{aligned} I_w(\mu) &= \left(\iint_{\Sigma_\varepsilon \times \Sigma_\varepsilon} + \iint_{\mathbf{C}^2 \setminus \Sigma_\varepsilon \times \Sigma_\varepsilon} \right) \log [|z - t| w(z) w(t)]^{-1} d\mu(z) d\mu(t) \\ &> \mu(\Sigma_\varepsilon)^2 I_w(\tilde{\mu}) + (V_w + 1)\left(1 - \mu(\Sigma_\varepsilon)^2\right), \end{aligned}$$

and since $I_w(\mu) < V_w + 1$, we get $I_w(\tilde{\mu}) < I_w(\mu)$.

Having thus established the claim (1.19) we can now prove assertion (b) of Theorem 1.3. By (1.19) there is a sequence $\{\mu_n\} \subseteq \mathcal{M}(\Sigma_\varepsilon)$ with

$$I_w(\mu_n) \to V_w \quad \text{as} \quad n \to \infty.$$

Each μ_n has support in the compact set Σ_ε; hence by Helly's theorem (Theorem 0.1.3) we can select a weak* convergent subsequence from $\{\mu_n\}$ and, without loss of generality, we can assume that $\{\mu_n\}$ itself converges to $\mu \in \mathcal{M}(\Sigma_\varepsilon)$ in the weak* topology on $\mathcal{M}(\Sigma_\varepsilon)$.

Since w is upper semi-continuous, there exists a sequence $\{w_m\}$ of continuous functions such that $w_{m+1} \le w_m$, $m = 1, 2, \ldots$, and $w_m(z) \to w(z)$ as $m \to \infty$ for every $z \in \Sigma_\varepsilon$. Thus, for every $(z, t) \in \Sigma_\varepsilon \times \Sigma_\varepsilon$, the continuous functions

$$G_m(z, t) := \log \left[\max\left(\frac{1}{m}, |z - t| \right) w_m(z) w_m(t) \right]^{-1}$$

converge monotone increasingly to

$$\log [|z - t| w(z) w(t)]^{-1}$$

(note that on Σ_ε the weight w is uniformly bounded away from zero and infinity), and we get from the monotone convergence theorem that

$$\begin{aligned} I_w(\mu) &= \lim_{m \to \infty} \iint G_m(z, t)\, d\mu(z) d\mu(t) \\ &= \lim_{m \to \infty} \left(\lim_{n \to \infty} \iint G_m(z, t)\, d\mu_n(z) d\mu_n(t) \right) \end{aligned}$$

$$\leq \lim_{m\to\infty}\left(\liminf_{n\to\infty}\iint \log[|z-t|w(z)w(t)]^{-1}d\mu_n(z)d\mu_n(t)\right)$$

$$= \lim_{n\to\infty} I_w(\mu_n) = V_w,$$

where in the second equality we used the continuity of G_m and the fact that if $\mu_n \to \mu$ in the weak* topology, then the product measures $\mu_n(z)\times\mu_n(t)$ converge to $\mu(z)\times\mu(t)$ in the weak* topology on $\mathcal{M}(\Sigma_\varepsilon\times\Sigma_\varepsilon)$. Thus, $\mu =: \mu_w$ is an extremal measure for the right-hand side of (1.13) or (1.19).

That μ_w has finite logarithmic energy follows from $I_w(\mu_w)<\infty$ and the upper boundedness of w on Σ_ε. To prove uniqueness, suppose that $\overline{\mu}\in\mathcal{M}(\Sigma)$ also satisfies $I_w(\overline{\mu}) = V_w$. From the preceding discussion we know that $\overline{\mu}\in\mathcal{M}(\Sigma_\varepsilon)$ and that $\overline{\mu}$ has finite energy. Now since $(\mu_w-\overline{\mu})(\mathbf{C})=0$,

$$J := \iint \log[|z-t|w(z)w(t)]^{-1}d\big(\tfrac{1}{2}(\mu_w-\overline{\mu})\big)(z)\,d\big(\tfrac{1}{2}(\mu_w-\overline{\mu})\big)(t)$$

$$= \iint \log\frac{1}{|z-t|}\,d\big(\tfrac{1}{2}(\mu_w-\overline{\mu})\big)(z)\,d\big(\tfrac{1}{2}(\mu_w-\overline{\mu})\big)(t),$$

which is the ordinary logarithmic energy integral of the compactly supported signed measure $(\mu_w-\overline{\mu})/2$ with total mass zero. Thus, from Lemma 1.8 we know that

$$J \geq 0 \tag{1.23}$$

with equality if and only if $\overline{\mu}=\mu_w$. But

$$I_w\left(\tfrac{1}{2}(\mu_w+\overline{\mu})\right) + J = \tfrac{1}{2}(I_w(\mu_w)+I_w(\overline{\mu})) = V_w$$

and $I_w(\frac{1}{2}(\mu_w+\overline{\mu}))\geq V_w$, so we have from (1.23) that $J=0$, proving $\overline{\mu}=\mu_w$. □

Proof of Theorem 1.3(c). In proving (b) we verified that $\mathcal{S}_w\subseteq\Sigma_\varepsilon$, and since Σ_ε is bounded, $\mathcal{S}_w$ must be compact and $\mathcal{S}_w\subseteq\Sigma_0$. Furthermore, μ_w has finite logarithmic energy (cf. (b)), so $\mathcal{S}_w=\operatorname{supp}(\mu_w)$ must have positive capacity. □

Proof of Theorem 1.3(d). Define

$$\mathcal{U}_w(z) := \int \log\frac{1}{|z-t|}\,d\mu_w(t) + Q(z), \quad z\in\Sigma. \tag{1.24}$$

Since logarithmic potentials are lower semi-continuous and so is Q, we can conclude that $\mathcal{U}_w$ is a lower semi-continuous extended real-valued function on Σ. Consequently, the set $\{z\in\Sigma \mid \mathcal{U}_w(z)\leq\alpha\}$ is closed for each $\alpha\in\mathbf{R}$. Now suppose, to the contrary, that the set

$$A := \{z\in\Sigma \mid \mathcal{U}_w(z) < F_w\}, \tag{1.25}$$

where F_w is defined in (1.14), has positive capacity. Then there exists a large positive integer n_0 such that the compact set

$$E_1 := \left\{ z \in \Sigma \,\middle|\, |z| \le n_0,\ \mathcal{U}_w(z) \le F_w - \frac{1}{n_0} \right\}$$

also has positive capacity. On the other hand, since

$$\int \mathcal{U}_w \, d\mu_w = \iint \log \frac{1}{|z-t|} \, d\mu_w(z) d\mu_w(t) + \int Q \, d\mu_w = F_w, \tag{1.26}$$

there exists a compact set $E_2 \subset \mathcal{S}_w$, disjoint from E_1, such that

$$\mathcal{U}_w(z) > F_w - \frac{1}{2n_0}, \quad z \in E_2,$$

and such that

$$m := \mu_w(E_2) > 0.$$

Now let σ be a positive measure supported on E_1 such that $I_w(\sigma)$ is finite and $\sigma(E_1) = m$. The existence of σ follows from the facts that $\operatorname{cap}(E_1) > 0$ and $E_1 \subset \Sigma_\varepsilon$ (cf. (1.18)) for some $\varepsilon > 0$. But then, for the signed measure σ_1 on Σ defined by

$$\sigma_1 := \sigma \quad \text{on} \quad E_1, \quad \sigma_1 := -\mu_w \quad \text{on} \quad E_2, \quad \sigma_1 := 0 \quad \text{elsewhere},$$

it can be easily verified that for $\eta > 0$ sufficiently small, the measure $\mu_w + \eta\sigma_1 \in \mathcal{M}(\Sigma)$ satisfies

$$I_w(\mu_w + \eta\sigma_1) < V_w. \tag{1.27}$$

In fact, here the left-hand side is

$$I_w(\mu_w) + 2\eta \int \mathcal{U}_w d\sigma_1 + O(\eta^2) \le I_w(\mu_w) - \frac{2\eta m}{2n_0} + O(\eta^2) < I_w(\mu_w)$$

for small η. As (1.27) contradicts the definition of V_w, the set A of (1.25) has capacity zero. Thus Theorem 1.3(d) follows. □

Proof of Theorem 1.3(e). Let $z_0 \in \mathcal{S}_w$ and suppose that

$$\mathcal{U}_w(z_0) > F_w,$$

where $\mathcal{U}_w$ is given by (1.24). Because of the lower semi-continuity of $\mathcal{U}_w$, there exists an open set $\mathcal{N}(z_0)$ around z_0 such that

$$\mathcal{U}_w(z) > F_w + \varepsilon, \quad z \in E := \mathcal{N}(z_0) \cap \mathcal{S}_w, \tag{1.28}$$

where ε is some positive number. Now since μ_w has finite logarithmic energy, the inequality of part (d) holds μ_w-almost everywhere (see Remark 1.7 above). Hence, from (1.26) and (1.28), we have

$$\begin{aligned} F_w = \int \mathcal{U}_w \, d\mu_w &= \int_E \mathcal{U}_w \, d\mu_w + \int_{\mathcal{S}_w \setminus E} \mathcal{U}_w \, d\mu_w \\ &\ge \mu_w(E)\{F_w + \varepsilon\} + [1 - \mu_w(E)]F_w, \end{aligned}$$

which implies that $\mu_w(E) = 0$. But this is absurd, because E is a nonempty relatively open subset of the support $\mathcal{S}_w$. Hence the assertion of (e) follows. □

The unicity of the extremal measure in the proof above was based on Lemma 1.8. This lemma will be used several times later in the book, and now we proceed with its proof.

Proof of Lemma 1.8. Let $z_1 \neq z_2$ be two points from the support of μ, and for large R form the integral

$$\mathcal{J}_R(z_1, z_2) := \frac{1}{2\pi} \int_{|t|\le R} \frac{1}{|t - z_1||t - z_2|} dm(t),$$

where m denotes 2-dimensional Lebesgue measure. The substitution $t \to t + z_2$ easily yields

$$\begin{aligned}
\mathcal{J}_R(z_1, z_2) &= \frac{1}{2\pi} \int_{|t|\le R} \frac{1}{|t||t - z_1 + z_2|} dm(t) + O\left(\frac{1}{R}\right) \\
&= \frac{1}{2\pi} \int_0^R \int_{-\pi}^{\pi} \frac{1}{\left|re^{i\varphi} - z_1 + z_2\right|} d\varphi dr + O\left(\frac{1}{R}\right) \\
&= \frac{1}{2\pi} \int_0^R \int_{-\pi}^{\pi} \frac{1}{\left|re^{i\varphi} - |z_1 - z_2|\right|} d\varphi dr + O\left(\frac{1}{R}\right),
\end{aligned}$$

where $O(1/R)$ is uniform in $z_1, z_2 \in \text{supp}(\mu)$, and where, at the last step, we used the symmetry of the integral. With $x := |z_1 - z_2|$ we have thus

$$\begin{aligned}
\mathcal{J}_R(z_1, z_2) &= \int_0^{R/x} \frac{1}{2\pi} \int_{-\pi}^{\pi} \frac{1}{\left|1 - ue^{i\varphi}\right|} d\varphi du + O\left(\frac{1}{R}\right) \\
&= \text{const.} + \int_2^{R/x} \left(\frac{1}{u} + O\left(\frac{1}{u^2}\right)\right) du + O\left(\frac{1}{R}\right), \\
&= \text{const.} + \log R - \log x + O\left(\frac{1}{R}\right),
\end{aligned}$$

where "const" denote certain constants not necessarily the same at each occurrence. On integrating this relation with respect to $d\mu(z_1)d\mu(z_2)$ and making use of the fact that $\mu(\mathbf{C}) = 0$ we get

$$\begin{aligned}
I(\mu) &= \iint \log \frac{1}{|z_1 - z_2|} d\mu(z_1) d\mu(z_2) \\
&= \iint \frac{1}{2\pi} \int_{|t|\le R} \frac{1}{|t - z_1||t - z_2|} dm(t) d\mu(z_1) d\mu(z_2) + O\left(\frac{1}{R}\right)
\end{aligned}$$

$$= \frac{1}{2\pi} \int_{|t| \le R} \left(\int \frac{1}{|t-z|} d\mu(z) \right)^2 dm(t) + O\left(\frac{1}{R}\right), \tag{1.29}$$

where the justification of the use of Fubini's theorem in the last step runs as follows: Let $|\mu|$ be the total variation of μ and $\|\mu\| = |\mu|(\mathbf{C})$ the total mass of $|\mu|$. Then the argument we applied to derive (1.29) yields

$$I(|\mu|) = \text{const} + \|\mu\|^2 \log R + \frac{1}{2\pi} \int_{|t| \le R} \left(\int \frac{1}{|t-z|} d|\mu|(z) \right)^2 dm(t) + O\left(\frac{1}{R}\right)$$

(note that in the present case $|\mu|(\mathbf{C}) \neq 0$, so there is no cancellation), which shows that

$$\int \frac{1}{|t-z|} d|\mu|(z)$$

exists for almost all t. Furthermore,

$$\iiint_{|t| \le R} \frac{1}{|t-z_1||t-z_2|} dm(t)\, d|\mu|(z_1)\, d|\mu|(z_2)$$

is finite. These facts show that the order of integration can be changed in (1.29).

Letting $R \to \infty$ we get from (1.29) that

$$I(\mu) = \frac{1}{2\pi} \int_{\mathbf{C}} \left(\int \frac{1}{|z-t|} d\mu(z) \right)^2 dm(t), \tag{1.30}$$

and this proves the nonnegativity of the energy integral.

If $I(\mu) = 0$, then in (1.30) the integral

$$\int \frac{1}{|z-t|} d\mu(z) \tag{1.31}$$

must be zero for almost all t; but this integral is continuous for large t, so we can conclude that (1.31) vanishes outside a compact set. Since for $t = Re^{i\varphi}$ we have for large R

$$\frac{1}{|z-t|} = R^{-1} \left(1 - \frac{z}{R} e^{-i\varphi}\right)^{-1/2} \left(1 - \frac{\bar{z}}{R} e^{i\varphi}\right)^{-1/2}$$

$$= R^{-1} \sum_{m=0}^{\infty} \sum_{n=0}^{\infty} (-1)^{m+n} \binom{-1/2}{m} \binom{-1/2}{n} z^m \bar{z}^n e^{-i(m-n)\varphi} R^{-m-n},$$

it follows that for large R and $k = 0, 1, \ldots$

$$\begin{aligned} 0 &= \frac{1}{2\pi} \int_{-\pi}^{\pi} e^{-ik\varphi} \int \frac{1}{|z - Re^{i\varphi}|} d\mu(z)\, d\varphi \\ &= (-1)^k \sum_{m=0}^{\infty} \binom{-1/2}{m} \binom{-1/2}{m+k} R^{-2m-k-1} \int z^m \bar{z}^{m+k} d\mu(z). \end{aligned}$$

Since this power series vanishes for all large R, we must have

$$\int z^m \bar{z}^{m+k} d\mu(z) = 0$$

for all $m, k \geq 0$. By taking complex conjugates we can deduce from this that

$$\int z^m \bar{z}^j d\mu(z) = 0$$

for all $m, j \geq 0$. But the monomials $\{z^m \bar{z}^j\}_{m,j=0}^{\infty}$ span a dense linear subset of the space of continuous functions on the support of μ (indeed, this span includes $x^k y^j$ for all $k, j \geq 0$). Hence we get

$$\int h \, d\mu = 0$$

for all continuous h, and this yields the claim that $\mu = 0$ if $I(\mu) = 0$. □

I.2 Minimum Principle, Dirichlet Problem

In this section we discuss some further properties of equilibrium measures that will be used several times later in the book.

First we establish a technical lemma.

Lemma 2.1. *Let E be a closed subset of* **C**, $z_0 \in$ **C**, *and S the set of radii r on the positive real half-line such that the circle*

$$C_r(z_0) := \{z \mid |z - z_0| = r\}$$

intersects E. If L is the linear measure of S, then $\mathrm{cap}(E) \geq L/4$. *In particular, if E has zero capacity, then S is nowhere dense on* **R**, *i.e. S is closed and its complement* **R** $\setminus S$ *is dense in* **R**.

Proof. Without loss of generality we may assume $z_0 = 0$. Let m denote the Lebesgue measure on **R**, and consider the mapping $T : E \to [0, L]$ defined by $T(z) = m(S \cap [0, |z|])$. It is immediate that T maps E onto $[0, L]$ and T is contractive, i.e.

$$|T(z) - T(w)| \leq |z - w|.$$

If σ is any measure on E, then we define a measure σ^* on $[0, L]$ with the stipulation that $\sigma^*(B) = \sigma(T^{-1}(B))$ for any Borel set $B \subseteq [0, L]$, where T^{-1} denotes inverse image. Next we show that every $\mu \in \mathcal{M}([0, L])$ can be obtained from some $\sigma \in \mathcal{M}(E)$ in this way. This is clear for a point mass and even for a discrete measure. Now let μ be arbitrary, and choose a sequence of discrete measures $\{\mu_n\}$ converging to μ in the weak* topology (for example,

$$\mu_n = \sum_{j=0}^{n} \mu([jL/n, (j+1)L/n])\delta_{jL/n}$$

is a possible choice). Then there are $\sigma_n \in \mathcal{M}(E)$ such that $\sigma_n^* = \mu_n$, and by Helly's theorem (Theorem 0.1.3) we can select from $\{\sigma_n\}$ a weak* convergent subsequence converging to some $\sigma \in \mathcal{M}(E)$, and without loss of generality we assume that the whole sequence converges to σ. We claim that $\mu = \sigma^*$. In fact, if $f \in C[0, L]$ is any continuous function, then by setting $F(z) = f(T(z))$, we can get a continuous F on E. This F can be continuously extended to the whole plane $\mathbf{C}$ with compact support, and we continue to denote the extension by F. Now

$$\int f\,d\sigma^* = \int F\,d\sigma = \lim_{n\to\infty} \int F d\sigma_n$$

by weak* convergence. However, here $\int F\,d\sigma_n = \int f\,d\mu_n$, so the preceding chain of equalities can be continued as

$$= \lim_{n\to\infty} \int f\,d\mu_n = \int f\,d\mu,$$

which shows that the integral of f against σ^* and μ are the same. Since this is true for every $f \in C[0, L]$, the claim $\mu = \sigma^*$ follows.

Now if $\mu = \sigma^*$, then for the corresponding energies, we get from the contractive property of T

$$\begin{aligned} I(\sigma) &= \iint \log \frac{1}{|z-t|}\,d\sigma(t)d\sigma(z) \\ &\le \iint \log \frac{1}{|T(z) - T(t)|}\,d\sigma(t)d\sigma(z) \\ &= \iint \log \frac{1}{|z-t|}\,d\mu(t)d\mu(z) = I(\mu), \end{aligned}$$

which, after taking the infimum for all $\sigma \in \mathcal{M}(\Sigma)$ yields

$$\operatorname{cap}(E) \ge \operatorname{cap}([0, L]) = L/4$$

(see the discussion after (1.6)). □

The same proof shows that, in general, the capacity of a set does not increase under a contractive mapping.

As a consequence of Lemma 2.1 we obtain

Lemma 2.2. *Let w be a continuous admissible weight on Σ. Then*

$$U^{\mu_w}(z_0) \ge -Q(z_0) + F_w$$

provided z_0 is an interior point of Σ. In particular, if $z_0 \in S_w$ is an interior point of Σ, then $U^{\mu_w}(z_0) = -Q(z_0) + F_w$.

This result supplements Theorem 1.3(f) for interior points.

Proof of Lemma 2.2. Without loss of generality we can assume $Q(z_0) < \infty$, for in the opposite case the conclusion is trivial. Let $\varepsilon > 0$ and choose a $\delta > 0$ such that for $|z - z_0| \le \delta$, we have $z \in \Sigma$ and

$$|Q(z) - Q(z_0)| < \varepsilon. \tag{2.1}$$

Set

$$E := \{z \in \Sigma \mid U^{\mu_w}(z) \le -Q(z_0) + F_w - \varepsilon, \ |z - z_0| \le \delta\}.$$

Then E is closed because U^{μ_w} is lower semi-continuous, and, by (2.1) and Theorem 1.3(d), it is of zero capacity. Thus we can apply Lemma 2.1 to choose an $0 < r < \delta$ with $E \cap C_r(z_0) = \emptyset$. This means that, for $z \in C_r(z_0)$,

$$U^{\mu_w}(z) \ge -Q(z_0) + F_w - \varepsilon,$$

and so from the superharmonicity of U^{μ_w} we can conclude

$$U^{\mu_w}(z_0) \ge -Q(z_0) + F_w - \varepsilon.$$

Letting $\varepsilon \to 0$ we arrive at the desired conclusion. □

In the classical case when Σ is compact and $w \equiv 1$, Lemma 2.2 implies that

$$U^{\mu_\Sigma}(z) = F_w = \log \frac{1}{\operatorname{cap}(\Sigma)} \tag{2.2}$$

for every interior point z of Σ. With this remark we can easily prove the following theorem of Evans (see [42]).

Lemma 2.3 (Evans' Theorem). *Let E be a bounded F_σ-set of zero capacity. Then there is a finite measure μ with compact support such that $U^\mu(z) = \infty$ for every $z \in E$. Furthermore, if $z_0 \notin E$ is fixed arbitrarily, then $U^\mu(z_0) < \infty$ can also be achieved.*

Recall that a set is called an F_σ-set if it is a countable union of closed sets. Then any F_σ-set is also a countable union of compact sets.

We remark that a measure μ as in Lemma 2.3 cannot exist for an E of positive capacity, because then, assuming E to be compact, we would get from (1.4) with the equilibrium measure μ_E of E that for any $\varepsilon > 0$ and any real number M

$$U^{\mu_E}(z) \le \varepsilon U^\mu(z) - M$$

for $z \in \operatorname{supp}(\mu_E)$, and so by the principle of domination (Theorem II.3.2) to be proven in the next chapter, the same holds for every z provided ε is sufficiently small (so that $\varepsilon\mu$ has total mass less than 1). But this is clearly impossible for any $z \notin \operatorname{supp}(\mu)$ and sufficiently large M.

Lemma 2.3 has a sharper version in which μ is supported on E (see Theorem III.1.11).

Proof of Lemma 2.3. We assume first that E is compact. Let E_δ be the closed δ-neighborhood of E, i.e. the set of those points whose distance from E is at most δ. We will construct μ of the form

$$\mu = \sum_{k=1}^{\infty} c_k \mu_{E_{\delta_k}} =: \sum_{k=1}^{\infty} c_k \mu_k$$

for some positive sequences $\{c_k\}$ and $\{\delta_k\}$. Then, by (2.2) mentioned before the lemma,

$$U^{\mu}(z) = \sum_{k=1}^{\infty} c_k U^{\mu_k}(z) = \sum_{k=1}^{\infty} c_k \log \frac{1}{\operatorname{cap}(E_{\delta_k})}$$

for every $z \in E$, and μ has total mass

$$\|\mu\| = \sum_{k=1}^{\infty} c_k.$$

If we can show that

$$\lim_{\delta \to 0} \operatorname{cap}(E_\delta) = 0, \tag{2.3}$$

then the choice

$$c_k = \left(\log \frac{1}{\operatorname{cap}(E_{\delta_k})} \right)^{-1/2}$$

will yield the desired μ for some sequence $\{\delta_k\}$ tending to zero sufficiently fast.

Suppose that (2.3) is not true. Then there is a constant K and a sequence $\delta_k \to 0$ such that

$$\log \frac{1}{\operatorname{cap}(E_{\delta_k})} \le K.$$

From the sequence $\{\mu_{E_{\delta_k}}\}$ we can select a subsequence converging to some ν in the weak* topology, and we may assume $\mu_k := \mu_{E_{\delta_k}} \xrightarrow{*} \nu$. But then with a computation very similar to the one in the proof of Theorem 1.3(b), we can write

$$\begin{aligned} I(\nu) &= \lim_{M\to\infty} \iint \min\left(M, \log\frac{1}{|z-t|}\right) d\nu(t) d\nu(z) \\ &= \lim_{M\to\infty} \lim_{k\to\infty} \iint \min\left(M, \log\frac{1}{|z-t|}\right) d\mu_k(t) d\mu_k(z) \\ &\le \liminf_{k\to\infty} I(\mu_k) \le K. \end{aligned}$$

Since obviously ν is supported on E, this contradicts the assumption $\operatorname{cap}(E) = 0$, and this contradiction proves the claim.

If $z_0 \notin E$, then this construction can be carried out in such a way that z_0 is outside the support of μ. Hence $U^{\mu}(z_0) < \infty$, and the lemma is proven for compact sets. If E is now an arbitrary F_σ-set of zero capacity, then E can be represented as $E = \cup_{j=1}^{\infty} H_j$, where each H_j is compact and of zero capacity.

Thus, for each H_j, there is a measure ν_j with support in a fixed compact set independent of j such that $U^{\nu_j}(z_0) < \infty$, but $U^{\nu_j}(z) = \infty$ for every $z \in H_j$. Now it is easy to see that for some positive sequence $\{\alpha_j\}$ tending sufficiently fast to zero, the measure $\mu := \sum \alpha_j \nu_j$ satisfies the conditions set forth in the lemma. □

Now we are ready to prove the so-called generalized minimum principle for superharmonic functions.

Theorem 2.4 (Generalized Minimum Principle). *Let $R \subseteq \overline{\mathbf{C}}$ be a domain and g a superharmonic function on R that is bounded from below and for which*

$$\liminf_{z \to z',\, z \in R} g(z) \geq m \tag{2.4}$$

is satisfied for quasi-every $z' \in \partial R$. Then

$$g(z) > m, \quad z \in R,$$

unless g is constant.

If $R = \mathbf{C}$, then obviously inequality (2.4) holds for quasi-every $z' \in \partial R = \{\infty\}$ for *every* m and we can conclude that *any lower bounded superharmonic function on* $\mathbf{C}$ *is constant.*

As a matter of fact, the lower boundedness can be weakened to

$$\liminf_{|z| \to \infty} \frac{g(z)}{\log |z|} \geq 0.$$

Indeed, then the minimum principle applied to $g(z) + \varepsilon \log |z|$ outside the unit disk shows for $\varepsilon \to 0$ that $g(z) \geq m$, where m is the minimum of g on the unit circle, so g is bounded from below. Thus g is constant as we have just observed.

A similar consequence shows that a Green function (see Sections I.4 and II.4) on a domain G can only exist if ∂G is of positive capacity:

Corollary 2.5. *If $G \subseteq \overline{\mathbf{C}}$ is a domain for which* $\operatorname{cap}(\partial G) = 0$ *and $a \in G$, then there is no nonnegative superharmonic function $g(z)$ on $G \setminus \{a\}$ that tends to infinity as $z \to a$.*

We now give the

Proof of Theorem 2.4. Without loss of generality, we assume $m = 0$ and that g is not constant. We start with the observation that superharmonicity is invariant under conformal mapping. Hence, by applying inversion around a point of R, we can assume that R contains the point infinity together with a neighborhood of it.

Next, we recall the minimum principle proved in Theorem 0.5.2 that if a superharmonic function attains its minimum at a point (on a domain), then it must be constant. Hence, we only have to prove that the infimum of g on R is at least 0.

Let m_g be this infimum, and let us assume that $m_g < 0$. Then

$$\liminf_{z \to \infty} g(z) \geq g(\infty) > m_g. \tag{2.5}$$

Let C be a circle containing ∂R in its interior D. By (2.5) and the minimum principle of Theorem 0.5.2 we can see that for some $0 > m_1 > m_g$

$$\inf_{z \notin D} g(z) > m_1. \tag{2.6}$$

Let $m_g < m_3 < m_2 < m_1$, and fix an arbitrary point $z_0 \in D \cap R$. Consider the set

$$E := \{z \in R \mid g(z) < m_2\}.$$

Using the lower semi-continuity of g we get from (2.6) that the closure of E is contained in $(D \cap R) \cup \partial R$. By (2.4) the set $\overline{E} \cap \partial R$ is compact and of zero capacity; hence by Lemma 2.3, there is a potential $U^{\mu}(z)$, where μ is a finite positive measure with compact support, that takes the value $+\infty$ everywhere on $\overline{E} \cap \partial R$ and $U^{\mu}(z_0) < \infty$. But then for every point z' of $\overline{E} \cap \partial R$ we have

$$\liminf_{z \to z',\, z \in R} U^{\mu}(z) = \infty;$$

hence for every $\varepsilon > 0$

$$\liminf_{z \to z',\, z \in R} (g(z) + \varepsilon U^{\mu}(z)) = \infty \tag{2.7}$$

(recall that g is lower bounded on R). At other points of $\partial(D \cap R)$ we have

$$\liminf_{z \to z',\, z \in R} g(z) \geq m_2.$$

Hence if γ is a lower bound of U^{μ} on $\overline{D}$, then for every $z' \in \partial(D \cap R)$

$$\liminf_{z \to z',\, z \in R} (g(z) + \varepsilon U^{\mu}(z)) \geq m_2 + \varepsilon\gamma \geq m_3 \tag{2.8}$$

if $0 < \varepsilon < (m_2 - m_3)/|\gamma|$.

From (2.7) and (2.8) we can deduce via Theorem 0.5.2 that

$$g(z) + \varepsilon U^{\mu}(z) \geq m_3$$

in $D \cap R$. If $\varepsilon \to 0$, then we arrive at $g(z_0) \geq m_3$, and since here $z_0 \in D \cap R$ was arbitrary, we obtain

$$\inf_{z \in D \cap R} g(z) \geq m_3. \tag{2.9}$$

But $m_g < m_3$, (2.6), and (2.9) contradict the definition of m_g, and this contradiction proves the generalized minimum principle. □

As another application of Lemma 2.3 we prove a technical lemma that allows us to recognize certain solutions of Dirichlet problems. To do this we introduce the concept of the Perron–Wiener–Brelot solution of the Dirichlet problem (see also Appendix A.2).

Consider a domain $R \subseteq \mathbf{C} \cup \{\infty\} =: \overline{\mathbf{C}}$ such that $\overline{\mathbf{C}} \setminus R$ has positive capacity, and suppose that f is a bounded Borel measurable function defined on ∂R. The upper and lower classes of functions corresponding to f and R are defined as

$$\mathcal{H}_f^{u,R} := \{g \mid g \text{ superharmonic and bounded below on } R, \liminf_{z\to x,\, z\in R} g(z) \geq f(x) \quad \text{for all} \quad x \in \partial R\}$$

and

$$\mathcal{H}_f^{l,R} := \{g \mid g \text{ subharmonic and bounded above on } R, \limsup_{z\to x,\, z\in R} g(z) \leq f(x) \quad \text{for all} \quad x \in \partial R\},$$

and the upper and lower solutions of the Dirichlet problem for the boundary function f are given by

$$\overline{H}_f^R(z) := \inf\left\{g(z) \,\middle|\, g \in \mathcal{H}_f^{u,R}\right\}, \quad z \in R,$$

and

$$\underline{H}_f^R(z) := \sup\left\{g(z) \,\middle|\, g \in \mathcal{H}_f^{l,R}\right\}, \quad z \in R.$$

If $\overline{H}_f^R \equiv \underline{H}_f^R$, then this function, denoted by H_f^R, is called the *Perron–Wiener–Brelot solution* of the Dirichlet problem on R for the boundary function f.

The following lemma will be useful in establishing several further results.

Lemma 2.6. *Let $R \subseteq \overline{\mathbf{C}}$ be a domain such that ∂R is of positive capacity, and let f be a bounded lower or upper semi-continuous function on ∂R. If u is a bounded harmonic function on R such that for quasi-every $z \in \partial R$*

$$\lim_{x\to z,\, x\in R} u(x) = f(z),$$

then u is the Perron–Wiener–Brelot solution of the Dirichlet problem on R with boundary function f.

The lemma is actually true for any bounded Borel function, but the above formulation is sufficient for our purposes.

Proof of Lemma 2.6. By applying a Möbius transformation if necessary, we may suppose that ∂R is compact. It is enough to show that $u \equiv \overline{H}_f^R$, because the proof of $u \equiv \underline{H}_f^R$ is identical (or apply the former relation to $-u$ and $-f$), which shows that both the upper and the lower solutions coincide with u, and so the solution to the Dirichlet problem in question exists and equals u.

Let $g \in \mathcal{H}_f^{u,R}$ be any upper function. Then $g - u$ is superharmonic, bounded below and, by the assumption of the lemma,

$$\liminf_{z\to x,\, z\in R} (g(z) - u(z)) \geq 0$$

for quasi-every $x \in \partial R$. Hence, it follows from Theorem 2.4 that $g(z) \geq u(z)$ for every $z \in R$. Since here $g \in \mathcal{H}_f^{u,R}$ is arbitrary, the inequality

$$\overline{H}_f^R(z) \geq u(z), \quad z \in R,$$

follows.

It remains to show that here the equality sign holds for every $z \in R$. Let $z_0 \in R$ be fixed. We have to show that $u(z_0)$ is the infimum of the values $g(z_0)$ of some upper functions $g \in \mathcal{H}_f^{u,R}$. We have assumed that

$$\liminf_{z \to x,\, z \in R} u(z) \geq f(x)$$

holds for quasi-every $x \in \partial R$. Let $E \subseteq \partial R$ be the set where the opposite inequality is true:

$$E := \left\{ x \in \partial R \,\middle|\, \liminf_{z \to x,\, z \in R} u(z) < f(x) \right\}.$$

Our first aim is to prove that E is an F_σ-set, i.e. the countable union of closed sets. This is immediate if f is upper semi-continuous, for then every set of the form

$$E_\varepsilon := \left\{ x \in \partial R \,\middle|\, \liminf_{z \to x,\, z \in R} u(z) \leq f(x) - \varepsilon \right\}$$

is closed, and clearly $E = \bigcup_{n=1}^\infty E_{1/n}$. If, however, f is lower semi-continuous, then f is the pointwise limit of an increasing sequence of continuous functions $\{f_m\}_{m=1}^\infty : f_m(x) \nearrow f(x)$ as $m \to \infty$ for every $x \in \partial R$. If we set

$$E^m := \left\{ x \in \partial R \,\middle|\, \liminf_{z \to x,\, z \in R} u(z) < f_m(x) \right\},$$

then we have just seen that E^m is an F_σ-set (note that f_m is continuous). But then $E = \bigcup_{m=1}^\infty E^m$ is also F_σ as we have claimed.

Thus, E is an F_σ-set of zero capacity; hence Lemma 2.3 guarantees the existence of a measure σ of compact support such that $U^\sigma(z) = \infty$ for every $z \in E$ and $U^\sigma(z_0) < \infty$.

Suppose now that R is bounded, and for an $\varepsilon > 0$ consider the function

$$g_\varepsilon(z) := u(z) + \varepsilon(U^\sigma(z) - m),$$

where m is the infimum of U^σ on ∂R. Since g_ε is superharmonic, the choice of E and σ yields that g_ε is an upper function: $g_\varepsilon \in \mathcal{H}_f^{u,R}$. Letting $\varepsilon \to 0$, we see that $u(z_0)$ is the infimum of the values $g_\varepsilon(z_0)$, $\varepsilon > 0$. This proves that $u = \overline{H}_f^R$ for bounded domains R.

If R is unbounded, then we have to modify the above g_ε as

$$g_\varepsilon(z) = u(z) + \varepsilon(U^\sigma(z) - m) - \varepsilon\|\sigma\| \left(U^{\mu_{\partial R}}(z) - \log \frac{1}{\operatorname{cap}(\partial R)} \right),$$

where $\mu_{\partial R}$ denotes the equilibrium measure of the set ∂R. In view of (1.4) we can again deduce that $g_\varepsilon \in \mathcal{H}_f^{u,R}$, and the rest is the same as above. □

I.3 The Extremal Measure

In this section we characterize the extremal (or equilibrium) measure introduced in Section I.1 via its potential. To do this we need the following theorem.

Principle of Domination. *Let μ and ν be two positive finite Borel measures with compact support on* $\mathbf{C}$, *and suppose that the total mass of ν does not exceed that of μ. Assume further that μ has finite logarithmic energy. If, for some constant c, the inequality*

$$U^{\mu}(z) \le U^{\nu}(z) + c \tag{3.1}$$

holds μ-almost everywhere, then it holds for all $z \in \mathbf{C}$.

In other words, we get the inequality (3.1) *everywhere* provided we have it on a "large" subset of supp(μ).

The principle of domination will be proved in Theorem 3.2 of Chapter II.

We now turn to the promised characterization of the extremal measure μ_w for the energy problem. For convenience let

$$\text{“inf”}_{z\in H}\ h(z)$$

denote the largest number L such that on H the real function h takes values smaller than L only on a set of zero capacity, and we similarly define "sup". With this shorthand notation parts (d) and (e) of Theorem 1.3 assert that

$$\text{“inf”}_{z\in\Sigma}\ (U^{\mu_w}(z) + Q(z)) = F_w \tag{3.2}$$

and

$$\sup_{z\in\operatorname{supp}(\mu_w)} (U^{\mu_w}(z) + Q(z)) = F_w, \tag{3.3}$$

where F_w is the modified Robin constant of (1.14). Now we show that (3.2) and (3.3) are extremal cases of inequalities involving general measures.

Theorem 3.1. *Let w be an admissible weight. Then, for any $\sigma \in \mathcal{M}(\Sigma)$ with compact support,*

$$\text{“inf”}_{z\in\Sigma}\ (U^{\sigma}(z) + Q(z)) \le F_w \tag{3.4}$$

and

$$\sup_{z\in\operatorname{supp}(\sigma)} (U^{\sigma}(z) + Q(z)) \ge F_w. \tag{3.5}$$

If equality holds both in (3.4) *and* (3.5), *then $\sigma = \mu_w$.*

Remark 3.2. The proof shows that on the left of (3.5) we can write "sup" provided σ has finite logarithmic energy.

Proof of Theorem 3.1. First consider (3.4), and suppose that for some L_1, and quasi-every $z \in \Sigma$

$$U^\sigma(z) + Q(z) \geq L_1.$$

Then we get from (3.3)

$$U^\sigma(z) \geq U^{\mu_w}(z) - F_w + L_1$$

for quasi-every $z \in \operatorname{supp}(\mu_w)$. Since μ_w has finite logarithmic energy, this inequality holds μ_w-almost everywhere (cf. Remark 1.7). Thus, by the principle of domination, the inequality holds for all z, and letting $z \to \infty$ we get $L_1 \leq F_w$.

The proof of (3.5) is similar. Suppose that for every $z \in \operatorname{supp}(\sigma)$

$$U^\sigma(z) + Q(z) \leq L_2, \tag{3.6}$$

where we need only to consider the case when $L_2 \leq F_w$. But then L_2 is finite, and since Q is bounded from below we get from (3.6) that σ has finite logarithmic energy. The inequalities (3.6) and (3.2) imply that

$$U^{\mu_w}(z) - F_w + L_2 \geq U^\sigma(z)$$

for quasi-every $z \in \operatorname{supp}(\sigma)$. Hence, as in the first part of the proof, we get $L_2 \geq F_w$.

If $L_1 = L_2 = F_w$, then the considerations above show that the potentials of μ_w and σ coincide everywhere, and so $\sigma = \mu_w$ follows from Lemma 1.8 (see also Corollary II.2.2). □

A useful consequence of Theorem 3.1 is the following:

Theorem 3.3. *Let w be an admissible weight. If $\sigma \in \mathcal{M}(\Sigma)$ has compact support and finite logarithmic energy, and*

$$U^\sigma(z) + Q(z)$$

coincides with a constant F quasi-everywhere on the support of σ and is at least as large as F quasi-everywhere on Σ, then $\sigma = \mu_w$ and $F = F_w$.

The same conclusion holds if we know that σ has finite logarithmic energy, $\operatorname{supp}(\sigma) \subseteq \mathcal{S}_w$*, and $U^\sigma(z) + Q(z)$ coincides with a constant F for quasi-every $z \in \mathcal{S}_w$.*

Proof. The assumptions imply that

$$\text{“}\inf_{z\in\Sigma}\text{”}\,(U^\sigma(z) + Q(z)) = F$$

and

$$\text{“}\sup_{z\in\operatorname{supp}(\sigma)}\text{”}\,(U^\sigma(z) + Q(z)) = F.$$

From Theorem 3.1 and Remark 3.2 we get first that $F_w \leq F \leq F_w$, i.e. $F = F_w$, and then that $\sigma = \mu_w$.

The second part of the theorem easily follows from the principle of domination by an argument similar to that used in Theorem 3.1. In fact, the assumptions and the principle of domination imply on the one hand that

$$U^{\sigma}(z) - F \le U^{\mu_w}(z) - F_w, \qquad z \in \mathbf{C},$$

and, on the other hand, the converse of this inequality. Thus, the equality sign must hold for every z, and for $|z| \to \infty$ we obtain $F = F_w$, which implies $U^{\sigma} \equiv U^{\mu_w}$ in view of what we have just established. Now $\sigma = \mu_w$ follows as before from Lemma 1.8. □

As simple applications of Theorem 3.3, we now determine the capacity and equilibrium distribution for a disk and for a line segment in the classical (unweighted) case.

Example 3.4. Let Σ be the closed disk $\overline{D_r(a)}$ of radius r or its circumference $C_r(a) := \{z \mid |z-a| = r\}$ with $w \equiv 1$. Set $d\sigma = ds/2\pi r$, where ds denotes arc measure on $C_r(a)$. Then, as shown in (0.5.5),

$$U^{\sigma}(z) = \begin{cases} \log \dfrac{1}{r} & \text{if } |z-a| \le r \\[2ex] \log \dfrac{1}{|z-a|} & \text{if } |z-a| > r, \end{cases}$$

and so U^{σ} is constant on Σ. Hence, by Theorem 3.3, we have $d\mu_{\Sigma} = ds/2\pi r$ and (cf. (2.2))

$$\log \frac{1}{r} = F_w = \log \frac{1}{\operatorname{cap}(\Sigma)};$$

that is, $\operatorname{cap}(\Sigma) = r$. □

Example 3.5. For a finite interval $\Sigma = [a, b] \subset \mathbf{R}$ with $w \equiv 1$ we show that

$$\operatorname{cap}(\Sigma) = \frac{b-a}{4} \quad \text{and} \quad d\mu_{\Sigma} = \frac{1}{\pi} \frac{dx}{\sqrt{(x-a)(b-x)}}, \quad x \in [a, b].$$

It suffices to consider only $\Sigma = [-1, 1]$. Set

$$d\sigma = \frac{1}{\pi} \frac{dx}{\sqrt{1-x^2}}, \qquad x \in [-1, 1].$$

Then $\sigma \in \mathcal{M}([-1, 1])$ and with the change of variable $x = \cos\theta$ we have

$$U^{\sigma}(z) = \frac{1}{\pi} \int_{-1}^{1} \log \frac{1}{|z-x|} \frac{dx}{\sqrt{1-x^2}} = \frac{1}{2\pi} \int_{-\pi}^{\pi} \log \frac{1}{|z - \cos\theta|} d\theta.$$

Now we apply the Joukowski transformation

$$z = \frac{1}{2}(\zeta + \zeta^{-1}),$$

which maps $|\zeta| > 1$ onto $\mathbf{C}\setminus[-1, 1]$ and maps the unit circle $|\zeta| = 1$ onto $[-1, 1]$ (covered twice). Its inverse is $h = z + \sqrt{z^2 - 1}$ with $\sqrt{z^2 - 1}$ denoting the branch that behaves like z near infinity. With $t = e^{i\theta}$ we compute

$$|z - \cos\theta| = \left|\frac{1}{2}(\zeta + \zeta^{-1}) - \frac{1}{2}(t + t^{-1})\right| = \frac{1}{2}|\zeta - t||\zeta^{-1} - t|.$$

Thus

$$U^\sigma(z) = \frac{1}{2\pi}\int_{-\pi}^{\pi} \log\frac{2}{|\zeta - t||\zeta^{-1} - t|}d\theta = \log 2 + U^{\mu_C}(\zeta) + U^{\mu_C}(\zeta^{-1}),$$

where $d\mu_C = d\theta/2\pi$ is, from the preceding example, the equilibrium distribution for the unit circle C. Consequently,

$$U^\sigma(z) = \log 2 + \log\frac{1}{|\zeta|} + \log 1 = \log 2 - \log|z + \sqrt{z^2 - 1}|.$$

In particular, for $z \in [-1, 1] = \operatorname{supp}(\sigma)$, we have $U^\sigma(z) = \log 2$, so that, by Theorem 3.3 and (2.2),

$$d\sigma = d\mu_{[-1,1]} \quad \text{and} \quad \operatorname{cap}([-1, 1]) = 1/2.$$

□

The next result, which will also be frequently used, gives a lower estimation for weighted polynomials.

Theorem 3.6. *Let w be an admissible weight and $P_n(z) = z^n + \cdots$ be a monic polynomial of degree n. Then*

$$\text{“}\sup_{z\in S_w}\text{”}\, [w(z)]^n |P_n(z)| \geq \exp(-nF_w).$$

Proof. Let $\tilde{w} := w_{|S_w}$, $\tilde{\Sigma} := S_w$. Then $\mu_{\tilde{w}} = \mu_w$ is obvious from the definitions. Furthermore, let σ_n be the discrete measure that has mass $1/n$ at every zero of P_n (counting multiplicity) so that

$$U^{\sigma_n}(z) = \frac{1}{n}\log\frac{1}{|P_n(z)|}.$$

We observe that in the proof of (3.4) we did not use the fact that σ had support in Σ, i.e. (3.4) holds for any σ with compact support. In particular, we can apply (3.4) for $\tilde{w}$, $\tilde{\Sigma}$, σ_n. If we raise the so obtained inequality to the n-th power, we obtain Theorem 3.6. □

Remark 3.7. Theorem 3.6 extends the familiar fact that for any monic polynomial $p(z) = z^n + \cdots$ and any compact set E, we have

$$\sup_{z\in E}|p(z)| \geq [\operatorname{cap}(E)]^n. \tag{3.7}$$

In the classical case when Σ is compact, $\text{cap}(\Sigma) > 0$ and $w \equiv 1$ on Σ, the relations (3.4) and (3.5) imply for any $\sigma \in \mathcal{M}(\Sigma)$

$$\text{“}\inf_{z\in\Sigma}\text{”}\ U^{\sigma}(z) \le \log\frac{1}{\text{cap}(\Sigma)} \tag{3.8}$$

and

$$\sup_{z\in\text{supp}(\sigma)} U^{\sigma}(z) \ge \log\frac{1}{\text{cap}(\Sigma)}, \tag{3.9}$$

with equality in both places if and only if $\sigma = \mu_\Sigma$. Equality in (3.8) alone is not sufficient for making the conclusion $\sigma = \mu_\Sigma$ as is shown by the following example: Let $\Sigma = \{z \mid |z| \le 1\}$ and $\sigma = \delta_0$, the unit measure placed at zero. In fact, in this case the equilibrium measure is the normalized arc measure on $\partial\Sigma$, $\text{cap}(\Sigma) = 1$, and

$$\mathring{U}^{\sigma}(z) = \log\frac{1}{|z|} \ge 0 = \log\frac{1}{\text{cap}(\Sigma)}$$

for all $z \in \Sigma$, i.e. in (3.8) equality holds but $\sigma \ne \mu_\Sigma$.

In contrast, if we have equality in (3.9), then we must have $\sigma = \mu_\Sigma$. This is immediate since if $U^{\sigma}(z) \le \log(1/\text{cap}(\Sigma))$ for every $z \in \text{supp}(\sigma)$, then for the logarithmic energy of σ we get

$$I(\sigma) = \int U^{\sigma} d\sigma \le \log\frac{1}{\text{cap}(\Sigma)} = I(\mu_\Sigma),$$

and so $\sigma = \mu_\Sigma$ follows from the extremality and unicity of μ_Σ. The following example demonstrates that in the weighted case we may have equality in (3.5) without having $\sigma = \mu_w$ (that equality in (3.4) does not imply $\sigma = \mu_w$ was shown above even for the unweighted case): Let $\Sigma = \{z \mid 1 \le |z| \le 2\}$ and $w(z) = |z|^{-1}$ for $z \in \Sigma$. If σ_1 and σ_2 denote the normalized arc measures on the inner and outer bounding circles of Σ, respectively, then

$$U^{\sigma_1}(z) = \log\frac{1}{|z|} = -Q(z)$$

for all $z \in \Sigma$ (recall Example 3.4). Hence Theorem 3.3 implies that $\mu_w = \sigma_1$ and $F_w = 0$. But for $z \in \text{supp}(\sigma_2) = \{z \mid |z| = 2\}$ we have

$$U^{\sigma_2}(z) = \log\frac{1}{2} = -Q(z) + F_w.$$

Thus for σ_2 the equality sign holds in (3.5), although $\sigma_2 \ne \sigma_1 = \mu_w$.

The assumption in Theorem 3.3 that

$$U^{\sigma}(z) + Q(z) \ge F_w$$

quasi-everywhere on Σ is essential; without it the conclusion need not hold. Indeed, there are many different measures μ with compact support and finite logarithmic energy such that

$$U^{\mu}(z) + Q(z) = \text{ const } = c$$

quasi-everywhere on $\text{supp}(\mu)$. In fact, if $\Sigma_1 \subseteq \Sigma$ is closed and w is not q.e. zero on Σ_1, and if we set $w_1 = w_{|_{\Sigma_1}}$, then μ_{w_1} will be such a measure. Note also that with this "restriction" procedure we get all such μ's (take $\Sigma_1 = \text{supp}(\mu)$ and apply Theorem 3.3). However, of all these μ's only μ_w will satisfy

$$U^{\mu}(z) + Q(z) \geq c$$

quasi-everywhere throughout Σ.

Theorem 3.3 also shows that if we change w outside $\mathcal{S}_w$ in such a way that (3.2) still holds, then for the obtained weight v we have $\mu_w = \mu_v$ (assuming v is admissible). Thus, in general, there are many essentially different weights for which the extremal measures coincide.

In the classical (unweighted) case $w \equiv 1$ the measure $\mu_w = \mu_\Sigma$ is always supported on the outer boundary of (the then compact set) Σ, i.e. on the boundary of the unbounded component of $\mathbf{C} \setminus E$, (see Corollary 4.5 in the next section). In the weighted case no such "structure theorem" can be given, for the next result says that "most" measures can arise as μ_w for some admissible w.

It easily follows from Theorem 1.3 (see Theorems 4.3 and 4.4 in the next section) that the potential U^{μ_w} is quasi-everywhere continuous on $\mathbf{C}$ and bounded on $\mathcal{S}_w = \text{supp}(\mu_w)$. Now we show that these are the only restrictions for the potential of a μ_w and hence for a μ_w.

Theorem 3.8. *Let $\mu \in \mathcal{M}(\mathbf{C})$ have compact support $\mathcal{S}$. Then $\mu = \mu_w$ (and $\mathcal{S} = \mathcal{S}_w$) for some admissible weight w (defined on $\Sigma := \mathbf{C}$) if and only if U^{μ} is bounded on $\mathcal{S}$ and continuous quasi-everywhere on* $\mathbf{C}$.

Proof. We only have to prove that the conditions are sufficient. The boundedness of U^{μ} on $\mathcal{S} = \text{supp}(\mu)$ implies that μ has finite logarithmic energy and so $\mathcal{S}$ is of positive capacity. Applying the principle of domination to μ we get that any upper bound of U^{μ} on $\mathcal{S}$ is an upper bound for it throughout $\mathbf{C}$ (cf. Corollary II.3.3). Hence U^{μ} is bounded from above on $\mathbf{C}$.

Suppose $\mathcal{S} \subseteq D_R := \{z \mid |z| < R\}$, and set

$$-Q(z) = \begin{cases} \limsup\limits_{x \to z} U^{\mu}(x) & \text{if } \ z \in \mathcal{S}, \\ U^{\mu}(z) & \text{if } \ z \in D_R \setminus \mathcal{S}, \\ U^{\mu}(z) - \log(|z|/R) & \text{if } \ z \notin D_R. \end{cases}$$

It is easily seen that $-Q$ is upper semi-continuous, bounded from above, and

$$\lim_{z \to \infty} [-Q(z) + \log |z|] = -\infty.$$

Thus $w(z) := \exp(-Q(z))$ is an admissible weight.

Clearly,

$$U^{\mu}(z) + Q(z) \geq 0$$

at every continuity point of U^μ and so this inequality holds quasi-everywhere on $\mathbf{C}$ by our assumptions. Furthermore, from the lower semi-continuity of U^μ we get

$$U^\mu(z) + Q(z) \le 0$$

for every $z \in \mathcal{S} = \text{supp}(\mu)$. Thus $\mu = \mu_w$ follows from Theorem 3.3. □

I.4 The Equilibrium Potential

In this section we examine the equilibrium potential

$$U^{\mu_w}(z) = \int \log \frac{1}{|z-t|} d\mu_w(t).$$

First we show that the results of the preceding section yield

Theorem 4.1. *Let $\mathcal{H}$ be the set of all superharmonic functions $g(z)$ on $\mathbf{C}$ that are harmonic for large $|z|$, and $g(z) + \log|z|$ is bounded from below near ∞. Then, for an admissible weight $w : \Sigma \to [0, \infty)$, the function*

$$U^{\mu_w}(z) - F_w$$

is the lower envelope of the functions g in $\mathcal{H}$ satisfying $g(z) \ge -Q(z)$ for quasi-every $z \in \Sigma$. Furthermore, the same conclusion holds if $g(z) \ge -Q(z)$ is required quasi-everywhere only on $\mathcal{S}_w$.

Proof. We start the proof with the observation that every $g \in \mathcal{H}$ is of the form

$$g(z) = u(z) + \int \log \frac{1}{|z-t|} d\nu(t),$$

where ν is compactly supported and u is harmonic on $\mathbf{C}$. This is an easy consequence of the Riesz decomposition theorem (Theorem II.3.1) and of the unicity theorem (Theorem II.2.1) to be proven in the next chapter.

The assumption of the lower boundedness of $g(z)+\log|z|$ near infinity implies that the function $u(z) + (1 - \nu(\mathbf{C}))\log|z|$ is bounded from below near ∞. Thus, by Corollary 0.3.7, u must be of the form $c_2 \log|z| + v(z)$, where v is harmonic at ∞. But then (recall that u is harmonic on the whole plane) the finiteness of u and the maximum principle forces $c_2 = 0$. Thus u is bounded near ∞ and so it is bounded on the whole plane; therefore, u is constant by Theorem 0.4.11.

So far we have shown that every $g \in \mathcal{H}$ is of the form

$$g(z) = \text{const} + \int \log \frac{1}{|z-t|} d\nu(t),$$

where ν is compactly supported. From the lower boundedness of $g(z) + \log|z|$ near infinity we then get $\nu(\mathbf{C}) \le 1$. Now if $g(z) \ge -Q(z)$ for quasi-every $z \in \mathcal{S}_w$, then Theorem 1.3(f) implies that

$$U^{\mu_w}(z) - F_w \leq g(z) \tag{4.1}$$

quasi-everywhere on $\mathcal{S}_w$, and so the desired conclusion follows from the principle of domination (Theorem II.3.2), according to which (4.1) is true everywhere (note also that $U^{\mu_w}(z) - F_w \in \mathcal{H}$ by Theorem 1.3(d)). □

Corollary 4.2. *$U^{\mu_w} - F_w$ and $-F_w$ are increasing uniformly continuous convex functions of $-Q := \log w$. More precisely, for admissible weights on Σ, we have:*

(a) *$w \leq v$ on Σ implies*

$$F_w \geq F_v, \tag{4.2}$$

and, for every $z \in \mathbf{C}$,

$$U^{\mu_w}(z) - F_w \leq U^{\mu_v}(z) - F_v. \tag{4.3}$$

(b) *If $|\log w - \log v| \leq \varepsilon$ on Σ, then*

$$|F_w - F_v| \leq \varepsilon$$

and, for every $z \in \mathbf{C}$,

$$|(U^{\mu_w}(z) - F_w) - (U^{\mu_v}(z) - F_v)| \leq \varepsilon. \tag{4.4}$$

(c) *If $w = \prod_{i=1}^{n} w_i^{\alpha_i}$ with $\alpha_i \geq 0$, $\sum_{i=1}^{n} \alpha_i = 1$, then*

$$F_w \geq \sum_{i=1}^{n} \alpha_i F_{w_i}, \tag{4.5}$$

and, for every $z \in \mathbf{C}$,

$$U^{\mu_w}(z) - F_w \leq \sum_{i=1}^{n} \alpha_i \left(U^{\mu_{w_i}}(z) - F_{w_i}\right). \tag{4.6}$$

According to Remark 1.2, we can assume without loss of generality that $\Sigma = \mathbf{C}$ for all the weights appearing in Corollary 4.2, i.e. the weights are defined everywhere. By the assumption $|\log w - \log v| \leq \varepsilon$ of part (b) we mean, more precisely, that $w(z) = 0$ if and only if $v(z) = 0$ and, if $w(z) > 0$, then $|\log[w(z)/v(z)]| < \varepsilon$.

Proof of Corollary 4.2. (a) Write $w = \exp(-Q)$, $v = \exp(-q)$. Then $w \leq v$ implies $-Q \leq -q$; hence (4.3) follows from Theorem 4.1. If we add $\log|z|$ to both sides of (4.3) and let $z \to \infty$, then we obtain (4.2).

(b) With the same notations as above we have

$$Q - \varepsilon \leq q \leq Q + \varepsilon.$$

For the weight $W(z) := \exp(-Q(z) - \varepsilon) = w(z)e^{-\varepsilon}$, we have $\mu_W = \mu_w$, $V_W = V_w + 2\varepsilon$ (cf. (1.13)), and so by (1.14)

$$U^{\mu_W}(z) = U^{\mu_w}(z), \quad F_W = F_w + \varepsilon.$$

Thus, by part (a), we have for all $z \in \mathbf{C}$

$$U^{\mu_w}(z) - F_w - \varepsilon = U^{\mu_W}(z) - F_W \leq U^{\mu_v}(z) - F_v,$$

and similarly it follows that

$$U^{\mu_v}(z) - F_v \leq U^{\mu_w}(z) - F_w + \varepsilon,$$

which proves (4.4). The first statement of (b) follows by letting $z \to \infty$ in (4.4).

(c) Let $w = \exp(-Q)$, $w_i = \exp(-Q_i)$. Then

$$-Q(z) = \sum_{i=1}^{n} \alpha_i(-Q_i(z)),$$

and we get from Theorem 1.3(d) that

$$\sum_{i=1}^{n} \alpha_i \left(U^{\mu_{w_i}}(z) - F_{w_i}\right) \geq -Q(z)$$

for quasi-every $z \in \Sigma = \mathbf{C}$. Now inequality (4.6) is a consequence of this and Theorem 4.1. Inequality (4.5) can be obtained from (4.6) by adding $\log|z|$ to both sides and letting $z \to \infty$. □

Next we consider the boundedness of U^{μ_w}.

Theorem 4.3. *Let w be admissible. Then the equilibrium potential U^{μ_w} is bounded on compact subsets of* $\mathbf{C}$.

Proof. U^{μ_w} is clearly bounded from below on compact subsets of $\mathbf{C}$, and now we show that it is bounded from above on all of $\mathbf{C}$. The principle of domination (Theorem II.3.2) implies that it is enough to show that U^{μ_w} is bounded from above on $\mathcal{S}_w$ (cf. also the maximum principle Corollary II.3.3), for then the same upper bound will serve on the whole complex plane. But the boundedness from above of U^{μ_w} on $\mathcal{S}_w$ is immediate from Theorem 1.3(e) and the lower boundedness of Q. □

Theorem 4.4. *U^{μ_w} is continuous at every $z \notin \mathcal{S}_w$ and at every $z \in \mathcal{S}_w$ where*

$$U^{\mu_w}(z) + Q(z) = F_w; \tag{4.7}$$

hence U^{μ_w} is continuous quasi-everywhere on $\mathbf{C}$. *Furthermore, $U^{\mu_w} + Q$ (considered as a function on $\mathcal{S}_w$) is continuous at $z \in \mathcal{S}_w$ if and only if (4.7) holds. In particular, $U^{\mu_w} + Q$ is continuous quasi-everywhere on $\mathcal{S}_w$ (considered as a function on $\mathcal{S}_w$) and, as a consequence, Q is continuous quasi-everywhere on $\mathcal{S}_w$ (considered as a function on $\mathcal{S}_w$).*

The last statement is actually a surprising fact about the positioning of $\mathcal{S}_w$, for Q may have "many" points of discontinuity on Σ. For more details on the continuity of the extremal potential see Theorems 4.8 and 5.1 below.

Note also that a general logarithmic potential can be discontinuous at "many" points of the support of the generating measure. In fact, if μ is a discrete measure with $\text{supp}(\mu) = [-1, 1]$, then U^μ takes the value $+\infty$ on a dense set, so it is discontinuous at every point where U^μ is finite. Thus, this potential is discontinuous at quasi-every point of $\text{supp}(\mu)$. Recall, however, Theorem 0.1.2, according to which U^μ is necessarily continuous on a dense subset of $\text{supp}(\mu)$.

Proof of Theorem 4.4. U^{μ_w} is clearly continuous outside $\mathcal{S}_w = \text{supp}(\mu_w)$. Now let $z \in \mathcal{S}_w$ be such that (4.7) holds. If $\varepsilon > 0$, then, by the upper semi-continuity of $-Q$, there is a $\delta > 0$ such that for $|z' - z| < \delta$ we have

$$-Q(z') \leq -Q(z) + \varepsilon.$$

Since for $|z' - z| < \delta,\ z' \in \mathcal{S}_w$, we get from Theorem 1.3(e)

$$U^{\mu_w}(z') \leq -Q(z') + F_w \leq -Q(z) + F_w + \varepsilon = U^{\mu_w}(z) + \varepsilon,$$

we can conclude that $U^{\mu_w}\big|_{\mathcal{S}_w}$ is upper semi-continuous on $\mathcal{S}_w$ at z . But U^{μ_w} is also lower semi-continuous and so $U^{\mu_w}\big|_{\mathcal{S}_w}$ is continuous at z. Hence the continuity of U^{μ_w} at z is guaranteed by Theorem II.3.5 to be proven in Chapter II, according to which the continuity of $U^{\mu_w}\big|_{\mathcal{S}_w}$ at z implies the continuity of U^{μ_w} at z.

Now suppose again that $z \in \mathcal{S}_w$ satisfies (4.7). Since $U^{\mu_w} + Q$ is lower semi-continuous on Σ and bounded from above by F_w on $\mathcal{S}_w$ (see Theorem 1.3(e)), it follows that $U^{\mu_w} + Q$ is continuous at z. Conversely, suppose that $U^{\mu_w} + Q$ is continuous at $z_0 \in \mathcal{S}_w$. Since μ_w has finite logarithmic energy, every neighborhood (relative to $\mathcal{S}_w$) of z_0 has positive capacity and so contains points $z \in \mathcal{S}_w$ that satisfy (4.7) (see Theorem 1.3(f)). Hence z_0 must satisfy (4.7). □

Corollary 4.5. *If Σ is a compact set of positive capacity and Ω is the unbounded component of $\overline{\mathbb{C}} \setminus \Sigma$, then the equilibrium measure μ_Σ of Σ is the same as the equilibrium measure $\mu_{\partial\Omega}$ of $\partial\Omega$. In particular, μ_Σ is supported on $\partial\Omega$. Furthermore, the equality*

$$U^{\mu_\Sigma}(z) = \log \frac{1}{\text{cap}(\Sigma)}$$

holds for quasi-every $z \in \Sigma$ and for all $z \notin \overline{\Omega}$, and we also have $\text{cap}(\Sigma) = \text{cap}(\partial\Omega)$.

Proof. First we prove that $\partial\Omega$ has positive capacity. In fact, in the opposite case, if we consider the negative $-U^{\mu_\Sigma}$ of the equilibrium potential on domains of the form

$$G_R := \{z \in \Omega \mid |z| < R\},$$

and apply the generalized minimum principle (Theorem 2.4, use also that equilibrium potentials are bounded on compact sets by Theorem 4.3), then on letting

$R \to \infty$ we can conclude that this potential is identically infinite, which is an obvious impossibility.

Thus, $\partial\Omega$ is of positive capacity, and $\mu_{\partial\Omega}$ exists. By Theorems 1.3(f), 4.4 and the generalized minimum principle (applied to connected components of $\mathbf{C}\setminus\overline{\Omega}$), the potential $U^{\mu_{\partial\Omega}}(z)$ coincides with $\log(1/\text{cap}(\partial\Omega))$ on $\mathbf{C}\setminus\overline{\Omega}$ (see also Theorem 4.3). Therefore we can invoke Theorem 3.3 according to which $\mu_\Sigma = \mu_{\partial\Omega}$, and all the assertions of the corollary are immediate consequences. □

In the classical theory, i.e. when Σ is compact, $\text{cap}(\Sigma) > 0$, and $w \equiv 1$ on Σ, the continuity of the equilibrium potential U^{μ_Σ} plays an important role. This is because U^{μ_Σ} and the Green function for the so-called outer domain relative to Σ are intimately connected. The *outer domain* Ω relative to Σ is the unbounded component of the complement $\overline{\mathbf{C}} \setminus \Sigma$ of Σ, and $\partial\Omega$ is called the *outer boundary* of Σ. The *polynomial convex hull* of Σ is defined to be $\mathbf{C} \setminus \Omega$ and is denoted by $\text{Pc}(\Sigma)$. It is easy to see that $\text{Pc}(\Sigma)$ is the union of Σ with the bounded components of $\overline{\mathbf{C}}\setminus\Sigma$; furthermore, the boundary $\partial\text{Pc}(\Sigma)$ of the polynomial convex hull coincides with the outer boundary of Σ.

The *Green function* of Ω with pole at infinity is the unique function $g_\Omega(z, \infty)$ on Ω with the following properties (see also Section II.4):

a) g_Ω is nonnegative and harmonic in $\Omega \setminus \{\infty\}$,

b) $$\lim_{|z|\to\infty} (g_\Omega(z,\infty) - \log|z|) = \log\frac{1}{\text{cap}(\Sigma)},$$

c) $$\lim_{z\to z',\, z\in\Omega} g_\Omega(z,\infty) = 0 \text{ for q.e. } z' \in \partial\Omega.$$

By Corollary 2.5 such a function can only exist if Σ is of positive capacity, which we shall assume henceforth. For such Σ the existence follows if we set

$$g_\Omega(z,\infty) = -U^{\mu_\Sigma}(z) + \log\frac{1}{\text{cap}(\Sigma)}. \tag{4.8}$$

In fact, a) follows from (1.4) and (1.5), b) is immediate from (4.8), while c) follows from Theorem 4.4. The unicity follows from the generalized minimum principle (Theorem 2.4), for if g^* denotes the right-hand side of (4.8), then the function $g_\Omega - g^*$ is harmonic and bounded below on $\Omega \cup \{\infty\}$ (Theorem 4.3) and vanishes at infinity; furthermore, it has boundary limit zero at quasi-every point of $\partial\Omega$. Hence by the minimum principle it must be identically zero.

One can extend $g_\Omega(z, \infty)$ by the stipulation

$$g_\Omega(z,\infty) = \begin{cases} 0, & z \in \text{Int}(\text{Pc}(\Sigma)) \\ \limsup_{z'\to z,\, z'\in\Omega} g_\Omega(z,\infty), & z \in \partial\text{Pc}(\Sigma) \end{cases}$$

(or by (4.8)) to the whole complex plane $\mathbf{C}$ by which it becomes a nonnegative *subharmonic* function.

In view of the representation (4.8) the continuity points of g_Ω and U^{μ_Σ} coincide, and since μ_Σ is supported on the outer boundary of Σ, both of these functions are continuous away from the outer boundary. A point z on this outer boundary $\partial\Omega$ is called a *regular* (boundary) *point* of Ω if $g_\Omega(z,\infty)$ is continuous at z; otherwise it is called *irregular.* It is easy to see (cf. Theorem 4.4) that $z \in \partial\Omega$ is a regular point if and only if

$$g_\Omega(z,\infty) = 0,$$

which is equivalent to

$$U^{\mu_\Sigma}(z) = \log\frac{1}{\operatorname{cap}(\Sigma)}.$$

In particular, the set of irregular points has zero capacity.

The regularity of a point plays an important role in solving Dirichlet's problem (cf. Section I.2) in Ω with prescribed boundary function on $\partial\Omega$. In fact, it turns out that $z \in \partial\Omega$ is a regular point if and only if for every bounded boundary function on $\partial\Omega$ that is continuous at z the solution of the Dirichlet problem (in Ω) is continuous at z (see the Appendix A.2). This is why regular points are usually called *regular points for the Dirichlet problem in* Ω. If every point of $\partial\Omega$ is regular, then we call Ω *regular* with respect to the Dirichlet problem.

A celebrated theorem of Wiener characterizes regular points as follows:

Theorem 4.6 (Wiener's Theorem). *Let* $0 < \lambda < 1$ *and set*

$$A_n(z) := \left\{x \,\middle|\, x \notin \Omega,\ \ \lambda^n \le |x-z| < \lambda^{n-1}\right\}.$$

Then $z \in \partial\Omega$ *is regular with respect to the Dirichlet problem in* Ω *if and only if*

$$\sum_{n=1}^{\infty}\frac{n}{\log(1/\operatorname{cap}(A_n(z)))} = \infty.$$

For a proof see the Appendix A.1. As an immediate consequence of Wiener's theorem, we have that every simply connected domain $\Omega \subset \overline{\mathbf{C}}$ is regular.

Next we establish an important representation for $U^{\mu_w} - F_w$ on $\mathbf{C}\setminus S_w$; namely that it coincides with the solution $H^R_{-Q}(z)$ of Dirichlet's problem (see Section I.2) with boundary values $-Q$ on any bounded component R of $\mathbf{C}\setminus S_w$, and with the solution of this Dirichlet problem minus the Green function on the unbounded component of $\mathbf{C}\setminus S_w$.

Theorem 4.7. *Let* w *be an admissible weight and* R *a bounded component of* $\overline{\mathbf{C}}\setminus S_w$. *Then*

$$U^{\mu_w}(z) - F_w \equiv H^R_{-Q}(z), \quad z \in R.$$

If, however, $R = \Omega$ *is the unbounded component of* $\overline{\mathbf{C}}\setminus S_w$, *then*

$$U^{\mu_w}(z) - F_w \equiv H^\Omega_{-Q}(z) - g_\Omega(z,\infty), \quad z \in \Omega,$$

where $g_\Omega(z,\infty)$ *denotes the Green function of* Ω *with pole at* ∞.

Note that in each case $\overline{\mathbf{C}} \setminus R$ is of positive capacity; in particular, g_Ω exists. We also remark that above we used the self-explanatory notation H^R_{-Q} instead of the more cumbersome expression $H^R_{-Q|_{\partial R}}$. The existence of the solution of the Dirichlet problem in R for the boundary function $-Q$ is part of the statement.

It is easy to see that in the classical unweighted case these results reduce to the ones discussed after Corollary 4.5 (cf. formula (4.8)).

Proof of Theorem 4.7. Let R be a bounded component of $\overline{\mathbf{C}} \setminus S_w$. By Theorem 1.3(f) and Theorem 4.4,

$$\lim_{z\to x}(U^{\mu_w}(z) - F_w) = -Q(x)$$

for quasi-every $x \in \partial R$. Thus, the claim follows from the boundedness of $U^{\mu_w}(z)$ on R (Theorem 4.3) and from Lemma 2.6.

The second statement concerning the unbounded component $R = \Omega$ of $\overline{\mathbf{C}} \setminus S_w$ can be similarly proved if we consider $U^{\mu_w}(z) + g_\Omega(z, \infty)$. In fact, by Theorem 4.4,

$$\lim_{x\to z,\, x\in R}[U^{\mu_w}(x) + g_\Omega(x,\infty) - F_w] = -Q(z)$$

for quasi-every $z \in \partial\Omega$. Furthermore, $U^{\mu_w}(z) + g_\infty(z,\infty) - F_w$ is harmonic and bounded on Ω (even at ∞) and so the rest of the proof is the same as before. □

Now we are in the position to make Theorem 4.4 more exact for continuous weights.

Theorem 4.8. *Let w be a continuous admissible weight on Σ. Then U^{μ_w} is continuous at $z_0 \in S_w$ if any one of the following conditions holds:*

(i) *z_0 is in the interior of Σ;*

(ii) *$z_0 \in \partial S_w \cap \partial\Sigma$ belongs to the boundary of at least two components of $\mathbf{C} \setminus S_w$;*

(iii) *$z_0 \in \partial S_w \cap \partial\Sigma$ does not belong to the boundary of any component of $\mathbf{C} \setminus S_w$;*

(iv) *$z_0 \in \partial S_w \cap \partial\Sigma$ belongs to the boundary of exactly one component R of $\mathbf{C} \setminus S_w$ and z_0 is a regular boundary point for the Dirichlet problem on R.*

In particular, if $\Sigma = \mathbf{C}$ and w is continuous, then U^{μ_w} is continuous everywhere.

We shall actually show that if w is continuous and z_0 is a regular boundary point for every component of $\mathbf{C} \setminus S_w$ that contains z_0 on its boundary, then U^{μ_w} is continuous at z_0. Furthermore, each of (ii)–(iv) implies this assumption.

To picture typical z_0's for which none of the conditions (i)–(iv) holds, consider the set S_w illustrated in Figure 4.1 (S_w consists of all the circles and of the two limit points z_0 and z_0'). Provided the radii of the small circles decrease sufficiently rapidly, the two limit points z_0 and z_0' do not satisfy any of (i)–(iv) above and, in

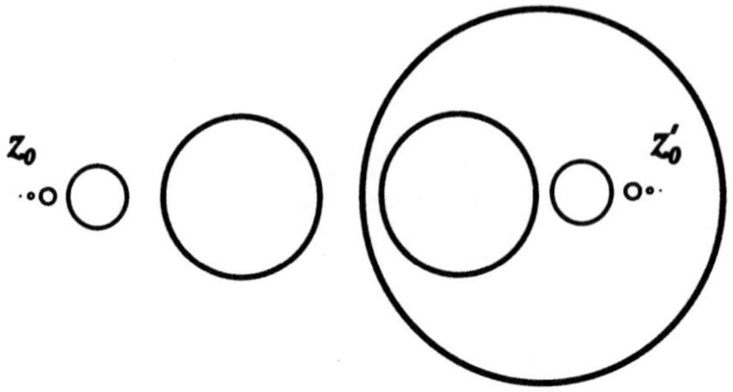

Fig. 4.1

fact, it may happen for some w that $\mathcal{S}_w$ is the above set and z_0 or z_0' are not points of continuity of U^{μ_w} (consider e.g. $w \equiv 1$ and the point z_0; see also Example 4.9 below).

Let us also mention that the continuity of U^{μ_w} at z_0 follows if z_0 is a regular boundary point of at least one component of $\mathbf{C} \setminus \Sigma$ containing z_0 on its boundary. This follows from Theorem 5.1 below which contains an extension of Theorem 4.8.

In the classical case $w \equiv 1$ if none of (i)–(iv) is satisfied, then z_0 is an irregular point of the outer boundary of $\mathcal{S}_w$ and we know that then U^{μ_w} is discontinuous at z_0 (see the discussion after Corollary 4.5). In the general case, however, U^{μ_w} may be continuous at z_0 even if all of the conditions (i)–(iv) fail:

Example 4.9. Let

$$\Sigma := \{0\} \cup \bigcup_{n=1}^{\infty} \left[2^{-n}, 2^{-n} + 2^{-n^3}\right],$$

μ_n the Lebesgue measure on $\left[2^{-n},\ 2^{-n} + 2^{-n^3}\right]$ and

$$\mu := \sum_{n=2}^{\infty} \mu_n + 2\left(1 - \sum_{n=2}^{\infty} 2^{-n^3}\right) \mu_1 .$$

Then U^μ is everywhere continuous and $\mathrm{supp}(\mu) = \Sigma$; hence by Theorem 3.8, $\mu = \mu_w$ and $\Sigma = \mathcal{S}_w$ for some admissible weight w. On the other hand, $z_0 = 0$ clearly does not satisfy any of (i)–(iii), and by Theorem 4.6 it does not satisfy (iv), either, since with $\lambda = 1/2$

$$\sum_{n=1}^{\infty} \frac{n}{\log(1/\mathrm{cap}(A_n))} = \sum_{n=1}^{\infty} \frac{n}{\log(4 \cdot 2^{n^3})} < \infty,$$

where we used that the capacity of an interval of length l is $l/4$. □

Now we give the

Proof of Theorem 4.8. Part (i) follows from Lemma 2.2 and Theorem 4.4.

We can simultaneously prove that each condition (ii)–(iv) implies continuity if we show that

A. U^{μ_w} *is continuous at* $z_0 \in \partial S_w$ *if* z_0 *is a regular boundary point for the Dirichlet problem for every component of* $\mathbf{C} \setminus S_w$ *containing* z_0 *on its boundary,* and

B. *Each of* (ii)–(iv) *implies the assumption in* A.

Assertion B is clear for (iii) and (iv) and first we show that it is true for (ii), as well. In fact, let R be any component of $\mathbf{C} \setminus S_w$ that contains z_0 on its boundary. Condition (ii) assumes that there exists at least one more component R_1 of $\mathbf{C} \setminus S_w$ with $z_0 \in \partial R_1$. But then the closure $\overline{R_1}$ of R_1 is a continuum (i.e. a connected closed set) in $\mathbf{C} \setminus R$ containing z_0 and hence, by Wiener's criterion (Theorem 4.6) and Lemma 2.1, z_0 is a regular boundary point for R.

Thus, it only remains to prove assertion A. Choose an arbitrary $\varepsilon > 0$ and then a $\delta > 0$ such that $|Q(z) - Q(z_0)| < \varepsilon$ for $z \in D_\delta(z_0) \cap \Sigma$, where

$$D_\delta(z_0) := \{z \mid |z - z_0| < \delta\}.$$

Let E_D be the set of points of discontinuity of U^{μ_w}. Then E_D is an F_σ set of zero capacity (cf. Theorem 4.4), and so, by Lemma 2.1, there are arbitrary small radii r such that

$$C_r(z_0) \cap (E \cup E_D) = \emptyset,$$

where E is the set

$$E := \{z \in \overline{D_\delta(z_0)} \cap \Sigma \mid U^{\mu_w}(z) \le -Q(z_0) + F_w - \varepsilon\}$$

(recall that by Theorem 1.3(d) E is a compact set of capacity zero). Let $r_1 < \delta$ be such a radius.

Let $R_1, \ldots, R_m$ be those components R of $\mathbf{C} \setminus S_w$ that have the following property: for some $z_R \in R \cap C_{r_1}(z_0)$,

$$U^{\mu_w}(z_R) \le -Q(z_0) + F_w - 2\varepsilon. \tag{4.9}$$

There are only finitely many such R's since otherwise there is a $z^* \in C_{r_1}(z_0)$ that is a limit point of z_R's associated with different R's; hence

$$\liminf_{z \to z^*} U^{\mu_w}(z) \le -Q(z_0) + F_w - 2\varepsilon.$$

However, between any two points z_R and $z_{R'}$ on $C_{r_1}(z_0)$ with different R and R' there must be at least one point of S_w on the circle $C_{r_1}(z_0)$. Hence, since $r_1 < \delta$ and $E \cap C_{r_1}(z_0) = \emptyset$, we get

$$\limsup_{z \to z^*} U^{\mu_w}(z) \ge -Q(z_0) + F_w - \varepsilon.$$

These inequalities mean that at $z^* \in C_{r_1}(z_0)$ the potential U^{μ_w} is discontinuous, contradicting the fact that $E_D \cap C_{r_1}(z_0) = \emptyset$.

Since the potential U^{μ_w} solves certain Dirichlet problems in each of $R_1, \ldots, R_m$ (cf. Theorem 4.7) it follows from the assumption in assertion A that there is an $r_0 < r_1$ such that for $r \le r_0$ and $1 \le j \le m$ either $D_r(z_0) \cap R_j = \emptyset$ or for $z \in D_r(z_0) \cap R_j$ we have

$$U^{\mu_w}(z) \ge -Q(z_0) + F_w - \varepsilon. \tag{4.10}$$

Choose $0 < r < r_0$ to satisfy $E \cap C_r(z_0) = \emptyset$. When $z \in C_r(z_0)$, we have several possibilities: If $z \in R_j$ for some j, then, as we have already seen, (4.10) is true. If $z \in C_r(z_0) \cap \mathcal{S}_w$, then by the definition of E, the relation $E \cap C_r(z_0) = \emptyset$ implies again (4.10). Finally, let $z \in R \cap C_r(z_0)$, where R is any component of $\mathbf{C} \setminus \mathcal{S}_w$ different from $R_1, \ldots, R_m$. Then z belongs to one of the components of $R \cap D_{r_1}(z_0)$, say to $\tilde{R}$. The boundary of the set $\tilde{R}$ consists of some part of the boundary of R and of some part of $C_{r_1}(z_0)$. It follows from the proof of Theorem 4.7 and from Theorem 4.7 itself that in $\tilde{R}$ the function U^{μ_w} is the solution of the Dirichlet problem with boundary values $-Q(t) + F_w$ on $\partial\tilde{R} \cap \partial R$ and $U^{\mu_w}(t)$ on $\partial\tilde{R} \cap C_{r_1}(z_0)$. Now, if $t \in \partial\tilde{R} \cap \partial R$, then $|t - z_0| < \delta$ and so

$$-Q(t) + F_w \ge -Q(z_0) + F_w - \varepsilon, \quad t \in \partial\tilde{R} \cap \partial R.$$

If $t \in \partial\tilde{R} \cap C_{r_1}(z_0) \subseteq R \cap C_{r_1}(z_0)$, then t does not belong to any of the components $R_1, \ldots, R_m$ for which (4.9) holds. Hence,

$$U^{\mu_w}(t) \ge -Q(z_0) + F_w - 2\varepsilon, \quad t \in \partial\tilde{R} \cap C_{r_1}(z_0).$$

Thus, the boundary function is not smaller than $-Q(z_0) + F_w - 2\varepsilon$, and so this number is a lower bound for the solution, as well, and we get for $z \in \tilde{R}$

$$U^{\mu_w}(z) \ge -Q(z_0) + F_w - 2\varepsilon. \tag{4.11}$$

Thus we have proved that (4.11) holds for all $z \in C_r(z_0)$ and so (4.11) follows for $z = z_0$ from the superharmonicity of U^{μ_w}. Letting $\varepsilon \to 0$ we get from Theorem 1.3(e) and Theorem 4.4 that U^{μ_w} is continuous at z_0, and the theorem is proved. □

I.5 Fine Topology and Continuity of Equilibrium Potentials

An alternative proof of Theorem 4.8 can be given that utilizes the *fine topology on* $\mathbf{C}$ (cf. [77, Chapter 10]). The fine topology $\mathcal{T}$ is the coarsest topology on $\mathbf{C}$ that makes every (logarithmic) potential continuous in $\mathcal{T}$.

The lower semi-continuity of logarithmic potentials implies that sets of the form

$$H = \bigcap_{i=1}^{m} \{z \mid U^{\mu_i}(z) < \alpha_i\} \cap G,$$

where G is an open set in Euclidean topology, form a basis for the fine topology. Thus, every set that is open in the Euclidean topology is also open in the fine topology; in other words, the fine topology is indeed finer than Euclidean topology.

A set E is called *thin* at z if z is not a limit point of E in the fine topology. Then it turns out that a set is of zero capacity if and only if it is a discrete set in $\mathcal{T}$, i.e. it is thin at every point z. In this context, if $\mathcal{S}_w$ is not thin at $z_0 \in \mathcal{S}_w$, then the set

$$\{z \in \mathcal{S}_w \mid U^{\mu_w}(z) = -Q(z) + F_w\} \tag{5.1}$$

is not thin at z_0 either. Since U^{μ_w} is continuous in the fine topology and the metric topology is rougher than the fine topology, we get from the continuity of Q at z_0 that

$$U^{\mu_w}(z_0) = -Q(z_0) + F_w. \tag{5.2}$$

Thus, by Theorem 4.4, U^{μ_w} is continuous at such a point z_0. Now it is not difficult to verify that each of the conditions (ii)–(iv) of Theorem 4.8 implies that $\mathcal{S}_w$ is not thin at z_0, and so the continuity of U^{μ_w} follows. In Theorem 4.8 the topological conditions (ii)–(iv) were stated instead of the thinness assumption because the former can be dealt with using elementary arguments.

To illustrate the power of the fine topology, the following variant of Theorem 4.8 is presented, which shows that with the assumptions of Theorem 4.8 the potential U^{μ_w} is continuous at $z_0 \in \mathcal{S}_w$ if any of the conditions Theorem 4.8(ii)–(iv) holds with $\mathbf{C} \setminus \mathcal{S}_w$ replaced by $\mathbf{C} \setminus \Sigma$.

Theorem 5.1. *Let w be a continuous admissible weight on Σ. Then U^{μ_w} is continuous at $z_0 \in \mathcal{S}_w$ if any one of the following conditions hold:*

(ii)' $z_0 \in \partial\mathcal{S}_w \cap \partial\Sigma$ *belongs to the boundary of at least two components of* $\mathbf{C} \setminus \Sigma$;

(iii)' $z_0 \in \partial\mathcal{S}_w \cap \partial\Sigma$ *does not belong to the boundary of any component of* $\mathbf{C} \setminus \Sigma$;

(iv)' $z_0 \in \partial\mathcal{S}_w \cap \partial\Sigma$ *belongs to the boundary of one component R of* $\mathbf{C} \setminus \Sigma$ *and z_0 is a regular boundary point for the Dirichlet problem on R;*

(v)

$$\sum_{n=1}^{\infty} \frac{n}{\log(1/\operatorname{cap}(A_n(z_0)))} = \infty,$$

where

$$A_n(z_0) = \{z \mid z \in \Sigma, \quad \lambda^{n+1} \le |z - z_0| < \lambda^n\}$$

for some fixed $0 < \lambda < 1$.

In particular, if w is continuous and $\Sigma = \mathbf{C}$, or every component of $\mathbf{C} \setminus \Sigma$ is regular with respect to the Dirichlet problem, then U^{μ_w} is continuous everywhere.

For the proof of Theorem 5.1 we need several lemmas. In each of them "open" etc. refers to Euclidean topology. The corresponding notions for the fine topology will always be preceded by the adjective "fine".

Lemma 5.2. *A set E is thin at a limit point z_0 of E if and only if there is a measure μ of compact support such that*

$$U^{\mu}(z_0) < \liminf_{z \to z_0,\, z \in E \setminus \{z_0\}} U^{\mu}(z). \tag{5.3}$$

Proof. The sufficiency is obvious, so we only have to prove the necessity of (5.3).

Suppose E is thin at z_0. As we have already mentioned, the lower semi-continuity of logarithmic potentials implies that sets of the form

$$H = \bigcap_{i=1}^{m} \{z \mid U^{\mu_i}(z) < \alpha_i\} \cap G,$$

where G is an open set in Euclidean topology, form a basis for the fine topology. Thus, there is an H of this form containing z_0 with $(H \cap E) \setminus \{z_0\} = \emptyset$. Choose an $\varepsilon > 0$ such that

$$U^{\mu_i}(z_0) < \alpha_i - \varepsilon, \quad i = 1, \dots, m, \tag{5.4}$$

and then a $\delta > 0$ such that for each i and $|z - z_0| < \delta$ we have

$$U^{\mu_i}(z) > U^{\mu_i}(z_0) - \frac{\varepsilon}{m}. \tag{5.5}$$

If now $\mu := \sum_{i=1}^{m} \mu_i$ and $z \in E \setminus \{z_0\}$, $|z - z_0| < \delta$, $z \in G$, then we can conclude from $H \cap E = \emptyset$ that for some $1 \le j \le m$

$$U^{\mu_j}(z) \ge \alpha_j > U^{\mu_j}(z_0) + \varepsilon,$$

and hence by (5.5)

$$U^{\mu}(z) \ge U^{\mu_j}(z_0) + \varepsilon + \sum_{i \ne j} \left(U^{\mu_i}(z_0) - \frac{\varepsilon}{m} \right) = U^{\mu}(z_0) + \frac{\varepsilon}{m}.$$

This verifies (5.3) for μ. □

Lemma 5.3. *If E is an F_σ-set of capacity zero, then for any $z_0 \in \mathbf{C}$ the set $(\mathbf{C} \setminus E) \cup \{z_0\}$ is a fine neighborhood of z_0.*

Proof. Without loss of generality we may and do assume E to be bounded. By Lemma 2.3 there is a measure μ such that its potential is $+\infty$ on $E \setminus \{z_0\}$, and at the same time $U^{\mu}(z_0) < \infty$ (note that $E \setminus \{z_0\}$ is again an F_σ-set). Hence

$$\{z \mid U^{\mu}(z) < U^{\mu}(z_0) + 1\}$$

is a fine neighborhood of z_0 contained in $(\mathbf{C} \setminus E) \cup \{z_0\}$. □

Lemma 5.4. *Suppose that E and z_0 satisfy the following property: for any $\delta > 0$ there is a neighborhood G of z_0 such that quasi-every boundary point of G belongs to $E \cap D_\delta(z_0)$. Then z_0 is a fine limit point of E.*

Proof. According to Lemma 5.2 we have to verify that if μ is a measure of compact support and

$$\liminf_{z \to z_0,\, z \in E \setminus \{z_0\}} U^{\mu}(z) \ge \alpha,$$

then

$$U^{\mu}(z_0) \geq \alpha. \tag{5.6}$$

For any $\varepsilon > 0$ choose $\delta > 0$ such that

$$U^{\mu}(z) \geq \alpha - \varepsilon$$

whenever $z \in (E \setminus \{z_0\}) \cap D_\delta(z_0)$, and consider the neighborhood G of z_0 guaranteed by the assumption for this δ. If $z \in \partial G \cap E$, then the lower semi-continuity of U^{μ} implies

$$\liminf_{z' \to z,\, z' \in G} U^{\mu}(z') \geq U^{\mu}(z) \geq \alpha - \varepsilon.$$

Since this is true for quasi-every $z \in \partial G$, we can deduce from the generalized minimum principle (Theorem 2.4) that $U^{\mu}(z_0) \geq \alpha - \varepsilon$, and letting $\varepsilon \to 0$ we get (5.6). □

Lemma 5.5. *Let E be a Borel set, $0 < \lambda < 1$, and*

$$E_n(z_0) = \{z \mid z \in E,\ \ \lambda^{n+1} \leq |z - z_0| < \lambda^n\}.$$

If

$$\sum_{n=1}^{\infty} \frac{n}{\log\left(1/\operatorname{cap}(E_n(z_0))\right)} = \infty, \tag{5.7}$$

then z_0 is a fine limit point of E.

We remark that the converse also holds, but to prove it one needs the deep theorem of Choquet on capacitability of Borel sets (see the notes to Section I.1 at the end of this chapter). However, for compact sets E the converse to Lemma 5.5 follows from Theorem 2.1 in Appendix A.2.

Corollary 5.6. *If G is a domain, then the boundary of G in the Euclidean and fine topologies are one and the same.*

First let us prove this corollary. Since the fine topology is finer than planar topology, all we have to show is that if $x \in \partial G$ (boundary of G in the Euclidean topology), then x is a fine limit point of G. But the connectedness of G yields that the circle $C_r(x)$ around x and of radius r has to intersect G for every small r. On invoking Lemma 2.1 we can conclude that for large n we have in (5.7)

$$\operatorname{cap}(E_n(z_0)) \geq \frac{\lambda^n - \lambda^{n+1}}{4} = \lambda^n \frac{1-\lambda}{4},$$

and so Lemma 5.5 can be applied to deduce the corollary.

Proof of Lemma 5.5. We can replace each set $E_n(z_0)$ by a compact subset of it, which we continue to denote by $E_n(z_0)$, so that (5.7) still holds. Hence we may assume that $E \cup \{z_0\}$ is closed.

Let $\delta > 0$, and consider the sets $E_\delta = (E \cup \{z_0\}) \cap \overline{D_\delta(z_0)}$. We distinguish two cases.

Case I. There is a $\delta_0 > 0$ such that z_0 is on the outer boundary of the set E_{δ_0} (that is, z_0 is on the boundary of the unbounded component of $\mathbf{C} \setminus E_{\delta_0}$). Then z_0 is on the outer boundary of E_δ for every $0 < \delta < \delta_0$. By (5.7) and Wiener's criterion (Theorem 4.6) for every such δ the point z_0 is a regular boundary point of the unbounded component of $\mathbf{C} \setminus E_\delta$, which is equivalent to the continuity at z_0 of the equilibrium potential $U^{\mu_\delta} := U^{\mu_{E_\delta}}$ of E_δ.

Now let us assume that for some measure μ we have

$$\liminf_{z \to z_0,\, z \in E \setminus \{z_0\}} U^\mu(z) \ge \alpha. \tag{5.8}$$

To prove the claim of the lemma we have to verify that (5.8) implies

$$U^\mu(z_0) \ge \alpha. \tag{5.9}$$

Without loss of generality, we may assume μ to have total mass at most 1. If $\varepsilon > 0$, then there is a $0 < \delta < \delta_0$ such that

$$U^\mu(z) \ge \alpha - \varepsilon$$

for every $z \in (E \setminus \{z_0\}) \cap \overline{D_\delta(z_0)}$. Hence, for the equilibrium measure $\mu_\delta = \mu_{E_\delta}$ of E_δ we have

$$U^\mu(z) \ge U^{\mu_\delta}(z) + \alpha - \varepsilon - \log \frac{1}{\operatorname{cap}(E_\delta)}$$

for quasi-every $z \in E_\delta \setminus \{z_0\}$, and this implies via the principle of domination (Theorem II.3.2) the same inequality for every $z \in \mathbf{C}$. In particular,

$$U^\mu(z_0) \ge U^{\mu_\delta}(z_0) + \alpha - \varepsilon - \log \frac{1}{\operatorname{cap}(E_\delta)}.$$

However, above we have established the continuity of U^{μ_δ} at $z_0 \in \operatorname{supp}(\mu_\delta)$ and this is equivalent to

$$U^{\mu_\delta}(z_0) = \log \frac{1}{\operatorname{cap}(E_\delta)};$$

hence

$$U^\mu(z_0) \ge \alpha - \varepsilon$$

follows. On letting $\varepsilon \to 0$ we obtain (5.9), which proves the claim in the first case.

Case II. For every $\delta > 0$, z_0 is in the interior of the polynomial convex hull $\operatorname{Pc}(E_\delta)$ of E_δ. In this case, for a $\delta > 0$, let G be the connected component of the interior of $\operatorname{Pc}(E_\delta)$ that contains z_0. It is clear that $\partial G \subseteq E_\delta \subseteq E$; hence in this case the claim that z_0 is a fine limit point of E follows from Lemma 5.4. □

With these preliminaries we can prove Theorem 5.1.

Proof of Theorem 5.1. First of all we claim that the hypotheses imply that z_0 is a fine limit point of Σ which is immediate for (v) by the preceding lemma. For (iv)' the claim follows from Wiener's criterion (Theorem 4.6) and Lemma 5.5. Lemma 5.5 can also be used for (ii)', since for every small $r > 0$ the circle $C_r(z_0)$ with center at z_0 of radius r intersects $E := \Sigma$; hence by Lemma 2.1

$$\operatorname{cap}(E_n(z_0)) \geq \frac{1}{4}\left(\lambda^n - \lambda^{n+1}\right) = \frac{\lambda - 1}{4}\lambda^n,$$

and so (5.7) holds. The same argument yields the claim for (iii)' provided $C_r(z_0) \cap \Sigma \neq \emptyset$ for every small $r > 0$. If, however, for some sequence $r_k \to 0+0$ we have $C_{r_k}(z_0) \cap \Sigma = \emptyset$, and yet z_0 does not belong to the boundary of any component of $\mathbf{C} \setminus \Sigma$ (cf. condition (iii)'), then for every $\delta > 0$, z_0 is in the interior of the polynomial convex hull of $\Sigma \cap \overline{D_\delta(z_0)}$, and so the claim that z_0 is a fine limit point of Σ follows from Lemma 5.4.

Thus, we have verified that z_0 is a fine limit point of Σ. Consider now the set

$$\Sigma' = \{z \mid z \in \Sigma,\ \ U^{\mu_w}(z) \geq -Q(z) + F_w\}.$$

By Theorem 1.3(d) the set $\Sigma \setminus \Sigma'$ is of zero capacity. Furthermore, it follows from the lower semi-continuity of U^{μ_w} and Q that $\Sigma \setminus \Sigma'$ is an F_σ-set. Hence, by Lemma 5.3, the set $(\mathbf{C} \setminus (\Sigma \setminus \Sigma')) \cup \{z_0\}$ is a fine neighborhood of z_0. If S' is any other fine neighborhood of z_0, then so is $S := S' \cap ((\mathbf{C} \setminus (\Sigma \setminus \Sigma')) \cup \{z_0\})$, and by what we have proved above, $S \setminus \{z_0\}$ must intersect Σ. This implies that $S' \setminus \{z_0\}$ must intersect Σ'. Keeping this in mind, the definition of Σ' and the continuity of U^{μ_w} in the fine topology implies

$$U^{\mu_w}(z_0) \geq -Q(z_0) + F_w$$

(recall that the fine topology contains Euclidean topology, hence Q is continuous in the fine topology, as well). Since $z_0 \in S_w$, the opposite inequality follows from Theorem 1.3(e), that is

$$U^{\mu_w}(z_0) = -Q(z_0) + F_w.$$

Now the continuity of U^{μ_w} at z_0 is a consequence of Theorem 4.4. □

I.6 Weighted Capacity

Let w be an admissible weight and V_w the corresponding energy defined in (1.13). The associated *weighted capacity* is defined as

$$c_w := \exp(-V_w). \tag{6.1}$$

In terms of F_w and Q we can write

$$c_w = \exp\left(-F_w - \int Q\, d\mu_w\right).$$

Notice that since V_w is finite, we have $0 < c_w < \infty$.

To allow the possibility that $c_w = 0$ we introduce the class of quasi-admissible weights.

Definition 6.1. w is called *quasi-admissible* if it satisfies properties (i) and (iii) of Definition 1.1, i.e. if it is nonnegative, upper semi-continuous and $|z|w(z) \to 0$ as $z \to \infty$, $z \in \Sigma$, when Σ is unbounded.

Now we extend the definition of c_w to quasi-admissible weights w by setting $c_w = 0$ if w is quasi-admissible but not admissible. We remark that this extension is in complete agreement with (6.1), for if w is quasi-admissible but not admissible, then the set

$$\Sigma_0 = \{z \mid w(z) > 0\}$$

is of zero capacity, and so the energy integral $I_w(\mu)$ is infinite for every $\mu \in \mathcal{M}(\Sigma)$, giving $V_w = \infty$.

Sometimes we want to talk about the weighted capacity of a closed subset E of Σ. Of course, this is covered by the above definitions since we only have to take the restriction $w|_E$ of w to E. However, to avoid cumbersome notation let us set

$$\operatorname{cap}(w, E) := c_{w|_E} \quad \text{and} \quad \mu(w, E) := \mu_{w|_E}.$$

Note that if E is compact, then

$$\operatorname{cap}(\chi_E, E) = \operatorname{cap}(1, E) = \operatorname{cap}(E),$$

where χ_E is the characteristic function for E.

Now we list some properties of the weighted capacity, and in doing so we may assume that $\Sigma = \mathbf{C}$ for all the weights.

Theorem 6.2. *Suppose that all the weights $w, w_1, \ldots$ below are quasi-admissible.*

(a) *If E is closed, then* $\operatorname{cap}(w, E) = 0$ *if and only if w vanishes quasi-everywhere on E.*

(b) *If $w_1 \le w_2$, then $c_{w_1} \le c_{w_2}$. In particular, if $E_1 \subseteq E_2$ are closed, then* $\operatorname{cap}(w, E_1) \le \operatorname{cap}(w, E_2)$.

(c) *If $w_n \ge w_{n+1}$, $n = 1, 2, \ldots$, and $v := \lim_{n\to\infty} w_n$, then v is quasi-admissible and*

$$c_v = \lim_{n\to\infty} c_{w_n}. \tag{6.2}$$

Furthermore, if $c_v > 0$ (i.e. v is admissible), then

$$\lim_{n\to\infty} \mu_{w_n} = \mu_v$$

in the weak topology of measures.*

In particular, if $E_{n+1} \subseteq E_n$, $n = 1, 2, \ldots$, are closed and $E := \bigcap_{n=1}^{\infty} E_n$, then

$$\operatorname{cap}(w, E) = \lim_{n\to\infty} \operatorname{cap}(w, E_n). \tag{6.3}$$

Furthermore, if the left-hand side of (6.3) is positive, then

$$\lim_{n\to\infty} \mu(w, E_n) = \mu(w, E) \tag{6.4}$$

in the weak topology.*

(d) *If* $w_n \le w_{n+1}$, $n = 1, 2, \dots$, *and* $v := \lim_{n\to\infty} w_n$ *is admissible (or agrees with an admissible weight q.e.), then*

$$c_v = \lim_{n\to\infty} c_{w_n}, \tag{6.5}$$

and, in the weak topology of measures,*

$$\lim_{n\to\infty} \mu_{w_n} = \mu_v. \tag{6.6}$$

(e) *Let M be a positive constant such that*

$$|z-t|w(z)w(t) \le M, \quad z,\ t \in \mathbf{C}.$$

Then $M \ge \mathrm{cap}(w, E)$ *for any closed set E. Moreover, if* E_n *are closed sets and* $E := \cup_{n=1}^{\infty} E_n$ *is closed, then*

$$\left[\log \frac{M}{\mathrm{cap}(w,E)}\right]^{-1} \le \sum_{n=1}^{\infty} \left[\log \frac{M}{\mathrm{cap}(w,E_n)}\right]^{-1}. \tag{6.7}$$

(f) *Let* $E_1, E_2, \dots$ *be disjoint closed sets, and m a nonnegative constant such that*

$$|z-t|w(z)w(t) \ge m, \qquad z \in E_n,\ z \in E_k, \tag{6.8}$$

for all $n \ne k$. *Then for* $E = \cup_n E_n$ *we have*

$$\left[\log^+ \frac{m}{\mathrm{cap}(w,E)}\right]^{-1} \ge \sum_{n=1}^{\infty} \left[\log^+ \frac{m}{\mathrm{cap}(w,E_n)}\right]^{-1}, \tag{6.9}$$

where $\log^+ = \max(\log, 0)$.

One can define the weighted capacity of nonclosed sets as the supremum of the weighted capacities of closed subsets of the given set. Then many of the above properties hold in this more general setting, but we shall not pursue this direction here.

Since statements (a) and (b) of the theorem are clear from the definitions, we proceed with the

Proof of Theorem 6.2(c). The quasi-admissibility of v is clear. The inequality

$$\lim_{n\to\infty} c_{w_n} \ge c_v$$

follows from part (b), and so (c) is true if the right-hand side of (6.2) is zero.

Thus, suppose that the right-hand side of (6.2) is positive, i.e. that

$$\lim_{n\to\infty} V_{w_n} =: V < \infty. \tag{6.10}$$

Using (6.10) we get from the proof of Theorem 1.3(b),(c) that there is an $\varepsilon > 0$ independent of n such that

$$\mathcal{S}_{w_n} \subseteq \{z \mid w_1(z) \geq \varepsilon\} =: K,$$

that is, all the measures $\mu_n := \mu_{w_n}$ have support in the compact set K. Thus, we can select from $\{\mu_n\}$ a weak* convergent subsequence and we may assume that the whole sequence $\{\mu_n\}$ converges to some $\mu \in \mathcal{M}(K)$ in the weak* sense. For every n, the weight w_n is an upper semi-continuous function on K; therefore, w_n is the limit of a decreasing sequence $\{\omega_m^{(n)}\}_{m=1}^{\infty}$ of positive continuous functions. Set

$$\omega_n := \min_{1 \leq i,\, j \leq n} \omega_i^{(j)}.$$

Then ω_n is a positive continuous function on K, and $\{\omega_n\}$ is monotone decreasing and converges to v. Setting

$$h_{M,m}(z,t) := \min\left(M, \log[|z-t|\omega_m(z)\omega_m(t)]^{-1}\right),$$

and

$$h_M(z,t) := \min\left(M, \log[|z-t|v(z)v(t)]^{-1}\right),$$

we get from the monotone convergence theorem that

$$\begin{aligned}
V &= \lim_{n\to\infty} V_{w_n} = \lim_{n\to\infty} \iint \log[|z-t|w_n(z)w_n(t)]^{-1} d\mu_n(z) d\mu_n(t) \\
&\geq \limsup_{n\to\infty} \iint h_{\infty,n}(z,t)\, d\mu_n(z) d\mu_n(t) \\
&\geq \lim_{M\to\infty} \limsup_{n\to\infty} \iint h_{M,n}(z,t)\, d\mu_n(z) d\mu_n(t) \\
&\geq \lim_{M\to\infty} \lim_{m\to\infty} \limsup_{n\to\infty} \iint h_{M,m}(z,t)\, d\mu_n(z) d\mu_n(t) \\
&= \lim_{M\to\infty} \lim_{m\to\infty} \iint h_{M,m}(z,t)\, d\mu(z) d\mu(t) \\
&= I_v(\mu) \geq V_v (\geq V),
\end{aligned}$$

and this shows that $\lim_{n\to\infty} c_{w_n} \leq c_v$, proving (6.2).

Above we verified that $I_v(\mu) = V_v$, i.e. $\mu = \mu_v$. This can be applied to any subsequence of $\{\mu_n\}$ and we get $\mu_n \to \mu_v$ in a standard way. □

Proof of Theorem 6.2(d). Since $c_{w_n} \leq c_v$ for all n, to prove (6.5) it is enough to observe that, by the monotone convergence theorem,

$$V_v = I_v(\mu_v) = \lim_{n\to\infty} I_{w_n}(\mu_v) \geq \lim_{n\to\infty} V_{w_n}.$$

In the proof of (6.6) we keep the notations of the proof of part (c) except that now $\{\omega_m\}$ is a monotone decreasing sequence of positive continuous functions converging to v (such ω_m's exist because we *assumed* the admissibility of v). Observe that now

$$\mathcal{S}_v \subseteq \{z \mid v(z) \geq \varepsilon\} =: K$$

for some $\varepsilon > 0$. Thus, similarly as above, the monotone convergence theorem yields with

$$h^*_{M,m}(z,t) := \min\left(M, \log[|z-t| w_m(z) w_m(t)]^{-1}\right),$$

that

$$\begin{aligned}
I_v(\mu) &= \lim_{m\to\infty} I_{\omega_m}(\mu) \\
&= \lim_{m\to\infty} \lim_{M\to\infty} \iint h_{M,m}(z,t)\, d\mu(z) d\mu(t) \\
&= \lim_{m\to\infty} \lim_{M\to\infty} \lim_{n\to\infty} \iint h_{M,m}(z,t)\, d\mu_n(z) d\mu_n(t) \\
&\leq \lim_{m\to\infty} \lim_{M\to\infty} \liminf_{n\to\infty} \iint h^*_{M,n}(z,t)\, d\mu_n(z) d\mu_n(t) \\
&\leq \lim_{m\to\infty} \lim_{n\to\infty} I_{w_n}(\mu_n) = \lim_{n\to\infty} V_{w_n} \leq V_v,
\end{aligned}$$

and $\mu = \mu_v$ and $\mu_n \to \mu_v$ follow in a standard way. □

Proof of Theorem 6.2(e). The fact that $M \geq \operatorname{cap}(w, E)$ is immediate from the definition of weighted capacity. In order to prove (6.7), we observe that with $\mu := \mu(w, E)$ we have for quasi-every $z \in \operatorname{supp}(\mu)$ (see Theorem 1.3(f))

$$\begin{aligned}
\log(M/\operatorname{cap}(w,E)) &= I_w(\mu) + \log M \\
&= \int \log[M/(|z-t| w(z) w(t))]\, d\mu(t) \\
&\geq \int_{E_n} \log[M/(|z-t| w(z) w(t))]\, d\mu(t).
\end{aligned}$$

This inequality holds $\mu|_{E_n}$-almost everywhere, and integrating it with respect to $\mu|_{E_n}$ we get that if $\mu(E_n) > 0$ (which implies that $\operatorname{cap}(w, E_n) > 0$, and so $\mu(w, E_n)$ exists)

$$\begin{aligned}
\log \frac{M}{\operatorname{cap}(w,E)} &\geq \mu(E_n)\left(I_{w|E_n}\left(\mu|_{E_n}/\mu(E_n)\right) + \log M\right) \\
&\geq \mu(E_n)\left(I_{w|E_n}(\mu(w,E_n)) + \log M\right) \\
&= \mu(E_n) \log(M/\operatorname{cap}(w, E_n)). \qquad (6.11)
\end{aligned}$$

Notice that (6.11) also holds when $\mu(E_n) = 0$. Finally (6.7) follows from (6.11) on dividing by $\log(M/\mathrm{cap}(w, E)) \cdot \log(M/\mathrm{cap}(w, E_n))$ and making use of the inequality $\sum_n \mu(E_n) \geq \mu(E) = 1$. □

Proof of Theorem 6.2(f). It is enough to prove the result just for two sets E_1 and E_2, for then induction gives the general case. We may also suppose that $m = 1$, for otherwise we can multiply w by $1/\sqrt{m}$, which changes the weighted capacity by the factor $1/m$. If $\mathrm{cap}(E, w) \geq 1$ or $\mathrm{cap}(E_j, w) = 0$ for some $j = 1, 2$, then there is nothing to prove, so suppose $0 < \mathrm{cap}(E, w) < 1$, and $0 < \mathrm{cap}(E_j, w) < 1$, $j = 1, 2$. If μ_j, $j = 1, 2$, denotes the equilibrium measure for the set E_j, then in the weighted energy of the measure

$$\mu = \frac{V_{w_2}}{V_{w_1} + V_{w_2}}\mu_1 + \frac{V_{w_1}}{V_{w_1} + V_{w_2}}\mu_2$$

the mutual energy

$$\int_{E_1}\int_{E_2} \log \frac{1}{|z-t|w(z)w(t)}\, d\mu_1(z)d\mu_2(t)$$

is non-positive by the assumption (6.8). Therefore,

$$I_w(\mu) \leq \left(\frac{V_{w_2}}{V_{w_1} + V_{w_2}}\right)^2 I_w(\mu_1) + \left(\frac{V_{w_1}}{V_{w_1} + V_{w_2}}\right)^2 I_w(\mu_2) = \frac{V_{w_1}V_{w_2}}{V_{w_1} + V_{w_2}},$$

and so

$$V_w \leq \frac{V_{w_1}V_{w_2}}{V_{w_1} + V_{w_2}},$$

which is exactly the inequality (6.9) for two summands. □

In the proof above we have appealed to the behavior of the capacity when the weight is multiplied by a constant. Let us record this as

Remark 6.3. If w is quasi-admissible and k is a nonnegative constant, then

$$c_{kw} = k^2 c_w.$$

Example 6.4. Suppose that E is compact, $w(z) \leq 1$ for $z \in E$ and $w(z) = 1$ for all z on the outer boundary $\partial_\infty E$ of E. Then

$$\mathrm{cap}(w, E) = \mathrm{cap}(E).$$

Indeed, from part (b) of Theorem 6.2 we have

$$\mathrm{cap}(w, E) \leq \mathrm{cap}(E).$$

The reverse inequality is trivial if $\mathrm{cap}(E) = 0$; so assume $\mathrm{cap}(E) > 0$. Let μ_E denote the (unweighted) equilibrium measure associated with the set E. As we have shown, $\mathrm{supp}(\mu_E)$ is part of the outer boundary $\partial_\infty E$ of E (Corollary 4.5). Hence

$$\begin{aligned}\log\frac{1}{\text{cap}(E)} &= \iint_{\partial_\infty E\times\partial_\infty E}\log|z-t|^{-1}d\mu_E(z)d\mu_E(t)\\ &= \iint_{\partial_\infty E\times\partial_\infty E}\log[|z-t|w(z)w(t)]^{-1}d\mu_E(z)d\mu_E(t)\\ &\geq I_{w|_E}(\mu(w,E)) = \log\frac{1}{\text{cap}(w,E)},\end{aligned}$$

which gives $\text{cap}(w, E) \geq \text{cap}(E)$. □

Next we prove the continuity of F_w and $U^{\mu_w}(z)$ in w with respect to monotone convergence.

Theorem 6.5. *Suppose that all the weights below are admissible and are defined on Σ.*

(a) *If $w_n \uparrow w$, then*

$$\lim_{n\to\infty} F_{w_n} = F_w, \tag{6.12}$$

and for every z

$$\lim_{n\to\infty} U^{\mu_{w_n}}(z) = U^{\mu_w}(z). \tag{6.13}$$

(b) *If $w_n \downarrow w$, then* (6.12) *holds. If, in addition, we know that the potential U^{μ_w} is everywhere continuous and*

$$U^{\mu_w}(z) - F_w \geq -Q(z) \quad on \quad \Sigma, \tag{6.14}$$

then (6.13) *also holds. In particular, this is the case if w is continuous and every component of $\mathbf{C}\setminus\Sigma$ is regular with respect to the Dirichlet problem.*

If, in (6.13), *the right-hand side is continuous on* $\mathbf{C}$, *then the convergence in* (6.13) *is uniform on* $\mathbf{C}$.

We note that in the case $w_n \downarrow w$ the convergence (6.13) may fail; in particular, the continuity assumption in part (b) cannot be dropped. This is shown by

Example 6.6. Let $\Sigma = \mathbf{R}$,

$$S = \{0\} \cup \bigcup_{n=1}^{\infty}\left[2^{-n}, 2^{-n} + 2^{-n^3}\right]$$

(cf. Example 4.9), $w(x) = 1$ on S and $w(x) = 0$ outside S, i.e. w is the characteristic function of S. We also set

$$S_n := S \cup [0,\ 1/n],$$

and let w_n be the characteristic function of S_n. Then $w_n \downarrow w$.

Since this corresponds to the classical setting, the potentials $U^{\mu_{w_n}}$ and U^{μ_w} are the equilibrium potentials of the sets S_n and S, respectively. S_n consists of finitely many intervals; hence its potential equals $\log(1/\mathrm{cap}(S_n))$ at every point of S_n. On the other hand, zero is an irregular point on the boundary of $\mathbf{C} \setminus S$; hence

$$U^{\mu_w}(0) < \log\left(\frac{1}{\mathrm{cap}(S)}\right).$$

Since $\mathrm{cap}(S_n) \to \mathrm{cap}(S)$ as $n \to \infty$, we finally can conclude

$$U^{\mu_w}(0) < \liminf_{n\to\infty} \log\left(\frac{1}{\mathrm{cap}(S_n)}\right) = \liminf_{n\to\infty} U^{\mu_{w_n}}(0),$$

i.e. (6.13) fails at 0. □

The preceding example showed that to be able to conclude (6.13) in part (b) the continuity of the potential U^{μ_w} cannot be dropped. Now we show that the other condition, namely (6.14), is also essential.

Example 6.7. Let $\mathbf{C} = \mathbf{R}$, w the characteristic function of $[-1, 1] \cup \{2\}$ and w_n the characteristic function of $K_n := [-1, 1] \cup [2, 2 + 1/n]$. Then U^{μ_w} is the equilibrium potential of the interval $[-1, 1]$; hence we get as in the preceding example (see also (1.8))

$$U^{\mu_w}(2) = -\log|2 + \sqrt{2^2 - 1}| + \log 2 \neq \log 2,$$

but

$$\log 2 = \lim_{n\to\infty} \log\left(\frac{1}{\mathrm{cap}(K_n)}\right) = \lim_{n\to\infty} U^{\mu_{w_n}}(2).$$

□

In the proof of Theorem 6.5 we shall need the following result which is called the *principle of descent* of the theory of logarithmic potentials.

Theorem 6.8 (Principle of Descent). *Let μ_n, $n = 1, 2, \ldots$, be probability measures all having support in a fixed compact subset of $\mathbf{C}$ and converging to some measure μ in the weak* topology. Suppose furthermore, that for each n a point z_n is given so that $z_n \to z^*$ for some $z^* \in \mathbf{C}$. Then*

$$U^{\mu}(z^*) \le \liminf_{n\to\infty} U^{\mu_n}(z_n). \tag{6.15}$$

Furthermore,

$$I(\mu) \le \liminf_{n\to\infty} I(\mu_n). \tag{6.16}$$

Recall that $I(\mu)$ denotes the (unweighted) energy of μ.

Let us remark that if U^μ is continuous on $\mathbf{C}$, then (6.15) implies that

$$U^\mu(z) \le \liminf_{n\to\infty} U^{\mu_n}(z)$$

uniformly on compact subsets of $\mathbf{C}$, and hence uniformly on $\mathbf{C}$.

Proof of Theorem 6.8. From the uniform continuity of the function

$$\min\left(M, \log\frac{1}{|t|}\right)$$

we easily get on applying the monotone convergence theorem that

$$\begin{aligned} U^\mu(z^*) &= \lim_{M\to\infty} \int \min\left(M, \log\frac{1}{|z^*-t|}\right) d\mu(t) \\ &= \lim_{M\to\infty}\lim_{n\to\infty} \int \min\left(M, \log\frac{1}{|z_n-t|}\right) d\mu_n(t) \\ &\le \lim_{M\to\infty}\liminf_{n\to\infty} U^{\mu_n}(z_n) = \liminf_{n\to\infty} U^{\mu_n}(z_n), \end{aligned}$$

and this is (6.15). The proof of (6.16) is similar if we note that $\mu_n \to \mu$ in the weak* topology implies that $\mu_n(z)\times\mu_n(t) \to \mu(z)\times\mu(t)$ in the weak* topology on measures on $\mathbf{C}\times\mathbf{C}$. □

Now we can proceed with the

Proof of Theorem 6.5(a). For simplicity let us write μ_n and F_n instead of μ_{w_n} and F_{w_n} and μ, F for μ_w, F_w. In Theorem 6.2(d) we proved that

$$\lim_{n\to\infty}\left(I(\mu_n) + 2\int Q_n\, d\mu_n\right) = I(\mu) + 2\int Q\, d\mu, \tag{6.17}$$

and that $\mu_n \to \mu$ in the weak* topology. Let $\{\omega_m\}$ be a monotone increasing sequence of continuous functions tending to Q (since Q is lower semi-continuous, such a sequence exists). From the proof of Theorem 1.3(a) it easily follows that the measures μ_n have support in a fixed compact set (note that $w_n \le w$), so the inequalities $Q_n \ge Q \ge \omega_m$ imply

$$\liminf_{n\to\infty}\int Q_n\, d\mu_n \ge \liminf_{n\to\infty}\int \omega_m\, d\mu_n = \int \omega_m\, d\mu,$$

which yields for $m\to\infty$

$$\liminf_{n\to\infty}\int Q_n\, d\mu_n \ge \int Q\, d\mu. \tag{6.18}$$

By the principle of descent, $\mu_n \to \mu$ implies that

$$\liminf_{n\to\infty} I(\mu_n) \geq I(\mu). \tag{6.19}$$

Now (6.18) and (6.19) together with (6.17) show that we must have

$$\lim_{n\to\infty} I(\mu_n) = I(\mu)$$

and

$$\lim_{n\to\infty} \int Q_n \, d\mu_n = \int Q \, d\mu.$$

But then (6.12) immediately follows from the representations

$$F_n = I(\mu_n) + \int Q_n \, d\mu_n \quad \text{and} \quad F = I(\mu) + \int Q \, d\mu.$$

To prove (6.13) we remark first of all that $w_n \leq w$ implies

$$U^{\mu_n}(z) - F_n \leq U^{\mu}(z) - F$$

(see Corollary 4.2(a)). Thus, taking into account the just proved relation (6.12) we can see that

$$\limsup_{n\to\infty} U^{\mu_n}(z) \leq U^{\mu}(z). \tag{6.20}$$

Now (6.13) follows from this and the principle of descent (Theorem 6.8) because $\mu_n \to \mu$ in the weak* topology. □

Proof of Theorem 6.5(b). Suppose that $w_n \downarrow w$, i.e. $Q_n \uparrow Q$. As in the preceding proof we can conclude from $w_n \leq w_1$ and the proof of Theorem 1.3(a) that, for some R, each μ_n has support in the disk $\overline{D_R} := \{z \mid |z| \leq R\}$. Again let $\{\omega_m\}$ be an increasing sequence of continuous functions tending to Q. Since $Q_n \uparrow Q \geq \omega_m$ and each Q_n is lower semi-continuous, we can get by simple compactness argument (Dini's theorem!) that for every $\varepsilon > 0$ there is an $n_\varepsilon = n_{\varepsilon,m}$ such that for $n \geq n_\varepsilon$ we have

$$Q_n(z) \geq \omega_m(z) - \varepsilon \quad \text{for} \quad z \in \overline{D_R} \cap \Sigma.$$

But then

$$\liminf_{n\to\infty} \int Q_n \, d\mu_n \geq \liminf_{n\to\infty} \int \omega_m \, d\mu_n - \varepsilon = \int \omega_m \, d\mu - \varepsilon.$$

For $m \to \infty$, $\varepsilon \to 0$ we can conclude (6.18), and from here on the proof of (6.12) coincides with the one given above in the monotone increasing case.

Now let us assume that U^{μ_w} is continuous and (6.14) holds for all $z \in \Sigma$. Since $(-Q_n) \downarrow (-Q)$ and each $-Q_n$ is upper semi-continuous, we can conclude as before from (6.14) and from the continuity of U^{μ_w} that, for every $\varepsilon > 0$, there is an n_ε such that we have

$$-Q_n(z) \leq U^{\mu_w}(z) - F_w + \varepsilon \quad \text{for} \quad z \in \overline{D_R} \cap \Sigma \text{ and } n \geq n_\varepsilon.$$

In particular, this is true for all $z \in S_n := \text{supp}(\mu_n)$, and since for such z's we also have

$$U^{\mu_n}(z) - F_n \leq -Q_n(z)$$

(see Theorem 1.3(e)) we can conclude

$$U^{\mu_n}(z) - F_n \leq U^{\mu_w}(z) - F_w + \varepsilon, \quad z \in S_n.$$

By invoking the principle of domination which will be proved in Theorem II.3.2 in Chapter II, we get the same inequality for all z, and letting first $n \to \infty$ and then $\varepsilon \to 0$ we can conclude the relation (6.20) (use also (6.12)), from which (6.13) is obtained as in part (a) via the principle of descent.

A careful examination of the above proof (see also the remark made after Theorem 6.8) yields the very last statement concerning the uniformity of the convergence in (6.13) (one shows this first for compact subsets of **C** from which the uniform convergence on **C** follows since

$$\lim_{|z|\to\infty} (U^{\mu_n}(z) - U^{\mu}(z)) = 0$$

uniformly in n – recall that the supports of the μ_n's lie in a disk $\overline{D_R}$). □

We conclude this chapter with the so-called lower envelope theorem.

Theorem 6.9 (Lower Envelope Theorem). *Let μ_n, $n = 1, 2, \ldots$, be a sequence of positive unit Borel measures all having support in a fixed compact set. If $\mu_n \to \mu$ in the weak* topology, then*

$$\liminf_{n\to\infty} U^{\mu_n}(z) = U^{\mu}(z) \tag{6.21}$$

*for quasi-every $z \in$ **C**.*

This is a supplement to the principle of descent (Theorem 6.8) which claims that the left-hand side in (6.21) is always at least as large as the right side.

For the proof of Theorem 6.9 and later results we shall appeal to the following lemma.

Lemma 6.10. *Let ν be a finite Borel measure of compact support with U^ν finite ν-almost everywhere. Then there exists an increasing sequence $\{\nu^{(i)}\}_{i=1}^{\infty}$ of Borel measures such that each of the measures $\nu^{(i)}$ has support in* supp(ν), $\|\nu^{(i)}\| \leq \|\nu\|$ *for all i,*

$$\lim_{i\to\infty} \|\nu - \nu^{(i)}\| = 0,$$

and each of the potentials $U^{\nu^{(i)}}$ is continuous on **C**. *Furthermore, $U^{\nu^{(i)}}(z) \to U^{\nu}(z)$ for all $z \in$* **C**, *with uniform convergence holding on any compact set where U^ν is continuous.*

We mention that $\nu^{(i)}$ is actually obtained by restricting ν to a compact set S_i.

Proof. By Lusin's continuity theorem [195], for every $i \geq 1$ there exists a compact set $S_i \subseteq \mathrm{supp}(\nu)$ such that $\nu(\mathbf{C} \setminus S_i) < 1/i$ and the potential U^ν (considered as a function on S_i only) is continuous on S_i. Without loss of generality we may assume $S_i \subset S_{i+1}$ for all i, so that the measures

$$\nu^{(i)} := \nu_{|S_i}$$

form an increasing sequence. Obviously, $\nu^{(i)}$ will be appropriate if we can show the continuity of the corresponding potentials. Since $\nu^{(i)}$ is supported on S_i, it is enough to prove the continuity of $U^{\nu^{(i)}}$ on the set S_i (see Theorem II.3.5). However,

$$U^{\nu^{(i)}} = U^\nu - U^{\nu - \nu^{(i)}},$$

and on the right U^ν is continuous on S_i while $U^{\nu-\nu^{(i)}}$ is lower semi-continuous on it; hence $U^{\nu^{(i)}}$ is upper semi-continuous on S_i. Since potentials of positive measures are lower semi-continuous everywhere, we deduce the continuity of $U^{\nu^{(i)}}$ on S_i.

From the definition of $\nu^{(i)}$ it follows that $\nu^{(i)} \to \nu$ in the weak*-topology, and also that $U^{\nu^{(i)}}(z) \to U^\nu(z)$ for $z \notin S := \mathrm{supp}(\nu)$. Furthermore, we can assume without loss of generality that the diameter of S is less than one so that $\log 1/|z-t| > 0$ for $z, t \in S$ (otherwise replace $\log 1/|z-t|$ by $\log M/|z-t|$ with $M > \mathrm{diam}(S)$). Then the sequence $U^{\nu^{(i)}}$ is increasing on S, and for $z \in S$ we have by the principle of descent that

$$U^\nu(z) \leq \liminf_{i\to\infty} U^{\nu^{(i)}}(z) \leq \limsup_{i\to\infty} U^{\nu^{(i)}}(z) \leq U^\nu(z);$$

hence $U^{\nu^{(i)}}(z) \to U^\nu(z)$ for all $z \in \mathbf{C}$. Finally the uniform convergence statement follows from the fact that $\nu^{(i)}$ is an increasing sequence. □

Applying Lemma 6.10 to the equilibrium measure for a set of positive capacity we immediately get

Corollary 6.11. *If S is a compact subset of $\mathbf{C}$ of positive capacity, then there is a measure ν such that $\|\nu\| > 0$, $\mathrm{supp}(\nu) \subseteq S$, and the potential U^ν is continuous everywhere.*

Proof of Theorem 6.9. The left-hand side of (6.21) is at least as large as the right side by the principle of descent. Hence if equality in (6.21) does not hold quasi-everywhere, there exists a compact set S of positive capacity such that

$$U^\mu(z) < \liminf_{n\to\infty} U^{\mu_n}(z) \quad \text{for} \quad z \in S. \tag{6.22}$$

Without loss of generality, we assume that U^{μ_n} is nonnegative on S for every n (recall that all μ_n are supported on a fixed compact set (necessarily containing S) which we can assume has diameter less than one). Let ν be the measure guaranteed by Corollary 6.11 for the set S. On integrating (6.22) with respect to ν and applying Fatou's lemma, we get

$$\int U^{\mu} d\nu < \int \liminf_{n\to\infty} U^{\mu_n} d\nu \leq \liminf_{n\to\infty} \int U^{\mu_n} d\nu. \tag{6.23}$$

But using that U^{ν} is continuous and $\mu_n \to \mu$ in the weak* topology we deduce from the Fubini–Tonelli theorem that

$$\liminf_{n\to\infty} \int U^{\mu_n} d\nu = \liminf_{n\to\infty} \int U^{\nu} d\mu_n = \int U^{\nu} d\mu = \int U^{\mu} d\nu,$$

which contradicts (6.23), and this contradiction proves the theorem. □

I.7 Notes and Historical References

Section I.1

It is likely that C.F. Gauss [57] was the first to study potentials with continuous external fields; therefore this part of potential theory is often referred to as the Gaussian variation problem. However, Gauss regarded the equilibrium not as a state but as a process to which the system descends. After the pioneering works of O. Frostman [51] on equilibrium measures, the emphasis shifted to more general kernels, although logarithmic potentials kept their importance because of their close relationship to polynomials and holomorphic functions. Frostman himself considered the energy problem in the presence of an external field, and proved the main theorem of this chapter Theorem 1.3 for the case when the Q is continuous and superharmonic (see [52]). Starting from the 1930's the Polish school headed by F. Leja investigated logarithmic potentials with continuous external fields because of their connections with the solutions to certain Dirichlet problems (see the book [123] and the other references to the works of Leja, J. Górski, W. Kleiner and J. Siciak). Independently several Japanese mathematicians such as S. Kametani [88], M. Ohtsuka [175], and N. Ninomiya [173],[174] studied the Gauss variational problem for generalized kernels in locally compact spaces and applied it to generalizing the balayage problem (for more references on the Japanese works see [175, p. 284]). G. Choquet likewise contributed substantially to the study of potentials for generalized kernels.

A new impulse came in the 1980's when E. A. Rakhmanov [189] and H. N. Mhaskar and E. B. Saff [157], [159] used potentials with external fields to study orthogonal polynomials with respect to exponential weights (see Chapter VII). Indeed, Saff pointed out in [196] that the theory of external fields provided a unified approach for treating incomplete polynomials (introduced by G. G. Lorentz [136]), ray sequences of Padé approximants to Markov functions (investigated in the dissertation of H. Stahl [210]) and orthogonal polynomials on **R** (particularly for the Freud weights [49]). The essential distinction between earlier works (say of the Polish school) and the newer treatments of Rakhmanov and Mhaskar-Saff lies not only in its greater generality, but in its emphasis on determining the support set S_w (cf. Theorem 1.3) from the given external field (especially when w is defined on an unbounded set or vanishes at some points). The determination of S_w is the essential step needed for recovering the extremal measure μ_w; see Chapter II.

Theorem 1.3 is essentially taken from Mhaskar-Saff [161], with a slight change in the notion of admissibility (cf. (1.10)). As we have already mentioned, O. Frostman [52] has verified the same theorem for superharmonic Q's. The proof is an adaptation of Frostman's argument.

The proof of Lemma 1.8 is taken from [80].

Sets of zero capacity are usually called *polar sets*. Under some regularity conditions (like compactness, or G_δ property) they are characterized by the fact that potentials can be equal to $+\infty$ on them (see Evan's theorem (Lemma 2.3)). It can also be shown (see e.g. the remark made before Lemma 2.2) that locally differentiable functions map polar sets into polar ones. Thus, "being of zero capacity" is conformally invariant.

Sets of zero capacity also play an important role in connection with removable singularities. For example, if D is a domain, $E \subset D$ is a closed set of zero capacity, and h is a bounded and harmonic function on $D \setminus E$, then h has a unique harmonic extension to the whole of D. For more on such removable singularities see [192, Section 3.6].

The *outer capacity* of a set E is defined as the infimum of the capacities of open sets containing E, as opposed to the definition made after (1.8) which might then be called the inner capacity. Now a theorem of Choquet [75, Section 5.8] says that every Borel set is capacitable in the sense that its outer and inner capacities are the same.

For the minimum energy problem described in Theorem 1.3, what happens if we impose a constraint on the measures μ? More precisely, suppose that σ is a given positive measure with $\text{supp}(\sigma) = \Sigma$ and $\|\sigma\| > 1$. Then by $\mathcal{M}^\sigma(\Sigma)$ we denote the set of measures $\mu \in \mathcal{M}(\Sigma)$ that are constrained by σ; that is

$$\mathcal{M}^\sigma(\Sigma) := \{\mu \in \mathcal{M}(\Sigma) \mid \mu \le \sigma\},$$

where $\mu \le \sigma$ means that $\sigma - \mu$ is a positive measure. The *constrained energy problem* with weight $w = e^{-Q}$ concerns the minimization

$$V_w^\sigma := \inf\{I_w(\mu) \mid \mu \in \mathcal{M}^\sigma(\Sigma)\}.$$

In the unweighted ($Q \equiv 0$) case (where we drop the subscript w), this problem was first introduced by Rakhmanov [190] who used it to deduce the asymptotic zero distribution of certain "ray sequences" of Chebyshev polynomials of a discrete variable. It was shown by P. Dragnev and E. B. Saff [36] that the unweighted constrained energy problem is equivalent to an unconstrained weighted energy problem. More precisely, suppose that Σ is compact and U^σ is continuous on Σ. Let $\lambda^\sigma \in \mathcal{M}^\sigma(\Sigma)$ be the extremal measure for the constrained problem, i.e.

$I(\lambda^\sigma) = V^\sigma$. If ν is the solution to the weighted energy problem on Σ with weight $w = \exp(U^\sigma/(||\sigma|| - 1))$, then $\lambda^\sigma = \sigma - (||\sigma|| - 1)\nu$.

Dragnev and Saff also investigate in [36] the more general constrained energy problem in the presence of an external field and obtain an analogue of Theorem 1.3 along with other characterizations. For example, they prove that if $w = e^{-Q}$ is an admissible weight on Σ and the constraint σ satisfies $\sigma(\Sigma_0) > 1$ and has finite energy on compact sets, then there exists a unique measure $\lambda = \lambda_w^\sigma \in \mathcal{M}^\sigma(\Sigma)$ such that $I_w(\lambda) = V_w^\sigma$. Furthermore, there exists a constant F_w^σ such that $U^\lambda + Q \geq F_w^\sigma$ holds $(\sigma - \lambda)$–a.e. and $U^\lambda + Q \leq F_w^\sigma$ holds for all $z \in \text{supp}(\lambda)$.

Section I.2

An alternative proof of Lemma 2.1 appears in the book of M. Tsuji [222, Section III.9]. After the proof of Lemma 2.1 we have mentioned that the capacity does not increase under a contractive mapping. More generally, if Φ is a mapping with $|\Phi(z) - \Phi(t)| \leq M|z-t|^\alpha$, then $\text{cap}(\Phi(K)) \leq M\text{cap}(K)^\alpha$.

Evans' theorem is from [42]. We shall return to it in Theorem III.1.11, where we prove that μ can actually be supported on E.

For the generalized minimum principle (Theorem 2.4), the assumption that g is lower bounded on R is essential; consider, for example, $g(z) = \log|z|$ on $R = \{z \mid 0 < |z| < 1\}$. Of course, this assumption is redundant if (2.4) holds for *every* $z' \in \partial R$ and R is bounded (see Theorem 0.5.2).

Sometimes the lower boundedness can be weakened provided we have a growth condition on the function around the point at infinity. As a typical example consider the strip $R := \{z = x + iy \mid 0 < x < 1\}$. In this case *if g is superharmonic on R, $|g(x+iy)| \leq Ae^{B|y|}$ for some A and $0 \leq B < \pi$, and*

$$\liminf_{z \to z'} g(z) \geq 0$$

for q.e. $z' \in \partial R$, then $g(z) \geq 0$ in R. The proof of this fact proceeds as follows: For a $B < b < \pi$ and $\varepsilon > 0$ consider the function

$$h_\varepsilon(z) = g(z) + \varepsilon \text{Re}\cos(b(z - 1/2)) = g(z) + \varepsilon \cos(b(x - 1/2))\cosh(by).$$

Then h_ε is superharmonic, bounded from below on R, and satisfies

$$\liminf_{z \to z'} h_\varepsilon(z) \geq 0$$

for q.e. $z' \in \partial R$. Thus, we deduce from the generalized minimum principle that $h_\varepsilon(z) \geq 0$ in R, and since this is true for every $\varepsilon > 0$, we can finally conclude that $g \geq 0$, as was claimed.

What we have just proven immediately implies the classical Three Lines Theorem: *If g is superharmonic on R, $|g(x+iy)| \leq Ae^{B|y|}$ for some A and $0 \leq B < \pi$, and*

$$\liminf_{z \to z'} g(z) \geq m_0$$

for q.e. $z' = iy' \in \partial R$, *and*

$$\liminf_{z\to z'} g(z) \geq m_1$$

for q.e. $z' = 1 + iy' \in \partial R$, *then*

$$g(x+iy) \geq (1-x)m_0 + xm_1.$$

To get this simply apply the preceding result to $g(z) - \mathrm{Re}((1-z)m_0 + zm_1)$.

In general, for domains R containing the point ∞ the lower boundedness in the minimum principle (to conclude $g \geq 0$ from $\liminf_{z\to z', z\in\mathbf{R}} g(z) \geq 0$ for q.e. $z' \in \partial R$) can be relaxed to $\limsup_{z\to\infty} g(z)/h(z) \leq 0$, where h is a finite-valued subharmonic function on R such that $\limsup_{z\to\infty} h(z) < 0$. For further such results of this Phragmén–Lindelöf type see [192, Section 2.3].

The Dirichlet problem may not be solvable in its original form, which is why we need the notion of a generalized solution. In fact, let R be the open unit disk without its center, and $f(z) = 0$ if $|z| = 1$ and $f(z) = 1$ if $z = 0$. For this boundary function there is no harmonic function on R that is continuous on its closure and agrees with f on the boundary; for such a function would actually be harmonic on the whole open unit disk, and would violate the maximum principle.

If ∂R is of zero capacity, then we cannot speak of the upper and lower solutions as defined in Section I.2, for in this case the definition would yield that the upper solution is identically $-\infty$, while the lower solution is identically $+\infty$. Note however, that if ∂R is a singleton, i.e. $R = \overline{\mathbf{C}} \setminus \{a\}$ for some a, then the Dirichlet problem is always solvable (by constants) in the strict sense.

Further discussion of the Dirichlet problem is given in Appendix A.2 and can be found in most textbooks on harmonic analysis (see e.g. the book of L. L. Helms [77]).

Section I.3

Theorem 3.1 extends classical extremal properties of conductor potentials (cf. [222, Theorem III.15]).

The basic fact, established in Example 3.4, that the equilibrium distribution for the disk in the classical case is simply normalized arc measure on the circumference also follows immediately from symmetry considerations and the facts that the extremal measure is unique and supported on the outer boundary (cf. Corollary 4.5). This last property, however, is not generally true in the presence of an external field.

The estimate of Theorem 3.6 is asymptotically sharp in the sense that the weighted Fekete polynomials $\Phi_n(z) = z^n + \cdots$ satisfy, for the sup norm over $\mathcal{S}_w$,

$$\lim_{n\to\infty} \|w^n \Phi_n\|_{\mathcal{S}_w}^{1/n} = \exp(-F_w);$$

see Section III.1.

Section I.4

The convexity property of Corollary 4.2(c) was observed by J. Siciak [203] who also considered extensions to plurisubharmonic functions in $\mathbf{C}^n$ (cf. [206]).

The continuity properties of equilibrium potentials (without external fields) were established by O. D. Kellogg [89], [90], G. Bouligand [22], [23], M. Brelot [25], [26], H. Cartan [29], and N. Wiener [233], [234], [235].

The notion of polynomial convex hull expresses its main property: $\mathrm{Pc}(K)$ is the set of those points z for which $|P(z)| \le \|P\|_K$ for all polynomials P. In one direction this is an immediate consequence of the maximum modulus theorem. In the other direction a more general result is true (see Corollary III.1.10 in Chapter III), namely there is a sequence of polynomials P_n of degree $n = 1, 2, \ldots$, such that

$$\lim_{n\to\infty} \left(\frac{|P_n(z)|}{\|P_n\|_K} \right)^{1/n} = \exp(g_{\overline{\mathbf{C}}\setminus \mathrm{Pc}(K)}(z, \infty)) > 1,$$

uniformly on compact subsets of $\overline{\mathbf{C}} \setminus \mathrm{Pc}(K)$, where the function g on the right is the Green function of the domain $\overline{\mathbf{C}} \setminus \mathrm{Pc}(K)$ (see Section II.4). In particlar, if U is a neighborhood of $\mathrm{Pc}(K)$, then there is a single polynomial P such that

$$\inf_{z\notin U} |P(z)| > \|P\|_K.$$

This is sometimes referred to as Hilbert's lemniscate theorem (see [192, Theorem 5.5.8]).

An open connected set Ω is said to have a *barrier* at a boundary point z_0 if for some $r > 0$ there exists a positive continuous superharmonic function $h(z)$ on $\Omega \cap D_r(z_0)$ such that $\lim_{z\to z_0} h(z) = 0$. Clearly, if z_0 is regular, then there exists a barrier at z_0 (take $h(z) = g_\Omega(z, \infty)$). Conversely, Bouligand (cf. Theorem I.13 of [222]) has shown that the existence of a barrier at z_0 implies that z_0 is a regular point. It follows from this fact (and also from Wiener's theorem (Theorem 4.6)) that the property of being a regular point is a *local* condition; it depends only on the behavior of the boundary of Ω in a neighborhood of the point.

The identities of Theorem 4.7 lead to numerical methods for solving Dirichlet's problem (cf. Section V.2).

Section I.5

The notion of fine topology was introduced by H. Cartan [29] and further developed by M. Brelot [26]. For further discussion, see the books by N. S. Landkof [111] and L. L. Helms [77].

Section I.6

Part (e) of Theorem 6.2 is essentially taken from Mhaskar-Saff [163, Theorem 3.3]. Notice that this part provides an analogue of subadditivity for weighted logarithmic capacity. In the unweighted case ($w \equiv 1$), Ch. Pommerenke [183] showed that

$$\text{cap}(E_1 \cup E_2) \le \text{cap}(E_1) + \text{cap}(E_2)$$

provided $E_1 \cup E_2$ is connected.

Part (f) of Theorem 6.2 and its proof follows [192, Theorem 5.1.4(b)].

For further discussion of the principle of descent, see M. Brelot [25], [26]. Theorem 6.9 is essentially also due to Brelot [25]. For an extension of these results to Riesz potentials, see [111, Theorem 3.8].

Chapter II. Recovery of Measures, Green Functions and Balayage

In this chapter we shall go deeper into the relationship between a measure and its potential.

In several applications one has a situation when the potential of a measure is given and one needs to determine the measure itself. For example, we shall develop in later chapters methods that allow us to determine the support of the extremal measure μ_w, and then the associated potential is given as the solution of certain Dirichlet problems (see Theorem 4.7 in the preceding chapter). Therefore, our first interest will be to determine a measure from its potential. This is possible, for we have several unicity theorems; one of them saying that if two potentials differ only by a harmonic function on a domain, then the two measures coincide there.

Concerning the recovery problem, first we establish Gauss' formula

$$\mu(R) = \frac{1}{2\pi} \int_{\partial R} \frac{\partial U^\mu}{\partial \mathbf{n}} \, ds$$

for the mass of the measure μ supported within a region R. Then we show that, in general,

$$d\mu = -\frac{1}{2\pi} \Delta U^\mu \, dm,$$

which is true in a distributional sense and also in an ordinary sense under suitable smoothness assumptions. A third possibility to recover μ on disks $D_r(z)$ is from the mean values

$$L(U^\mu; z_0, r) = \frac{1}{2\pi} \int_{-\pi}^{\pi} U^\mu(z_0 + re^{i\theta}) \, d\theta$$

of the potential via the formula

$$\mu(D_r(z_0)) = -r \frac{d}{dr} L(U^\mu; z_0, r).$$

We shall also frequently encounter the situation when the measure μ is supported on an arc γ, for which the recovery takes the form

$$d\mu = -\frac{1}{2\pi} \left(\frac{\partial U^\mu}{\partial \mathbf{n}_-}(s) + \frac{\partial U^\mu}{\partial \mathbf{n}_+}(s) \right) ds$$

with $\mathbf{n}_\pm$ denoting the normals to γ. This formula will allow us in later chapters to derive smoothness properties of the extremal measure μ_w from smoothness on w.

One of the most useful theorems concerning potentials is the principle of domination claiming that the inequality

$$U^\mu(z) \leq U^\nu(z) + c$$

holds on the whole plane provided it is true μ-almost everywhere (at least for the case when μ is of finite energy and has no less total mass than ν). Its importance is clear: in establishing an inequality for potentials one needs only to establish the inequality almost everywhere on the support, which may be a considerably easier problem. This principle of domination contains as a special case the maximum theorem for potentials claiming that the supremum of a potential U^μ is attained on the support of μ, which in turn leads to the continuity theorem stating that a potential U^μ that is continuous on the support of μ as a function on that support is actually also continuous as a function on the whole plane. With the same technique that is used in the proof of the basic unicity theorem and of the principle of domination we are going to prove the Riesz decomposition theorem asserting that any superharmonic function can be represented as the sum of a logarithmic potential and a harmonic function.

After that we shall turn to a powerful technique called balayage (the French word for "sweeping"), which consists of sweeping out the mass from a region G without altering the potential outside G. To be more precise, the balayage problem consists of finding another measure $\widehat{\nu}$ supported on ∂G such that it has the same mass as ν and the potentials U^ν and $U^{\widehat{\nu}}$ coincide on ∂G (quasi-everywhere). The determination of $\widehat{\nu}$ is neatly facilitated by solving the weighted energy problem with ∂G as the conductor and $Q = -U^\mu$ as the external field. With this choice $\widehat{\nu} = \mu_w$, where $w = \exp(-Q)$. This approach easily yields the main properties of balayage measures such as the nonincreasing behavior of the potential under the transition $\nu \to \widehat{\nu}$ or the fact that this mapping preserves integrals against harmonic functions h:

$$\int h \, d\widehat{\nu} = \int h \, d\nu.$$

The solution to the balayage problem immediately leads to the construction of Green functions. In fact, the Green function for a bounded domain G with pole at $a \in G$ is given by

$$g_G(z, a) = \log \frac{1}{|z-a|} - \int_{\partial G} \frac{1}{|z-t|} \, d\widehat{\delta_a}(t),$$

where δ_a is the Dirac mass at a. We shall use this representation in conjunction with the aforementioned recovery formulas to derive several representations for balayage measures such as

$$\widehat{\delta_a}(s) = \frac{1}{2\pi} \frac{\partial g_G(s, a)}{\partial \mathbf{n}} ds.$$

This combined with Green's formula shows the importance of the balayage measure $\widehat{\delta_a}$ in representing harmonic functions from their boundary values. This is why $\widehat{\delta_a}$ is often referred to as harmonic measure, a topic which is discussed in some detail in Appendix A.3.

II.1 Recovering a Measure from Its Potential

In this section we consider the problem of how to find a signed measure μ from its potential. We do not present the results under the most general conditions because the details tend to be technical. Our aim is more to illustrate the possible theorems and to have them in a form that is convenient for applications.

We shall always assume that the measure μ is finite and of compact support on the complex plane. The results will be of local character, but obviously local determination of μ is equivalent to its complete determination.

We start with the determination of the measure from its logarithmic potential on two-dimensional domains. For clearer notation, let m denote two-dimensional Lebesgue measure on $\mathbf{C}$ and, as usual, Δ denotes the Laplacian operator

$$\Delta F := \frac{\partial^2 F}{\partial x^2} + \frac{\partial^2 F}{\partial y^2}.$$

Most of our results are based on Green's formula (cf. [31, p. 280] or [75, Theorem 1.9]):

Let R be a bounded open set with C^1 boundary ∂R, and let u and v be twice continuously differentiable functions defined in a neighborhood of $\overline{R}$, the closure of R. Then

$$\int_R (v\Delta u - u\Delta v)\, dm = -\int_{\partial R} \left(v\frac{\partial u}{\partial \mathbf{n}} - u\frac{\partial v}{\partial \mathbf{n}} \right) ds, \tag{1.1}$$

where $\partial/\partial\mathbf{n}$ denotes differentiation in the direction of the inner normal of R, and ds indicates integration with respect to the arc length on ∂R. In particular, if both u and v are harmonic in R, then

$$\int_{\partial R} v\frac{\partial u}{\partial \mathbf{n}} = \int_{\partial R} u\frac{\partial v}{\partial \mathbf{n}}. \tag{1.2}$$

As a typical result which immediately follows from Green's formula we state

Theorem 1.1 (Gauss' Theorem). *Let R be a bounded open set with C^1 boundary, and μ a signed measure with compact support that is disjoint from ∂R. Then*

$$\mu(R) = \frac{1}{2\pi}\int_{\partial R} \frac{\partial U^\mu}{\partial \mathbf{n}}\, ds.$$

Proof. First assume that $\mu = \delta_{z_0}$ (unit mass at z_0) with $z_0 \in R$. Let us remove a small circle $C_r(z_0)$ around z_0 together with its interior from R. We can apply Green's formula to the resulting domain R_r and to $u = U^\mu$ and $v \equiv 1$. By the harmonicity of these functions on R_r, Green's formula takes the form

$$\int_{\partial R_r} \frac{\partial U^{\delta_{z_0}}}{\partial \mathbf{n}} ds = 0. \tag{1.3}$$

The boundary of R_r consists of the boundary of R and of $C_r(z_0)$. But on $C_r(z_0)$ the normal derivative of $U^{\delta_{z_0}}(z) = \log 1/|z - z_0|$ equals $-1/r$; hence (1.3) yields

$$1 = \frac{1}{2\pi} \int_{\partial R} \frac{\partial U^{\delta_{z_0}}}{\partial \mathbf{n}} ds, \tag{1.4}$$

i.e. the theorem follows in this special case.

Similar considerations show that if $\mu = \delta_{z_0}$ but $z_0 \notin \overline{R}$, then

$$0 = \frac{1}{2\pi} \int_{\partial R} \frac{\partial U^{\delta_{z_0}}}{\partial \mathbf{n}} ds. \tag{1.5}$$

Theorem 1.1 now follows for arbitrary μ by integrating (1.4) and (1.5) against μ. □

Of similar flavor is the next result, which however refers only to disk domains $D_r(z_0) := \{z \mid |z - z_0| < r\}$.

Theorem 1.2. *Let μ be a finite positive measure of compact support on the plane. Then for any z_0 and $r > 0$ the mean value*

$$L(U^\mu; z_0, r) = \frac{1}{2\pi} \int_{-\pi}^{\pi} U^\mu(z_0 + re^{i\theta})\, d\theta$$

exists as a finite number, and $L(U^\mu; z_0, r)$ is a nonincreasing function of r that is absolutely continuous (actually Lip 1*) on any closed subinterval of $(0, \infty)$. Furthermore,*

$$\lim_{r\to 0} L(U^\mu; z_0, r) = U^\mu(z_0). \tag{1.6}$$

If r is a value for which $\frac{d}{dr} L(U^\mu; z_0, r)$ exists, then

$$\mu(D_r(z_0)) = -r \frac{d}{dr} L(U^\mu; z_0, r). \tag{1.7}$$

In particular, this is true for all $r > 0$ with at most countably many exceptions.

Proof. Let $z_0 = 0$, and let us write $L(r)$ instead of $L(U^\mu; 0, r)$. On applying formula (0.5.5) from Chapter 0 we can see that

$$L(r) = \int \log_r(z)\, d\mu(z), \tag{1.8}$$

where

$$\log_r(z) := \begin{cases} \log 1/r & \text{if } |z| \le r \\ \log 1/|z| & \text{if } |z| > r. \end{cases}$$

Since $\log_r$ is decreasing in r, we get that $L(r)$ is nonincreasing. The very definition of superharmonicity gives that $L(r) \le U^{\mu}(0)$. Furthermore, it follows from the lower semi-continuity of U^{μ} that

$$\liminf_{r \to 0+0} L(r) \ge U^{\mu}(0),$$

which yields (1.6).

From the representation (1.8) we get for $h > 0$

$$L(r+h) - L(r) = -\log \frac{r+h}{r} \int_{|z| \le r} d\mu(z) - \int_{r < |z| \le r+h} \log \frac{r+h}{|z|} \, d\mu(z),$$

which in absolute value is less than $\|\mu\| \log[(r+h)/r] \le \|\mu\| h/r$, and this proves the absolute continuity of $L(r)$ on every finite subinterval $[a, b] \subset (0, \infty)$. If we divide the preceding formula by h and let h tend to zero through positive values we get from the dominated convergence theorem that the limit is

$$-\frac{1}{r} \mu(\overline{D_r(0)}).$$

Thus, the right derivative of $L(r)$ always exists and equals the preceding expression.

A similar argument gives

$$\frac{L(r-h) - L(r)}{-h} = \frac{1}{h} \log \frac{r-h}{r} \int_{|z| \le r-h} d\mu(z) - \int_{r-h < |z| \le r} \frac{1}{h} \log \frac{r}{|z|} \, d\mu(z),$$

from which we can again deduce via the monotone convergence theorem and the bounded convergence theorem that the left derivative of $L(r)$ exists and equals

$$-\frac{1}{r} \mu(D_r(0)).$$

Thus, the (two-sided) derivative exists and equals the preceding number for every r for which the circle $\{z \mid |z| = r\}$ has zero μ-measure, which is true for all but at most countably many r. □

Theorem 1.3. *If in a region R the potential U^{μ} has continuous second partial derivatives, then μ is absolutely continuous in R with respect to two-dimensional Lebesgue measure m, and on R we have the formula*

$$d\mu = -\frac{1}{2\pi} \Delta U^{\mu} \, dm.$$

Conversely, if in a neighborhood of a point z_0 the measure μ has the form $d\mu = f(t)\,dm(t)$, and at z_0 the function f satisfies a Lip α, $\alpha > 0$, *condition, then ΔU^μ exists at z_0 and*

$$f(z_0) = -\frac{1}{2\pi}\Delta U^\mu(z_0).$$

The Lip α condition at z_0 means that, in a neighborhood of z_0,

$$|f(z) - f(z_0)| \le L|z - z_0|^\alpha$$

is satisfied for some constant L.

Proof of Theorem 1.3. First let us suppose that in the region R the potential U^μ has continuous second partial derivatives, and we prove that then $d\mu_{|R}$ coincides with $-(2\pi)^{-1}\Delta U^\mu dm_{|R}$.

Let R_0 be a simply connected subdomain of R with continuously differentiable boundary curve ∂R_0 such that the closure of R_0 is contained in R. We are going to prove that the sum

$$\int_{R_0} (\Delta U^\mu(t)) \log\frac{1}{|z - t|}\,dm(t) + 2\pi U^\mu(z) \tag{1.9}$$

is harmonic on R_0. Then, by Theorem 2.1 below, the signed measure $(2\pi)^{-1}\Delta U^\mu dm_{|R_0} + \mu$ does not have mass in R_0; hence

$$d\mu_{|R_0} = -\frac{1}{2\pi}\Delta U^\mu dm_{|R_0}.$$

Since here R_0 is arbitrary with the above properties, this will yield the desired result.

Let $z_0 \in R_0$ be arbitrary and, for small $r > 0$, let R_r be the domain that we obtain from R_0 by deleting the circle $C_r(z_0)$ with center at z_0 and radius r together with its interior $D_r(z_0)$. With $u = U^\mu$ and $v(t) = \log 1/|z_0 - t|$ we apply Green's identity (1.1). Since v is harmonic in R_r, the left-hand side in (1.1) coincides with the first term in (1.9) minus the corresponding integral over the disk $D_r(z_0)$, which is of the order $O(r^2 \log 1/r)$. The right-hand side can be written as an integral over ∂R_0 plus an integral over $C_r(z_0)$. Since for $t \ne z_0$ both v and $\partial v/\partial \mathbf{n}$ are harmonic in R_0 as a function of z_0, the contribution of the first integral (the one over ∂R_0) is harmonic in R_0. The integral over $C_r(z_0)$ consists of two terms. The term that is the integral of $v\partial u/\partial \mathbf{n}$ is of the order $O(r \log 1/r)$. The other term is

$$\int_{C_r(z_0)} u\frac{\partial v}{\partial \mathbf{n}}\,ds,$$

and here $\partial v/\partial \mathbf{n}$ is identically equal to $-1/r$ on $C_r(z_0)$; hence this integral tends to $-2\pi u(z_0)$ as $r \to 0$. Thus, in the limit when r tends to zero we get that (1.9) is in fact a harmonic function in R_0 and this proves the first part of the theorem.

Turning now to the second part let us assume that in a neighborhood of a point z_0 the measure μ is of the form $f(t)dm(t)$ and f satisfies a Lip α, $\alpha > 0$, condition at z_0. We have to prove that under these assumptions $f(z_0) = -(2\pi)^{-1}\Delta U^\mu(z_0)$. Let $D_r(z_0)$ be a neighborhood of z_0. The part of μ lying outside $D_r(z_0)$ contributes to U^μ with a term that is harmonic in $D_r(z_0)$; hence we can assume that μ vanishes outside $D_r(z_0)$ and in $D_r(z_0)$ it is of the form $f(t)dm(t)$. If f is constant throughout $D_r(z_0)$, then for $z \in D_r(z_0)$ we get from (0.5.5)

$$U^\mu(z) = 2\pi\left(\frac{r^2}{2}\log\frac{1}{r} + \frac{r^2}{4} - \frac{|z-z_0|^2}{4}\right)f(z_0),$$

in which case the claim is obvious. By subtracting $f(z_0)$ from f we may thus suppose that f vanishes at z_0. This and the Lip α condition implies then that

$$|f(z)| \le L|z-z_0|^\alpha. \tag{1.10}$$

Without loss of generality we assume $z_0 = 0$, and let us write $z = \zeta_1 + i\zeta_2$ and $t = \tau_1 + i\tau_2$. It is obvious that $\partial U^\mu(z)/\partial\zeta_1$ is given by

$$-\int_{D_r(0)} f(t)\frac{\zeta_1 - \tau_1}{|z-t|^2}\,dm(t)$$

(note that the integral converges because $|\zeta_1 - \tau_1| \le |z - t|$); hence

$$\begin{aligned} -\frac{\partial^2 U^\mu}{\partial\zeta_1^2}(0) &= \lim_{\zeta_1\to 0}\int_{D_r(0)} f(t)\left(\frac{\zeta_1-\tau_1}{|\zeta_1-t|^2} - \frac{-\tau_1}{|t|^2}\right)\frac{1}{\zeta_1}\,dm(t) \\ &= \lim_{\zeta_1\to 0}\int_{D_r(0)} f(t)\left(\frac{\tau_2^2-\tau_1^2+\zeta_1\tau_1}{|\zeta_1-t|^2|t|^2}\right)dm(t). \end{aligned} \tag{1.11}$$

We shall consider here the limit for positive ζ_1's; the left-hand limit can be analogously calculated. Thus, in what follows we assume $\zeta_1 > 0$, and let

$$\varphi(\zeta_1, t) := \frac{\tau_2^2-\tau_1^2+\zeta_1\tau_1}{|\zeta_1-t|^2|t|^2} = \frac{\tau_2^2-\tau_1^2+\zeta_1\tau_1}{((\zeta_1-\tau_1)^2+\tau_2^2)(\tau_1^2+\tau_2^2)}.$$

On applying the inequality

$$|\tau_1(\tau_1-\zeta_1)| \le \frac{1}{2}(\tau_1^2 + (\zeta_1-\tau_1)^2),$$

we can see that

$$|\varphi(\zeta_1,t)| \le \frac{1}{2}\left(\frac{1}{(\zeta_1-\tau_1)^2+\tau_2^2} + \frac{1}{\tau_1^2+\tau_2^2}\right).$$

Thus, if

$$\Phi(\zeta_1, t) := \max\left(\min\left(\varphi(\zeta_1,t), \frac{9}{\tau_1^2+\tau_2^2}\right), \frac{-9}{\tau_1^2+\tau_2^2}\right),$$

then it follows that Φ agrees with φ for each of the following parameter values:

$\zeta_1 \geq 2\tau_1$ (because then $(\zeta_1 - \tau_1)^2 + \tau_2^2 \geq \tau_1^2 + \tau_2^2$),
$\tau_1 \geq \frac{3}{2}\zeta_1$ (because then $(\zeta_1 - \tau_1)^2 + \tau_2^2 \geq \frac{1}{9}(\tau_1^2 + \tau_2^2)$),
$\frac{1}{2}\zeta_1 < \tau_1 < \frac{3}{2}\zeta_1$ and $|\tau_2| \geq \zeta_1$ (because $(\zeta_1 - \tau_1)^2 + \tau_2^2 \geq \frac{1}{4}(\tau_1^2 + \tau_2^2)$).

Hence, Φ and φ can only differ in the region $R(\zeta_1)$ determined by

$$\tfrac{1}{2}\zeta_1 < \tau_1 < \tfrac{3}{2}\zeta_1, \quad |\tau_2| < \zeta_1,$$

in which case $|\varphi - \Phi| \leq |\varphi|$, and for such values we have with $\xi = \zeta_1 - \tau_1$

$$|\varphi(\zeta_1, t)| \leq \frac{1}{\tau_1^2 + \tau_2^2} + \frac{(3/2)\zeta_1}{|\zeta_1/2|^2}\frac{|\xi|}{\xi^2 + \tau_2^2} \leq \frac{4}{\zeta_1^2} + 6|\zeta_1|^{-1}\frac{|\xi|}{\xi^2 + \tau_2^2},$$

from which we deduce

$$\int_{R(\zeta_1)} |\varphi|\, dm \leq \left| \int_{-\zeta_1}^{\zeta_1} \int_{-\zeta_1/2}^{\zeta_1/2} \left(\frac{4}{\zeta_1^2} + 6|\zeta_1|^{-1} \frac{|\xi|}{\xi^2 + \tau_2^2} \right) d\xi\, d\tau_2 \right| \leq 8 + 12\pi.$$

In view of (1.10), this implies

$$\int_{R(\zeta_1)} |f\varphi|\, dm = o(1)$$

as $\zeta_1 \to 0$.

What we have verified so far is that

$$\lim_{\zeta_1 \to 0} \int_{D_r(0)} f(t)\varphi(\zeta_1, t)\, dm(t) = \lim_{\zeta_1 \to 0} \int_{D_r(0)} f(t)\Phi(\zeta_1, t)\, dm(t). \tag{1.12}$$

But by (1.10), we have independently of ζ_1

$$|f(t)\Phi(\zeta_1, t)| \leq 9L|t|^{\alpha - 2},$$

and here the right-hand side is an integrable function (with respect to m). Hence we can deduce from (1.11) and (1.12) via the dominated convergence theorem that

$$\begin{aligned} -\frac{\partial^2 U^\mu}{\partial \zeta_1^2}(0) &= \lim_{\zeta_1 \to 0} \int_{D_r(0)} f(t)\Phi(\zeta_1, t)\, dm(t) \\ &= \int_{D_r(0)} \left(\lim_{\zeta_1 \to 0} f(t)\Phi(\zeta_1, t) \right) dm(t) \\ &= \int_{D_r(0)} f(t) \frac{\tau_2^2 - \tau_1^2}{(\tau_1^2 + \tau_2^2)^2}\, dm(t). \end{aligned}$$

In a similar fashion one can prove that

$$-\frac{\partial^2 U^\mu}{\partial \zeta_2^2}(0) = \int_{D_r(0)} f(t) \frac{\tau_1^2 - \tau_2^2}{(\tau_1^2 + \tau_2^2)^2}\, dm(t),$$

and so

$$\Delta U^\mu(0) = \frac{\partial^2 U^\mu}{\partial \zeta_1^2}(0) + \frac{\partial^2 U^\mu}{\partial \zeta_2^2}(0) = 0 = -2\pi f(0)$$

as we have claimed. □

Next we turn to the problem of determining the measure from its potential on arcs of its support. In other words, we shall suppose that the portion of the support of μ in a region is a $C^{1+\delta}$-curve γ for some $\delta > 0$ (which means that $\gamma = \gamma(t)$, $0 < t < 1$ is continuously differentiable and its derivative satisfies a Lip δ condition:

$$|\gamma'(t_1) - \gamma'(t_1)| \le L|t_1 - t_2|^\delta),$$

and then we want to determine μ on γ from U^μ. Note that in this case Theorem 1.3 cannot be applied.

To avoid pathological cases we shall always assume when we are speaking of a $C^{1+\delta}$-curve that its derivative does not vanish at any point. We shall think of γ as an oriented curve and in this respect we can talk about the left and right-hand sides of γ, left and right neighborhoods of points on γ and of the left and right normals to γ. Also, if $z_0, z_1 \in \gamma$, then we can speak of z_0 preceding z_1 in the orientation of γ, and then by $\gamma_{[z_0,z_1]}$ we denote the portion of γ lying between z_0 and z_1. We say that D is a domain attached to γ from the left if D contains a left neighborhood of every point of (an arc of) γ (see Figure 1.1). In a similar manner one can speak of domains attached from the right to γ and, of course, a domain may be simultaneously attached to γ from both sides.

We shall also need the concept of nontangential limit from one side to a point of γ. Let us suppose that H is a function defined in a left neighborhood D of a point $z_0 \in \gamma$. If θ is a fixed positive number, then the set of points z lying in D satisfying the condition $\operatorname{dist}(z, z_0) \le \theta \operatorname{dist}(z, \gamma)$ is a "sector" (or cone) $S^{(\theta)}$ with vertex at z_0 (see Figure 1.1). We say that H has nontangential limit $H(z_0 + 0)$ from the left at z_0 if for each such sector, i.e. for each θ,

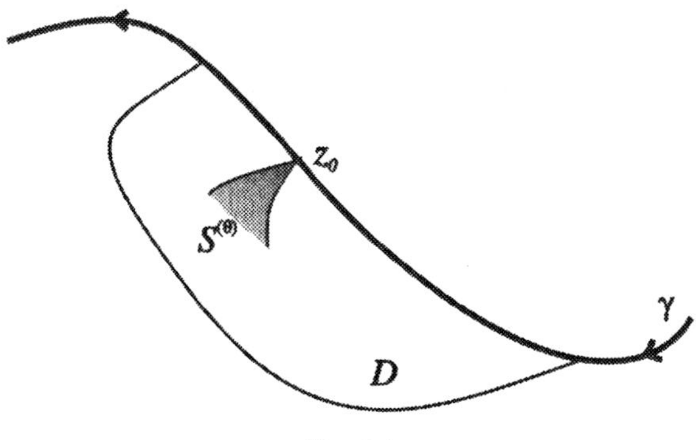

Fig. 1.1

$$\lim_{z\to z_0,\, z\in S^{(\theta)}} H(z) = H(z_0+0).$$

The nontangential limit $H(z_0-0)$ is similarly defined provided H is given in a right neighborhood of z_0.

With these notations, we are ready to prove

Theorem 1.4. *Suppose that the intersection of* supp(μ) *with a domain $\mathcal{D}$ is a simple $C^{1+\delta}$-curve γ for some $\delta > 0$, and assume that the potential U^μ of μ is bounded on γ. Let H_+ (resp. H_-) be analytic functions in a domain $\mathcal{D}_+$ (resp. $\mathcal{D}_-$) attached to γ from the left (resp. right) with real part equal to U^μ. Then if $z_0, z_1 \in \gamma$ and z_0 precedes z_1 on γ, the μ-measure of the arc $\gamma_{[z_0,z_1]}$ of γ lying between z_0 and z_1 is given by*

$$\mu(\gamma_{[z_0,z_1]}) = \frac{1}{2\pi i}\{H_+(z_1+0) - H_+(z_0+0) - H_-(z_1-0) + H_-(z_0-0)\}.$$

Naturally, the existence of the nontangential limits on the right is contained in the statement of the theorem.

Proof of Theorem 1.4. The part of μ lying outside $\mathcal{D}$ generates a potential that is harmonic in a neighborhood of γ. This fact easily implies that we can ignore this part of the measure in the proof. Hence, without loss of generality, we assume that μ is supported on γ (more precisely on its closure).

The assumptions imply that the analytic functions

$$H_\pm(z) - \int \log\frac{1}{z-t}\,d\mu(t)$$

have zero real parts, and hence are constant in $\mathcal{D}_\pm$, respectively. Thus we assume without loss of generality that

$$H_\pm(z) = \int \log\frac{1}{z-t}\,d\mu(t).$$

Let z_0 and z_1 be two points on γ, and for $\theta > 0$, let $S_{0,\pm}^{(\theta)}$ denote the "sectors" lying in $\mathcal{D}_\pm$ determined by the condition dist$(z, z_0) \le \theta$dist(z,γ), with $S_{1,\pm}^{(\theta)}$ similarly defined for z_1. For an $h > 0$ choose points $A_h \in S_{0,+}^{(\theta)}$, $B_h \in S_{0,-}^{(\theta)}$ the distance of which to z_0 is smaller than h, and let the points D_h, C_h be similarly chosen with respect to z_1 (see Figure 1.2). Let us also connect D_h and A_h in $\mathcal{D}_+$ by a smooth Jordan curve γ_+ which lies close to γ, and similarly let γ_- connect B_h and C_h in $\mathcal{D}_-$. Finally let γ^* be the positively oriented closed curve consisting of these two curves $\gamma_\pm$ and the four segments A_hz_0, z_0B_h, C_hz_1, z_1D_h.

Clearly, we can write

$$H_+(D_h) - H_+(A_h) - H_-(C_h) + H_-(B_h) = \int_\gamma\int_{\gamma^*}\frac{1}{u-t}\,du\,d\mu(t)$$

$$-\left(\int_\gamma\int_{A_hz_0} + \int_\gamma\int_{z_0B_h} + \int_\gamma\int_{C_hz_1} + \int_\gamma\int_{z_1D_h}\right)\frac{1}{u-t}\,du\,d\mu(t). \tag{1.13}$$

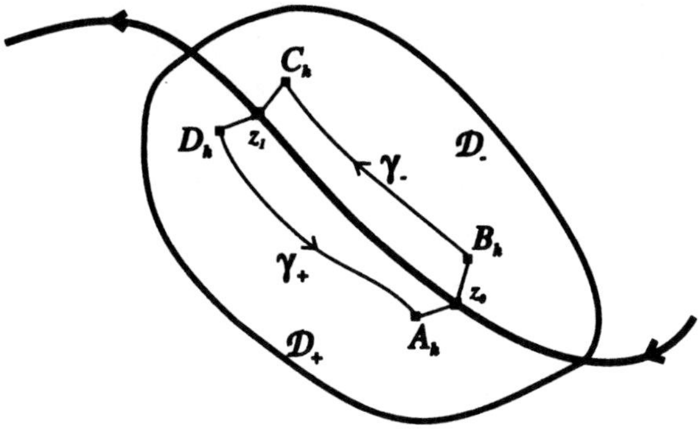

Fig. 1.2

Since

$$\int_{A_h z_0} \frac{1}{|u-t|} d|u| \leq \begin{cases} Ch/|t-z_0| & \text{if} \quad |t-z_0| \geq h \\ C + C \log h/|t-z_0| & \text{if} \quad |t-z_0| < h, \end{cases}$$

with a constant C independent of h and $t \in \gamma$ that may however depend on the fixed θ (here is where we need that the point A_h lies in the sector $S_{0,+}^{(\theta)}$), and since we have assumed that $U^{\mu}(z_0)$ is finite, i.e. $\log 1/|t-z_0|$ is integrable with respect to μ, we can deduce from this and analogous reasoning concerning $z_0 B_h$, $C_h z_1$, and $z_1 D_h$ that the value of the inner integral

$$\int_{\gamma^*} \frac{1}{u-t} du \tag{1.14}$$

in the above formula is integrable with respect to $d\mu(t)$. Furthermore,

$$\lim_{h \to 0} \left(\int_\gamma \int_{A_h z_0} + \int_\gamma \int_{z_0 B_h} + \int_\gamma \int_{C_h z_1} + \int_\gamma \int_{z_1 D_h} \right) \frac{1}{u-t} du\, d\mu(t) = 0 \tag{1.15}$$

independently of how the points A_h, B_h and C_h, D_h tend to z_0 and z_1, respectively, provided they stay in the corresponding sectors defined above, i.e. assuming that the convergence is nontangential.

Making use of the fact that the value of (1.14) is $2\pi i$ if t is inside γ^* and it is 0 otherwise, we finally get from (1.13) and (1.15) that

$$\lim_{h \to 0} (H_+(D_h) - H_+(A_h) - H_-(C_h) + H_-(B_h)) = 2\pi i \mu(\gamma_{[z_0, z_1]}).$$

The existence of the limit on the left shows that the nontangential limits $H_\pm(z \pm 0)$ exist at every point of γ, and obviously we get the required formula

$$H_+(z_1 + 0) - H_+(z_0 + 0) - H_-(z_1 - 0) + H_-(z_0 - 0) = 2\pi i \mu(\gamma_{[z_0, z_1]}).$$

□

Finally, we prove a theorem which is a kind of combination of the preceding two ones, again for determination of μ on arcs.

Theorem 1.5. *Suppose that the intersection of* supp(μ) *with a domain* $\mathcal{D}$ *is a simple* $C^{1+\delta}$*-curve* γ *for some* $\delta > 0$, *let* ds *denote the arc length on* γ *and* $\mathbf{n}_+(z)$ *resp.* $\mathbf{n}_-(z)$ *the left (resp. right) normal to* γ *at a point* $z \in \gamma$.

(i) *Suppose that the potential* U^μ *is in* Lip 1 *in a neighborhood of* γ. *Then on* γ *the measure* μ *is locally absolutely continuous with respect to* ds, *and we have the representation*

$$d\mu = -\frac{1}{2\pi}\left(\frac{\partial U^\mu}{\partial \mathbf{n}_-}(s) + \frac{\partial U^\mu}{\partial \mathbf{n}_+}(s)\right) ds.$$

(ii) *Conversely, if on* γ *the measure* μ *is of the form* $d\mu = f(s)\,ds$, *where* f *is continuous at* $z_0 \in \gamma$ *or even if* z_0 *is merely a Lebesgue point of* f *with respect to* ds, *then the normal derivatives* $\partial U^\mu/\partial \mathbf{n}_-$ *and* $\partial U^\mu/\partial \mathbf{n}_+$ *exist at* z_0 *and*

$$f(z_0) = -\frac{1}{2\pi}\left(\frac{\partial U^\mu}{\partial \mathbf{n}_-}(z_0) + \frac{\partial U^\mu}{\partial \mathbf{n}_+}(z_0)\right).$$

The Lebesgue point property with respect to ds at z_0 mentioned in the second half of the theorem means that

$$\lim_{z \to z_0,\, z \in \gamma} \frac{1}{s(\gamma_{[z_0,z]})} \int_{\gamma_{[z_0,z]}} |f(s) - f(z_0)|\, ds = 0,$$

where $s(\gamma_{[z_0,z]})$ denotes the arc measure of the arc of γ lying between z_0 and z.

Proof of Theorem 1.5. We begin the proof of (i) with the following observation. If U^μ is in Lip 1 in a neighborhood on γ, then the directional derivatives of U^μ in the directions $\mathbf{n}_\pm(z)$ are continuous at z for ds-almost every $z \in \gamma$. In fact, let e.g. $D \subseteq \mathcal{D}$ be a simply connected domain (with smooth boundary) attached to some arc γ' of γ from the left with the property that U^μ is in Lip 1 on D. Let h be an analytic function on D with real part equal to U^μ. From the Lip 1 property of U^μ it follows that U^μ has uniformly bounded directional derivatives in D; hence the Cauchy-Riemann formulae imply that h' is bounded in D. Thus, we can apply Fatou's theorem according to which h' has nontangential limits at z from the left for almost every $z \in \gamma'$ (for this classical result when D is on a disk see e.g. [82, p. 38], the case of general D then can be obtained by a conformal mapping since such maps are conformal up to γ' [151, Theorem II.2.24]). It is easy to see that at every such z the normal derivative $\partial U^\mu/\partial \mathbf{n}_+$ exists and is equal to the limit of the directional derivative of U^μ at z' in the direction $\mathbf{n}_+(z)$ as $z' \to z$ nontangentially. Since D is any subdomain of $\mathcal{D}$ with the above properties, the claim concerning the continuity of directional derivatives follows at ds-almost every point of γ.

To prove the first part of the theorem under the assumptions stated there, let A and B be two points on γ, and connect B and A by two other smooth Jordan curves γ_+ and γ_- going on opposite sides of and lying sufficiently close to γ (see Figure 1.3). Let z_0 be any inner point of the arc $\gamma_{[A,B]}$, and construct a small circle $C_r(z_0)$ around z_0 that intersects γ in the points C and D. Consider the following simple curves Γ_+ and Γ_- : Γ_+ consists of γ_+, the arcs $\gamma_{[A,C]}$ and $\gamma_{[D,B]}$ of γ and that arc $C_r(z_0)_{[C,D]}$ of the circle $C_r(z_0)$ that lies on the same side of γ as γ_+, and

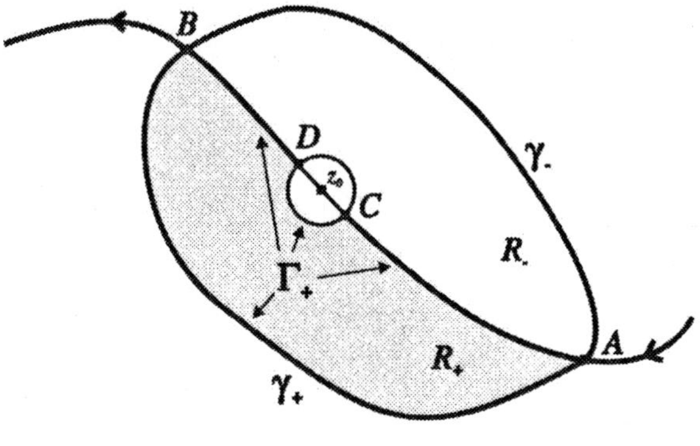

Fig. 1.3

the definition of Γ_- is symmetric. Furthermore, let $R_\pm$ be the domains enclosed by $\Gamma_\pm$.

With $u = U^\mu$ and $v(z) = \log 1/|z_0 - z|$ we can easily get from Green's identity (1.1) the formula

$$\int_{R_+} (v\Delta u - u\Delta v)dm = -\int_{\Gamma_+} \left(v\frac{\partial u}{\partial \mathbf{n}} - u\frac{\partial v}{\partial \mathbf{n}} \right) ds \tag{1.16}$$

where $\partial/\partial\mathbf{n}$ denotes differentiation in the direction of the inner normal of R_+, and ds indicates integration with respect to the arc length on Γ_+. In fact, if u was differentiable on $\gamma_{[A,B]}$, then (1.16) would be just Green's formula (1.1). Thus, (1.16) can be obtained by exhausting R_+ from the inside by similar domains and making use of the continuity and boundedness (implied by the Lip 1 property of U^μ) of the derivatives involved (see the beginning of the proof).

Both u and v are harmonic in R_+, so the left-hand side in (1.16) is zero. Thus,

$$\int_{\gamma_{[A,B]}} v\frac{\partial u}{\partial \mathbf{n}} ds - \int_{\gamma_{[C,D]}} v\frac{\partial u}{\partial \mathbf{n}} ds$$
$$= \int_{\Gamma_+} u\frac{\partial v}{\partial \mathbf{n}} ds - \int_{\gamma_+} v\frac{\partial u}{\partial \mathbf{n}} ds - \int_{C_r(z_0)_{[C,D]}} v\frac{\partial u}{\partial \mathbf{n}} ds.$$

The corresponding formula for the region R_- is

$$\int_{\gamma_{[A,B]}} v\frac{\partial u}{\partial \mathbf{n}} ds - \int_{\gamma_{[C,D]}} v\frac{\partial u}{\partial \mathbf{n}} ds$$
$$= \int_{\Gamma_-} u\frac{\partial v}{\partial \mathbf{n}} ds - \int_{\gamma_-} v\frac{\partial u}{\partial \mathbf{n}} ds - \int_{C_r(z_0)_{[C,D]}} v\frac{\partial u}{\partial \mathbf{n}} ds,$$

where, of course, now $\mathbf{n}$ means inner normal with respect to R_-, and the arcs $C_r(z_0)_{[C,D]}$ in these two formulae are opposite arcs on $C_r(z_0)$. Adding these two formulae together and making use of the fact that on γ the two derivatives in the first terms on the right-hand sides are opposite values ($\partial v/\partial \mathbf{n}$ means differentiation in the direction of the inner normals of R_+ and R_-, respectively, and on γ these are opposite directions), we obtain

$$\begin{aligned}\int_{\gamma_{[A,B]}} \left(v\frac{\partial u}{\partial \mathbf{n}_-} + v\frac{\partial u}{\partial \mathbf{n}_+} \right) ds &- \int_{\gamma_{[C,D]}} \left(v\frac{\partial u}{\partial \mathbf{n}_-} + v\frac{\partial u}{\partial \mathbf{n}_+} \right) ds \\ = \int_{\gamma_+ \cup \gamma_-} \left(u\frac{\partial v}{\partial \mathbf{n}} - v\frac{\partial u}{\partial \mathbf{n}} \right) ds &+ \int_{C_r(z_0)} \left(u\frac{\partial v}{\partial \mathbf{n}} - v\frac{\partial u}{\partial \mathbf{n}} \right) ds,\end{aligned} \tag{1.17}$$

where on the left $\partial u/\partial \mathbf{n}_-$ and $\partial u/\partial \mathbf{n}_+$ are the derivatives of u in the direction of the right and left normal to γ, while the differentiations on the right-hand side are with respect to the inner normal of the region $R_+ \cup R_-$.

From Green's formula it is obvious that (1.17) remains true if z_0 is not on γ provided $C_r(z_0)$ is in one of the regions determined by γ and γ_+ or γ and γ_- (in this case the second term on the left is missing).

The first term on the right of (1.17) is harmonic as a function of z_0 in the region determined by γ_+ and γ_-, while, in view of

$$\frac{\partial v}{\partial \mathbf{n}}(t) = -\frac{1}{r}, \qquad t \in C_r(z_0),$$

the second integral can be written as

$$-2\pi u(z_0) + o(1) + O(r \log 1/r)$$

because the Lip 1 property of $u = U^\mu$ implies that it has bounded directional derivatives. For the same reason the second term on the left is of order $O(r \log 1/r)$. Hence, as $r \to 0$ we arrive at the formula (replace z_0 by z)

$$U^\mu(z) = -\frac{1}{2\pi} \int_{\gamma_{[A,B]}} \left(\frac{\partial U^\mu}{\partial \mathbf{n}_-}(s) + \frac{\partial U^\mu}{\partial \mathbf{n}_+}(s) \right) \log \frac{1}{|z-s|}\, ds + g(z),$$

where g is a function harmonic in a neighborhood of the open arc $\gamma_{(A,B)}$. Thus, by Theorem 2.1 below in the next section, on $\gamma_{(A,B)}$ the measure μ is given by the formula

$$d\mu = -\frac{1}{2\pi} \left(\frac{\partial U^\mu}{\partial \mathbf{n}_-}(s) + \frac{\partial U^\mu}{\partial \mathbf{n}_+}(s) \right) ds$$

as we have claimed.

This proves the first part of the theorem.

Now we turn to the second half, and suppose first that f is continuous at z_0. Without loss of generality we assume that $z_0 = 0$, the x-axis is tangent to γ at 0, and that the direction of the positive y-axis is the direction of $\mathbf{n}_+$ at this point. Then the fact that γ is of class $C^{1+\delta}$ around 0 implies that it has a parametric representation $(t, \gamma(t))$ with parameter values in an interval around 0 for which $\gamma(t) = O(t^{1+\delta})$ as $t \to 0$, and we may assume $|\gamma(t)| \le Lt^{1+\delta}$ for $t \in [-a, a]$ with an $a > 0$. Since the contribution to U^μ of the part of μ that lies outside the arc $\gamma^* := \{\gamma(t) \mid t \in [-a, a]\}$ is harmonic around 0, and the normal derivatives of this contribution at 0 in the directions $\mathbf{n}_+$ and $\mathbf{n}_-$ are each other's negatives, we can assume that μ actually lives on γ^*, and then we can write γ instead of γ^*. Then for $z = x + iy$ the potential of μ is given by

$$U^\mu(z) = \int_{-a}^{a} \log\left(((x-t)^2 + (y-\gamma(t))^2\right)^{-1/2} \sqrt{1+(\gamma'(t))^2} f((t,\gamma(t)))\,dt.$$

For the sake of simpler notation, let

$$F(t) := \sqrt{1+(\gamma'(t))^2} f((t,\gamma(t))).$$

Now consider the partial derivative of U^μ with respect to y on the positive y-axis. It is equal to

$$\begin{aligned}
\frac{\partial U^\mu}{\partial y}(iy) &= -\int_{-a}^{a} (y-\gamma(t))\left(t^2+(y-\gamma(t))^2\right)^{-1} F(t)\,dt \\
&= -\int_{-a}^{a} y(t^2+y^2)^{-1} F(t)\,dt \\
&\quad + \int_{-a}^{a} y\left((t^2+y^2)^{-1} - \left(t^2+(y-\gamma(t))^2\right)^{-1}\right) F(t)\,dt \\
&\quad + \int_{-a}^{a} \gamma(t)\left(t^2+(y-\gamma(t))^2\right)^{-1} F(t)\,dt \\
&=: I_1 + I_2 + I_3.
\end{aligned}$$

Since $\gamma(t)/t^2$ is integrable on $[-a, a]$, we get from the dominated convergence theorem that

$$\lim_{y\to 0+} I_3 = \int_{-a}^{a} \gamma(t)\left(t^2+\gamma(t)^2\right)^{-1} F(t)\,dt.$$

The integral I_1 can be written as

$$\begin{aligned}
-\int_{-\infty}^{\infty} y(t^2+y^2)^{-1} F(0)\,dt \quad &+ \quad \left(\int_{-\infty}^{-a} + \int_{a}^{\infty}\right) y(t^2+y^2)^{-1} F(0)\,dt \\
&- \quad \int_{-a}^{a} y(t^2+y^2)^{-1}(F(t)-F(0))\,dt.
\end{aligned}$$

Here the first term is $-\pi F(0) = -\pi f(0)$, the second term tends to zero with y, and if for a prescribed $\varepsilon > 0$ we split the third integral for integration over $[-\eta, \eta]$ and the rest, where η is so small that on $[-\eta, \eta]$ we have $|F(t) - F(0)| \le \varepsilon$, then the same can be seen for the third term. Hence,

$$\lim_{y\to 0+} I_1 = -\pi f(0).$$

Finally, in I_2 the function F is multiplied by

$$y\left((t^2+y^2)^{-1}-\left(t^2+(y-\gamma(t))^2\right)^{-1}\right)=\frac{y(\gamma(t)^2-2\gamma(t)y)}{(t^2+y^2)(t^2+(y-\gamma(t))^2)}$$

which is easily seen to be at most as large in absolute value as

$$Cy\frac{\gamma(t)^2/t^2+1}{t^2+y^2}$$

with an absolute constant C. Since here $|\gamma(t)/t|\le Lt^\delta$, $\delta>0$, the same analysis as applied for I_1 shows that

$$\lim_{y\to 0+} I_2=0.$$

What we have shown up to now is that

$$\lim_{y\to 0+}\frac{\partial U^\mu}{\partial y}(iy)=\int_{-a}^{a}\gamma(t)\left(t^2+\gamma(t)^2\right)^{-1}F(t)\,dt-\pi f(0).$$

This easily implies via the mean value theorem that the partial derivative of U^μ with respect to $y>0$ exists at 0 and equals the right-hand side of the preceding formula, in other words

$$\frac{\partial U^\mu}{\partial \mathbf{n}_+}(0)=\int_{-a}^{a}\gamma(t)\left(t^2+\gamma(t)^2\right)^{-1}F(t)\,dt-\pi f(0). \tag{1.18}$$

When we form the derivative in the direction $\mathbf{n}_-$, the above argument gives

$$\begin{aligned}\frac{\partial U^\mu}{\partial \mathbf{n}_+}(0) &= -\lim_{y\to 0-}\frac{\partial U^\mu}{\partial y}(iy)\\ &= -\int_{-a}^{a}\gamma(t)\left(t^2+\gamma(t)^2\right)^{-1}F(t)\,dt-\pi f(0).\end{aligned} \tag{1.19}$$

The only point that must be made clear is that the sign of $\pi f(0)$ is again negative because for negative y the integral

$$-\int_{-\infty}^{\infty}y(t^2+y^2)^{-1}dt$$

equals π (cf. the derivation of (1.18)).

The second half of the theorem in the case when f is continuous at z_0 follows now by adding together (1.18) and (1.19).

If we merely assume that $z_0=0$ is a Lebesgue point of f with respect to ds, then 0 is a Lebesgue point of F with respect to dt. Then we follow the preceding proof and the only change we have to make is in the proof of

$$\lim_{y\to 0+}\int_{-a}^{a}y(t^2+y^2)^{-1}(F(t)-F(0))\,dt=0 \tag{1.20}$$

and in the proof of the analogous relation in the estimate of I_2. In fact, the rest holds word for word because (see the consideration below) $\gamma(t)t^{-2}F(t)$ is in L^1 if $\gamma(t) = O(t^{1+\delta})$, $\delta > 0$, and 0 is a Lebesgue point for F.

As for (1.20), the Lebesgue point property of F implies that if $\varepsilon > 0$ is given, then there is an $\eta > 0$ such that for $0 < t < \eta$ we have

$$\int_0^t |F(u) - F(0)|\,du \le \varepsilon t.$$

Integration by parts yields for $y > 0$, $y \to 0$,

$$\int_0^a y(t^2 + y^2)^{-1}|F(t) - F(0)|\,dt = \int_0^y + \int_y^\eta + \int_\eta^a$$

$$\le \frac{1}{y}\int_0^y |F(t) - F(0)|\,dt$$

$$+ \left(y(\eta^2 + y^2)^{-1} \int_0^\eta |F(t) - F(0)|\,dt - \frac{1}{2y}\int_0^y |F(t) - F(0)|\,dt \right.$$

$$\left. + \int_y^\eta 2ty(t^2 + y^2)^{-2} \int_0^t |F(u) - F(0)|\,du\,dt \right) + O(y)$$

$$= o(1) + \int_y^\eta 2ty(t^2 + y^2)^{-2}\varepsilon t\,dt \le o(1) + 2\varepsilon\pi$$

which, together with an analogous estimate for the integral over the interval $[-a, 0]$, proves (1.20). □

The above proof for the second part (ii) of the theorem shows that actually more is true, namely the limits of the partial derivatives of U^μ in the directions $\mathbf{n}_\pm$ exist at z_0 if we approach z_0 along the corresponding normals.

II.2 The Unicity Theorem

In this section we prove the following unicity result that has already been used in the preceding section.

Theorem 2.1 (Unicity Theorem). *Suppose that the positive measures μ and ν have compact support and in a region $D \subseteq \mathbf{C}$ the potentials U^μ and U^ν satisfy*

$$U^\mu(z) = U^\nu(z) + u(z)$$

almost everywhere with respect to two-dimensional Lebesgue measure, where the function u is harmonic in D. Then in D the measures μ and ν coincide, i.e.

$$\mu_{|D} = \nu_{|D}.$$

The requirement that the measures have compact support is not needed, but this is a convenient way to ensure the existence of the logarithmic potentials.

As an immediate consequence of the theorem we get the following unicity result.

Corollary 2.2. *If* μ *and* ν *are compactly supported measures and the potentials* U^μ *and* U^ν *coincide almost everywhere with respect to two-dimensional Lebesgue measure, then* $\mu = \nu$.

It will be convenient to use in the proof the the *convolution* notation

$$h * g(z) := \int h(z-t)g(t)\,dm(t), \tag{2.1}$$

where m denotes two dimensional Lebesgue measure, and similarly if h is a function bounded from one side on compact subsets of $\mathbf{C}$ and μ is a compactly supported measure, then we define the convolution with μ *from the right* by

$$h * \mu(z) := \int h(z-t)\,d\mu(t). \tag{2.2}$$

With this notation the potential U^μ coincides with $-\log|\cdot| * \mu$. In what follows, the functions used in convolution will always be bounded from one side on compact subsets of $\mathbf{C}$ and at most one of them will have unbounded support, and these properties easily imply via Fubini's theorem that the convolution is commutative and associative, and this is so even if one of the factors in a multiple convolution is a measure of compact support. We will use these facts in the proof without further mention.

Proof of Theorem 2.1. For $\varepsilon > 0$ let

$$D_\varepsilon := \{z \in D \mid \operatorname{dist}(z, \partial D) > \varepsilon\},$$

$$\chi_\varepsilon(z) := \begin{cases} 1 & \text{if } |z| \le \varepsilon \\ 0 & \text{if } |z| > \varepsilon \end{cases} \tag{2.3}$$

and

$$\log_\varepsilon(z) := \min\left(\log\frac{1}{|z|}, \log\frac{1}{\varepsilon}\right) = \begin{cases} \log 1/|z| & \text{if } |z| \ge \varepsilon \\ \log 1/\varepsilon & \text{if } 0 \le |z| \le \varepsilon. \end{cases}$$

If we also set

$$\mu_\varepsilon(B) := \int \mu(B-t)\chi_\varepsilon(t)\,dm(t),$$

for every Borel set B, then for this measure it can be verified that

$$f * \mu_\varepsilon = (f * \mu) * \chi_\varepsilon = (f * \chi_\varepsilon) * \mu.$$

In a similar fashion we define the measure ν_ε.

Now the assumptions in the theorem imply

$$U^{\mu_\varepsilon}(z) = U^\mu * \chi_\varepsilon(z) = U^\nu * \chi_\varepsilon(z) + u * \chi_\varepsilon(z) = U^{\nu_\varepsilon}(z) + u * \chi_\varepsilon(z) \quad (2.4)$$

for *every* $z \in D_\varepsilon$, and of course $u * \chi_\varepsilon$ is again harmonic in D_ε. Since $\log|\cdot| * \chi_\varepsilon$ is continuous on $\mathbf{C}$, the same is true of

$$U^{\mu_\varepsilon} = -(\log|\cdot| * \chi_\varepsilon) * \mu \quad \text{and} \quad U^{\nu_\varepsilon} = -(\log|\cdot| * \chi_\varepsilon) * \mu.$$

Let us now integrate the equality (2.4) on a circle $\{\xi \mid |z-\xi| = \eta\}$, $\eta \le \varepsilon$. Since (see (0.5.5))

$$\frac{1}{2\pi}\int_{-\pi}^{\pi} \log\frac{1}{|z+\eta e^{i\varphi}-t|}\,d\varphi = \log_\eta(z-t),$$

we get again from Fubini's theorem that on the one hand

$$\frac{1}{2\pi}\int_{-\pi}^{\pi} U^{\mu_\varepsilon-\nu_\varepsilon}(z+\eta e^{i\varphi})\,d\varphi = \log_\eta * (\mu_\varepsilon - \nu_\varepsilon)(z),$$

and on the other hand for $z \in D_{2\varepsilon}$ (recall $\eta \le \varepsilon$)

$$\frac{1}{2\pi}\int_{-\pi}^{\pi} U^{\mu_\varepsilon-\nu_\varepsilon}(z+\eta e^{i\varphi})\,d\varphi = \frac{1}{2\pi}\int_{-\pi}^{\pi} u * \chi_\varepsilon(z+\eta e^{i\varphi})\,d\varphi$$

$$= u * \chi_\varepsilon(z) \quad = \quad U^{\mu_\varepsilon-\nu_\varepsilon}(z) = \log_0 * (\mu_\varepsilon - \nu_\varepsilon)(z),$$

where in the second step we used the mean value property for the harmonic function $u * \chi_\varepsilon$ (note that $\log_0 * (\mu_\varepsilon - \nu_\varepsilon)$ is well defined because of the continuity of U^{μ_ε} and U^{ν_ε} mentioned above).

By subtracting the last two formulae we arrive at

$$(\log_0 - \log_\eta) * (\mu_\varepsilon - \nu_\varepsilon)(z) = 0$$

for all $z \in D_{2\varepsilon}$ and $0 < \eta \le \varepsilon$. On applying this with η and an $\eta_1 \le \varepsilon$ and making another subtraction we can conclude

$$(\log_{\eta_1} - \log_\eta) * (\mu_\varepsilon - \nu_\varepsilon) \equiv 0$$

in $D_{2\varepsilon}$. Finally, if we divide this equation by $\log(\eta/\eta_1)$ and take the limit as $\eta_1 \to \eta - 0$, we get from the bounded convergence theorem that

$$\chi_\eta * (\mu_\varepsilon - \nu_\varepsilon)(z) = 0, \quad z \in D_{2\varepsilon}, \quad (2.5)$$

because

$$\lim_{\eta_1\to\eta-0} \frac{1}{\log(\eta/\eta_1)}\left(\log_{\eta_1}(z) - \log_\eta(z)\right) = \chi_\eta(z),$$

and the expression on the left-hand side is between zero and one for all $0 < \eta_1 < \eta$.

Now let h be any continuous function that vanishes outside a compact subset of D. Then there is an $\varepsilon_0 > 0$ such that h vanishes outside a compact subset of $D_{3\varepsilon_0}$. Choose $0 < \eta \leq \varepsilon \leq \varepsilon_0$ arbitrarily, multiply (2.5) by h and integrate the result with respect to two dimensional Lebesgue measure. The result can be written as

$$\int h * \chi_\eta \, d(\mu_\varepsilon - \nu_\varepsilon) = 0,$$

and since here $(h * \chi_\eta)/\pi\eta^2$ uniformly tends to h as $\eta \to 0$, we can conclude

$$\int h \, d(\mu_\varepsilon - \nu_\varepsilon) = 0,$$

that is

$$\int h * \chi_\varepsilon \, d\mu = \int h * \chi_\varepsilon \, d\nu.$$

By dividing here by $\pi\varepsilon^2$ and applying the preceding reasoning we finally arrive at

$$\int h \, d\mu = \int h \, d\nu.$$

Since this is true for all h with compact support in D, we get that $\mu|_D = \nu|_D$, and the proof is complete. □

II.3 Riesz Decomposition Theorem and Principle of Domination

In this section we prove two important theorems concerning logarithmic potentials. We start with the so-called Riesz decomposition theorem, which asserts that superharmonic functions are essentially potentials. Let us recall from Section 0.5 that an extended real-valued function f defined on a domain $\mathcal{D}$ is called superharmonic in $\mathcal{D}$ if f is not identically $+\infty$ and satisfies

(a) $f(z) > -\infty$ for all $z \in \mathcal{D}$,
(b) f is lower semi-continuous in $\mathcal{D}$,
(c) $f(z) \geq \frac{1}{2\pi}\int_{-\pi}^{\pi} f(z + re^{i\varphi})\, d\varphi$ for all $z \in \mathcal{D}$ and $r > 0$ for which the disk $\{z' \mid |z' - z| \leq r\}$ is contained in $\mathcal{D}$.

We have seen in Theorem 0.5.6 that logarithmic potentials are superharmonic. Now for superharmonic functions the following converse result holds.

Theorem 3.1 (Riesz Decomposition Theorem). *If f is a superharmonic function on a domain $\mathcal{D}$, then there exists a positive measure λ on $\mathcal{D}$ such that on every subdomain $D \subseteq \mathcal{D}$ for which $\overline{D} \subseteq \mathcal{D}$ we have the representation*

$$f(z) = u_D(z) + \int_D \log \frac{1}{|z - t|} \, d\lambda(t), \quad z \in D, \tag{3.1}$$

where u_D is a harmonic function on D.

Notice that for smooth functions this result immediately follows from Theorem 1.3. In fact, if f is three times continuously differentiable and

$$d\lambda = -\frac{1}{2\pi}\Delta f(t)\,dm(t),$$

then the subharmonicity of f easily yields (cf. Theorem 0.5.5) that λ is a positive measure and, by Theorem 1.3,

$$\Delta(f - U^{\lambda}) = 0$$

in $\mathcal{D}$, i.e. $f - U^{\lambda}$ is harmonic (in case U^{λ} does not exist, one has to apply the same reasoning on compact subsets of $\mathcal{D}$).

Proof of Theorem 3.1. In what follows all integrals are with respect to two-dimensional Lebesgue measure m. First we remark that the lower semi-continuity of f and $f(z) > -\infty$, $z \in \mathcal{D}$, imply that f is bounded from below on compact subsets of $\mathcal{D}$. Hence, all the integrals below are well defined with possibly $+\infty$ value.

Let $z_0 \in \mathcal{D}$ be such that $f(z_0) < \infty$. If $r > 0$ is such that the disk

$$D_r(z_0) = \{z \mid |z - z_0| < r\}$$

is contained in $\mathcal{D}$, then property (c) for f implies

$$\frac{1}{2\pi r^2}\int_{D_r(z_0)} f \le f(z_0);$$

hence $f(z) < \infty$ for almost all $z \in D_r(z_0)$ (with respect to two-dimensional Lebesgue measure). We can repeat the same argument for almost every point $z \in D_r(z_0)$ instead of z_0, and by a process of continuation via chains of overlapping disks we can conclude that $f(z) < \infty$ for almost all $z \in \mathcal{D}$. Since f is integrable on any disk around every such z that is contained in $\mathcal{D}$, we have

$$f \in L^1(D) \tag{3.2}$$

for all $D \subseteq \mathcal{D}$ with $\overline{D} \subseteq \mathcal{D}$, i.e. f is integrable with finite integral on compact subsets of $\mathcal{D}$.

Let $C_0(\mathcal{D})$ denote the space of continuous functions with compact support in $\mathcal{D}$ and let $C_0^2(\mathcal{D})$ be the subspace of two times continuously differentiable functions (considered as functions of two variables) of $C_0(\mathcal{D})$. Our next aim is to show that

$$\int\left(-\frac{1}{2\pi}\Delta h\right) f \ge 0 \tag{3.3}$$

for all nonnegative $h \in C_0^2(\mathcal{D})$, where, as usual, Δ denotes the Laplacian. To do this, we first recall from Theorem 0.5.5 that if g is twice continuously differentiable and superharmonic on $\mathcal{D}$, then for all $z \in \mathcal{D}$

$$\Delta g(z) \le 0. \tag{3.4}$$

Next we have to smooth out f, and to this end consider the functions χ_ε from (2.3) and form

$$q_\varepsilon(z) = \left(\frac{1}{\pi\varepsilon^2}\right)^3 \chi_\varepsilon * \chi_\varepsilon * \chi_\varepsilon(z).$$

This q_ε is supported in the disk $D_{3\varepsilon}(0)$, is two times continuously differentiable and has integral 1 (with respect to Lebesgue measure). Now let

$$g_\varepsilon(z) := f * q_\varepsilon(z), \quad z \in \mathcal{D}_{3\varepsilon} := \{z \in \mathcal{D} \,|\, \mathrm{dist}\,(z, \partial\mathcal{D}) > 3\varepsilon\}.$$

Since f is integrable on compact subsets of $\mathcal{D}$, it follows from the differentiability of q_ε that g_ε is twice continuously differentiable in $\mathcal{D}_{3\varepsilon}$.

Now let $h \in C_0^2(\mathcal{D})$ be nonnegative, and choose $\varepsilon_0 > 0$ so that the support of h is a compact subset of $\mathcal{D}_{3\varepsilon_0}$. Then integration by parts yields, for $0 < \varepsilon < \varepsilon_0$,

$$\int \left(-\frac{1}{2\pi}\Delta h\right) g_\varepsilon = \int h \left(-\frac{1}{2\pi}\Delta g_\varepsilon\right) \geq 0, \tag{3.5}$$

where we used (3.4) for g_ε and the fact that the superharmonicity of f implies the superharmonicity of g_ε (apply Fubini's theorem). But the left-hand side in (3.5) can also be written as

$$\int \left(\left(-\frac{1}{2\pi}\Delta h\right) * q_\varepsilon\right) f,$$

and here $(\Delta h) * q_\varepsilon$ uniformly tends to Δh as $\varepsilon \to 0$. Hence (3.3) follows from (3.5) and (3.2) by letting ε tend to zero in (3.5).

Thus, the functional

$$Lh = \int \left(-\frac{1}{2\pi}\Delta h\right) f \tag{3.6}$$

is linear and nonnegative on $C_0^2(\mathcal{D})$, and so the Riesz representation theorem for positive functionals on locally compact spaces [195, Theorem 2.14] can be applied to deduce the existence of a measure λ such that

$$Lh = \int h \, d\lambda. \tag{3.7}$$

Strictly speaking, the Riesz representation theorem for positive functionals can be applied only with some precaution because L in (3.6) is defined only on $C_0^2(\mathcal{D})$ and not on $C_0(\mathcal{D})$, which would be needed in the Riesz representation theorem. But a nonnegative functional from $C_0^2(\mathcal{D})$ can always be extended via the Hahn–Banach theorem to a nonnegative functional on $C_0(\mathcal{D})$. Since L is finite for all $h \in C_0^2(\mathcal{D})$ (and hence for all $h \in C_0(\mathcal{D})$, as well) we also know that λ is finite on every compact subset of $\mathcal{D}$.

Now let $D \subseteq \mathcal{D}$ be an open set whose closure $\overline{D}$ is contained in $\mathcal{D}$. In place of h we apply (3.7) to a function of the form $h * q_\varepsilon$ where $h \in C_0^2(\mathcal{D})$ and ε is sufficiently small (eventually, ε will go to zero). It is easy to see that the Laplacian and convolution commute, so we get from the Fubini-Tonelli theorem for the measures $\lambda^* := \lambda\big|_D$, $d\lambda_\varepsilon^* := q_\varepsilon * \lambda^* \, dm$

$$\int h\left(-\frac{1}{2\pi}\Delta(q_\varepsilon * f)\right) = \int\left(-\frac{1}{2\pi}\Delta h\right) q_\varepsilon * f = \int\left(-\frac{1}{2\pi}\Delta(h * q_\varepsilon)\right) f$$

$$= \int (h * q_\varepsilon)\, d\lambda^* = \int h(q_\varepsilon * \lambda^*)$$

$$= \int h\left(-\frac{1}{2\pi}\Delta U^{\lambda_\varepsilon^*}\right) = \int h\left(-\frac{1}{2\pi}\Delta(q_\varepsilon * U^{\lambda^*})\right),$$

where at the third equality we used (3.7), at the fifth equality we applied the second part of Theorem 1.3, and where we also used that, because λ^* is a compactly supported finite measure, its logarithmic potential U^{λ^*} exists (with the possibility of taking $+\infty$ values, as well). Since both $\Delta(q_\varepsilon * U^{\lambda^*})$ and $\Delta(q_\varepsilon * f)$ are continuous functions, we deduce that

$$\Delta(q_\varepsilon * f - q_\varepsilon * U^{\lambda^*})$$

is orthogonal onto every $h \in C_0^2(D_{3\varepsilon})$, which implies

$$\Delta\left(q_\varepsilon * f - q_\varepsilon * U^{\lambda^*}\right)(z) = 0$$

for all $z \in D_{3\varepsilon}$, i.e.

$$q_\varepsilon * f(z) = q_\varepsilon * U^{\lambda^*}(z) + u_\varepsilon(z), \tag{3.8}$$

where u_ε is a harmonic function on $D_{3\varepsilon}$.

Now we let ε tend to zero in (3.8). By the lower semi-continuity of f we get

$$\liminf_{\varepsilon \to 0+} q_\varepsilon * f(z) \geq f(z).$$

On the other hand, property (c) of superharmonicity yields by the circular symmetry of q_ε that for every $\varepsilon > 0$

$$q_\varepsilon * f(z) \leq f(z),$$

and this together with the preceding lower estimate shows

$$\lim_{\varepsilon \to 0+} q_\varepsilon * f(z) = f(z)$$

for all z. The same is true for U^{λ^*} because U^{λ^*} is also superharmonic, and so we get that, in (3.8), $u_\varepsilon(z)$ has a limit $u(z)$ as $\varepsilon \to 0+$ for every $z \in D$ where both $f(z)$ and $U^{\lambda^*}(z)$ is finite, i.e. for almost every $z \in D$. Furthermore,

$$f(z) = U^{\lambda^*}(z) + u(z) \quad \text{for a.e. } z \in D. \tag{3.9}$$

Now we take the convolution of both sides of (3.8) with q_η for a small η, use that the harmonicity of u_ε implies $u_\varepsilon * q_\eta(z) = u_\varepsilon(z)$ for all $z \in D_{3\eta+3\varepsilon}$, and let $\varepsilon \to 0$. What we obtain is that for almost all $z \in D_{3\eta}$

$$q_\eta * f(z) = q_\eta * U^{\lambda^*}(z) + u(z),$$

which, when compared to (3.8) shows that on $D_{3\eta}$ the functions u_η and u coincide. In particular, u is uniformly bounded on compact subsets of $D_{3\eta}$ (wherever it is defined) and we get from (3.9) and the lower semi-continuity of the functions f and U^{λ^*} that they become $+\infty$ simultaneously on $D_{3\eta}$. By extending u by setting $u = u_\eta$ we conclude that

$$f(z) = U^{\lambda^*}(z) + u(z)$$

for every $z \in D_{3\eta}$; furthermore, here $u = u_\eta$ is harmonic on $D_{3\eta}$. Finally, for $\eta \to 0$ we get the desired formula (3.1) on all of D. □

Next we prove the so-called principle of domination which is one of the fundamental theorems concerning potentials.

Theorem 3.2 (Principle of Domination). *Let μ and ν be two positive finite Borel measures with compact support on* **C**, *and suppose that the total mass of ν does not exceed that of μ. Assume further that μ has finite logarithmic energy. If, for some constant c, the inequality*

$$U^\mu(z) \le U^\nu(z) + c$$

holds μ-almost everywhere, then it holds for all $z \in$ **C**.

As an immediate consequence we get the *maximum principle* (for logarithmic potentials).

Corollary 3.3 (Maximum Principle for Potentials). *If μ is a positive finite Borel measure with compact support and $U^\mu(z) \le M$ for all $z \in \mathrm{supp}(\mu)$, then $U^\mu(z) \le M$ for all $z \in$* **C**.

In fact, the hypothesis of the corollary implies that μ has finite logarithmic energy, and so the conclusion follows from Theorem 3.2 by setting $\nu \equiv 0$ and $c = M$.

Corollary 3.4. *If K is a compact subset of* **C** *of positive capacity and μ_K is its equilibrium measure, then*

$$U^{\mu_K}(z) \le \log \frac{1}{\mathrm{cap}(K)}$$

holds for every $z \in$ **C**.

In fact, we know this inequality for every $z \in \mathrm{supp}(\mu_K)$ (see Theorem I.1.3), and we only have to apply the preceding corollary.

Proof of Theorem 3.2. We have, by assumption, that the set of points z satisfying

$$U^\nu(z) + c \ge U^\mu(z)$$

is a Borel set of the form $\mathbf{C} \setminus E$ with $\mu(E) = 0$. Set

$$U(z) := \min(U^\nu(z) + c, U^\mu(z)).$$

From properties (a) – (c) of superharmonic functions discussed above it is immediate that the minimum of two superharmonic functions is again superharmonic; hence U is superharmonic.

Now we recall the Riesz decomposition theorem (with $\mathcal{D} = \mathbf{C}$) according to which there exists a positive Borel measure λ such that, for every $r > 0$,

$$U(z) = u_r(z) + \int_{D_r} \log \frac{1}{|z-t|}\, d\lambda(t), \quad z \in D_r, \tag{3.10}$$

where

$$D_r := \{z \mid |z| < r\},$$

and u_r is harmonic in D_r. We now distinguish two cases:

Case I. $c \geq 0$ or $\nu(\mathbf{C}) < \mu(\mathbf{C})$. By replacing c by $c+\varepsilon$ for some small $\varepsilon > 0$ (and then letting $\varepsilon \to 0$) we can assume without loss of generality that either $c > 0$ (and $\nu(\mathbf{C}) \leq \mu(\mathbf{C})$) or $\nu(\mathbf{C}) < \mu(\mathbf{C})$. But then, as

$$\nu(\mathbf{C})U^{\mu}(z) - \mu(\mathbf{C})U^{\nu}(z) \to 0 \quad \text{as} \quad z \to \infty,$$

we get for large $|z|$, say for $|z| > R$, that $U(z)$ coincides with $U^{\mu}(z)$, and hence $U^{\lambda}(z)$ is harmonic for $|z| > R$. According to Theorem 2.1, the harmonicity of a potential in a region prevents any mass from being present in the region, i.e. $\lambda(\mathbf{C} \setminus \overline{D}_R) = 0$, and so λ has compact support contained in $\overline{D}_R$. Thus, for $r > R$, we can write (3.10) as

$$U(z) = u_r(z) + \int \log \frac{1}{|z-t|}\, d\lambda(t), \quad z \in D_r. \tag{3.11}$$

This tells us that for $\rho > r > R$ the function u_ρ coincides with u_r on D_r, and so

$$u(z) := \lim_{r \to \infty} u_r(z)$$

exists and is harmonic on $\mathbf{C}$, and we get from (3.11) that

$$U(z) = u(z) + \int \log \frac{1}{|z-t|}\, d\lambda(t), \quad z \in \mathbf{C}. \tag{3.12}$$

If $\lambda(\mathbf{C}) < \mu(\mathbf{C})$, then we have

$$\begin{aligned} \lim_{z \to \infty} u(z) &= \lim_{z \to \infty} \left(U(z) - U^{\lambda}(z)\right) \\ &= \lim_{z \to \infty} \int \log \frac{1}{|z-t|} (d\mu(t) - d\lambda(t)) = -\infty, \end{aligned}$$

which yields via the maximum principle for harmonic functions that $u \equiv -\infty$, which is absurd. Thus, we must have $\lambda(\mathbf{C}) = \mu(\mathbf{C})$, since $\lambda(\mathbf{C}) > \mu(\mathbf{C})$ is likewise impossible. If, however, $\lambda(\mathbf{C}) = \mu(\mathbf{C})$, then it follows as before that

$$\lim_{z \to \infty} u(z) = \lim_{z \to \infty} \left(U^{\mu}(z) - U^{\lambda}(z)\right) = 0,$$

and hence $u \equiv 0$. We have thus proved that

$$U(z) = U^{\lambda}(z), \quad z \in \mathbf{C}. \tag{3.13}$$

Now on $\mathbf{C} \setminus E$ the functions U^{μ} and $U = U^{\lambda}$ coincide; furthermore, we have $\mu(E) = 0$. Below we shall show that both μ and λ have finite logarithmic energy. Hence for the logarithmic energy of the signed measure $\mu - \lambda$ with total mass equal to zero we get

$$\begin{aligned} \iint \log \frac{1}{|z-t|}(d\mu(t) - d\lambda(t))(d\mu(z) - d\lambda(z)) \qquad (3.14) \\ = \int \left(U^{\mu}(z) - U^{\lambda}(z)\right)(d\mu(z) - d\lambda(z)) \\ = \int_E \left(U^{\mu}(z) - U^{\lambda}(z)\right)(d\mu(z) - d\lambda(z)) \\ = -\int_E \left(U^{\mu}(z) - U^{\lambda}(z)\right) d\lambda(z) \le 0 \end{aligned}$$

because $U^{\mu}(z) \ge U(z) = U^{\lambda}(z)$ for $z \in \mathbf{C}$.

Having this inequality, we would like to bring into play Lemma I.1.8 for the signed measure $\mu - \lambda$. To do so it suffices to show that both μ and λ are of finite logarithmic energy. For μ this fact is given, and since $U^{\lambda} \le U^{\mu}$, the same is true for λ :

$$I(\lambda) := \int U^{\lambda}\, d\lambda \le \int U^{\mu}\, d\lambda = \int U^{\lambda}\, d\mu \le \int U^{\mu}\, d\mu = I(\mu) < \infty,$$

where, in the second equality, we used the Fubini-Tonelli theorem. Thus, according to Lemma 1.8, $(\mu - \lambda)(\mathbf{C}) = 0$ implies that the logarithmic energy of $\mu - \lambda$ is nonnegative, and it is zero if and only if $\mu = \lambda$. In view of (3.14) this yields $\mu = \lambda$, which implies that for all $z \in \mathbf{C}$ (cf. (3.13))

$$U^{\mu}(z) = U^{\lambda}(z) = U(z) \le U^{\nu}(z) + c,$$

and the proof of this case is complete.

Case II. $c < 0$ and $\nu(\mathbf{C}) = \mu(\mathbf{C})$. (Note that one of the assumptions in Theorem 3.2 is $\nu(\mathbf{C}) \le \mu(\mathbf{C})$). We shall show that this is impossible.

Suppose not. Reasoning as in Case I we get now that $U(z)$ agrees with $U^{\nu}(z)+c$ for large z, the measure λ again has compact support, (3.12) is valid for some harmonic function u, $\lambda(\mathbf{C}) = \nu(\mathbf{C}) = \mu(\mathbf{C})$ and

$$\lim_{z\to\infty}(u(z) - c) = \lim_{z\to\infty}\left\{\left((U^{\nu}(z) + c) - U^{\lambda}(z)\right) - c\right\} = 0,$$

that is, $u \equiv c$. Thus, similarly as above,

$$\iint \log \frac{1}{|z-t|}(d\mu(t) - d\lambda(t))(d\mu(z) - d\lambda(z))$$

$$= \int \left(U^{\mu}(z) - U^{\lambda}(z)\right)(d\mu(z) - d\lambda(z))$$

$$= \int (U^{\mu}(z) - (U(z) - c))\,(d\mu(z) - d\lambda(z))$$

$$= \int (U^{\mu}(z) - U(z))\,(d\mu(z) - d\lambda(z))$$

$$= \int_E (U^{\mu}(z) - U(z))\,(d\mu(z) - d\lambda(z))$$

$$= -\int_E (U^{\mu}(z) - U(z))\,d\lambda(z) \le 0,$$

and we can conclude again that $\mu = \lambda$.

Since μ was assumed to have finite logarithmic energy and $U^{\mu}(z) = U(z)$ holds μ-almost everywhere, there is a $z_0 \in \mathbf{C}$ such that

$$-\infty < U(z_0) = U^{\mu}(z_0) < \infty.$$

But then $\mu = \lambda$ and (3.12) with $u \equiv c$ yield

$$U(z_0) = U^{\mu}(z_0) = U^{\lambda}(z_0) > U^{\lambda}(z_0) + c = U(z_0),$$

and this contradiction shows that Case II is indeed impossible. □

As an application of the maximum principle we prove

Theorem 3.5 (Continuity Theorem). *Let μ have compact support and $z_0 \in$ supp(μ). If the potential U^{μ}, considered as a function on* supp(μ), *is continuous at z_0, then U^{μ} (considered as a function on* **C**) *is continuous at z_0.*

Proof. Since potentials are lower semi-continuous, it is enough to show that U^{μ} is continuous from above at z_0. If $U^{\mu}(z_0) = \infty$, then there is nothing to prove, so assume $U^{\mu}(z_0) < \infty$.

Let $\delta > 0$, $D_\delta := \{z \mid |z - z_0| < \delta\}$, $\mu_\delta := \mu|_{D_\delta}$ and $\mu_\delta^* := \mu - \mu_\delta$. Since $U^{\mu}(z_0) < \infty$, the measure μ has no mass point at z_0, and hence $U^{\mu_\delta}(z_0) \to 0$ as $\delta \to 0$. Thus, given $\varepsilon > 0$, there is a $0 < \delta_0 < 1/2$ such that for $0 < \delta \le \delta_0$ we have

$$U^{\mu_\delta}(z_0) < \varepsilon.$$

Since $U^{\mu_\delta^*}$ is continuous at z_0, the assumed continuity of $U^{\mu}\big|_{\mathrm{supp}(\mu)}$ at z_0 implies that for some $0 < \delta_1 < \delta_0$

$$U^{\mu_{\delta_0}}(z) \le \varepsilon$$

whenever $z \in D_{\delta_1} \cap \text{supp}(\mu)$. But then, because of

$$0 \le U^{\mu_{\delta_1}}(z) \le U^{\mu_{\delta_0}}(z), \quad z \in D_{\delta_1},$$

we get

$$U^{\mu_{\delta_1}}(z) \le \varepsilon$$

for all $z \in \text{supp}(\mu_{\delta_1})$, and so the maximum principle for potentials (Corollary 3.3) implies the same inequality for all $z \in \mathbf{C}$. This and the continuity of $U^{\mu^*_{\delta_1}}$ at z_0 yield

$$\limsup_{z \to z_0} U^{\mu}(z) \le U^{\mu}(z_0),$$

which completes the proof. □

II.4 Green Functions and Balayage Measures

In this section we discuss two important and extremely useful concepts of potential theory: Green functions and the concept of balayage ("sweeping out") measure.

Let us begin with the Green function of an unbounded domain with pole at infinity, which we use at many places in the book. Let G be a domain (connected open set) on $\overline{\mathbf{C}}$ that contains the point infinity, and assume that $\overline{\mathbf{C}} \setminus G$ is of positive capacity. The latter condition is equivalent to $\text{cap}(\partial G) > 0$. The Green function $g_G(z, \infty)$ of G with pole at ∞ is defined as the unique function with the properties

$$\begin{aligned} &\text{(i)} \quad g_G(z,\infty) \text{ is nonnegative and harmonic in } G \setminus \{\infty\},\\ &\text{(ii)} \quad \lim_{|z|\to\infty} (g_G(z,\infty) - \log|z|) = \log\frac{1}{\text{cap}(\partial G)},\\ &\text{(iii)} \quad \lim_{z\to x,\, z\in G} g_G(z,\infty) = 0 \text{ for quasi-every } x \in \partial G. \end{aligned} \tag{4.1}$$

The existence and uniqueness of $g_G(z, \infty)$ were established in Section I.4, where we saw that

$$g_G(z,\infty) := \log\frac{1}{\text{cap}(\partial G)} - U^{\mu_{\partial G}}(z), \tag{4.2}$$

(cf. (I.4.8)) with $\mu_{\partial G}$ denoting the equilibrium measure for ∂G.

Instead of (ii) we can request only the boundedness of $g_G(z, \infty) - \log|z|$ around infinity, but then we also have to assume the boundedness of g_G away from ∞. In other words, an alternative definition is that $g_G(z, \infty)$ is the unique function with the properties

(i') $g_G(z, \infty)$ is harmonic in $G \setminus \{\infty\}$ and is bounded as z stays away from ∞,

(ii') $g_G(z, \infty) - \log|z|$ is bounded around ∞,

(iii) $\lim_{z\to x,\, z\in G} g_G(z, \infty) = 0$ for quasi-every $x \in \partial G$.

In fact, to show that this is an equivalent definition it is enough to prove that any g_G^* with properties (i'), (ii') and (iii) is identical to g_G with properties (i), (ii) and (iii). But this is an immediate consequence of the generalized minimum principle, for $g_G - g_G^*$ is *bounded* and harmonic on G with zero boundary limit quasi-everywhere. Hence by the generalized minimum principle (applied to $g_G - g_G^*$ and to $g_G^* - g_G$) we obtain that $g_G \equiv g_G^*$.

The notion of a Green function $g_G(z, a)$ with pole at some finite point a is similar. Let again $G \subset \overline{\mathbf{C}}$ be a domain such that $\operatorname{cap}(\partial G) > 0$, and a a finite point in G. The Green function $g_G(z, a)$ of G with pole at a is defined as the unique function on G satisfying the following properties:

(i) $g_G(z, a)$ is nonnegative and harmonic in $G \setminus \{a\}$ and bounded as z stays away from a,

(ii) $g_G(z, a) - \log \dfrac{1}{|z - a|}$ is bounded in a neighborhood of a, (4.3)

(iii) $\lim_{z \to x,\, z \in G} g_G(z, a) = 0$ for quasi-every $x \in \partial G$.

Naturally, in (ii) a must be excluded from the neighborhood in question, and

$$g_G(z, a) - \log \frac{1}{|z - a|}$$

has a limit at a (see below). We also remark that the existence of a function satisfying (4.3) requires that $\operatorname{cap}(\partial G) > 0$ as can be seen from Corollary I.2.5.

Both the unicity and the existence of $g_G(z, a)$ can be based on inversion with center at a: if G' is the domain that we obtain from G under the mapping $z \to 1/(z - a)$, then the formula

$$g_G(z, a) := g_{G'}\left(\frac{1}{z - a}, \infty\right) \tag{4.4}$$

establishes a relationship by which questions concerning $g_G(z, a)$ can be transformed into related ones concerning $g_{G'}(z, \infty)$.

For simply connected domains, the Green function is related to the conformal map ψ of G onto the unit disk that maps a into the origin via the formula $g_G(z, a) = \log 1/|\psi(z)|$. Based on this (or the unicity established above) one can easily verify the following examples of Green functions: if G is the unit disk, then

$$g_G(z, a) = \log \left| \frac{1 - z\bar{a}}{z - a} \right|,$$

while for the exterior of the unit disk the same formula holds if a is finite, and for $a = \infty$ we have

$$g_G(z, \infty) = \log |z|.$$

If G is the complement of $[-1, 1]$, then

$$g_G(z,a) = \log\left|\frac{1-\overline{\varphi(a)}\varphi(z)}{\varphi(z)-\varphi(a)}\right|, \qquad \varphi(z) := z + \sqrt{z^2-1},$$

and finally if G is the right-half plane $\operatorname{Re} z > 0$, then

$$g_G(z,a) = \log\left|\frac{z+\overline{a}}{z-a}\right|.$$

An alternative approach to the existence of Green functions can be based on the notion of balayage measures which we now discuss. After that we shall briefly return to Green functions and their properties.

Let $G \subset \overline{\mathbf{C}}$ be an open set such that its boundary ∂G is a compact subset of $\mathbf{C}$ of positive capacity. Let ν be a (finite) Borel measure on G by which we mean $\nu(\overline{\mathbf{C}} \setminus G) = 0$. The balayage (or "sweeping out") problem consists of finding another measure $\widehat{\nu}$ *supported on* ∂G such that $\|\widehat{\nu}\| = \|\nu\|$ and the potentials U^ν and $U^{\widehat{\nu}}$ coincide on ∂G quasi-everywhere. $\widehat{\nu}$ is called the *balayage measure* associated with ν when we sweep out ν from G onto ∂G. We shall see that if G is unbounded, then $\widehat{\nu}$ with the above property may not exist because during this "sweeping out" process the potential increases by a constant. For unbounded regions we shall consider balayage in this sense.

Sometimes we also want to take the balayage of a measure ν for which $\nu(\overline{\mathbf{C}} \setminus G) \neq 0$. In such a case we write $\nu = \nu|_G + \nu|_{\overline{\mathbf{C}}\setminus G}$, and sweep out only the part $\nu|_G$ lying on G, i.e. we set $\widehat{\nu} := \widehat{\nu|_G} + \nu|_{\overline{\mathbf{C}}\setminus G}$.

It is also customary to speak of taking balayage *onto* a compact set K of positive capacity by which we mean taking balayage *out of* $\overline{\mathbf{C}} \setminus K$ onto $\partial(\overline{\mathbf{C}} \setminus K)$.

Before discussing the balayage problem for a general open set G we consider the special case when G is connected and the (compact) support of ν is contained in G. This special case is sufficient for our purposes in this book, but we shall return to the general problem for completeness.

The case of connected G splits naturally into two subcases: G is bounded or G contains the point infinity. We begin with the former one.

Theorem 4.1. *Let G be a bounded domain, and ν a measure with compact support in G. Then there exists a unique measure $\widehat{\nu}$ supported on ∂G such that $\|\widehat{\nu}\| = \|\nu\|$, the potential $U^{\widehat{\nu}}$ is bounded on ∂G, and*

$$U^{\widehat{\nu}}(z) = U^\nu(z) \tag{4.5}$$

for quasi-every $z \in \partial G$.

This $\widehat{\nu}$ has the following additional properties:

(a) $U^{\widehat{\nu}}(z_0) = U^\nu(z_0)$ *if $z_0 \notin \overline{G}$ or if $z_0 \in \partial G$ and z_0 is a regular boundary point of G.*

(b) $U^{\widehat{\nu}}(z) \le U^\nu(z)$ *for every $z \in \mathbf{C}$.*

(c) *If h is a continuous function on $\overline{G}$ that is harmonic on G, then*

$$\int h\, d\widehat{\nu} = \int h\, d\nu. \tag{4.6}$$

Note that in (4.6) the integral on the left is taken on ∂G.

Assertion (a) implies that if G is regular with respect to the Dirichlet problem, then $\widehat{\nu}$ solves the balayage problem in the *strict sense*; namely the potentials $U^{\widehat{\nu}}$ and U^{ν} coincide everywhere outside G.

Example 4.2. One has to be cautious with the unicity of $\widehat{\nu}$, namely we cannot drop the requirements that $U^{\widehat{\nu}}$ is bounded on ∂G and $\|\widehat{\nu}\| = \|\nu\|$. In fact, let G be the punctured disk $\{z \mid 0 < |z| < 1\}$, and ν the normalized arc measure on the circle $\{z \mid |z| = 1/2\}$ with the normalization $\|\nu\| = 1$. If m is the normalized arc measure on the unit circumference, then for any $0 \le a \le 1$ the measure

$$\sigma = a\delta_0 + (1-a)m$$

satisfies $\|\sigma\| = 1 = \|\nu\|$ and

$$U^{\sigma}(z) = 0 = U^{\nu}(z) \tag{4.7}$$

for quasi-every $z \in \partial G$. However, only the choice $a = 0$ gives a σ $(= \widehat{\nu})$ for which the potential is bounded on ∂G.

Similarly, for any $a > 0$ the measure $\sigma = am$ satisfies the boundedness condition and (4.7), but only the choice $a = 1$ yields $\|\sigma\| = 1 = \|\nu\|$. □

Example 4.3. If G is the disk $D_R(0)$ and $z_0 \in D_R(0)$, then it follows from Poisson's formula (cf. Section 0.4) that the balayage of the point mass δ_{z_0} is given by

$$d\widehat{\delta_{z_0}}(t) = \frac{1}{2\pi} P(t, z_0)\, d\theta, \qquad t = Re^{i\theta}, \tag{4.8}$$

where $P(t, z)$ is the Poisson kernel. □

Proof of Theorem 4.1. Without loss of generality we assume that $\|\nu\| = 1$.

Let $\Sigma = \partial G$, $Q(z) = -U^{\nu}(z)$ for $z \in \Sigma$, and consider the weighted energy problem with $w = e^{-Q}$. We set $\widehat{\nu} = \mu_w$. Then $\|\widehat{\nu}\| = 1$, and $U^{\widehat{\nu}}$ is bounded on ∂G (Theorem I.4.3). To prove (4.5) we first remark that by Theorem I.1.3(d) and (e)

$$U^{\widehat{\nu}}(z) \ge U^{\nu}(z) + F_w \tag{4.9}$$

for quasi-every $z \in \Sigma$, and

$$U^{\widehat{\nu}}(z) \le U^{\nu}(z) + F_w \tag{4.10}$$

for every $z \in S_w = \operatorname{supp}(\widehat{\nu})$. If we apply the principle of domination (Theorem 3.2) to the latter inequality we conclude (4.10) for every $z \in \mathbf{C}$, which, together with (4.9), yields

$$U^{\widehat{\nu}}(z) = U^{\nu}(z) + F_w \tag{4.11}$$

for quasi-every $z \in \Sigma = \partial G$. Now let us integrate (4.11) with respect to the equilibrium measure $\mu_{\partial G}$. Since sets of capacity zero have zero $\mu_{\partial G}$– and $\widehat{\nu}$–measures, it follows from the property

$$U^{\mu_{\partial G}}(z) = \log \frac{1}{\operatorname{cap}(\partial G)}$$

for $z \in G$ and for quasi-every $z \in \partial G$ that

$$\begin{aligned} \log \frac{1}{\operatorname{cap}(\partial G)} &= \int U^{\mu_{\partial G}} d\widehat{\nu} = \int U^{\widehat{\nu}} d\mu_{\partial G} \\ &= \int U^{\nu} d\mu_{\partial G} + F_w = \int U^{\mu_{\partial G}} d\nu + F_w = \log \frac{1}{\operatorname{cap}(\partial G)} + F_w. \end{aligned} \tag{4.12}$$

This gives $F_w = 0$, and so (4.11) proves (4.5).

The unicity of $\widehat{\nu}$ with the three properties in the theorem follows from the principle of domination and the unicity result Corollary 2.2.

It remains to prove properties (a) – (c).

Property (b) was proved above; namely we verified (4.10) for all z and we also know that $F_w = 0$.

Next we prove (a). If $z_0 \in \partial G$ and z_0 is a regular boundary point of G, then $U^{\widehat{\nu}}(z_0) = U^{\nu}(z_0)$ follows from our construction and Theorems I.4.4 and I.5.1, according to which $U^{\widehat{\nu}}$ is continuous at z_0 and continuity implies $U^{\widehat{\nu}}(z_0) = U^{\nu}(z_0)$.

Now let $z_0 \notin \overline{G}$. Then z_0 belongs to a connected component Γ of $\overline{\mathbf{C}} \setminus \overline{G}$. Since $\partial\Gamma \subseteq \partial G$, and for quasi-every $z \in \partial G$ we have just verified the equality $U^{\widehat{\nu}}(z) = U^{\nu}(z)$ and the continuity of $U^{\widehat{\nu}}$ at z, it follows from the generalized minimum principle (Theorem I.2.4) that the function $U^{\widehat{\nu}} - U^{\nu}$, which is harmonic on Γ, must identically vanish on Γ (recall also that the potential $U^{\widehat{\nu}}$ is bounded on compact subsets of $\mathbf{C}$, which implies the boundedness of $U^{\widehat{\nu}} - U^{\nu}$ on Γ). Hence $U^{\widehat{\nu}}(z_0) = U^{\nu}(z_0)$ as we claimed.

Finally, we prove (c). First we verify the formula

$$\widehat{\nu} = \int \widehat{\delta_t}\, d\nu(t), \tag{4.13}$$

where δ_t denotes the unit mass at t. The equality in (4.13) is understood in the sense that for every Borel set $E \subseteq \partial G$

$$\widehat{\nu}(E) = \int \widehat{\delta_t}(E)\, d\nu(t), \tag{4.14}$$

which is easily seen to be equivalent to

$$\int f\, d\widehat{\nu} = \int \left(\int f\, d\widehat{\delta_t} \right) d\nu(t) \tag{4.15}$$

for every nonnegative Borel function f. Then (4.15) also holds for every lower bounded Borel function f.

To prove (4.13), let σ denote the right-hand side. Then $\|\sigma\| = \|\nu\|$, and for every regular boundary point $z \in \partial G$ we get from (a)

$$\begin{aligned} U^{\sigma}(z) &= \int \left(\int \log \frac{1}{|z-u|} d\widehat{\delta_t}(u) \right) d\nu(t) \\ &= \int U^{\widehat{\delta_t}}(z)\, d\nu(t) = \int U^{\delta_t}(z)\, d\nu(t) \\ &= \int \log \frac{1}{|z-t|} d\nu(t) = U^{\nu}(z); \end{aligned}$$

hence this relation holds true for quasi-every $z \in \partial G$. By (b), if $t \in \operatorname{supp}(\nu)$, then for $z \in \partial G$

$$U^{\widehat{\delta_t}}(z) \le U^{\delta_t}(z) \le \log \frac{1}{\operatorname{dist}(\operatorname{supp}(\nu), \partial G)}, \tag{4.16}$$

which, together with the previous computation yields the boundedness of U^{σ} on ∂G. Hence, by the unicity part of the theorem, $\sigma = \widehat{\nu}$ as we claimed in (4.13).

We use formula (4.13) to reduce (c) to the special case $\nu = \delta_t$, $t \in G$. In fact, suppose (c) has been verified in this special case. Then we obtain from (4.15)

$$\begin{aligned} \int h\, d\widehat{\nu} &= \int \left(\int h\, d\widehat{\delta_t} \right) d\nu(t) \\ &= \int \left(\int h\, d\delta_t \right) d\nu(t) = \int h\, d\nu, \end{aligned} \tag{4.17}$$

i.e. then (c) holds for every ν.

Thus, in proving (c) we can assume that $\nu = \delta_{t_0}$ for some $t_0 \in G$. Let $G_1 \subseteq G_2 \subseteq \cdots$ be an increasing sequence of open sets such that each of them contains t_0, has closure contained in G, and $G = \bigcup G_n$. Let $\widehat{\nu}_n$ denote the balayage of $\nu = \delta_{t_0}$ onto ∂G_n. We claim that $\widehat{\nu}_n \to \widehat{\nu}$ in the weak* topology. In fact, let σ be any weak* limit point of the sequence $\{\widehat{\nu}_n\}$, say $\widehat{\nu}_{n_k} \to \sigma$. Inequality (4.16) and the principle of descent (Theorem I.6.8) show that U^{σ} is bounded on ∂G. Since

$$\|\sigma\| = \int 1\, d\sigma = \lim_{n\to\infty} \int 1\, d\widehat{\nu}_n = \lim_{n\to\infty} \|\widehat{\nu}_n\| = 1,$$

and by the lower envelope theorem (Theorem I.6.9) and part (a) of the theorem,

$$U^{\sigma}(z) = \liminf_{k\to\infty} U^{\widehat{\nu}_{n_k}}(z) = \liminf_{k\to\infty} U^{\nu}(z) = U^{\nu}(z)$$

for quasi-every $z \in \partial G$, the equality $\sigma = \widehat{\nu}$ is a consequence of the unicity of the balayage measure. This is true for any weak* limit point; hence the whole sequence $\{\widehat{\nu}_n\}$ converges to $\widehat{\nu}$ in weak* sense.

Now suppose that property (c) has been established for every G_n in place of G. Then it also follows for G:

$$\int h\, d\widehat{\nu} = \lim_{n\to\infty} \int h\, d\widehat{\nu}_n = \lim_{n\to\infty} \int h\, d\nu_n = \int h\, d\nu. \tag{4.18}$$

Thus, it is enough to establish (c) for every G_n.

Until now we have not specified how to choose G_n, but the above procedure allows us to replace G by some G_n's with smooth boundary. Hence, by choosing G_n to have C^2 boundary, we can assume that the boundary of G consists of a finite number of C^2 curves, and that h is harmonic on $\overline{G}$.

Now consider the potential

$$U(z) := U^{\delta_{t_0} - \widehat{\delta_{t_0}}}(z).$$

By (a), $U(z) = 0$ for $z \notin G$. Let

$$\overline{D_r} = \{z \mid |z - t_0| \le r\}$$

be a closed neighborhood of t_0, and for some small $r > 0$ set $G_r = G \setminus \overline{D_r}$. We can apply Green's formula (1.1) to the functions h and U to deduce

$$\int_{\partial G_r} \left(h \frac{\partial U}{\partial \mathbf{n}} - U \frac{\partial h}{\partial \mathbf{n}} \right) ds = 0, \tag{4.19}$$

where $\partial/\partial\mathbf{n}$ denotes differentiation in the direction of the inner normal of G_r. The second term in the integrand vanishes on ∂G, and in absolute value is $O(\log 1/r)$ on ∂D_r; hence for $r \to 0$ this term vanishes completely.

The first term equals

$$h(z)\left(-\frac{1}{r} - \frac{\partial U^{\widehat{\delta_{t_0}}}(z)}{\partial \mathbf{n}} \right)$$

on ∂D_r, and so it follows from the mean value theorem for h that if $r \to 0$, then (4.19) becomes

$$\int_{\partial G} h \frac{\partial U}{\partial \mathbf{n}} = 2\pi h(t_0). \tag{4.20}$$

Since the potential U vanishes on $\mathbf{C} \setminus G$, it follows from Theorem 1.5 that

$$-\frac{1}{2\pi} \frac{\partial U}{\partial \mathbf{n}} \tag{4.21}$$

gives the density of the measure $\delta_{t_0} - \widehat{\delta_{t_0}}$ on ∂G with respect to the arc measure ds on ∂G (the required Lip 1 property is easy to establish, consider e.g. the construction of $\widehat{\delta_{t_0}}$ and Theorem I.4.7); hence (4.20) is simply

$$\int h\, d\widehat{\delta_{t_0}} = h(t_0),$$

which is exactly (c) for $\nu = \delta_{t_0}$. □

We continue now with the case of an unbounded region.

Theorem 4.4. *Let G be a domain on $\overline{\mathbf{C}}$ containing the point infinity, such that $\mathbf{C}\setminus G$ is of positive capacity, and let ν be a measure with compact support in $G \setminus \{\infty\}$. Then there exists a unique measure $\widehat{\nu}$ supported on ∂G such that $\|\widehat{\nu}\| = \|\nu\|$, the potential $U^{\widehat{\nu}}$ is bounded on ∂G, and for some constant c*

$$U^{\widehat{\nu}}(z) = U^{\nu}(z) + c \tag{4.22}$$

for quasi-every $z \in \partial G$. Here

$$c = \int_G g_G(t, \infty)\, d\nu(t), \tag{4.23}$$

where g_G denotes the Green function of G with pole at ∞.

This $\widehat{\nu}$ has the following additional properties:

(a) *$U^{\widehat{\nu}}(z_0) = U^{\nu}(z_0) + c$ for $z_0 \notin \overline{G}$ and for $z_0 \in \partial G$ provided z_0 is a regular boundary point of G.*

(b) *$U^{\widehat{\nu}}(z) \leq U^{\nu}(z) + c$ for every $z \in \mathbf{C}$.*

(c) *If h is a continuous function on $\overline{G}$ that is harmonic on G, then*

$$\int h\, d\widehat{\nu} = \int h\, d\nu.$$

Proof. We follow the proof of Theorem 4.1. Equation (4.12) now takes the form

$$\begin{aligned} \log \frac{1}{\operatorname{cap}(\partial G)} &= \int U^{\mu_{\partial G}} d\widehat{\nu} = \int U^{\widehat{\nu}} d\mu_{\partial G} \\ &= \int U^{\nu} d\mu_{\partial G} + F_w = \int U^{\mu_{\partial G}} d\nu + F_w, \end{aligned}$$

and

$$F_w = \int g_G(t, \infty)\, d\nu(t)$$

follows from here and the representation (4.2). Thus, (4.11) implies the first part of the theorem.

The proofs of (a) – (c) are identical with the corresponding parts in Theorem 4.1; one only has to use (4.22) – (4.23) instead of (4.5). □

Before we consider the balayage problem for general open sets, as an application of the balayage concept we prove the following unicity theorem.

Definition 4.5. A Borel measure μ is called *C-absolutely continuous* if $\mu(E) = 0$ for every Borel set E of zero capacity.

For example, we know that measures with finite energy are C-absolutely continuous. Now we prove

Theorem 4.6. *Let μ_1 and μ_2 be two C-absolutely continuous measures with compact support on* $\mathbf{C}$. *If $\|\mu_1\| = \|\mu_2\|$ and $U^{\mu_1}(z) = U^{\mu_2}(z) + c$ with some constant c for quasi-every $z \in \operatorname{supp}(\mu_1) \cup \operatorname{supp}(\mu_2)$, then $\mu_1 = \mu_2$.*

Proof. It is enough to prove $U^{\mu_1}(z_0) = U^{\mu_2}(z_0) + c$ for $z_0 \notin S := \operatorname{supp}(\mu_1) \cup \operatorname{supp}(\mu_2)$, for then $z_0 \to \infty$ yields $c = 0$, and so the two potentials U^{μ_1} and U^{μ_2} coincide quasi-everywhere, and we can invoke Corollary 2.2 (we remark also that by Lemma III.4.6 in the next chapter every Borel set of capacity zero has zero two dimensional Lebesgue measure).

Thus let $z_0 \notin S$, and let G be that connected component of $\mathbf{C} \setminus S$ that contains z_0. We sweep out δ_{z_0} onto ∂G, and if $\widehat{\delta_{z_0}}$ is the balayage measure, we obtain

$$U^{\widehat{\delta_{z_0}}}(z) = \log \frac{1}{|z - z_0|} + d \tag{4.24}$$

for quasi-every $z \notin G$, where $d = 0$ if G is bounded and $d = g_G(z_0, \infty)$ in the opposite case (see Theorems 4.1 and 4.4). Since $S \subseteq \overline{\mathbf{C}} \setminus G$ and μ_1 and μ_2 are C-absolutely continuous, we get from (4.24) by integration with respect to μ_j, $j = 1, 2$,

$$U^{\mu_j}(z_0) = \int U^{\widehat{\delta_{z_0}}} d\mu_j - \|\mu_j\| d = \int_S U^{\mu_j} d\widehat{\delta_{z_0}} - \|\mu_j\| d. \tag{4.25}$$

This so-called Riesz formula immediately implies $U^{\mu_1}(z_0) = U^{\mu_2}(z_0) + c$, for by the assumption $\|\mu_1\| = \|\mu_2\|$, and for $j = 1$ and $j = 2$ the integrands differ $\widehat{\delta_{z_0}}$-almost everywhere by the constant c. □

Next we can turn to the balayage problem out of general open sets.

Theorem 4.7. *Let G be an open subset of $\overline{\mathbf{C}}$. Assume that ∂G is a compact subset of* $\mathbf{C}$ *of positive capacity. Further let ν be a finite Borel measure on G (i.e. $\nu(\overline{\mathbf{C}} \setminus G) = 0$) with compact support in* $\mathbf{C}$. *Then there exists a unique C-absolutely continuous measure $\widehat{\nu}$ on ∂G and a constant c for which $\|\widehat{\nu}\| = \|\nu\|$ and*

$$U^{\widehat{\nu}}(z) = U^{\nu}(z) + c \tag{4.26}$$

for quasi-every $z \notin G$. Here

$$c = \int_{\Omega} g_{\Omega}(t, \infty)\, d\nu(t), \tag{4.27}$$

where Ω denotes the unbounded component of G. Furthermore, this $\widehat{\nu}$ has the following properties:

(a) (4.26) *holds for every $z \notin \overline{G}$ and also for every $z \in \partial G$ which is a regular boundary point for every component of G containing z on its boundary.*

(b) *$U^{\widehat{\nu}}(z) \le U^{\nu}(z) + c$ for every z.*

(c) *If h is a continuous function on $\overline{G}$ that is harmonic in G, then*

$$\int h\, d\widehat{\nu} = \int h\, d\nu.$$

Note that (4.26) holds if $z \in \partial G$ and

- z is not a boundary point of any connected component of G, or
- z is a boundary point for at least two connected components of G, or
- z is on the boundary of exactly one component of G and it is a regular boundary point for this component.

In fact, in these cases, (a) can be applied (see the proof of Theorem I.4.8).

Proof of Theorem 4.7. The unicity of $\widehat{\nu}$ follows from Theorem 4.6.

To prove the existence, first suppose that $\mathrm{supp}(\nu) \subseteq G$. Let $G_1, \ldots, G_k$ be those connected components of G that intersect $\mathrm{supp}(\nu)$, and let ν_k be the restriction of ν to G_k. Finally, let $\widehat{\nu}_k$ be the balayage measure that we obtain by sweeping out ν_k from G_k onto ∂G_k in the sense of Theorems 4.1 and 4.4. It is immediate from Theorems 4.1 and 4.4 that the measure

$$\widehat{\nu} := \widehat{\nu}_1 + \cdots + \widehat{\nu}_k$$

satisfies all the requirements in Theorem 4.7.

Now let ν be arbitrary. Without loss of generality we can assume that ∂G is contained in the disk $D_{1/2} = \{z \mid |z| \leq 1/2\}$. For an $n \geq 4$ we set

$$K_n := \{z \in G \mid \mathrm{dist}(z, \partial G) \geq 1/n\},$$

and let $\mu_n = \nu_{|K_n}$ be the restriction of ν to K_n. Then μ_n has compact support in G; hence we can form the balayage $\widehat{\mu_n}$ of μ_n out of G onto ∂G. Since

$$\widehat{\mu_{n+1}} = \widehat{\mu_n} + \widehat{\nu_{|K_{n+1}\setminus K_n}} \geq \widehat{\mu_n},$$

the sequence $\{\widehat{\mu_n}\}$ is increasing together with $\{\mu_n\}$. We set

$$\widehat{\nu} := \lim_{n\to\infty} \widehat{\mu_n}.$$

Each $\widehat{\mu_n}$ was C-absolutely continuous (actually each of them has finite logarithmic energy because its potential is bounded on its support), from which the C-absolute continuity of $\widehat{\nu}$ easily follows. Since $\mathrm{supp}(\widehat{\mu_n}) \subseteq \overline{D_{1/2}} = \{z \mid |z| \leq 1/2\}$, and for $z, t \in \overline{D_{1/2}}$ we have $\log 1/|z - t| \geq 0$, every property of $\widehat{\nu}$ listed in Theorem 4.7 follows from the analogous property of $\widehat{\mu_n}$ and the monotone convergence theorem. For example, the proof of (4.26) runs as follows:

$$\begin{aligned} U^{\widehat{\nu}}(z) &= \lim_{n\to\infty} U^{\widehat{\mu_n}}(z) = \lim_{n\to\infty}\left(U^{\mu_n}(z) + \int_\Omega g_\Omega(t,\infty)\,d\mu_n(t)\right) \\ &= U^{\nu}(z) + \int_\Omega g_\Omega(t,\infty)\,d\nu(t) \end{aligned}$$

for quasi-every $z \in \partial G$. □

As an application of balayage let us consider the important case of an external field given by a potential.

Example 4.8. Let $\Sigma \subset \mathbf{C}$ be compact, and let the external field Q on Σ be of the form $-aU^{\nu}(z)$, where ν is a positive measure of compact support disjoint from Σ having total mass 1, and $0 \le a \le 1$. Then for $w(x) = \exp(-Q(x))$ we have

$$\mu_w = a\widehat{\nu} + (1-a)\mu_\Sigma, \tag{4.28}$$

where $\widehat{\nu}$ denotes the balayage of ν onto Σ, and μ_Σ is the equilibrium measure of Σ.

In fact, this is a consequence of the properties of balayage measures and of Theorem I.3.3, and actually for $a = 1$ this is how we defined the balayage.

□

Up to now we have assumed that the support of ν is a bounded subset of $\mathbf{C}$. We shall need property (c) for $\nu = \delta_\infty$. So what would be the balayage of δ_∞ in the sense of (c) in case $\infty \in G$? We can easily answer this question by considering m_R, the normalized arc measure on the circle $C_R = \{z \mid |z| = R\}$ for large R. In fact, if C_R contains ∂G in its interior, then $\widehat{m_R} = \mu_{\partial G}$, because both of these measures have bounded potential on ∂G, and the potential of each of them is constant quasi-everywhere on ∂G (see the unicity in the first part of Theorem 4.4). Thus, by applying Theorem 4.4(c) to m_R we can see that

$$\int h\, dm_R = \int h\, d\mu_{\partial G},$$

which, by the mean value property of h, implies

$$h(\infty) = \int h\, d\mu_{\partial G}. \tag{4.29}$$

Thus, in this sense the balayage of δ_∞ onto ∂G is the equilibrium measure $\mu_{\partial G}$.

Let us also mention the formula

$$\widehat{\nu} = \int \widehat{\delta_t}\, d\nu(t), \tag{4.30}$$

which is a consequence of (4.13) and the construction above.

Now we return to the Green function of a domain G with pole at a finite $a \in G$. Let $\widehat{\delta_a}$ be the balayage of δ_a onto ∂G, and let $c_a = 0$ if G is bounded and $c_a = g_G(a, \infty)$ if G contains the point infinity. By Theorems 4.1 and 4.4 the function

$$\begin{aligned} g(z) &:= U^{\delta_a}(z) - U^{\widehat{\delta_a}}(z) + c_a \\ &= \log\frac{1}{|z-a|} - \int_{\partial G} \log\frac{1}{|z-t|}\, d\widehat{\delta_a}(t) + c_a \end{aligned} \tag{4.31}$$

satisfies all the requirements (i) – (iii) in (4.3); hence $g(z) = g_G(z, a)$. Thus,

$$g_G(z,a) = \log\frac{1}{|z-a|} - \int_{\partial G}\log\frac{1}{|z-t|}\,d\widehat{\delta_a}(t) + c_a, \tag{4.32}$$

where c_a is a nonnegative constant (equal to $g_G(a,\infty)$ if G contains the point infinity and 0 otherwise).

Using this representation we are now going to prove the symmetry of the Green function:

Theorem 4.9. *For all $a, z \in G$ we have*

$$g_G(z,a) = g_G(a,z). \tag{4.33}$$

Proof. We may assume that G is regular with respect to the Dirichlet problem. In fact, we can exhaust G by an increasing sequence of regular domains $G_1 \subseteq G_2 \subseteq \cdots$ with closure in G. If $(\widehat{\delta_a})_n$ denotes the balayage of δ_a onto ∂G_n, then it was verified in the proof of Theorem 4.1(c) that $(\widehat{\delta_a})_n \to \widehat{\delta_a}$ in the weak* sense; hence for every $z, a \in G$ we have by the representation (4.31)

$$\lim_{n\to\infty} g_{G_n}(z,a) = g_G(z,a).$$

Hence the symmetry of every $g_{G_n}(z,a)$ implies that of $g_G(z,a)$. Thus, we assume the regularity of G, which is equivalent to the continuity of the equilibrium potential $U^{\mu_{\partial G}}$.

If G is bounded, then $c_a = 0 = c_z$, and we obtain from (4.31) and Theorem 4.1(c)

$$\begin{aligned} g_G(z,a) &= \log\frac{1}{|z-a|} - U^{\widehat{\delta_a}}(z) = \log\frac{1}{|a-z|} - \int U^{\widehat{\delta_a}}\,d\widehat{\delta_z} \\ &= \log\frac{1}{|a-z|} - \int U^{\widehat{\delta_z}}\,d\widehat{\delta_a} = \log\frac{1}{|a-z|} - U^{\widehat{\delta_z}}(a) = g_G(a,z). \end{aligned}$$

If, however, G is unbounded but a and z are finite, then with $\Sigma = \partial G$

$$\begin{aligned} g_G(z,a) &= \log\frac{1}{|z-a|} + g_G(a,\infty) - U^{\widehat{\delta_a}}(z) \qquad (4.34) \\ &= \log\frac{1}{|a-z|} + \log\frac{1}{\operatorname{cap}(\partial G)} - U^{\mu_\Sigma}(a) \\ &\quad - \int \left(U^{\widehat{\delta_a}} - U^{\mu_\Sigma}\right) d\widehat{\delta_z} - U^{\mu_\Sigma}(z). \end{aligned}$$

Here the second and fifth terms on the right give $g_G(z,\infty)$, while the fourth term is equal to

$$\begin{aligned} -\int \left(U^{\widehat{\delta_a}} - U^{\mu_\Sigma}\right) d\widehat{\delta_z} &= -\int U^{\widehat{\delta_z}}\,d(\widehat{\delta_a} - \mu_\Sigma) \\ &= -\int \left(U^{\widehat{\delta_z}} - U^{\mu_\Sigma}\right) d(\widehat{\delta_a} - \mu_\Sigma), \end{aligned}$$

where at the last step we used that U^{μ_Σ} is constant on Σ, and $\widehat{\delta_a}$ and μ_Σ are probability measures on Σ. But here

$$\int \left(U^{\widehat{\delta_z}} - U^{\mu_\Sigma}\right) d\mu_\Sigma = \int U^{\mu_\Sigma}\, d(\widehat{\delta_z} - \mu_\Sigma) = 0$$

for the same reasons; hence (4.34) can be continued as

$$\begin{aligned} g_G(z,a) &= \log\frac{1}{|a-z|} - U^{\mu_\Sigma}(a) - \int \left(U^{\widehat{\delta_z}} - U^{\mu_\Sigma}\right) d\widehat{\delta_a} + g_G(z,\infty) \\ &= \log\frac{1}{|a-z|} - U^{\mu_\Sigma}(a) - \left(U^{\widehat{\delta_z}}(a) - U^{\mu_\Sigma}(a)\right) + g_G(z,\infty) \\ &= g_G(a,z). \end{aligned}$$

If one of a or z is infinite, say $z = \infty$, then we get the symmetry by letting $z \to \infty$ in the formula

$$g_G(z,a) = \log\frac{1}{|z-a|} - \int_{\partial G} \log\frac{1}{|z-t|} d\widehat{\delta_a}(t) + g_G(a,\infty) \tag{4.35}$$

(see (4.31)): in the limit the first two terms on the right cancel each other; hence by continuity

$$g_G(\infty,a) = \lim_{z\to\infty} g_G(z,a) = g_G(a,\infty).$$

□

On applying (4.32) in case of a bounded domain we get

$$\log\frac{1}{|z-a|} = g_G(z,a) + \int_{\partial G} \log\frac{1}{|z-t|}\, d\widehat{\delta_a}(t) \tag{4.36}$$

for all $z \in G$ and $a \in G$. Since every $z \in \partial G$ is a fine limit point of G, this equality extends to $z \in \partial G$. If G is a regular domain, then the Green function is continuous on ∂G and vanishes on ∂G, so for $z \in \partial G$ the first term on the right is actually zero. Finally, by setting $g_G(z,a) = 0$ for $z \notin G$, (4.36) still holds for $z \notin \overline{G}$ since in this case the function $\log 1/|z-a|$ is harmonic on $\overline{G}$, and we can apply part (c) of Theorem 4.7. Thus, (4.36) is true for all $z \in \mathbf{C}$ and on integrating this equality against a measure $\lambda(z)$, interchanging the order of integration, and applying the symmetry of the Green function we obtain

$$U^\lambda(a) = \int_G g_G(a,z)\, d\lambda(z) + \int_{\partial G} U^\lambda(t)\, d\widehat{\delta_a}(t). \tag{4.37}$$

The formula

$$u(a) = \int_{\partial G} u(t)\, d\widehat{\delta_a}(t)$$

is also true in view of Theorem 4.7(c) for all functions u harmonic in a neighborhood of $\overline{G}$, so we can obtain from the last two equalities and the Riesz decomposition theorem the so-called Poisson–Jensen formula.

Theorem 4.10 (Poisson–Jensen Formula). *Let G be a bounded regular domain and f a superharmonic function on a neighborhood of $\overline{G}$. Then for $a \in G$*

$$f(a) = \int_G g_G(a, z)\, d\lambda(z) + \int_{\partial G} f(t)\, d\widehat{\delta_a}(t), \tag{4.38}$$

where λ is the measure associated with f in the Riesz decomposition theorem.

As a consequence of formula (4.32) we obtain

Theorem 4.11. *If G is a domain with compact and C^2 boundary, then for every $a \in G$ the balayage of the point mass δ_a at a onto ∂G is given by the formula*

$$d\widehat{\delta_a}(s) = \frac{1}{2\pi} \frac{\partial g_G(s, a)}{\partial \mathbf{n}} ds, \tag{4.39}$$

where ds is the arc length on ∂G and $\mathbf{n}$ denotes inner normal relative to G.

Proof. From the proof of Theorem 4.1 it follows that the balayage measure in question is given by

$$d\widehat{\delta_a}(s) = \frac{1}{2\pi} \frac{\partial U^{\delta_a - \widehat{\delta_a}}(s)}{\partial \mathbf{n}} ds$$

(use formulas (4.20) and (4.21)). Now all we have to do is to apply (4.32). □

Formula (4.39) implies that if u is harmonic on G and continuous on $\overline{G}$, then

$$u(a) = \frac{1}{2\pi} \int_{\partial G} u(s) \frac{\partial g_G(s, a)}{\partial \mathbf{n}} ds, \quad a \in G \tag{4.40}$$

which is a generalization of the classical Poisson formula (0.4.1).

Strictly speaking the domain $G = \overline{\mathbf{C}} \setminus [-1, 1]$ does not satisfy the hypotheses of Theorem 4.11, but formula (4.39) remains valid in this case in the form

$$d\widehat{\delta_a}(x) = \frac{1}{2\pi} \left(\frac{\partial g_G(x, a)}{\partial \mathbf{n}_+} + \frac{\partial g_G(x, a)}{\partial \mathbf{n}_-} \right) dx, \tag{4.41}$$

where of course $\mathbf{n}_\pm$ denote the two normals to the interval $(-1, 1)$. In fact, this immediately follows from the preceding theorem and the exhaustion process applied in the proof of Theorem 4.1. Now (4.41) yields for any measure ν supported on $\overline{\mathbf{C}} \setminus [-1, 1]$ the formula

$$\frac{d\widehat{\nu}}{dx}(x) = \frac{1}{2\pi} \int \left(\frac{\partial g_G(x, a)}{\partial \mathbf{n}_+} + \frac{\partial g_G(x, a)}{\partial \mathbf{n}_-} \right) d\nu(a) \tag{4.42}$$

for the balayage of ν onto $[-1, 1]$.

If a is real, or ν is supported on the real line, then in view of the symmetry of the Green function with respect to the real line these formulas take the form

$$d\widehat{\delta_a}(x) = \frac{1}{\pi} \frac{\partial g_G(x, a)}{\partial \mathbf{n}} dx, \tag{4.43}$$

and

$$\frac{d\widehat{\nu}}{dx}(x) = \frac{1}{\pi}\int \frac{\partial g_G(x,a)}{\partial \mathbf{n}} d\nu(a), \tag{4.44}$$

where $\mathbf{n}$ denotes either the upper or the lower normal.

We now make these formulas more concrete. Let $\varphi(z) := z + \sqrt{z^2-1}$ be the conformal map of the domain $G = \overline{\mathbf{C}} \setminus [-1, 1]$ onto the exterior G' of the unit disk with $\sqrt{z^2-1}$ positive for real $z > 1$. Then $g_G(z, a) = g_{G'}(\varphi(z), \varphi(a))$ and since

$$g_{G'}(\zeta, a') = -\log\left|\frac{\zeta - a'}{\overline{a'}\zeta - 1}\right|,$$

it follows that

$$g_G(z, a) = -\log\left|\frac{\varphi(z) - \varphi(a)}{\overline{\varphi(a)}\varphi(z) - 1}\right|. \tag{4.45}$$

Now for real a the (upper) normal derivative in (4.44) is the same as the limit of the real part of i times the derivative of the analytic function

$$-\log\left(\frac{\varphi(z) - \varphi(a)}{\varphi(a)\varphi(z) - 1}\right)$$

as z approaches the point a from the upper half-plane. The latter derivative is

$$\frac{\varphi'(z)}{\varphi(z) - 1/\varphi(a)} - \frac{\varphi'(z)}{\varphi(z) - \varphi(a)} = \frac{-\varphi'(z)(\varphi(a) - 1/\varphi(a))}{(\varphi(z) - \varphi(a))(\varphi(z) - 1/\varphi(a))},$$

and here the denominator is $2\varphi(z)(z - a)$, while

$$\varphi'(z) = \frac{\varphi(z)}{\sqrt{z^2-1}}, \qquad \varphi(a) - \frac{1}{\varphi(a)} = 2\sqrt{a^2-1},$$

where we have to keep in mind that the last square root on the right-hand side is positive for $a > 1$ and negative for $a < -1$. Thus, for a measure ν supported on $\mathbf{R}$ for which there is no mass on the interval $[-1, 1]$, we obtain

$$\frac{d\widehat{\nu}}{dx}(x) = \frac{1}{\pi}\int \frac{\left|\sqrt{a^2-1}\right|}{|x-a|\sqrt{1-x^2}} d\nu(a), \quad x \in [-1, 1]. \tag{4.46}$$

More generally, if ν is a measure supported on $\mathbf{R}$ and having no mass on the compact interval $[\alpha, \beta]$, then the balayage of ν on $[\alpha, \beta]$ is given by

$$\frac{d\widehat{\nu}}{dx}(x) = \frac{1}{\pi}\int \frac{\left|\sqrt{(a-\alpha)(a-\beta)}\right|}{|x-a|\sqrt{(x-\alpha)(\beta-x)}} d\nu(a), \quad x \in [\alpha, \beta]. \tag{4.47}$$

From these representations one can immediately derive

Corollary 4.12. *Let $I \subset \mathbf{R}$ be a closed interval and let $\widehat{\nu}$ be the balayage onto I of a finite measure ν with compact support disjoint from I. Then $\widehat{\nu}$ is absolutely continuous (with respect to Lebesgue measure) and its density is a C^∞ function inside I. The same conclusion holds if the support of ν lies on $\mathbf{R}$ and $\nu(I) = 0$.*

Proof. Indeed, the first part of the corollary is a consequence of (4.41) if we use the formula (4.45) for the Green function, while the very last statement is immediate from (4.46). □

As an application of balayage we conclude this section by proving the following unicity theorem.

Theorem 4.13 (Carleson's Unicity Theorem). *Let K be a compact set of positive capacity, and Ω the unbounded component of $\overline{\mathbf{C}} \setminus K$. If μ and ν are two unit measures supported on $\partial\Omega$, and if the potentials U^μ and U^ν coincide in Ω, then $\mu = \nu$.*

Proof. We start the proof by the observation (see Corollary I.5.6) that $\partial\Omega$ is also the boundary of Ω in the fine topology. Hence, the continuity of logarithmic potentials in the fine topology and the assumptions of the theorem imply that

$$U^\mu(z) = U^\nu(z) \qquad \text{for all } z \in \overline{\Omega}. \tag{4.48}$$

If we can show that the same equality holds true in $\mathbf{C} \setminus \overline{\Omega}$, then $U^\mu \equiv U^\nu$, and the conclusion $\mu = \nu$ follows from Corollary 2.2.

Let $z_0 \in \mathbf{C}\setminus\overline{\Omega}$, and let R be that (connected) component of $\mathbf{C}\setminus\overline{\Omega}$ that contains z_0. It follows from Wiener's criterion Theorem I.4.6 that R is a regular domain for the Dirichlet problem. Hence, if $\widehat{\delta_{z_0}}$ is the balayage of δ_{z_0} onto ∂R, then

$$U^{\widehat{\delta_{z_0}}}(z) = U^{\delta_{z_0}}(z), \qquad z \in \partial R.$$

On integrating this identity with respect to μ we obtain (cf. (4.25))

$$U^\mu(z_0) = \int U^{\delta_{z_0}}\,d\mu = \int U^{\widehat{\delta_{z_0}}}\,d\mu = \int U^\mu\,d\widehat{\delta_{z_0}}.$$

If we write the same chain of equalities for the measure ν instead of μ, then the last integral will be the same because of (4.48) (recall that μ and ν are supported on $\partial\Omega$). Thus, the starting values $U^\mu(z_0)$ and $U^\nu(z_0)$ are also the same and this is what we had to prove. □

II.5 Green Potentials

We have seen in the preceding section that if G is a bounded domain and ν is a finite positive measure on G (i.e. $\nu(\overline{\mathbf{C}} \setminus G) = 0$), then the potential $U^{\widehat{\nu}}$ of its balayage measure on ∂G satisfies

$$U^\nu(z) - U^{\widehat{\nu}}(z) = 0 \ \text{ for quasi} - \text{every } z \notin G \tag{5.1}$$

(cf. Theorem 4.7). But what can be said about the difference $U^\nu - U^{\widehat{\nu}}$ in G? To answer this question we return to the representation (4.31) for the Green function $g_G(z, \zeta)$ of G:

$$g_G(z,\zeta) = \log\frac{1}{|z-\zeta|} - \int_{\partial G}\log\frac{1}{|z-t|}d\widehat{\delta_\zeta}(t), \tag{5.2}$$

where $\widehat{\delta_\zeta}$ is the balayage of the unit mass at ζ onto ∂G. Integrating this equation with respect to $d\nu(\zeta)$ and recalling from (4.13) and (4.29) that

$$\widehat{\nu} = \int \widehat{\delta_\zeta}\, d\nu(\zeta),$$

we obtain

$$\int g_G(z,\zeta)\, d\nu(\zeta) = U^\nu(z) - U^{\widehat{\nu}}(z), \quad z \in G. \tag{5.3}$$

The function on the left-hand side plays an essential role in the study of superharmonic functions on G. It is called the *Green potential* of ν and is denoted by U_G^ν.

More generally, if G is any domain possessing a Green function, we set

$$U_G^\nu(z) := \int g_G(z,\zeta)\, d\nu(\zeta), \quad z \in G, \tag{5.4}$$

for any positive measure ν on G. For example, when G is the open unit disk $D_1(0)$ (or the exterior of the closed unit disk), then

$$U_G^\nu(z) = \int \log\left|\frac{1-\overline{\zeta}z}{z-\zeta}\right| d\nu(\zeta), \tag{5.5}$$

and when G is the right half-plane $\{z \mid \operatorname{Re} z > 0\}$,

$$U_G^\nu(z) = \int \log\left|\frac{z+\overline{\zeta}}{z-\zeta}\right| d\nu(\zeta). \tag{5.6}$$

The following properties of Green potentials are easy consequences of the properties of Green functions and balayage measures.

Theorem 5.1. *If the domain $G \subset \overline{\mathbf{C}}$ possesses a Green function and ν is a finite positive measure on G with compact support in $\mathbf{C}$, then*

(i) $U_G^\nu \geq 0$ *in* G;

(ii) U_G^ν *is superharmonic in G and harmonic in* $G \setminus \operatorname{supp}(\nu)$;

(iii) *If G is bounded or $\infty \in \partial G$, then $U^\nu - U_G^\nu$ is harmonic in G; in the former case, this difference equals $U^{\widehat{\nu}}$, where $\widehat{\nu}$ is the balayage of ν onto ∂G. If $\infty \in G$, then $U^\nu - U_G^\nu$ is harmonic in $G \setminus \{\infty\}$ and equals $U^{\widehat{\nu}} - \int g_G(\zeta,\infty)\, d\nu(\zeta)$;*

(iv) *If ν has compact support in G, then $\lim_{z\to x, z\in G} U_G^\nu(z) = 0$ for q.e. $x \in \partial G$.*

We remark that in part (iii), we tacitly assume that the difference $U^\nu - U_G^\nu$ is suitably defined at points where both potentials are infinite.

Proof. Property (i) is immediate from the nonnegativity of $g_G(z,\zeta)$. Property (ii) can be easily verified directly or regarded as a consequence of the representations described in (iii).

Assertion (iii) for bounded domains G follows from (5.3) and the fact that $U^{\widehat{\nu}}$ is harmonic in G. If $\infty \in G$, the right-hand side of formula (5.2) must be modified by the addition of the function $g_G(\zeta,\infty)$ (cf. (4.31)), which leads to the representation

$$U_G^\nu(z) = U^\nu(z) - U^{\widehat{\nu}}(z) + \int g_G(\zeta,\infty)\,d\nu(\zeta), \tag{5.7}$$

where $\widehat{\nu}$ is the balayage of ν onto ∂G. For the case when $\infty \in \partial G$, select a point $z_0 \in G$ with $z_0 \notin \text{supp}(\nu)$. Under the inversion $z' = 1/(z-z_0)$, the domain G is mapped to a domain G' with $\infty \in G'$ and $\nu'(\zeta') = \nu(z_0 + 1/\zeta')$ is a finite measure with compact support in **C**. Since

$$g_G(z,\zeta) = g_{G'}\left(\frac{1}{z-z_0}, \frac{1}{\zeta - z_0}\right) = g_{G'}(z',\zeta'),$$

we have from (5.7) applied to G' that

$$U_G^\nu(z) = U_{G'}^{\nu'}(z') = U^{\nu'}(z') - U^{\widehat{\nu'}}(z') + \text{ const.} \tag{5.8}$$

But

$$U^{\nu'}(z') = \int \log\left[\frac{|z-z_0||\zeta - z_0|}{|z-\zeta|}\right] d\nu(\zeta) = U^\nu(z) + \|\nu\| \log|z-z_0| + \text{const.},$$

and

$$U^{\widehat{\nu'}}(z') = \int_{\partial G'} \log \frac{1}{|1-\zeta'(z-z_0)|}\, d\widehat{\nu'}(\zeta') + \|\nu\|\log|z-z_0|;$$

hence from (5.8) we see that

$$U^\nu(z) - U_G^\nu(z) = \int_{\partial G'} \log\frac{1}{|1-\zeta'(z-z_0)|}\, d\widehat{\nu'}(\zeta') + \text{const.},$$

which is harmonic in G.

Assertion (iv) follows from the representations in (iii) and the properties of the balayage measure (for the $\infty \in \partial G$ case, also apply the transformation $z \to 1/(z-z_0)$ as before). For example, if G is bounded, then since U^ν is continuous in a neighborhood of ∂G and $U^\nu(x) = U^{\widehat{\nu}}(x)$ for q.e. $x \in \partial G$, we have

$$\begin{aligned}
\limsup_{z\to x, z\in G} U_G^\nu(z) &= \limsup_{z\to x, z\in G} [U^\nu(z) - U^{\widehat{\nu}}(z)] \\
&= U^\nu(x) - \liminf_{z\to x, z\in G} U^{\widehat{\nu}}(z) \\
&\le U^\nu(x) - U^{\widehat{\nu}}(x) = 0,
\end{aligned}$$

for q.e. $x \in \partial G$. Together with the nonnegativity of U_G^ν we deduce that (iv) holds. □

Remark 5.2. If we do not assume that ν has finite mass, then U_G^ν can be identically infinite on G. However, if ν is a measure on G that is finite on compact subsets of G, then U_G^ν is either identically infinite on G or is superharmonic in G, as the following argument shows. Let G_n be an increasing sequence of open sets whose compact closures lie in G and $\cup G_n = G$. Set $g_n(z,\zeta) := \min\{g_G(z,\zeta), n\}$. Then for each $n = 1, 2, \ldots$,

$$u_n(z) := \int_{G_n} g_n(z,\zeta)\, d\nu(\zeta)$$

is continuous on G; indeed, if $z_j \to z_0 \in G$, then $g_n(z_j,\zeta) \to g_n(z_0,\zeta)$ and since

$$0 \le \int_{G_n} g_n(z_j,\zeta)\, d\nu(\zeta) \le n\nu(G_n) < \infty,$$

we can deduce from the Lebesgue dominated convergence theorem that $u_n(z_j) \to u_n(z_0)$ as $j \to \infty$. Furthermore, $u_n \le u_{n+1}$ on G and, by the monotone convergence theorem,

$$\lim_{n\to\infty} u_n(z) = U_G^\nu(z), \quad z \in G.$$

Thus U_G^ν is lower semi-continuous on G. The function U_G^ν also satisfies the super-mean value inequality property in G since it is the limit of the increasing sequence of superharmonic functions $U_G^{\nu_n}$, where $\nu_n := \nu|_{\overline{G}_n}$. Hence, if U_G^ν is not identically infinite, then it is superharmonic in G.

From the representations for Green potentials in Theorem 5.1 (iii) it is easy to deduce analogues of many of the basic results previously obtained for logarithmic potentials, such as the principle of descent, the lower envelope theorem, the continuity theorem, and the unicity theorem. For example, we prove

Theorem 5.3 (Unicity Theorem for Green Potentials). *Suppose the domain $G \subset \overline{\mathbf{C}}$ possesses a Green function and μ, ν are finite positive measures on G with compact supports in $\mathbf{C}$. If U_G^μ and U_G^ν satisfy*

$$U_G^\mu(z) = U_G^\nu(z) + h(z) \tag{5.9}$$

almost everywhere in G with respect to two-dimensional Lebesgue measure, where h is harmonic in G, then $\mu = \nu$.

Proof. Assume that G is bounded; the proof in the contrary case is similar. Then from (5.9) and Theorem 5.1 (iii) we have

$$U^\mu(z) - U^{\widehat{\mu}}(z) = U^\nu(z) - U^{\widehat{\nu}}(z) + h(z)$$

or

$$U^{\mu+\widehat{\nu}}(z) = U^{\widehat{\mu}+\nu}(z) + h(z)$$

m_2-almost everywhere on G, where $\widehat{\mu}, \widehat{\nu}$ are the balayage measures of μ, ν respectively onto ∂G. Hence by the unicity result for logarithmic potentials (Theorem 2.1), we obtain

$$\mu = (\mu + \widehat{\nu})\big|_G = (\widehat{\mu} + \nu)\big|_G = \nu.$$

□

As with logarithmic potentials, the notion of energy plays a crucial role in the study of Green potentials. Suppose that the domain $G \subset \overline{\mathbf{C}}$ possesses a Green function and that μ and ν are two positive measures on G. Then the *mutual Green energy* of μ and ν is defined by

$$(\mu, \nu)_e := \int U_G^\mu \, d\nu = \int U_G^\nu \, d\mu. \tag{5.10}$$

Notice that $(\mu, \nu)_e = (\nu, \mu)_e \geq 0$ and that $(\mu, \nu)_e$ may be infinite. The quantity

$$||\mu||_e^2 := (\mu, \mu)_e = \int\int g_G(z, \zeta) \, d\mu(z) \, d\mu(\zeta) \tag{5.11}$$

is called the *Green energy* of μ. If μ is a finite positive measure with compact support in G, then it is easy to see from the representations in Theorem 5.1 (iii) that μ has finite Green energy if and only if μ has finite logarithmic energy.

Lemma 5.4. *Suppose μ, ν are finite positive measures on G having compact support in* **C**. *If μ and ν have finite Green energies, then*

(i) $(\mu, \nu)_e \leq ||\mu||_e ||\nu||_e < \infty$;
(ii) $\mu + \nu$ *has finite Green energy.*

Proof. We give the proof only for the case when G is bounded. Assume at first that μ and ν have compact supports in G. Since μ has finite Green energy, U_G^μ is finite μ-a.e. Furthermore, μ is a finite measure, and so Lusin's continuity theorem can be applied to U_G^μ on $S := \operatorname{supp}(\mu)$. Thus, for each $k = 1, 2, \ldots,$ there exists a compact subset $S_k \subset S$ such that the restriction of U_G^μ to S_k is continuous on S_k and $\mu(S \setminus S_k) < 1/k$. Let μ_k be the restriction of μ to S_k. Then, as in the proof of Lemma I. 6.10, it follows that $U_G^{\mu_k}$ restricted to S_k is continuous on S_k, and hence $U_G^{\mu_k}$ is continuous on G (here we use the analogue of the continuity theorem for logarithmic potentials, Theorem 3.5).

Next, observe that for each k,

$$0 \leq (\mu_k, \nu)_e = \int U_G^{\mu_k} \, d\nu \leq M_k \nu(\mathbf{C}) < \infty,$$

where $M_k := \max\{U_G^{\mu_k}(z) | \, z \in \operatorname{supp}(\nu)\}$. Thus for each $\varepsilon > 0$, the Green energy of the measure $\mu_k + \varepsilon\nu$,

$$||\mu_k + \varepsilon\nu||_e^2 = ||\mu_k||_e^2 + 2\varepsilon(\mu_k, \nu)_e + \varepsilon^2 ||\nu||_e^2,$$

is finite. Consequently, for the signed measure $\lambda := \mu_k - \varepsilon\nu$, the integral

$$J_{k,\varepsilon} := \int g_G(z, \zeta) \, d\lambda(z) \, d\lambda(\zeta)$$

is absolutely convergent. We now show that $J_{k,\varepsilon} \geq 0$.

Let $\widehat{\mu}_k$ and $\widehat{\nu}_\varepsilon$ denote, respectively, the balayage measures of μ_k and $\varepsilon\nu$ onto ∂G, and set $\widehat{\lambda} := \widehat{\mu}_k - \widehat{\nu}_\varepsilon$. Then, from Theorem 5.1 (iii), we have

$$J_{k,\varepsilon} = \int \left[\int g_G(z,\zeta)\, d\lambda(z) \right] d\lambda(\zeta) = \int [U^{\lambda}(\zeta) - U^{\widehat{\lambda}}(\zeta)]\, d\lambda(\zeta)$$

$$= \int U^{\lambda-\widehat{\lambda}}(\zeta)\, d\lambda(\zeta) = \int U^{\lambda-\widehat{\lambda}}(\zeta)\, d(\lambda - \widehat{\lambda})(\zeta), \tag{5.12}$$

where the last equality follows from the fact that $U^{\lambda-\widehat{\lambda}}(\zeta) = 0$ q.e. on ∂G, and hence $|\widehat{\lambda}|$ - a.e., since $|\widehat{\lambda}| \le \widehat{\mu}_k + \widehat{\nu}_\varepsilon$ has finite logarithmic energy. Now observe that $(\lambda - \widehat{\lambda})(\mathbf{C}) = 0$ and so Lemma I.1.8 applies to the last integral in (5.12). Thus $J_{k,\varepsilon} \ge 0$.

Next, from the definition of λ, we have the representation

$$J_{k,\varepsilon} = ||\mu_k||_e^2 - 2\varepsilon(\mu_k, \nu)_e + \varepsilon^2||\nu||_e^2.$$

We have shown that this quadratic in ε is nonnegative for $\varepsilon \ge 0$ and since it is clearly nonnegative for $\varepsilon < 0$, it follows that

$$(\mu_k, \nu)_e \le ||\mu_k||_e||\nu||_e \le ||\mu||_e||\nu||_e.$$

Thus, from Theorem 0.1.4, we obtain

$$(\mu, \nu)_e = \int U_G^\nu\, d\mu \le \liminf_{k\to\infty} \int U_G^\nu\, d\mu_k \le ||\mu||_e||\nu||_e,$$

which proves assertion (i) for the case when μ and ν have compact supports in G. The general case then follows by taking an increasing sequence of compact subsets that exhaust G.

Finally, since $||\mu+\nu||_e^2 = ||\mu||_e^2+2(\mu,\nu)_e+||\nu||_e^2$, assertion (ii) is an immediate consequence of (i). □

We now use Lemma 5.4 to show that the mutual Green energy provides us with an inner product on a linear space of signed measures.

Definition 5.5. Suppose $G \subset \overline{\mathbf{C}}$ is a domain possessing a Green function. Let $\mathcal{E}^+$ note the set of finite positive Borel measures on G having compact support in $\mathbf{C}$ and finite Green energy. By Lemma 5.4, $\mathcal{E}^+$ is closed under addition of measures. Furthermore, it is closed under multiplication by nonnegative (real) scalars. Let

$$\mathcal{E} := \{\sigma \mid \sigma = \mu - \nu, \quad \mu \in \mathcal{E}^+, \ \nu \in \mathcal{E}^+\}.$$

Then $\mathcal{E}$ is a linear vector space over the reals. If $\sigma_i = \mu_i - \nu_i \in \mathcal{E}$, with $\mu_i, \nu_i \in \mathcal{E}^+$, $i = 1, 2$, the *mutual Green energy* $(\sigma_1, \sigma_2)_e$ is defined by

$$(\sigma_1, \sigma_2)_e := \int\int g_G(z,\zeta)\, d\sigma_1(z)\, d\sigma_2(\zeta)$$

$$= (\mu_1, \mu_2)_e - (\mu_1, \nu_2)_e - (\nu_1, \mu_2)_e + (\nu_1, \nu_2)_e, \tag{5.13}$$

which is consistent with the previously defined mutual energy for positive measures.

From Lemma 5.4, the integral in (5.13) is absolutely convergent and the mutual energy is independent of the representations for σ_1 and σ_2. It is also clear that $(\cdot,\cdot)_e$ is a bilinear form on $\mathcal{E}\times\mathcal{E}$ and we now show that it is an inner product.

Theorem 5.6. *If $\sigma\in\mathcal{E}$, then*

$$(\sigma,\sigma)_e=\int g_G(z,\zeta)\,d\sigma(z)\,d\sigma(\zeta)\geq 0, \tag{5.14}$$

with equality if and only if σ is the zero measure. Consequently, $(\cdot,\cdot)_e$ is an inner product on $\mathcal{E}\times\mathcal{E}$.

Notice that here we do *not* assume that $\sigma(G)=0$.

Proof. Write $\sigma=\mu-\nu$ with $\mu,\nu\in\mathcal{E}^+$. Then, by Lemma 5.4 (i), we have

$$\begin{aligned}(\sigma,\sigma)_e=||\mu||_e^2-2(\mu,\nu)_e+||\nu||_e^2&\geq||\mu||_e^2-2||\mu||_e||\nu||_e+||\nu||_e^2\\&=(||\mu||_e-||\nu||_e)^2\geq 0,\end{aligned}$$

which proves (5.14).

Next suppose that $(\sigma,\sigma)_e=0$. Let $\tau\in\mathcal{E}$ and $c\in\mathbf{R}$. Then $\sigma+c\tau\in\mathcal{E}$ and so from (5.14) (with σ replaced by $\sigma+c\tau$) we have

$$(\sigma+c\tau,\sigma+c\tau)_e=(\sigma,\sigma)_e+2c(\sigma,\tau)_e+c^2(\tau,\tau)_e=2c(\sigma,\tau)_e+c^2(\tau,\tau)_e\geq 0.$$

Since the last inequality holds for all $c\in\mathbf{R}$, we deduce that $(\sigma,\tau)_e=0$; that is,

$$(\mu,\tau)_e=(\nu,\tau)_e\quad\text{for all }\tau\in\mathcal{E}. \tag{5.15}$$

For a fixed $x\in G$, let $\tau=\tau_r$ be normalized Lebesgue measure on the circle centered at x with radius r whose closed interior is contained in G. Clearly, $\tau_r\in\mathcal{E}$, and since the Green potential U_G^μ is superharmonic in G, it follows from formula (1.6) in Theorem 1.2 that

$$U_G^\mu(x)=\lim_{r\to 0}\int U_G^\mu(z)\,d\tau_r(z)=\lim_{r\to 0}(\mu,\tau_r)_e. \tag{5.16}$$

(As remarked in the notes for Section II.1 at the end of this chapter, equation (1.6) is valid not only for logarithmic potentials, but for all superharmonic functions.) As (5.16) also holds with μ replaced by ν, we deduce from (5.15) that $U_G^\mu(x)=U_G^\nu(x)$ for all $x\in G$. Hence by Theorem 5.3 we have $\mu=\nu$; that is $\sigma=\mu-\nu=0$. □

Since $(\cdot,\cdot)_e$ is an inner product on $\mathcal{E}\times\mathcal{E}$,

$$||\sigma||_e:=\sqrt{(\sigma,\sigma)_e} \tag{5.17}$$

serves as a norm on $\mathcal{E}$. Consequently, the triangle inequality and the Cauchy-Schwarz inequality are valid.

Corollary 5.7. *If $\sigma, \lambda \in \mathcal{E}$, then*

$$||\sigma + \lambda||_e \leq ||\sigma||_e + ||\lambda||_e \text{ and } |(\sigma, \lambda)_e| \leq ||\sigma||_e ||\lambda||_e.$$

We remark that with the mutual Green energy as an inner product, $\mathcal{E}$ is not a Hilbert space because it lacks the completeness property. However, H. Cartan has shown that $\mathcal{E}^+$ is complete with respect to the energy norm provided we also include in $\mathcal{E}^+$ infinite positive measures of finite Green energy that are finite on compact subsets of G (see e.g. Helms [77], Chapter 11).

The principle of domination for Green potentials takes the following form.

Theorem 5.8 (Principle of Domination). *Let ν be a positive Borel measure on a domain $G \subset \overline{\mathbb{C}}$ having a Green function and suppose that ν is finite on compact subsets of G. Assume further that ν has finite Green energy. If, for some nonnegative superharmonic function f on G, the inequality*

$$U_G^{\nu}(z) \leq f(z) \tag{5.18}$$

holds ν-almost everywhere, then it holds for all $z \in G$.

Proof. Let G_n be an increasing sequence of open sets with compact closures $\overline{G}_n \subset G$ and $G = \cup G_n$. Set $\nu_n = \nu|_{\overline{G}_n}$ and observe from (5.18) that

$$U_G^{\nu_n}(z) \leq f(z) \tag{5.19}$$

holds ν_n - a.e. for each n. Furthermore, ν_n has finite Green energy, and so $U_G^{\nu_n}$ is finite ν_n - a.e.

Let n be fixed and set $S := \text{supp}(\nu_n)$. Then there exists a subset $E \subset S$ such that $\nu_n(E) = 0$ and (5.19) holds for all $z \in S \setminus E$. Because ν_n is a finite measure and $U_G^{\nu_n}$ is finite ν_n - a.e., Lusin's continuity theorem can be applied to $U_G^{\nu_n}$ on S. Thus there exists a compact subset $S_k \subset S$ such that the restriction of $U_G^{\nu_n}$ to S_k is continuous on S_k and $\nu_n(S \setminus S_k) < 1/k$, $k = 1, 2, \ldots$. Since $\nu_n(E) = 0$, we can further choose a compact subset $S_k^* \subset S_k \setminus E$ so that $\nu_n(S_k \setminus S_k^*) < 1/k$. Let ν_k^* be the restriction of ν_n to S_k^*. Then, as in the proof of Lemma 5.4, it follows that $U_G^{\nu_k^*}$ is continuous on G. Furthermore,

$$U_G^{\nu_k^*}(z) \leq f(z) \quad \text{for all } z \in S_k^*. \tag{5.20}$$

Next, let $H_k(z) := \min(f(z), U_G^{\nu_k^*}(z))$ for all $z \in G$. Then H_k is nonnegative and superharmonic in G, and we shall show that $H_k \equiv U_G^{\nu_k^*}$. Indeed, from Theorem 5.1(iv), it follows that

$$\liminf_{z \to x, z \in G} [H_k(z) - U_G^{\nu_k^*}(z)] \geq 0 \quad \text{for q.e. } x \in \partial G.$$

Also, from the lower semi-continuity of H_k and the continuity of $U_G^{\nu_k^*}$, we have for all $x \in \partial S_k^*$,

$$\liminf_{z\to x, z\in G\setminus S_k^*}[H_k(z) - U_G^{\nu_k^*}(z)] \geq H_k(x) - U_G^{\nu_k^*}(x) = 0,$$

where the last equality follows from (5.20). Since $H_k - U_G^{\nu_k^*}$ is superharmonic and bounded from below on $G \setminus S_k^*$, the generalized minimum principle (Theorem I.2.4) yields

$$H_k(z) - U_G^{\nu_k^*}(z) \geq 0 \quad \text{for } z \in G \setminus S_k^*.$$

Hence, from the definition of H_k, we have $H_k = U_G^{\nu_k^*}$ on $G \setminus S_k^*$ which, together with (5.20), yields $H_k = U_G^{\nu_k^*}$ in G.

We have thus shown that (5.20) holds for all $z \in G$, and since $\nu_k^* \to \nu_n$ as $k \to \infty$, inequality (5.19) is valid in G by the principle of descent. Finally, letting $n \to \infty$ in (5.19), we deduce that (5.18) also holds everywhere in G. □

As an immediate consequence, we get the following *maximum principle for Green potentials.*

Corollary 5.9. *Let G and ν be as in Theorem* 5.8. *If $U_G^\nu(z) \leq M$ holds ν almost everywhere, then $U_G^\nu(z) \leq M$ for all $z \in G$.*

Indeed, the hypotheses of the corollary imply that the measure ν_n in the preceding proof has finite Green energy and satisfies $U_G^{\nu_n}(z) \leq M$, ν_n-almost everywhere.

We now turn to the weighted energy problem for Green potentials. Let G be a domain possessing a Green function and E a closed subset of G. If $w = e^{-Q}$ is an admissible weight on E (cf. Definition I.1.1), then the *weighted Green energy* for a positive measure μ supported on E is defined by

$$I_w^G(\mu) := \int\int g_G(z,\zeta)\,d\mu(z)\,d\mu(\zeta) + 2\int Q\,d\mu = \|\mu\|_e^2 + 2\int Q\,d\mu. \tag{5.21}$$

The analogue of Theorem I.1.3 is then the following.

Theorem 5.10. *Let $w = e^{-Q}$ be an admissible weight on the closed subset $E \subset G$ and set*

$$V_w^G := \inf\{I_w^G(\mu) \mid \mu \in \mathcal{M}(E)\},$$

where $\mathcal{M}(E)$ denotes the set of probability measures on E. Then the following hold.

(i) *V_w^G is finite.*
(ii) *There is a unique measure $\mu_w^G \in \mathcal{M}(E)$ such that*

$$I_w^G(\mu_w^G) = V_w^G.$$

Moreover, μ_w^G has compact support S_w^G and finite Green energy.

(iii) *Setting*

$$F_w^G := V_w^G - \int Q\, d\mu_w^G = \|\mu_w^G\|_e^2 + \int Q\, d\mu_w^G, \tag{5.22}$$

we have

$$U_G^{\mu_w^G}(z) + Q(z) \geq F_w^G \quad q.e.\ on\ E. \tag{5.23}$$

(iv) *For all* $z \in S_w^G$,

$$U_G^{\mu_w^G}(z) + Q(z) \leq F_w^G. \tag{5.24}$$

The measure $\mu_w^G$is called the *Green equilibrium* (or *extremal*) *measure* associated with w. In the unweighted case, when E is compact, $\text{cap}(E) > 0$, and $Q \equiv 0$ on E, we replace the subscript w by E and call μ_E^G the *Green equilibrium measure for the set* E. For this measure we have the following additional properties.

Theorem 5.11. *Let* $E \subset G$ *be compact with* $\text{cap}(E) > 0$, *and let* μ_E^G *be the Green equilibrium measure for* E. *Then*

(i) $U_G^{\mu_E^G}(z) = V_E^G$ *q.e. on* E;

(ii) $U_G^{\mu_E^G}(z) \leq V_E^G$ *for all* $z \in G$.

The constant $C(E, \partial G) := 1/V_E^G$ is called the *capacity of the condenser* $(E, \partial G)$; see also Sections VIII.2 and VIII.3.

The proofs of the above results follow via the same reasoning used in the case of logarithmic potentials. Although we omit the details, we remark that in establishing the uniqueness of the extremal measure we appeal to Theorem 5.6.

To determine Green equilibrium measures, the following result is useful. It is the analogue of Theorem I.3.3, but its proof is slightly different.

Theorem 5.12. *Let* $w = e^{-Q}$ *be an admissible weight on the closed subset* $E \subset G$. *If* $\nu \in \mathcal{M}(E)$ *has compact support and finite Green energy, and the function*

$$U_G^\nu(z) + Q(z)$$

coincides with a constant F *quasi-everywhere on* $\text{supp}(\nu)$ *and is at least as large as* F *quasi-everywhere on* E, *then* $\nu = \mu_w^G$ *and* $F = F_w^G$.

Proof. Set $\mu := \mu_w^G$, $S := \text{supp}(\mu_w^G)$, and $F' := F_w^G$. We first show that $F' = F$. The hypotheses together with Theorem 5.10 imply that quasi-everywhere on S

$$U_G^\nu(z) + Q(z) - F \geq U_G^\mu(z) + Q(z) - F',$$

and since Q is finite on S,

$$U_G^\nu(z) + F' \geq U_G^\mu(z) + F \quad \text{q.e. on } S. \tag{5.25}$$

Let μ_S^G be the Green equilibrium measure for the set S. Then integrating (5.25) with respect to μ_S^G yields

$$(\nu, \mu_S^G)_e + F' \geq (\mu, \mu_S^G)_e + F.$$

But, from Theorem 5.11,

$$(\mu, \mu_S^G)_e = \int U_G^{\mu_S^G} d\mu = V_S^G$$

and

$$(\nu, \mu_S^G)_e = \int U_G^{\mu_S^G} d\nu \le V_S^G;$$

hence $V_S^G + F' \ge V_S^G + F$, that is $F' \ge F$. Reversing the roles of μ and ν, we obtain by the same reasoning that $F \ge F'$. Thus $F = F'$.

To prove that $\mu = \nu$, observe that (5.25) becomes $U_G^\nu \ge U_G^\mu$ quasi-everywhere (and hence μ - a.e.) on $S = \text{supp}(\mu)$. Thus, from the principle of domination (Theorem 5.8) this inequality holds everywhere in G. Likewise, it follows that $U_G^\mu \ge U_G^\nu$ everywhere in G, and so from the unicity theorem, $\nu = \mu$. □

We now present two examples of Green equilibrium measures for the so-called hyperbolic case when G is the open unit disk.

Example 5.13. Let $G = D_1(0)$ be the open unit disk, E the concentric closed disk of radius $r < 1$, and $w \equiv 1$. Then μ_E^G is simply the normalized Lebesgue measure $d\sigma_r = d\theta/2\pi$ on the circumference $|z| = r$. Indeed, from Example I.3.4 and (5.5) we have for all $z \in E$

$$U_G^{\sigma_r}(z) = \int \log \frac{1}{|z-\zeta|} d\sigma_r(\zeta) + \int \log|1 - \zeta\bar{z}|\, d\sigma_r(\zeta) = \log\frac{1}{r} + 0 = \log\frac{1}{r},$$

and so $U_G^{\sigma_r}$ is constant on E. Hence, by Theorem 5.12, $\mu_E^G = \sigma_r$. □

Example 5.14. Let $G = D_1(0)$, $E = [-a, a] \subset \mathbf{R}$ with $0 < a < 1$, and $w \equiv 1$. Then

$$d\mu_E^G(x) = \frac{1}{2K} \frac{dx}{\sqrt{(a^2 - x^2)(1 - a^2x^2)}}, \quad x \in [-a, a], \tag{5.26}$$

where

$$K := \int_0^1 \frac{dt}{\sqrt{(1-t^2)(1-a^4t^2)}}.$$

To verify this formula, one can easily check that the measure μ defined by the right-hand side of (5.26) is a unit measure on $E = [-a, a]$ and that its Green potential U_G^μ is continuous on $G = D_1(0)$. We shall show that U_G^μ is constant on $[-a, a]$, so that $\mu = \mu_E^G$ will follow from Theorem 5.12. For this purpose, we first observe that

$$U_G^\mu(z) = \int_{-a}^{a} \log\left|\frac{1-tz}{z-t}\right| d\mu(t) = U^\mu(z) + \int_{-a}^{a} \log|1 - tz|\, d\mu(t)$$

$$= U^\mu(z) - \int_{s^{-1}\in[-a,a]} \log\frac{1}{|z-s|} d\mu(1/s) + \int_{s^{-1}\in[-a,a]} \log\frac{1}{|s|} d\mu(1/s),$$

and so

$$U_G^{\mu}(z) = U^{\sigma}(z) + \text{const.}, \tag{5.27}$$

where the signed measure σ is given by

$$d\sigma(t) = h(t)dt = \begin{cases} \frac{1}{2K}[|(a^2 - t^2)(1 - a^2t^2)|]^{-\frac{1}{2}}dt, & t \in [-a, a] \\ \frac{-1}{2K}[|(a^2 - t^2)(1 - a^2t^2)|]^{-\frac{1}{2}}dt, & t \in (-\infty, -\frac{1}{a}] \cup [\frac{1}{a}, \infty). \end{cases}$$

Let $\Gamma = (-\infty, -1/a] \cup [-a, a] \cup [1/a, \infty)$, oriented with increasing t, and set

$$F(z) := \int_{\Gamma} \log \frac{1}{z - t} h(t)dt, \quad \operatorname{Im}(z) > 0.$$

For a suitable branch of the logarithm, F is analytic in the upper-half plane and we have

$$U^{\sigma}(z) = \operatorname{Re} F(z) \quad \text{for } \operatorname{Im}(z) > 0. \tag{5.28}$$

Furthermore,

$$F'(z) = \int_{\Gamma} \frac{h(t)}{t - z} dt, \tag{5.29}$$

is analytic in $\mathbf{C} \setminus \Gamma$. To evaluate this last integral, let $f(z)$ denote the branch of $(a^2 - z^2)^{1/2}$ analytic in $\mathbf{C} \setminus [-a, a]$ that behaves like $-iz$ as $z \to \infty$, and set

$$H(z) := \frac{1}{iz\, f(z) f(1/z)}, \quad \text{for } z \in \mathbf{C} \setminus \Gamma.$$

Clearly H is analytic in $\mathbf{C} \setminus \Gamma$ and it is straightforward to verify that for each $t \in \Gamma$

$$\lim_{z \to t+i0} H(z) = h(t), \quad \lim_{z \to t-i0} H(z) = -h(t), \tag{5.30}$$

where $z \to t + i0$ ($z \to t - i0$) denotes that z approaches t from the upper-half (lower-half) plane. Since $H(z) \to 0$ as $|z| \to \infty$, it then follows from the Cauchy integral formula (after deforming the contour of integration so as to consist of the "upper half" and "lower half" of Γ) that

$$H(z) = \frac{2}{2\pi i} \int_{\Gamma} \frac{h(t)}{t - z} dt, \quad \text{for } \operatorname{Im}(z) > 0,$$

and so from (5.29) we have

$$F'(z) = \pi i H(z), \quad \text{for } \operatorname{Im}(z) > 0.$$

Finally, we observe from this formula and (5.30) that $F'(z)$ has a continuous extension to the closed upper-half plane and that

$$\lim_{z \to t+i0} \operatorname{Re} F'(z) = \operatorname{Re}\{\pi i h(t)\} = 0, \quad \text{for } t \in \Gamma.$$

Consequently, for each $x \in [-a, a]$,

$$\lim_{z\to x+i0} \operatorname{Re} F(z) = \lim_{z\to x+i0} \operatorname{Re}\left(\int_0^z F'(\zeta)d\zeta + F(0)\right) = \operatorname{Re} F(0),$$

so that from (5.28), (5.27), and the continuity of the potential it follows that $U_G^{\mu}(x)$ is constant for $x \in [-a, a]$. □

The next example applies to arbitrary domains G possessing a Green function and is the analogue of Example II.4.8.

Example 5.15. Let ν be a unit measure with compact support $E \subset G$ such that U_G^{ν} is continuous on G. If $w = \exp(c\, U_G^{\nu})$ on E, where $0 < c < 1$, then

$$\mu_w^G = c\nu + (1-c)\mu_E^G.$$

Indeed $\lambda := c\nu + (1-c)\mu_E^G \in \mathcal{M}(E)$ has finite Green energy and, by Theorem 5.11,

$$U_G^{\lambda} + Q = U_G^{\lambda} - c\, U_G^{\nu} = (1-c)U_G^{\mu_E^G}$$

is constant for q.e. $z \in E$. Hence $\lambda = \mu_w^G$ follows from Theorem 5.12. □

In Theorem 5.10, the infimum of the weighted energy was taken only over probability measures supported on E. If, instead, we allow measures of arbitrary mass, we obtain the following result which is known as the Gauss-Frostman theorem.

Theorem 5.16. *In Theorem* 5.10, *assume that E is compact and set*

$$v_w := \inf_{\tau} J(\tau), \quad J(\tau) := \|\tau\|_e^2 + 2\int Q\, d\tau, \tag{5.31}$$

where the infimum is taken over all positive Borel measures τ *with* $\operatorname{supp}(\tau) \subset E$. *Then* v_w *is finite and there exists a unique measure* $\tau_0 \in \mathcal{E}^+$ *on E such that* $J(\tau_0) = v_w$. *Furthermore,*

$$U_G^{\tau_0}(z) + Q(z) \geq 0 \ \ q.e. \ on\ E, \tag{5.32}$$

and

$$U_G^{\tau_0}(z) + Q(z) \leq 0 \ \ for\ all\ z \in \operatorname{supp}(\tau_0). \tag{5.33}$$

Recall that the set $\mathcal{E}^+$ is described in Definition 5.5.

Proof. Since the zero measure is allowed, we have $v_w \leq 0$. To see that $v_w > -\infty$, let

$$m := \min_{z\in E} Q(z), \quad \kappa := \min_{E\times E} g_G(z,\zeta) > 0$$

(both minima exist because of lower semi-continuity). Then for any positive measure τ we have

$$J(\tau) \geq \kappa[\tau(E)]^2 + 2m\tau(E) = \kappa\left(\tau(E) + \frac{m}{\kappa}\right)^2 - \frac{m^2}{\kappa} \geq -\frac{m^2}{\kappa}, \tag{5.34}$$

and so v_w is finite.

If $m \geq 0$, then clearly $v_w = 0$ and the zero measure is a minimizing measure. If $m < 0$, then it follows from (5.34) that $J(\tau) > 0 \geq v_w$ whenever $\tau(E) > -2m/\kappa$. Hence

$$v_w = \inf\{J(\tau) \mid \tau(E) \leq -2m/\kappa, \quad \operatorname{supp}(\tau) \subset E\}.$$

With respect to the weak* topology of measures on E, $J(\tau)$ is lower semi-continuous and hence attains its infimum over the set of measures τ with $\tau(E) \leq -2m/\kappa$. Thus there exists a τ_0 on E such that $J(\tau_0) = v_w$ and, clearly, $\tau_0 \in \mathcal{E}^+$.

Next, let $|\varepsilon| < 1$. Since $J((1+\varepsilon)\tau_0) \geq J(\tau_0)$, we deduce that

$$\varepsilon^2 ||\tau_0||_e^2 + 2\varepsilon \left(||\tau_0||_e^2 + \int Q \, d\tau_0 \right) \geq 0.$$

Since the sign of ε is arbitrary, it follows that

$$||\tau_0||_e^2 + \int Q \, d\tau_0 = 0. \tag{5.35}$$

Furthermore, if τ_0 is not the zero measure, then $\tau_0^* := \tau_0/\tau_0(E)$ is the probability measure on E that minimizes the weighted Green energy for the weight $w^* := e^{-Q/\tau_0(E)}$, that is, $\tau_0^* = \mu_{w^*}^G$. Hence, from Theorem 5.10, it follows that

$$U_G^{\tau_0}(z) + Q(z) \geq \tau_0(E) F_{w^*}^G \quad \text{q.e. on } E \tag{5.36}$$

and

$$U_G^{\tau_0}(z) + Q(z) \leq \tau_0(E) F_{w^*}^G \quad \text{for all } z \in \operatorname{supp}(\tau_0). \tag{5.37}$$

But from (5.22) and (5.35) we have

$$F_{w^*}^G = ||\tau_0/\tau_0(E)||_e^2 + \int \frac{Q}{\tau_0(E)} \, d\tau_0/\tau_0(E) = 0.$$

Thus, if $\tau_0(E) > 0$, inequalities (5.32) and (5.33) follow, respectively, from (5.36) and (5.37).

For the case when τ_0 is the zero measure, we only need to verify (5.32), that is, to show $Q(z) \geq 0$ q.e. on E. In this case, for every measure $\tau \in \mathcal{E}^+$ on E, we have

$$J(\delta\tau) = \delta^2 ||\tau||_e^2 + 2\delta \int Q d\tau \geq v_w = 0$$

for every $\delta > 0$, and so

$$\int Q d\tau \geq 0 \quad \text{for all } \tau \in \mathcal{E}^+. \tag{5.38}$$

If the subset of E where Q is negative has positive capacity, then for some positive interger n, the compact set $E_n := \{z \in E \mid Q(z) \leq -1/n\}$ also has positive capacity. But then, for the Green equilibrium measure $\mu_{E_n}^G$ for the set E_n, we have $\int Q d\mu_{E_n}^G < 0$, which contradicts (5.38).

To complete the proof, it remains to show that τ_0 is unique. If τ_1 is another minimizing measure, then $\tau_1 \in \mathcal{E}^+$ and, as above, the inequalities (5.32) and (5.33) hold for τ_1 as well. Thus, for quasi-every $z \in \operatorname{supp}(\tau_1)$,

$$U_G^{\tau_1}(z) + Q(z) \leq 0 \leq U_G^{\tau_0} + Q(z),$$

that is, $U_G^{\tau_1} \leq U_G^{\tau_0}$ q.e. on the support of τ_1 (note that Q is finite q.e. on $\operatorname{supp}(\tau_1)$). By the principle of domination (Theorem 5.8), this inequality holds everywhere in G, and so, by interchanging τ_1 and τ_0, we deduce that $U_G^{\tau_1} = U_G^{\tau_0}$ for all z in G. Hence, from the unicity theorem, we get that $\tau_1 = \tau_0$. □

Remark 5.17. If $Q(z)$ is a negative constant function on E, say $Q(z) \equiv -c$, with $c > 0$, then the minimizing measure τ_0 is just $c\mu_E^G / V_E^G$. Indeed, from Theorem 5.11, the inequalities (5.32) and (5.33) also hold for the measure $c\mu_E^G / V_E^G$, and so $\tau_0 = c\mu_E^G / V_E^G$ follows by the argument used in the last step of the above proof.

II.6 Notes and Historical References

Many of the theorems in this section are folklore, for which it is easier to find a formulation and proof than the original source.

Section II.1

The function $L(U^\mu; z_0, r)$ (see Theorem 1.2) is, in fact, a concave function of $\log r$ which follows from the observation that the function $\log_r(z) := \min(\log 1/r, \log 1/|z|)$ is a concave function of $\log |z|$ (cf. (1.8)).

Theorem 1.2 holds, with only slight modification, for arbitrary superharmonic functions f in place of U^μ; that is for

$$L(f; z_0, r) := \frac{1}{2\pi} \int_{-\pi}^{\pi} f(z_0 + re^{i\theta})\, d\theta.$$

This observation follows from the Riesz decomposition theorem (see Theorem 3.1 in Section II.3).

Along with $L(f; z_0, r)$ it is also customary to consider

$$m(f; z_0, r) := \inf_{|z-z_0|=r} f(z).$$

Like $L(f; z_0, r)$, the function $m(f, z_0, r)$ is a non-increasing concave function of $\log r$. In fact, set

$$m_f(z) := \inf_t f(z_0 + e^{it}(z - z_0)).$$

Since f is uniformly lower semi-continuous on compact subsets of its domain of definition, a compactness argument gives that m_f is also lower semi-continuous. But then one can easily see that m_f, being the infimum of superharmonic functions, is itself superharmonic. Finally,

$$m(f; z_0, r) = L(m_f; z_0, r),$$

so the concavity follows from that of L.

The formula (see Theorem 1.3)

$$\mu = -\frac{1}{2\pi} \Delta U^\mu \, dm,$$

holds true for arbitrary μ of compact support provided the right-hand side is understood in the distributional sense; see the notes below concerning the Riesz decomposition theorem in Section II.3.

Theorems 1.4 and 1.5 are analogous to results known for the Cauchy transform

$$\tilde{\mu}(z) := \int \frac{d\mu(t)}{t - z};$$

see, for example, the Stieltjes-Perron inversion formula [78, Theorem 12.10d], and the Sokhotskyi formulas [79, Section 14.1].

Section II.2

An alternative proof for Theorem 2.1 follows from Theorem 1.2. In fact, the latter one easily implies that μ and ν take the same value on $D_r(z_0)$ for almost all r provided $\overline{D_r(z_0)} \subseteq D$. This implies the same conclusion for every such r and $\mu = \nu$ easily follows.

For the analogue of Corollary 2.2 for Cauchy transforms see J. Garnett [55, Section II.1].

The smoothing technique used in the proofs of Theorem 2.1 and Theorem 3.1 is fundamental in potential theory, for it replaces superharmonic functions, which need not be continuous, by smooth ones. Furthermore, as a consequence of Theorem 1.2 and the previously made remark to Section II.1, the symmetric averages U^{μ_ε}, g_ε, etc. used in these proofs behave nicely in the sense that they are decreasing functions of ε. For example, one can show with the same smoothing technique the following result of Brelot and Cartan: if f is an extended real-valued function on a domain D such that f is bounded from below on compact subsets of D, and the inequality

$$f(z) \geq \frac{1}{2\pi} \int_{-\pi}^{\pi} f(z + re^{it}) \, dt$$

holds for every disk $\overline{D_r(z)} \subset D$, then its *lower regularization*

$$f^*(z) := \lim_{\delta \to 0} \inf_{z' \in D_\rho(z)} f(z')$$

is superharmonic, and coincides with f quasi-everywhere (in particular, almost everywhere). See e.g. [192, pp. 51, 62].

Section II.3

Theorem 3.1 is due to F. Riesz [193] and holds more generally for superharmonic functions in $\mathbf{R}^m$; see also [75, Section 3.5]. The Riesz decomposition theorem can be made more precise by claiming that

$$\lambda = -\frac{1}{2\pi}\Delta f\, dm,$$

where m is two-dimensional Lebesgue measure, and the Laplace operator Δ is understood in the distributional sense, i.e. in the sense (suggested by Green's formula) that

$$\int_{\mathcal{D}} h\, d\lambda = -\frac{1}{2\pi}\int_{\mathcal{D}} f \cdot \Delta h\, dm$$

for every C^∞ function h with compact support in $\mathcal{D}$. In fact, this is what the proof gives.

Note that in the principle of domination (Theorem 3.2), the constant c cannot, in general, be replaced by an arbitrary function $u(z)$ harmonic on $\mathbf{C}$; indeed, if μ is the normalized Lebesgue measure on the circle $|z| = 1$, $\nu = 0$, and $u(x+iy) = 1-x$, then $U^\mu(z) \le U^\nu(z) + u(z)$ for $|z| = 1$, but the last inequality fails for $z = x$ large and positive.

Corollary 3.3 is due to A. J. Maria [149].

Corollary 3.4 is part of O. Frostman's fundamental theorem [51].

Sometimes the continuity theorem (Theorem 3.5) is stated as

$$\limsup_{z'\to z} U^\mu(z') = \limsup_{z'\to z,\ z'\in \operatorname{supp}(\mu)} U^\mu(z') \tag{6.1}$$

for every $z \in \operatorname{supp}(\mu)$. This is somewhat more general than the assertion in Theorem 3.5; however, the proof of the latter also proves (6.1).

Section II.4

Our treatment of balayage (a notion due to H. Poincaré) based on the weighted equilibrium problem is a very natural approach to the balayage concept. A detailed discussion of the theory of balayage for logarithmic potentials appears in the book of de La Vallée-Poussin [32]. See Landkof [111, Chapter IV] and Helms [77] for further discussion of balayage.

We remark that in the definition of Green function (cf. (4.1) and (4.3)), the requirement in part (i) that g_G be nonnegative is redundant and is mentioned only for emphasis.

In the Poisson–Jensen formula (4.38) the second integral is the solution to the Dirichlet problem on G with boundary function f on ∂G (see Appendix A.3). Thus, the first term on the right is a measure of "how much" f differs from a harmonic function.

Theorem 4.13 is implicit in the paper of L. Carleson [28] and has been extended by A. Cornea [30].

Section II.5

The treatment of Green potentials and Green energy in this section is based primarily on the work of O. Frostman [52]; see also Helms [77, Chapter 11].

The Green equilibrium measure μ_E^G for the compact set E described in Theorem 5.11 can be characterized in terms of balayage measures as follows. Suppose that ∂G is a compact subset of $\mathbf{C}$. Then μ_E^G is the unique measure $\mu \in \mathcal{M}(E)$ with the property that when its balayage $\widehat{\mu}$ on ∂G is balayaged back onto E, then the resulting measure is μ itself; i.e. $\widehat{\widehat{\mu}} = \mu$. Indeed, any such measure μ satisfies $U^\mu - U^{\widehat{\mu}} = \text{const.}$ q.e. on E and so $\mu = \mu_E^G$ follows from Theorem 5.12.

The Green equilibrium measure μ_E^G is also related to the following minimum logarithmic energy problem for *signed* measures. Let $F := \overline{\mathbf{C}} \setminus G$, where we assume that $\infty \in G$. Let $\mathcal{M}(E, F)$ denote the set of all signed measures σ of the form $\sigma = \sigma_1 - \sigma_2$, where $\sigma_1 \in \mathcal{M}(E)$, $\sigma_2 \in \mathcal{M}(F)$ and set

$$V_{E,F} := \inf_{\sigma \in \mathcal{M}(E,F)} \int\int \log \frac{1}{|z-t|} d\sigma(z) d\sigma(t).$$

Then (cf. Theorem VIII.2.6 and Corollary VIII.2.7), there exists a unique measure $\sigma^* = \sigma_1^* - \sigma_2^* \in \mathcal{M}(E, F)$ for which $V_{E,F}$ is attained; furthermore, U^{σ^*} is constant q.e. on E and constant q.e. on F. From Theorem 5.12, it can be seen that $\mu_E^G = \sigma_1^*$ and that $V_E^G = V_{E,F} = 1/C(E, F)$.

An alternative method for deriving the equilibrium distribution $\mu = \mu_E^G$ in Example 5.14 is to start with the known function $w = \varphi(z)$ that maps $G \setminus E = D_1(0) \setminus [-a, a]$ onto an annulus $\{w \mid \rho < |w| < 1\}$; namely

$$\varphi(z) = \rho \exp\left\{\frac{\pi}{2iK}\left[\int_0^{z/a} \frac{dt}{\sqrt{(1-t^2)(1-a^4t^2)}} - K\right]\right\},$$

where

$$\rho = \exp\left\{-\frac{\pi}{4}\frac{K'}{K}\right\}, \quad K' := \int_0^1 \frac{dt}{\sqrt{(1-t^2)(1-(1-a^4)t^2)}}$$

(cf. [166], Section VI.3). Then for some constant c, we have $U_{D_1(0)}^\mu(z) = c \log|\varphi(z)|^{-1}$, and μ can be recovered via the normal derivative technique described in Section IV.2.

Chapter III. Weighted Polynomials

Logarithmic potentials are intimately connected with polynomials on the complex plane. Indeed, if P_n is a monic polynomial, then $\log(1/|P_n|)$ is the potential of the counting measure on the zeros of P_n. In a similar fashion, potentials with external fields are closely related to weighted polynomials. In this chapter we shall utilize this relationship.

We begin with the weighted analogue of Fekete points. These are points for which the expression

$$\prod_{1\leq i<j\leq n} |z_i - z_j| w(z_i) w(z_j)$$

attains its maximum value for all possible choices of the $z_i, z_j \in \Sigma$. The $(n(n-1)/2)$-th root of this maximum converges to a limit δ_w, called the weighted transfinite diameter associated with w. Since Fekete points minimize the weighted energy expression for discrete measures, it is expected that their asymptotic distribution is the equilibrium distribution μ_w, which we show to be the case.

An associated concept is that of the weighted Chebyshev number. The weighted Chebyshev polynomials are the extremal polynomials minimizing the supremum norm $\|w^n T_n\|_\Sigma$ among all monic polynomials of degree n. The n-th root of this minimum tends to a limit t_w, called the weighted Chebyshev constant associated with $w = \exp(-Q)$. In the classical setting, i.e. when Σ is compact and w is identically 1 on Σ, the three quantities: logarithmic capacity, transfinite diameter and Chebyshev constant all coincide. In the weighted case we have the analogous formula

$$c_w = \delta_w = t_w \exp\left(-\int Q \, d\mu_w\right),$$

which reduces to the classical one if Q is identically zero. We shall also consider the asymptotic behavior of the Chebyshev polynomials and their zeros. The zeros accumulate on the polynomial convex hull of the support S_w of the equilibrium measure, and in the case when S_w has empty interior and connected complement, the asymptotic distribution of the zeros is the equilibrium distribution μ_w. Actually, we shall verify that the same property is shared by all monic polynomial sequences $\{P_n\}$ for which the norms $\|w^n P_n\|^{1/n}$ are asymptotically minimal. The asymptotic behavior of all such polynomials in the unbounded component of $\mathbf{C} \setminus S_w$ is given (modulo spurious zeros) by the equilibrium potential:

$$\lim_{n\to\infty} |P_n(z)|^{1/n} = \exp(-U^{\mu_w}(z)).$$

Another important topic in this chapter is the problem of determining where the supremum norm of weighted polynomials $w^n P_n$ live. A weighted variant of the Bernstein–Walsh lemma yields that (even if the set Σ is unbounded) these norms actually live on a fixed compact set which turns out to be the support $\mathcal{S}_w$ of μ_w. We shall also verify that $\mathcal{S}_w$ is the smallest compact set with this property. In some sense the same is true for the L^p norms of weighted polynomials, i.e. in many cases they essentially live on the set $\mathcal{S}_w$.

The investigation of the norm of weighted polynomials is naturally connected to the investigation of the function

$$\Phi(z) := \sup_{\deg(P_n)\leq n} |P_n(z)|^{1/n}, \qquad \|P_n w^n\|_\Sigma \leq 1.$$

Historically this function has played a significant role in solving Dirichlet problems and finding conformal mappings. We shall identify Φ quasi-everywhere as

$$\Phi(z) = \exp(-U^{\mu_w}(z) + F_w).$$

III.1 Weighted Fekete Points, Transfinite Diameter and Fekete Polynomials

In this section we discuss a discretized version of the weighted energy problem. In the classical case this was done by M. Fekete [44]; so here we speak of *weighted* Fekete points, transfinite diameter and Fekete polynomials. Fekete's idea was to look for points in a compact set Σ that are as far apart as possible in the sense of the geometric mean of the distances between the points. Thus, consider the Vandermonde determinant

$$V(x_1, x_2, \ldots, x_n) := \begin{vmatrix} 1 & 1 & \cdots & 1 \\ x_1 & x_2 & \cdots & x_n \\ x_1^2 & x_2^2 & \cdots & x_n^2 \\ \vdots & \vdots & \ddots & \vdots \\ x_1^{n-1} & x_2^{n-1} & \cdots & x_n^{n-1} \end{vmatrix},$$

where each x_i belongs to Σ. Then V has the form

$$V(x_1, \ldots, x_n) = \prod_{1\leq i<j\leq n} (x_j - x_i)$$

and any system of points $\{x_1, \ldots, x_n\} \subseteq \Sigma$ maximizing

$$|V(x_1, \ldots, x_n)|$$

is called an n-th *Fekete set* for Σ; the points x_i in a Fekete set are called *Fekete points*. If the $n(n-1)/2$-th root of this maximum is denoted by δ_n, then the sequence $\{\delta_n\}_{n=2}^{\infty}$ has a limit $\delta(\Sigma)$ which is called the *transfinite diameter* of Σ. A remarkable theorem of Fekete [44] asserts that the transfinite diameter of any compact set coincides with its logarithmic capacity. For example, if Σ is the unit disk or unit circle, then the n-th roots of unity form an n-th Fekete set and in this case $\delta_n = n^{1/(n-1)}$, which tends to 1, the capacity of Σ.

It easily follows from their definition that Fekete points are "almost ideal" points for Lagrange interpolation. In fact, each basic interpolating polynomial

$$L_{n-1,k}(z) := \prod_{j=1,\, j\neq k}^{n} (z - x_j) \Big/ \prod_{j=1,\, j\neq k}^{n} (x_k - x_j)$$

satisfies $|L_{n-1,k}(z)| \leq 1$ for all $z \in \Sigma$ by the choice of the Fekete points $\{x_j\}$; hence the Lebesgue constant is at most n :

$$L = \sup_{z\in\Sigma} \sum_{k=1}^{n} |L_{n-1,k}(z)| \leq n.$$

In particular, if P is a polynomial of degree at most $n-1$, then, for the sup norms

$$\|P\|_\Sigma \leq n\|P\|_{\{x_k\}_{k=1}^n}.$$

In the weighted case F. Leja [112]–[125] investigated analogues of Fekete points. His definition is analogous to the one given above; for completeness we give it for quasi-admissible weights. Thus, let w be a quasi-admissible weight on the closed set Σ, i.e. $w \geq 0$ is upper semi-continuous and $|z|w(z) \to 0$ as $|z| \to \infty$, $z \in \Sigma$, if Σ is unbounded. For an integer $n \geq 2$ we set

$$\delta_n^w := \sup_{z_1,\dots,z_n\in\Sigma} \left\{ \prod_{1\leq i<j\leq n} |z_i - z_j| w(z_i) w(z_j) \right\}^{2/n(n-1)}. \tag{1.1}$$

The supremum defining δ_n^w is obviously attained for some set

$$\mathcal{F}_n = \{z_1, \dots, z_n\} \subseteq \Sigma.$$

These $\mathcal{F}_n$ are called n-th *weighted-Fekete sets* associated with w, or more briefly *w-Fekete sets*, and the points $z_1, \dots, z_n$ in $\mathcal{F}_n$ are called *weighted Fekete points*. As we will shortly see, $\{\delta_n^w\}$ are closely related to the weighted capacity which can be anticipated from the fact that $\log(1/\delta_n^w)$ is a sort of substitute for the "discrete minimal energy".

For fixed n, the sets $\mathcal{F}_n$ need not be unique; however the results below apply to any choice of them. Since w is fixed, we will often drop the superscript w from δ_n^w.

Theorem 1.1. *Let w be a quasi-admissible weight. Then $\{\delta_n^w\}_{n=2}^\infty$ is a decreasing sequence.*

Thus the limit

$$\delta_w := \lim_{n\to\infty} \delta_n^w \tag{1.2}$$

exists. The constant δ_w is called the *weighted transfinite diameter* of Σ with respect to w.

Proof of Theorem 1.1. Let $z_1, \dots, z_{n+1} \in \Sigma$ be so chosen that

$$\delta_{n+1}^{n(n+1)/2} = \prod_{1\le i<j\le n+1} |z_i - z_j| w(z_i) w(z_j). \tag{1.3}$$

For any fixed $1 \le k \le n+1$,

$$\delta_{n+1}^{n(n+1)/2} \le \left(\prod_{1\le i\le n+1,\, i\ne k} |z_i - z_k| w(z_i) w(z_k) \right) \delta_n^{n(n-1)/2}.$$

Multiplying these inequalities for $k = 1, \dots, n+1$ and making use of (1.3) we get

$$\delta_{n+1}^{n(n+1)^2/2} \le \delta_{n+1}^{n(n+1)} \delta_n^{n(n^2-1)/2},$$

from which the statement immediately follows. □

In the classical unweighted case, it easily follows from the maximum modulus theorem for analytic functions that Fekete points lie on the outer boundary of Σ. In the weighted case the situation is more subtle, and it will be of importance for some later results to locate the weighted Fekete points. With the notation of Theorem I.1.3, we prove

Theorem 1.2. *Let w be an admissible weight. Then the set*

$$\mathcal{S}_w^* := \{z \in \Sigma \mid U^{\mu_w}(z) + Q(z) \le F_w\} \tag{1.4}$$

is compact and $\mathcal{F}_n \subseteq \mathcal{S}_w^*$, for each $n = 2, 3, \dots$.

We note that, by Theorem I.1.3(e), $\mathcal{S}_w \subseteq \mathcal{S}_w^*$. The set $\mathcal{S}_w^*$ will appear several times later in this book.

Proof of Theorem 1.2. That $\mathcal{S}_w^*$ is compact follows from the facts that $U^{\mu_w}(z) + Q(z)$ is lower semi-continuous and, in case Σ is unbounded,

$$\lim_{z\to\infty} [U^{\mu_w}(z) + Q(z)] = \lim_{z\to\infty} \left[\log\frac{1}{|z|} + Q(z)\right] = \infty$$

(recall that $|z|w(z) \to 0$ as $z \to \infty$, $z \in \Sigma$).

To prove that $\mathcal{F}_n \subseteq \mathcal{S}_w^*$ we use the following easy consequence of Theorem I.4.1 (see Theorem 2.1 in the next section): If $P_n(z)$ is a polynomial of degree at most n, then

$$|w(z)^n P_n(z)| \le \|w^n P_n\|_\Sigma \exp(n(F_w - U^{\mu_w}(z) - Q(z))) \tag{1.5}$$

for all $z \in \Sigma$, where $\|\cdot\|_\Sigma$ denotes the supremum norm on Σ.

Now let $\mathcal{F}_n = \left\{z_1^{(n)}, z_2^{(n)}, \ldots, z_n^{(n)}\right\}$, and set

$$P_{n-1,j}(z) := \prod_{i=1, i\neq j}^{n} \left(z - z_i^{(n)}\right).$$

Then, from the definition of $\mathcal{F}_n$, we have

$$\|w^{n-1}P_{n-1,j}\|_\Sigma = \left|w\left(z_j^{(n)}\right)^{n-1} P_{n-1,j}\left(z_j^{(n)}\right)\right|.$$

But if $z \notin S_w^*$, then (1.5) implies that

$$\left|w(z)^{n-1}P_{n-1,j}(z)\right| < \left\|w^{n-1}P_{n-1,j}\right\|_\Sigma .$$

This shows that $z_j^{(n)} \notin S_w^*$ is impossible and so $\mathcal{F}_n \subseteq S_w^*$. □

Next we show that the weighted transfinite diameter and the weighted capacity coincide. Furthermore, we shall see that the asymptotic distribution of Fekete points is the equilibrium distribution μ_w.

To this end we define for a finite set H the *normalized counting measure* on H:

$$\nu_H(A) := \frac{1}{|H|}\sum_{x\in H\cap A} 1 = \frac{|H\cap A|}{|H|}, \tag{1.6}$$

where A is any Borel subset of $\mathbf{C}$ and $|H|$ denotes the cardinality of H.

Theorem 1.3. *Let w be a quasi-admissible weight. Then*

$$\delta_w = c_w,$$

where c_w is the weighted capacity. Furthermore, if w is admissible (*i.e. if $c_w > 0$*), *then for any w-Fekete sets $\{\mathcal{F}_n\}_{n=2}^\infty$ we have*

$$\lim_{n\to\infty} \nu_{\mathcal{F}_n} = \mu_w$$

in the weak topology of measures.*

Note that the second statement is precisely that the Fekete points have limit distribution μ_w.

Proof of Theorem 1.3. First we consider the case $c_w > 0$, i.e. we assume that w is admissible. If $z_1, \ldots, z_n \in \Sigma$, then

$$\frac{n(n-1)}{2}\log\frac{1}{\delta_n^w} \le \sum_{1\le i<j\le n} \log\left[|z_i - z_j|w(z_i)w(z_j)\right]^{-1}. \tag{1.7}$$

If $\sigma \in \mathcal{M}(\Sigma)$ is an arbitrary probability measure on Σ, then integrating both sides of (1.7) with respect to $d\sigma(z_1)\cdots d\sigma(z_n)$ we get (cf. Section I.I.1)

$$\log \frac{1}{\delta_n^w} \le I_w(\sigma).$$

Hence

$$\log \frac{1}{\delta_n^w} \le V_w,$$

that is, $c_w \le \delta_n^w$, $n = 2, 3 \ldots$, and so

$$c_w \le \delta_w$$

follows.

We now proceed with the proof of the inequality $\delta_w \le c_w$ for the case of admissible weights w. From Theorem 1.2 we know that the normalized counting measures $\nu_n := \nu_{\mathcal{F}_n}$ are all supported in the compact set $\mathcal{S}_w^*$; hence from any subsequence $\{\nu_{n_k}\}$ we can extract a subsequence $\{\nu_{n'_k}\} =: \{\overline{\nu}_s\}$ converging to some σ in the weak* topology on $\mathcal{M}(\mathcal{S}_w^*)$. Let $\{w_m\}_{m=1}^\infty$ be a decreasing sequence of admissible continuous weight functions converging to w pointwise (the upper semi-continuity of w guarantees the existence of such w_m's). Furthermore, set

$$h_{M,m}(z,t) := \min\{M, \log[|z-t|w_m(z)w_m(t)]^{-1}\},$$

$$h_M(z,t) := \min\{M, \log[|z-t|w(z)w(t)]^{-1}\}.$$

Then $h_{M,m} \le h_M$ and since w is bounded from below by a positive constant on $\mathcal{S}_w^*$ (recall that by Theorem I.4.3 the equilibrium potential U^{μ_w} is bounded on compact subsets of $\mathbf{C}$), we get

$$\begin{aligned}
I_w(\sigma) &:= \iint \log[|z-t|w(z)w(t)]^{-1} d\sigma(z)d\sigma(t) \\
&= \lim_{M\to\infty} \lim_{m\to\infty} \iint h_{M,m}(z,t)\, d\sigma(z)d\sigma(t) \\
&= \lim_{M\to\infty} \lim_{m\to\infty} \lim_{s\to\infty} \iint h_{M,m}(z,t)\, d\overline{\nu}_s(z)d\overline{\nu}_s(t) \\
&\le \lim_{M\to\infty} \limsup_{s\to\infty} \iint h_M(z,t)\, d\overline{\nu}_s(z)d\overline{\nu}_s(t) \\
&\le \lim_{M\to\infty} \limsup_{s\to\infty} \left(\frac{1}{s}M + \frac{s(s-1)}{s^2} \log\left(1/\delta_s^w\right) \right) \\
&= \log(1/\delta_w);
\end{aligned}$$

that is, $V_w \le \log(1/\delta_w)$, which is equivalent to $\delta_w \le c_w$. Hence $\delta_w = c_w$.

As a further consequence we have $I_w(\sigma) = V_w$ and so, from the unicity of the extremal measure μ_w, it follows that $\sigma = \mu_w$. Thus, from every subsequence of $\{\nu_n\}$ we can select a subsequence converging to μ_w in the weak* sense and so the whole sequence $\{\nu_n\} \equiv \{\nu_{\mathcal{F}_n}\}$ converges to μ_w in the weak* topology. The proof of Theorem 1.3 is therefore complete if w is admissible, i.e. $c_w > 0$.

Now suppose that $c_w = 0$, i.e. $\mathrm{cap}(\Sigma_0) = 0$, where $\Sigma_0 = \{z \mid w(z) > 0\}$. We must show that $\delta_w = 0$. With

$$\chi_\varepsilon(z) := \begin{cases} 1 & \text{if} \quad |z| \leq \varepsilon \\ 0 & \text{if} \quad |z| > \varepsilon \end{cases},$$

we set $\Sigma^{(\varepsilon)} := \Sigma \cup \overline{D}_\varepsilon$, $\overline{D}_\varepsilon := \{z \,|\, |z| \leq \varepsilon\}$, and $w_\varepsilon(z) := w(z) + \varepsilon\chi_\varepsilon(z)$ for $z \in \Sigma^{(\varepsilon)}$ with the obvious agreement that $w(z) = 0$ if $z \in \overline{D}_\varepsilon \setminus \Sigma$. Then w_ε is an admissible weight for any $\varepsilon > 0$, and so, according to what has already been proved, $\delta_{w_\varepsilon} = c_{w_\varepsilon}$. We also have $w \leq w_\varepsilon$; hence $\delta_w \leq \delta_{w_\varepsilon}$ for every $\varepsilon > 0$. Thus it is enough to show that

$$\lim_{\varepsilon \to 0} c_{w_\varepsilon} = 0$$

in order to prove that $\delta_w = 0$. But this is a consequence of Theorem I.6.2(c). □

Remark 1.4. The following observation is of some importance: the proof of Theorem 1.3 can be modified to show that the same conclusion concerning the limit distribution holds not only for Fekete sets but also for "asymptotically extremal" sets $\mathcal{T}_n$ consisting of n points from Σ, for which

$$\left(\prod_{z,\, t \in \mathcal{T}_n,\, z \neq t} |z - t| w(z) w(t)\right)^{1/n(n-1)} \to \delta_w \quad \text{as} \quad n \to \infty.$$

The next result will play a useful role in several subsequent theorems. In particular, it implies that, for admissible w,

$$\int Q \, d\nu_{\mathcal{F}_n} \to \int Q \, d\mu_w \quad \text{as} \quad n \to \infty.$$

Lemma 1.5. *Let w be admissible, $\tau_n \in \mathcal{M}(S_w^*)$, $n = 1, 2, \ldots$, where S_w^* is the set defined in* (1.4), *and suppose that $\{\tau_n\}$ converges to some $\mu \in \mathcal{M}(S_w)$ in the weak* topology of measures, where μ is assumed to have finite logarithmic energy. Then*

$$\lim_{n \to \infty} \int Q \, d\tau_n = \int Q \, d\mu. \tag{1.8}$$

In particular, (1.8) *holds if the assumption $\tau_n \in \mathcal{M}(S_w^*)$ is replaced by $\tau_n \in \mathcal{M}(S_w)$, $n = 1, 2, \ldots$.*

Of course (1.8) requires proof because Q may not be continuous.

Proof of Lemma 1.5. Since $\tau_n \to \mu$ in the weak* topology, we have by Theorem 0.1.4

$$\int Q \, d\mu \leq \liminf_{n \to \infty} \int Q \, d\tau_n. \tag{1.9}$$

On the other hand, for $z \in \mathcal{S}_w^*$ we have (cf. (1.4))

$$Q(z) \leq F_w - U^{\mu_w}(z),$$

and so

$$\limsup_{n\to\infty} \int Q\, d\tau_n \leq \limsup_{n\to\infty} \int (F_w - U^{\mu_w})\, d\tau_n. \tag{1.10}$$

Since U^{μ_w} is lower semi-continuous, we again get from Theorem 0.1.4 that

$$\limsup_{n\to\infty} \int (F_w - U^{\mu_w})\, d\tau_n \leq \int (F_w - U^{\mu_w})\, d\mu. \tag{1.11}$$

But $Q = F_w - U^{\mu_w}$ quasi-everywhere on $\mathcal{S}_w$ (cf. Theorem I.1.3(f)) and hence, since μ has finite energy, this equality holds μ-a.e. Thus

$$\int Q d\mu = \int (F_w - U^{\mu_w})\, d\mu,$$

and we get from (1.10) and (1.11)

$$\limsup_{n\to\infty} \int Q\, d\tau_n \leq \int Q\, d\mu.$$

Combining this with (1.9) completes the proof. □

Corollary 1.6. *If w is admissible, then*

$$\lim_{n\to\infty} \int Q\, d\nu_{\mathcal{F}_n} = \int Q\, d\mu_w.$$

Proof. By Theorem 1.2, each measure $\nu_{\mathcal{F}_n}$ is supported on $\mathcal{S}_w^*$ and, by Theorem 1.3, $\nu_{\mathcal{F}_n} \to \mu_w$. Hence the corollary follows from Lemma 1.5. □

Next we consider some quantities and points related to weighted Fekete points. Let $\mathcal{T}_n = \{t_1, \ldots, t_n\} \subseteq \Sigma$ be any n-point subset of Σ and, for $1 \leq i \leq n$, let

$$\Delta^{(i)}(\mathcal{T}_n, w) := \prod_{k=1,\, k\neq i}^{n} |t_i - t_k| w(t_i) w(t_k).$$

The following theorem is of interest when compared to Theorem 1.3.

Theorem 1.7. *Let w be an admissible weight and $\{\mathcal{F}_n\}_{n=2}^{\infty}$ w-Fekete sets. Then for the quantities*

$$\tilde{\delta}_n^w := \min_{1\leq i\leq n} \left\{\Delta^{(i)}(\mathcal{F}_n, w)\right\}^{1/(n-1)}$$

and

$$\tilde{\tilde{\delta}}_n^w := \sup_{\mathcal{T}_n \subseteq \Sigma,\, |\mathcal{T}_n| = n} \min_{1\leq i\leq n} \left\{\Delta^{(i)}(\mathcal{T}_n, w)\right\}^{1/(n-1)},$$

we have

$$\lim_{n\to\infty} \tilde{\delta}_n^w = \lim_{n\to\infty} \tilde{\tilde{\delta}}_n^w = \delta_w (= c_w). \tag{1.12}$$

Proof. The inequalities

$$\tilde{\delta}_n^w \le \tilde{\tilde{\delta}}_n^w \le \delta_n^w \tag{1.13}$$

easily follow from the definitions; for the second inequality observe that for any $\mathcal{T}_n \subseteq \Sigma$

$$\min_{1\le i\le n} \left\{\Delta^{(i)}(\mathcal{T}_n, w)\right\}^{1/(n-1)} \le \left\{ \prod_{s,t\in\mathcal{T}_n,\, s\ne t} |s-t|w(s)w(t)\right\}^{1/n(n-1)}.$$

From Corollary 1.6 we get

$$\lim_{n\to\infty}\left(\prod_{t\in\mathcal{F}_n} w(t)\right)^{1/n} = \lim_{n\to\infty}\exp\left(-\frac{1}{n}\sum_{t\in\mathcal{F}_n} Q(t)\right) \tag{1.14}$$

$$= \lim_{n\to\infty}\exp\left(-\int Q\,d\nu_{\mathcal{F}_n}\right) = \exp\left(-\int Q\,d\mu_w\right).$$

Note also that if, say,

$$\Delta^{(1)}(\mathcal{F}_n, w) := \min_{1\le i\le n} \Delta^{(i)}(\mathcal{F}_n, w)$$

and $\mathcal{F}_n = \left\{t_1^{(n)}, \ldots, t_n^{(n)}\right\}$, then, by the definition of Fekete sets,

$$\Delta^{(1)}(\mathcal{F}_n, w)^{1/(n-1)} = \sup_{z\in\Sigma}\left\{[w(z)]^{n-1}\prod_{i=2}^n \left|z - t_i^{(n)}\right|\right\}^{1/(n-1)}$$

$$\times\left\{\prod_{i=2}^n w\left(t_i^{(n)}\right)\right\}^{1/(n-1)}.$$

In Theorem I.3.6 we verified that the first term on the right is at least as large as $\exp(-F_w) = c_w \exp\left(\int Q\,d\mu_w\right)$ (actually this is true for the weighted norm of any monic polynomial). Hence by (1.14),

$$\liminf_{n\to\infty} \Delta^{(1)}(\mathcal{F}_n, w)^{1/(n-1)} \ge c_w,$$

and this together with (1.13) and Theorem 1.3 proves (1.12). □

In many extremal problems polynomials arise that have asymptotically minimal (weighted) norms. If Σ is a compact set and $T_n(z) = z^n + \cdots$ are polynomials that have the smallest possible supremum norm among all monic polynomials of degree $n = 1, 2, \ldots$, then in many questions related to extremal polynomials these T_n, the so-called Chebyshev polynomials for Σ, can serve as a prototype for comparison; hence the knowledge of $\{T_n\}$ or their asymptotic properties is essential. However, it is usually very hard to determine T_n, and for comparison purposes any other sequence of polynomials with asymptotically minimal norm could serve. Now Fekete points provide a convenient way for generating such asymptotically minimal polynomials.

Let $\mathcal{F}_n = \left\{t_1^{(n)}, \ldots, t_n^{(n)}\right\}$, $n = 2, 3, \ldots$, be w-Fekete sets. The polynomials

$$\Phi_n^w(z) \equiv \Phi_n(z) := \prod_{i=1}^{n}\left(z - t_i^{(n)}\right) \tag{1.15}$$

are called *Fekete polynomials* associated with w. For these polynomials we immediately get from Theorems 1.2 and 1.3 the following result.

Theorem 1.8. *For the Fekete polynomials associated with an admissible weight w we have*

$$\lim_{n\to\infty} \left|\Phi_n^w(z)\right|^{1/n} = \exp\left(\int \log|z-t|\,d\mu_w(t)\right)$$

uniformly on compact subsets of $\mathbf{C} \setminus S_w^*$, where S_w^* is defined in (1.4).

We now prove a norm estimate for weighted Fekete polynomials that will be of importance in Section III.2.

Theorem 1.9. *Let w be an admissible weight and Φ_n Fekete polynomials associated with w of respective degrees $n = 2, 3, \ldots$. Then*

$$\lim_{n\to\infty} \left\|w^n \Phi_n\right\|_\Sigma^{1/n} = \lim_{n\to\infty} \left\|w^n \Phi_n\right\|_{S_w}^{1/n} = \exp(-F_w),$$

where $\|\cdot\|_\Sigma$ and $\|\cdot\|_{S_w}$ denote the supremum norms on Σ and S_w, respectively.

If w is quasi-admissible with $c_w = 0$, Σ is compact and $w > 0$ in infinitely many points of Σ, then

$$\lim_{n\to\infty} \|w^n \Phi_n\|_\Sigma^{1/n} = 0.$$

Recall that by Theorem I.3.6 for any sequence $P_n(z) = z^n + \cdots$, $n = 1, 2, \ldots$, of monic polynomials we have

$$\liminf_{n\to\infty} \left\|w^n P_n\right\|_{S_w}^{1/n} \geq \exp(-F_w),$$

and hence the theorem shows that the polynomials Φ_n have the asymptotic minimality property discussed previously.

Proof of Theorem 1.9. As we have just mentioned, for an admissible w,

$$\liminf_{n\to\infty} \left\|w^n \Phi_n\right\|_\Sigma^{1/n} \geq \liminf_{n\to\infty} \left\|w^n \Phi_n\right\|_{S_w}^{1/n} \geq \exp(-F_w) = c_w \exp\left(\int Q\,d\mu_w\right).$$

Let $\mathcal{F}_n$ be the set of zeros of Φ_n and $\nu_{\mathcal{F}_n}$ the associated normalized counting measure. From the definition of the quantities $\delta_n^w =: \delta_n$ (cf. (1.1)) we get for $z \in \Sigma$,

$$
\begin{aligned}
|[w(z)]^n \Phi_n(z)|^{1/n} & \qquad\qquad (1.16)\\
&= \exp\left(\frac{1}{n}\sum_{t\in\mathcal{F}_n} Q(t)\right)\left(\delta_n^{-n(n-1)/2}\right)^{1/n} \times \\
&\qquad \times \left(\prod_{s,t\in\mathcal{F}_n\cup\{z\},\, s\neq t} |s-t|w(s)w(t)\right)^{1/2n} \\
&\le \exp\left(\int Q d\nu_{\mathcal{F}_n}\right)\left(\delta_n^{-n(n-1)/2}\delta_{n+1}^{n(n+1)/2}\right)^{1/n} \\
&\le \exp\left(\int Q d\nu_{\mathcal{F}_n}\right)\delta_{n+1},
\end{aligned}
$$

where we used that $\delta_{n+1} \le \delta_n$ (see Theorem 1.1). On taking the supremum for all $z \in \Sigma$ and applying Corollary 1.6 we deduce that

$$
\limsup_{n\to\infty} \|w^n\Phi_n\|_{S_w}^{1/n} \le \limsup_{n\to\infty} \|w^n\Phi_n\|_{\Sigma}^{1/n} \le c_w \exp\left(\int Q d\mu_w\right),
$$

and this completes the proof for admissible w.

To prove the second assertion of Theorem 1.9, it follows from the hypotheses that $\delta_n > 0$ for $n = 2, 3, \ldots$, and, from Theorem 1.3, that $\delta_n \to 0$ as $n \to \infty$. From definition (1.1) for δ_n we obtain

$$
\begin{aligned}
\exp\left(\int Q d\nu_{\mathcal{F}_n}\right) &= \frac{1}{\delta_n^{1/2}}\left(\prod_{s,t\in\mathcal{F}_n, s\neq t} |s-t|\right)^{1/(2n(n-1))} \\
&\le \frac{1}{\delta_n^{1/2}}[\mathrm{diam}(\Sigma)]^{1/2}.
\end{aligned}
$$

Using this estimate in (1.16) we can write

$$
\|w^n\Phi_n\|_{\Sigma}^{1/n} \le \frac{\delta_{n+1}}{\delta_n^{1/2}}[\mathrm{diam}(\Sigma)]^{1/2} \le \delta_n^{1/2}[\mathrm{diam}(\Sigma)]^{1/2},
$$

and the desired result follows. □

Combining Theorems 1.8 and 1.9 we immediately get

Corollary 1.10. *The Fekete polynomials Φ_n associated with an admissible weight w satisfy*

$$
\lim_{n\to\infty}\left(\frac{|\Phi_n(z)|}{\|w^n\Phi_n\|_{\Sigma}}\right)^{1/n} = \exp(F_w - U^{\mu_w}(z)),
$$

uniformly on compact subsets of $\mathbf{C} \setminus S_w^*$, where S_w^* is defined in (1.4).

With the help of Theorem 1.9 we can prove the following sharper version of Lemma I.2.3, in which the measure is supported on the set E.

Theorem 1.11 (Evans' Theorem). *Let E be a bounded F_σ-set of zero capacity. Then there is a finite measure μ on E (i.e. $\mu(\mathbf{C} \setminus E) = 0$) such that $U^\mu(z) = \infty$ for every $z \in E$.*

Proof. It is enough to consider the compact case. We suppose that E contains infinitely many points, since otherwise the proof is trivial. For $w \equiv 1$ on $\Sigma = E$, let $\mathcal{F}_n$, Φ_n and $\nu_{\mathcal{F}_n}$ be as in the proof of Theorem 1.9. Then, with $\nu_n := \nu_{\mathcal{F}_n}$, we have

$$U^{\nu_n}(z) = \frac{1}{n} \log \frac{1}{|\Phi_n(z)|},$$

and so

$$U^{\nu_n}(z) \geq \frac{1}{n} \log \frac{1}{\|\Phi_n\|_E}, \quad z \in E. \tag{1.17}$$

Since $\operatorname{cap}(E) = 0$, we have by Theorem 1.9 that the right-hand side of (1.17) tends to $+\infty$ as $n \to \infty$. Hence we can choose integers $1 < n_1 < n_2 < \cdots$ so that

$$\frac{1}{n_k} \log \frac{1}{\|\Phi_{n_k}\|_E} \geq 2^k, \quad k = 1, 2, \ldots.$$

Now set

$$\mu := \sum_{k=1}^{\infty} \frac{1}{2^k} \nu_{n_k}.$$

Clearly $\operatorname{supp}(\mu) \subseteq E$, $\|\mu\| = 1$ and from (1.17) we see that $U^\mu(z) = \infty$ for $z \in E$. □

We conclude this section by showing that Fekete points pin down weighted polynomials in the sense that the norm of a weighted polynomial $P_n w^n$ on Σ cannot be much larger than its norm on $\mathcal{F}_{n+1}$.

Theorem 1.12. *Let w be admissible on Σ and $\mathcal{F}_{n+1} = \{z_0, z_1, \ldots, z_n\}$ be an $(n+1)$-th Fekete set associated with w. Then for every polynomial P_n of degree at most n we have*

$$\left\| w^n P_n \right\|_\Sigma \leq (n+1) \left\| w^n P_n \right\|_{\mathcal{F}_{n+1}}.$$

Proof. It follows from Lagrange's interpolation formula that for polynomials P_n with $\deg(P_n) \leq n$:

$$(w^n P_n)(z) = \sum_{i=0}^{n} (w^n P_n)(z_i) L_{n,i}(z), \tag{1.18}$$

where

$$L_{n,i}(z) := \prod_{j \neq i} \frac{(z - z_j) w(z)}{(z_i - z_j) w(z_i)}.$$

From the extremality property of the Fekete points, we have $\|L_{n,i}\|_\Sigma = 1$, and the result follows from (1.18) and the triangle inequality. □

III.2 Where Does the Sup Norm of a Weighted Polynomial Live?

The weighted polynomials in the title of this section are of the form

$$w(z)^n P_n(z), \quad \deg(P_n) \le n,$$

which essentially differ from the usual definition of a weighted polynomial because here the weight varies together with the degree. We shall see in later chapters that this is what is needed in many applications. If w is an admissible weight on Σ, then the supremum norm behavior of these weighted polynomials is roughly as follows: the supremum norm actually "lives" on a subset of Σ that is independent of n and P_n, and the behavior outside this subset is typically exponentially small. In this section we determine this essential support and show that it is exactly $\mathcal{S}_w$.

First we verify

Theorem 2.1. *Let $w : \Sigma \to [0, \infty)$ be an admissible weight. If P_n is a polynomial of degree at most n and*

$$|w(z)^n P_n(z)| \le M \quad \text{for q.e.} \quad z \in \mathcal{S}_w, \tag{2.1}$$

then for all $z \in \mathbf{C}$

$$|P_n(z)| \le M \exp\left(n(-U^{\mu_w}(z) + F_w)\right). \tag{2.2}$$

Furthermore, (2.1) *implies*

$$|w(z)^n P_n(z)| \le M \quad \text{for q.e.} \quad z \in \Sigma. \tag{2.3}$$

Remark 2.2. It immediately follows from (2.2) that a (not identically zero) weighted polynomial $w(z)^n P_n(z)$, $\deg(P_n) \le n$, can attain its maximum modulus at z_0 only for z_0 in the set

$$\mathcal{S}_w^* = \{z \in \Sigma \mid U^{\mu_w}(z) + Q(z) \le F_w\}$$

which was introduced in Theorem 1.2.

The first part of Theorem 2.1 is the weighted analogue of the well-known *Bernstein–Walsh lemma* ([229, p. 77]): If Σ is compact, and $\text{cap}(\Sigma) > 0$, then for any polynomial of degree at most n and for any $z \in \mathbf{C}$

$$|P_n(z)| \le \exp(n g_\Omega(z, \infty)) \|P_n\|_\Sigma, \tag{2.4}$$

where

$$g_\Omega(z, \infty) = -U^{\mu_\Sigma}(z) + \log \frac{1}{\text{cap}(\Sigma)}$$

is the Green function of the unbounded component Ω of $\mathbf{C} \setminus \Sigma$ with pole at infinity that was introduced in Section I.3 (see also Section II.4). With $w \equiv 1$, Theorem 2.1 reduces to this statement; however, notice that (2.1) is required to hold only quasi-everywhere.

The second part of the theorem is the assertion that the (essential) norm of $w^n P_n$ "lives" on $\mathcal{S}_w$.

Proof of Theorem 2.1. Notice that the function

$$g(z) := \frac{1}{n} \log \frac{M}{|P_n(z)|}$$

is superharmonic on **C**. Furthermore, near infinity, $g(z)$ is harmonic and $g(z) + \log|z|$ is bounded from below. By assumption, $g(z) \geq -Q(z)$ quasi-everywhere on $\mathcal{S}_w$, and hence the first statement of Theorem 2.1 immediately follows from Theorem I.4.1.

The second statement is a consequence of (2.2) and Theorem I.1.3(d). □

We can reformulate the second part of Theorem 2.1 by saying that

$$\|w^n P_n\|_\Sigma^* = \|w^n P_n\|_{\mathcal{S}_w}^*, \quad \deg P_n \leq n, \tag{2.5}$$

where $\|f\|_K^*$ denotes the "sup" from Section I.2, i.e. the smallest number that is an upper bound for $|f|$ quasi-everywhere on K. The next theorem shows that $\mathcal{S}_w$ is the smallest set with this property.

Theorem 2.3. *Let $w : \Sigma \to [0, \infty)$ be an admissible weight and $S \subseteq \Sigma$ a closed set. If, for every $n = 1, 2, \ldots$ and every polynomial P_n with $\deg P_n \leq n$,*

$$\|w^n P_n\|_\Sigma^* = \|w^n P_n\|_S^*,$$

then $\mathcal{S}_w \subseteq S$.

Proof. We assume that $\Sigma = \mathbf{C}$, since for $z \notin \Sigma$ we can set $w(z) = 0$ (cf. Remark I.1.2). Define

$$\overline{w}(z) := \lim_{\delta \to 0+} \|w\|_{\overline{D_\delta(z)}}^*, \tag{2.6}$$

where $\overline{D_\delta(z)} := \{t \mid |z - t| \leq \delta\}$. We claim that

A. *$\overline{w}$ is admissible, $\overline{w} \leq w$, and for quasi-every z*

$$\overline{w}(z) = w(z).$$

Consequently, $\mu_{\overline{w}} = \mu_w$ and $\mathcal{S}_{\overline{w}} = \mathcal{S}_w$.

B. *For any polynomial P_n,*

$$\|w^n P_n\|_S^* \leq \|\overline{w}^n P_n\|_S \leq \|\overline{w}^n P_n\|_\Sigma = \|w^n P_n\|_\Sigma^*. \tag{2.7}$$

Hence, the hypotheses of Theorem 2.3 imply that

$$\|\overline{w}^n P_n\|_S = \|\overline{w}^n P_n\|_\Sigma$$

for every polynomial P_n of $\deg \leq n$, $n = 1, 2, \ldots$.

Thus, with the transition $w \to \overline{w}$, we can replace $\|\cdot\|^*$ by the ordinary supremum norm $\|\cdot\|$. Thereby we avoid the unwanted effect of certain negligible sets of zero capacity (in contrast, from the point of view of Fekete sets, such zero capacity sets may not be negligible).

We first show that assertion B is a consequence of assertion A and the definition (2.6). Assuming $\overline{w}(z) = w(z)$ for q.e. z we immediately get

$$\|w^n P_n\|_S^* \le \|\overline{w}^n P_n\|_S, \quad \|w^n P_n\|_\Sigma^* \le \|\overline{w}^n P_n\|_\Sigma.$$

Thus, to establish (2.7), it remains to show that

$$|\overline{w}(z)^n P_n(z)| \le \|w^n P_n\|_\Sigma^*, \quad \text{for all } z. \tag{2.8}$$

Let z be fixed and $\varepsilon > 0$ given. Then there exists a $\delta_\varepsilon > 0$ such that

$$|P_n(z)| < |P_n(\zeta)| + \varepsilon, \quad \text{if} \quad |z - \zeta| \le \delta_\varepsilon.$$

Set

$$A_\varepsilon := \{\zeta \mid w(\zeta)^n |P_n(\zeta)| \le \|w^n P_n\|_\Sigma^* + \varepsilon\}.$$

Then $\operatorname{cap}(\mathbf{C} \setminus A_\varepsilon) = 0$, and so there exists a $\hat{\zeta} \in A_\varepsilon \cap \overline{D_{\delta_\varepsilon}(z)}$ such that

$$\|w^n\|_{\overline{D_{\delta_\varepsilon}(z)}}^* \le w(\hat{\zeta})^n + \varepsilon.$$

Thus (cf. (2.6))

$$\begin{aligned} |\overline{w}(z)^n P_n(z)| &\le \|w^n\|_{\overline{D_{\delta_\varepsilon}(z)}}^* |P_n(z)| \\ &\le (w(\hat{\zeta})^n + \varepsilon)(|P_n(\hat{\zeta})| + \varepsilon) \\ &= w(\hat{\zeta})^n |P_n(\hat{\zeta})| + \varepsilon |P_n(\hat{\zeta})| + \varepsilon w(\hat{\zeta})^n + \varepsilon^2 \\ &\le \|w^n P_n\|_\Sigma^* + \varepsilon(1 + |P_n(\hat{\zeta})| + w(\hat{\zeta})^n + \varepsilon). \end{aligned}$$

Since P_n and w^n are bounded near z, inequality (2.8) follows on letting $\varepsilon \to 0$, which completes the justification of claim B.

To establish assertion A, we note that since w is upper semi-continuous, we have $\overline{w} \le w$ everywhere. That $\overline{w}$ is upper semi-continuous is an easy consequence of definition (2.6). Thus, the only nontrivial claim that must be proved is that w and $\overline{w}$ coincide quasi-everywhere.

Let

$$V^{(\varepsilon)} := \{z \mid w(z) > \overline{w}(z) + \varepsilon\}.$$

If $w = \overline{w}$ is not true quasi-everywhere, then $\operatorname{cap}\left(V^{(\varepsilon)}\right) > 0$ for some $\varepsilon > 0$. Define

$$\tau_0 := \inf\{\tau \mid \operatorname{cap}(\{z \in V^{(\varepsilon)} \mid \overline{w}(z) < \tau\}) > 0\}.$$

Then the set

$$V^+ := \left\{z \in V^{(\varepsilon)} \,\middle|\, \overline{w}(z) < \tau_0 + \frac{\varepsilon}{4}\right\}$$

has positive capacity, while the set

$$V^- := \left\{ z \in V^{(\varepsilon)} \,\middle|\, \overline{w}(z) < \tau_0 - \frac{\varepsilon}{4} \right\}$$

is of zero capacity. For each $z \in V^+$ there is a δ_z such that $|z-t| < \delta_z$ and

$$w(t) > \overline{w}(z) + \frac{\varepsilon}{4}$$

can simultaneously happen only for t's in a set of zero capacity. The disks

$$\{t \mid |z-t| < \delta_z\}, \quad z \in V^+,$$

cover V^+; hence we can select a countable subcover from them. Since the countable union of sets of zero capacity is of zero capacity, we conclude that the set

$$\left\{ t \in V^+ \,\middle|\, w(t) > \tau_0 + \frac{\varepsilon}{2} \right\}$$

is of zero capacity. Thus the set

$$\left\{ z \in V^+ \,\middle|\, \tau_0 - \frac{\varepsilon}{4} \le \overline{w}(z) < \tau_0 + \frac{\varepsilon}{4}, \quad w(z) \le \tau_0 + \frac{\varepsilon}{2} \right\}$$

must be of positive capacity which is absurd because this set is actually empty – the conditions that define it are contradictory (note that $V^+ \subseteq V^{(\varepsilon)}$).

With these preliminaries we can now easily prove Theorem 2.3. Consider the above $\overline{w}$ and Fekete sets (cf. the preceding section) $\overline{\mathcal{F}}_n$ associated with $\overline{w}$. We claim that for each n we can choose $\overline{\mathcal{F}}_n \subseteq S$. This will already prove the theorem because the normalized counting measures $\overline{\nu}_n$ (cf. (1.6)) associated with $\overline{\mathcal{F}}_n$ then have support in S and converge in the weak* topology to the measure $\mu_{\overline{w}} = \mu_w$ (Theorem 1.3); hence $S_w = \text{supp}(\mu_w) \subseteq S$.

Let $\overline{\mathcal{F}}_n = \{t_1, \ldots, t_n\}$ be any $\overline{w}$-Fekete set and suppose that $t_1, \ldots, t_{s-1}$ are in S but $t_s, \ldots, t_n \notin S$. Consider the polynomial

$$P_{n-1}(z) := \prod_{k \neq s} (z - t_k).$$

By the choice of $\overline{w}$-Fekete points

$$\max_{z \in \Sigma} |P_{n-1}(z)| \overline{w}(z)^{n-1} = |P_{n-1}(t_s)| \overline{w}(t_s)^{n-1}.$$

Assertion B above applied to P_{n-1} shows that there is a $t_s^* \in S$ with

$$\max_{z \in \Sigma} |P_{n-1}(z)| \overline{w}(z)^{n-1} = |P_{n-1}(t_s^*)| \overline{w}(t_s^*)^{n-1}.$$

Thus, the set $\{t_1, \ldots, t_{s-1}, t_s^*, t_{s+1}, \ldots, t_n\}$ is also a $\overline{w}$-Fekete set and this set has s points in S. We can continue this process and eventually arrive at a $\overline{w}$-Fekete set contained in S. □

From the proof we have

Corollary 2.4. *Let w be an admissible weight and $\overline{w}$ the weight constructed from it in* (2.6). *Then $w = \overline{w}$ quasi-everywhere on Σ, and for any polynomial P_n*

$$\|\overline{w}^n P_n\|_\Sigma = \|\overline{w}^n P_n\|_{S_w}, \quad \deg P_n \le n. \tag{2.9}$$

Furthermore, if $S \subseteq \Sigma$ is closed and has the property that for any polynomial P_n

$$\|\overline{w}^n P_n\|_\Sigma = \|\overline{w}^n P_n\|_S, \quad \deg P_n \le n, \quad n = 1, 2, \ldots, \tag{2.10}$$

then $\mathcal{S}_w \subseteq S$.

Remark 2.5. Note that by (2.7) and (2.5) we have

$$\|\overline{w}^n P_n\|_\Sigma = \|w^n P_n\|_\Sigma^*$$

and

$$\|\overline{w}^n P_n\|_{S_w} = \|w^n P_n\|_{S_w}^*.$$

The hypotheses of the next result imply that $\overline{w} = w$, and so Corollary 2.4 yields

Corollary 2.6. *If w is a continuous admissible weight on Σ, and Σ is of positive capacity at every point $z_0 \in \Sigma$ (i.e. the set $\{z \mid |z - z_0| < \delta,\ z \in \Sigma\}$ is of positive capacity for every $\delta > 0$), then $\|w^n P_n\|_\Sigma = \|w^n P_n\|_{S_w}$ for any polynomial P_n with* $\deg P_n \le n$. *Furthermore, if $S \subset \Sigma$ is closed and has the property $\|w^n P_n\|_\Sigma = \|w^n P_n\|_S$ for every polynomial P_n with* $\deg P_n \le n$, *$n = 1, 2 \ldots$, then $\mathcal{S}_w \subset S$.*

Theorems 2.1 and 2.3 do not rule out the possibility that a weighted polynomial $w^n P_n$ has a (essential) maximum outside $\mathcal{S}_w$. What they say is that $w^n P_n$ must take the same (essential) maximum somewhere on $\mathcal{S}_w$, as well. As an example, consider the annulus

$$\Sigma = \{z \mid 1 \le |z| \le 2\}$$

and let $w(z) = 1/|z| = \exp(-\log|z|)$. For this weight, $Q(z) = \log|z|$ coincides on Σ with the negative of the potential of the normalized Lebesgue measure on the unit circumference. Hence, by Theorem I.3.3, μ_w is this Lebesgue measure and $\mathcal{S}_w$ is the unit circumference. On the other hand, $w(z)^n|z^n|$ has a global maximum at every point of Σ.

Having settled the problem that the essential supremum norm of weighted polynomials lives on $\mathcal{S}_w$, let us consider the question of where the actual supremum norms of weighted polynomials live. Let

$$R_w := \{z \in \Sigma \mid U^{\mu_w}(z) + Q(z) < F_w\}. \tag{2.11}$$

This is a bounded σ-compact set (that is, the union of countably many compact sets) having zero capacity because $U^{\mu_w} + Q$ is lower semi-continuous, $R_w \subseteq \mathcal{S}_w^*$, where $\mathcal{S}_w^*$ is the compact set defined in (1.4), and $U^{\mu_w} + Q(z) \ge F_w$ q.e. on Σ (cf. Theorem I.1.3(d)). The closure of this set has to be added to $\mathcal{S}_w$ to get the smallest closed set where all weighted polynomials take their supremum.

Theorem 2.7. *Let w be an admissible weight. Then for any $n = 1, 2, \ldots$ and any polynomial P_n of degree at most n*

$$\|w^n P_n\|_\Sigma = \|w^n P_n\|_{S_w \cup \overline{R_w}}. \tag{2.12}$$

Furthermore, if S is any closed set with the property that

$$\|w^n P_n\|_\Sigma = \|w^n P_n\|_S, \quad \deg P_n \leq n, \quad n = 1, 2, \ldots, \tag{2.13}$$

then $\mathcal{S}_w \cup \overline{R_w} \subseteq S$.

Thus, the supremum norm lives on $\mathcal{S}_w \cup \overline{R_w}$. Before proving Theorem 2.7 we state the following result which should be compared to Theorem 1.2.

Theorem 2.8. *If w is admissible, then for each $n \geq 2$ there are w-Fekete sets $\mathcal{F}_n$ with $\mathcal{F}_n \subseteq \mathcal{S}_w \cup \overline{R_w}$. Furthermore, if S is any closed set such that for each n there is a w-Fekete set $\mathcal{F}_n \subseteq S$, then $\mathcal{S}_w \cup \overline{R_w} \subseteq S$.*

Thus, $\mathcal{S}_w \cup \overline{R_w}$ is also the smallest closed set that contains Fekete sets for all orders n.

Proof of Theorem 2.7. First we establish (2.12). Suppose that

$$w(z)^n |P_n(z)| \leq M \quad \text{on} \quad \mathcal{S}_w \cup \overline{R_w}.$$

Then, by Theorem 2.1,

$$w(z)^n |P_n(z)| \leq M \exp(n(-U^{\mu_w}(z) - Q(z) + F_w))$$

everywhere on Σ. However, outside $\mathcal{S}_w \cup \overline{R_w}$ the right-hand side is at most M and this proves (2.12).

To verify the second part of Theorem 2.7 we have to show that (2.13) implies $\mathcal{S}_w \subseteq S$ and $\overline{R_w} \subseteq S$. To prove these inclusions we can follow the same argument that was used to prove Theorem 2.3. In fact, exactly as there, we get that if (2.13) holds, then for every n there is a w-Fekete set $\mathcal{F}_n \subseteq S$. But (cf. (1.6)) $\nu_{\mathcal{F}_n} \to \mu_w$ in the weak* sense (see Theorem 1.3); therefore, $\mathcal{S}_w = \text{supp}(\mu_w) \subseteq S$.

Now let $z_0 \in R_w \setminus \mathcal{S}_w$ be arbitrary. We have to show that $z_0 \in S$. If this is not the case, then, for some $\delta > 0$, $\text{dist}(z_0, \mathcal{F}_n) \geq \delta$ holds for all n (as before we assume $\mathcal{F}_n \subseteq S$). Let $\mathcal{F}_n = \left\{z_1^{(n)}, \ldots, z_n^{(n)}\right\}$ where, anticipating the application of Theorem 1.7, we assume that the points are ordered in such a way that among

$$\Delta^{(i)}(\mathcal{F}_n, w) = \prod_{k=1,\, k\neq i}^{n} \left|z_i^{(n)} - z_k^{(n)}\right| w\left(z_i^{(n)}\right) w\left(z_k^{(n)}\right), \quad 1 \leq i \leq n,$$

$\Delta^{(n)}(\mathcal{F}_n, w)$ is the smallest. Consider the polynomials

$$\varphi_{n-1}(z) := \prod_{k=1}^{n-1} \left(z - z_k^{(n)}\right).$$

If ν_{n-1} denotes the normalized counting measure on the set

$$\left\{z_1^{(n)}, \ldots, z_{n-1}^{(n)}\right\},$$

then, since $\nu_{\mathcal{F}_n} \to \mu_w$, we also have $\nu_{n-1} \to \mu_w$ in the weak* sense, and this shows that

$$\lim_{n\to\infty} w(z_0)|\varphi_{n-1}(z_0)|^{1/(n-1)} = \exp(-U^{\mu_w}(z_0) - Q(z_0)) > \exp(-F_w)$$

because $z_0 \in R_w$. On the other hand, since $z_k^{(n)} \in \mathcal{S}_w^*$ (cf. Theorem 1.2) and Q is bounded on $\mathcal{S}_w^*$, we have from Corollary 1.6 that

$$\begin{aligned}\lim_{n\to\infty}\left(\prod_{k=1}^{n-1} w\left(z_k^{(n)}\right)\right)^{1/(n-1)} &= \lim_{n\to\infty} \exp\left(-\int Q d\nu_{n-1}\right)\\ &= \exp\left(-\int Q d\mu_w\right).\end{aligned}$$

Moreover, by Theorem 1.7,

$$\lim_{n\to\infty}\left(\Delta^{(n)}(\mathcal{F}_n, w)\right)^{1/(n-1)} = c_w = \exp(-V_w).$$

Consequently,

$$\begin{aligned}\lim_{n\to\infty} w\left(z_n^{(n)}\right)\left|\varphi_{n-1}\left(z_n^{(n)}\right)\right|^{1/(n-1)} &= \exp(-V_w)\exp\left(\left(\int Q d\mu_w\right)\right.\\ &= \exp(-F_w).\end{aligned}$$

Thus, for large n,

$$w(z_0)^{n-1}|\varphi_{n-1}(z_0)| > w\left(z_n^{(n)}\right)^{n-1}\left|\varphi_{n-1}\left(z_n^{(n)}\right)\right|,$$

which is a contradiction because, by the definition of w-Fekete sets, we must have

$$w\left(z_n^{(n)}\right)^{n-1}\left|\varphi_{n-1}\left(z_n^{(n)}\right)\right| = \max_{z\in\Sigma}\left\{w(z)^{n-1}|\varphi_{n-1}(z)|\right\}.$$

The contradiction obtained shows that $z_0 \in S$, and since S is closed, we get $\overline{R_w} \subseteq S$. □

Proof of Theorem 2.8. The first statement was established in the preceding proof.

Let us now suppose that S is closed and for each n there is an $\mathcal{F}_n \subseteq S$. Then following the second part of the preceding proof with these $\mathcal{F}_n$'s we get $\mathcal{S}_w \cup \overline{R_w} \subseteq S$ exactly as before. □

The last theorem in this section establishes a close relationship between weighted polynomials and the equilibrium potential. Actually, the result to be proven below is a partially sharpened form of Theorem I.4.1, because in it we represent $U^{\mu_w} - F_w$ q.e. as a lower envelope of discrete potentials from $\mathcal{H}$ (cf. Theorem I.4.1).

Theorem 2.9. *Let w be an admissible weight on Σ. Then, for any $z \in \mathbf{C}$ where the potential U^{μ_w} is continuous, there holds*

$$\exp(-U^{\mu_w}(z) + F_w) = \sup\left\{ |P_n(z)|^{1/n} \,\middle|\, \deg P_n = n, \; n = 1, 2 \ldots, \right.$$
$$\left. w(x)^n |P_n(x)| \le 1 \text{ for q.e. } x \in \Sigma \right\}. \quad (2.14)$$

In particular, (2.14) *holds quasi-everywhere.*

We mention that equality (2.14) cannot hold at any discontinuity point of $U^{\mu_w}(z)$. In fact, the left-hand side is upper, while the right-hand side is lower semi-continuous; furthermore, the left-hand side is always at least as large as the right one (Theorem 2.1). Thus, if equality holds in (2.14) for a z, then U^{μ_w} must be continuous there.

In Section III.5 we shall show that, in fact

$$\exp(-U^{\mu_w}(z) + F_w)$$

coincides quasi-everywhere with the supremum of the expression $|P_n(z)|^{1/n}$, where P_n, $\deg(P_n) \le n$, $n = 1, 2, \ldots$, are polynomials satisfying

$$w(x)^n |P_n(x)| \le 1$$

everywhere on Σ. This is a much stronger requirement than having $w^n|P_n| \le 1$ only quasi-everywhere, so the same supremum as in (2.14) is attained on a narrower class of polynomials. On the other hand, Theorem 2.9 is a more precise result with regard to where equality takes place in (2.14), namely exactly at the continuity points of U^{μ_w}.

Proof of Theorem 2.9. As already mentioned, the right-hand side of (2.14) is at most as large as the left-hand side. Thus we only have to prove that if U^{μ_w} is continuous at z_0, then

$$\exp(-U^{\mu_w}(z_0) + F_w) \le \sup\left\{ |P_n(z_0)|^{1/n} \,\middle|\, \deg P_n = n, \right. \quad (2.15)$$
$$\left. n = 1, 2, \ldots, \; w(x)^n |P_n(x)| \le 1 \quad \text{for q.e. } x \in \Sigma \right\}.$$

If $z_0 \notin S_w$, then by choosing

$$P_n = \frac{\tilde{\Phi}_n}{\|w^n \tilde{\Phi}_n\|_{S_w}},$$

where $\tilde{\Phi}_n$ are the Fekete polynomials associated with the restricted weight $\tilde{w} := w|_{S_w}$ (so that $\Sigma_{\tilde{w}} = S_w$, $S_{\tilde{w}} = S_w$, $\mu_{\tilde{w}} = \mu_w$, and $S_{\tilde{w}}^* \subseteq S_w$) we obtain (2.15) from Corollary 1.10.

Next we prove (2.15) for every $z_0 \in S_w$ where the potential U^{μ_w} is continuous. For $\delta > 0$ sufficiently small, we set

$$\Sigma^{(\delta)} := \{x \in \Sigma \mid |x - z_0| \geq \delta\},$$

$w_\delta := w|_{\Sigma^{(\delta)}}$ and let μ_δ be the extremal measure corresponding to w_δ. By Theorem I.6.2(d) we get that as $\delta \to 0+$, $\mu_\delta \to \mu_w$ in the weak* topology; furthermore (see Corollary I.4.2(a))

$$U^{\mu_\delta}(z) - F_{w_\delta} \leq U^{\mu_w}(z) - F_w \quad \text{and} \quad -F_{w_\delta} \leq -F_w, \tag{2.16}$$

and the left-hand sides increase monotonically as δ decreases. We also know from Theorem I.6.5(a) that

$$\lim_{\delta \to 0+} F_{w_\delta} = F_w$$

and

$$\lim_{\delta \to 0+} U^{\mu_\delta}(z) = U^{\mu_w}(z)$$

for all z. Now choose a $\delta_0 > 0$ such that for $0 < \delta \leq \delta_0$ we have

$$\left|F_{w_\delta} - F_w\right| < \varepsilon \tag{2.17}$$

and

$$|U^{\mu_\delta}(z_0) - U^{\mu_w}(z_0)| < \varepsilon. \tag{2.18}$$

By the continuity of U^{μ_w} and U^{μ_δ} at z_0 (note that $z_0 \notin \text{supp}(\mu_\delta)$), we can choose $\rho > 0$ such that

$$|U^{\mu_\delta}(z) - U^{\mu_w}(z)| < \varepsilon \quad \text{if} \quad |z - z_0| \leq \rho, \quad \delta = \delta_0.$$

Now let $\delta \leq \delta_0$ and $|z - z_0| \leq \rho$ be arbitrary. Then, using also the monotonicity of the left-hand sides in (2.16), we get

$$U^{\mu_\delta}(z) = U^{\mu_\delta}(z) - F_{w_\delta} + F_{w_\delta} \leq U^{\mu_w}(z) - F_w + F_w + \varepsilon = U^{\mu_w}(z) + \varepsilon$$

and

$$\begin{aligned} U^{\mu_\delta}(z) &= U^{\mu_\delta}(z) - F_{w_\delta} + F_{w_\delta} \geq U^{\mu_{\delta_0}}(z) - F_{w_{\delta_0}} + F_w - \varepsilon \\ &\geq U^{\mu_w}(z) - \varepsilon - F_w - \varepsilon + F_w - \varepsilon \geq U^{\mu_w}(z) - 3\varepsilon. \end{aligned}$$

That is, independently of $\delta \leq \delta_0$ and $|z - z_0| \leq \rho$, we have

$$|U^{\mu_\delta}(z) - U^{\mu_w}(z)| \leq 3\varepsilon. \tag{2.19}$$

Now let $\delta < \min(\delta_0, \rho)$ be fixed, and consider the Fekete polynomials $\tilde{\Phi}_n^{(\delta)}$, $n = 2, 3, \ldots$, associated with the restricted weight $\tilde{w}_\delta := w_\delta|_{S_{w_\delta}}$. We get from Theorem 1.9 that

$$\left|\tilde{\Phi}_n^{(\delta)}(z) w(z)^n\right|^{1/n} \leq (1 + \varepsilon) \exp(-F_{w_\delta})$$

uniformly in $z \in \operatorname{supp}(\mu_\delta)$ for every large n, say $n \geq n_0$, and so

$$P_n(z) := \left(\frac{\exp(F_{w_\delta})}{1+\varepsilon}\right)^n \tilde{\Phi}_n^{(\delta)}(z)$$

satisfies $|P_n(z)w(z)^n| \leq 1$ for $z \in \operatorname{supp}(\mu_\delta)$ and $n \geq n_0$. This implies by Theorem 2.1 that for $n \geq n_0$ and all $z \in \Sigma$,

$$|P_n(z)w(z)^n|^{1/n} \leq \exp\left(-U^{\mu_\delta}(z) + F_{w_\delta} - Q(z)\right).$$

If $z \in \Sigma^{(\delta)}$, then of course (see Theorem I.1.3(d))

$$-U^{\mu_\delta}(z) + F_{w_\delta} - Q(z) \leq 0$$

quasi-everywhere. If, however, $z \in \Sigma \setminus \Sigma^{(\delta)}$, then $|z - z_0| < \rho$, and we get from (2.17) and (2.19) that q.e.

$$-U^{\mu_\delta}(z) + F_{w_\delta} - Q(z) \leq -U^{\mu_w}(z) + F_w - Q(z) + 4\varepsilon \leq 4\varepsilon.$$

Thus, the polynomials $P_n^*(z) := e^{-4n\varepsilon} P_n(z)$, $n \geq n_0$, satisfy

$$|P_n^*(z)w(z)^n|^{1/n} \leq 1 \quad \text{for q.e. } z \in \Sigma.$$

On the other hand, we get from Theorem 1.8 applied to the restriction of w_δ to $\operatorname{supp}(\mu_\delta)$ that

$$\begin{aligned}
\lim_{n\to\infty} |P_n^*(z_0)|^{1/n} &= \lim_{n\to\infty} \frac{\exp(F_{w_\delta} - 4\varepsilon)}{1+\varepsilon} \left|\tilde{\Phi}_n^{(n)}(z_0)\right|^{1/n} \\
&\geq \frac{e^{F_w - 5\varepsilon}}{1+\varepsilon} \exp(-U^{\mu_\delta}(z_0)) \\
&\geq \frac{e^{-6\varepsilon}}{1+\varepsilon} \exp(-U^{\mu_w}(z_0) + F_w),
\end{aligned}$$

where we used (2.17) and (2.18) again. Since $\varepsilon > 0$ is arbitrary, (2.15) follows. □

III.3 Weighted Chebyshev Polynomials

The name "*Chebyshev polynomial*" usually refers to the polynomials

$$T_n(x) = \frac{1}{2^n}\left\{\left(x + \sqrt{x^2-1}\right)^n + \left(x - \sqrt{x^2-1}\right)^n\right\}, \quad n \geq 1,$$

which have minimal supremum norm on $[-1, 1]$ among all monic polynomials $P_n(x) = x^n + \cdots$. But if Σ is an arbitrary compact subset of $\mathbf{C}$ containing infinitely many points, then we can analogously define the n-th Chebyshev polynomial T_n corresponding to Σ as the (unique) monic polynomial $P_n(z) = z^n + \cdots$

that minimizes the supremum norm on Σ. If the minimum value is t_n, then t_n is called the n-th *Chebyshev number* of Σ. In only a very few cases do we exactly know the Chebyshev polynomials and numbers; such a case is the unit circle for which $T_n(z) = z^n$ and $t_n = 1$ and the interval $[-1, 1]$ for which T_n is given by the previously stated formula. It is well known however (see e.g. [222, Theorem III.26]), that the sequence $\{t_n^{1/n}\}_{n=1}^{\infty}$ converges and its limit is $\mathrm{cap}(\Sigma)$. If we call the limit the *Chebyshev constant* of Σ, then in the classical theory the three important quantities associated with a compact set: its logarithmic capacity, transfinite diameter and Chebyshev constant, all coincide. We shall see below that there is a slight deviation from this in the weighted case.

As we have already remarked in connection with Fekete polynomials in Section III.1, the solutions of many extremal problems behave like the Chebyshev polynomials and the Chebyshev polynomials can serve as a good standard for comparison. In this section we discuss some properties of Chebyshev polynomials and constants in the weighted case, the definitions of which are as follows.

For an admissible weight w on the closed set Σ, the numbers

$$t_n^w := \inf\{\ \|w^n P\|_\Sigma \ \ | P(z) = z^n + \cdots\} \tag{3.1}$$

are called the *Chebyshev numbers* corresponding to w. It is easily seen that the infimum on the right is attained for a polynomial $T_n(z) = T_n^w(z)$ which is called an n-th *Chebyshev polynomial* corresponding to w. Of course, if $w \equiv 1$ on Σ and Σ is compact, then these definitions coincide with the classical definitions of Chebyshev polynomials and numbers corresponding to compact sets. Our main concern will be the n-th root behavior of the sequences $\{t_n^w\}$, $\{|T_n(z)|\}$ and the zero distribution of the polynomials T_n.

We also introduce the *restricted Chebyshev numbers*

$$\tilde{t}_n^w := \inf\{\ \|w^n P\|_\Sigma \ \ | P(z) = z^n + \cdots, \ \text{ with all zeros in } \Sigma\ \}, \tag{3.2}$$

and the corresponding extremal polynomials $\tilde{T}_n \equiv \tilde{T}_n^w$. The first result describes the n-th root behavior of these Chebyshev numbers.

Theorem 3.1. *Let w be admissible. Then*

$$\exp(-F_w) \le (t_n^w)^{1/n} \le (\tilde{t}_n^w)^{1/n}, \quad n = 1, 2, \ldots, \tag{3.3}$$

and

$$\lim_{n\to\infty} (t_n^w)^{1/n} = \lim_{n\to\infty} (\tilde{t}_n^w)^{1/n} = \exp(-F_w). \tag{3.4}$$

Thus, if the *Chebyshev constants* associated with w are defined as

$$t_w := \lim_{n\to\infty} (t_n^w)^{1/n} \qquad \text{and} \qquad \tilde{t}_w := \lim_{n\to\infty} (\tilde{t}_n^w)^{1/n},$$

then

$$t_w = \tilde{t}_w = \exp(-F_w) = c_w \exp\left(\int Q\, d\mu_w\right). \tag{3.5}$$

Note that if $w \equiv 1$, then $Q \equiv 0$, and we get $t_w = c_w$; i.e., the above mentioned result that the capacity and the Chebyshev constant of Σ are the same. In the general weighted case, however, the additional factor $\exp\left(\int Q\, d\mu_w\right)$ appears in the expression for t_w.

If w vanishes quasi-everywhere then we can apply

Corollary 3.2. *If w is quasi-admissible but not admissible, then*

$$t_w = \tilde{t}_w = 0.$$

Proof of Theorem 3.1. That (3.3) is true follows immediately from Theorem I.3.6.

To prove (3.4) it suffices to show that

$$\limsup_{n\to\infty} (\tilde{t}_n^w)^{1/n} \le \exp(-F_w). \tag{3.6}$$

But from Theorem 2.8 we know that there exist w-Fekete sets $\mathcal{F}_n$ with $\mathcal{F}_n \subseteq \mathcal{S}_w \cup \overline{R_w} \subseteq \Sigma$ for each $n = 2, 3, \ldots$. Hence if we let Φ_n be the associated (monic) Fekete polynomials, we get from Theorem 1.9 that

$$\limsup_{n\to\infty} (\tilde{t}_n^w)^{1/n} \le \limsup_{n\to\infty} \|w^n \Phi_n\|_\Sigma^{1/n} = \exp(-F_w).$$

□

Proof of Corollary 3.2. The quantities t_w and $\tilde{t}_w$ are obviously monotone increasing functions of the weight w. Consider now the weights $w_\varepsilon := w + \varepsilon\chi_\varepsilon$,

$$\chi_\varepsilon(z) := \begin{cases} 1 & \text{if } |z| \le \varepsilon \\ 0 & \text{if } |z| > \varepsilon, \end{cases}$$

used in the proof of Theorem 1.3. In that proof we verified that $c_{w_\varepsilon} \to 0$ as $\varepsilon \to 0$ and that μ_{w_ε} is supported in $\overline{D}_\varepsilon = \{z \mid |z| \le \varepsilon\}$. These facts easily imply that $F_{w_\varepsilon} \to \infty$ as $\varepsilon \to 0$, and so we get from Theorem 3.1

$$t_w \le \tilde{t}_w \le \lim_{\varepsilon\to 0} \tilde{t}_{w_\varepsilon} = \lim_{\varepsilon\to 0} \exp(-F_{w_\varepsilon}) = 0.$$

□

Example 3.3. Let q_m be a monic polynomial of degree m. Then for the *lemniscate set*

$$E_a := \{z \mid |q_m(z)| \le a\}$$

we have

$$\operatorname{cap}(E_a) = a^{1/m}. \tag{3.7}$$

Furthermore, the Chebyshev polynomials for E_a ($w \equiv 1$) of respective degrees km, $k = 1, 2, \ldots$, are given by $T_{km} = q_m^k$. Indeed, the last assertion is an immediate consequence of the maximum modulus principle applied to ratios of the form $(z^{mk} + \cdots)/q_m^k(z)$ in the exterior of E_a. From this we obtain $t_{mk} = a^k$ and Theorem 3.1 asserts that as $k \to \infty$, we have $t_{mk}^{1/mk} = a^{1/m} \to \operatorname{cap}(E_a)$. □

Formula (3.7) has an extension: if $q = q_m$ is a monic polynomial of degree m, then for every compact set K

$$\operatorname{cap}(q^{-1}(K)) = \operatorname{cap}(K)^{1/m}. \tag{3.8}$$

In fact, it is enough to prove this for polynomially convex K of positive capacity. But then $q^{-1}(\mathbf{C} \setminus K) = \mathbf{C} \setminus q^{-1}(K)$, and this set is connected. Indeed, if z_0 is a point of this set, then there is a polynomial p such that $|p(q(z_0))| > \|p\|_K$, because $q(z_0)$ is not in the polynomial convex hull of K (see e.g. Corollary 1.10). Therefore, for the polynomial $P(z) = p(q(z))$ we have $|P(z_0)| > \|P\|_{q^{-1}(K)}$; hence z_0 cannot belong to the polynomial convex hull of $q^{-1}(K)$. In other words, it must belong to the unbounded component of $\mathbf{C} \setminus q^{-1}(K)$, and since this is true for every $z \in \mathbf{C} \setminus q^{-1}(K)$, we conclude the connectedness of $\mathbf{C} \setminus q^{-1}(K)$. But then the unicity of the Green function shows that

$$\frac{1}{m} g_{\overline{\mathbf{C}} \setminus K}(q(z), \infty) = g_{\overline{\mathbf{C}} \setminus q^{-1}(K)}(z, \infty),$$

and checking this equality at ∞ we arrive at (3.8).

In some cases it is still true that if T_n is the n-th Chebyshev polynomial for K then $T_n(q)$ is the mn-th Chebyshev polynomial for $q^{-1}(K)$, but not always (see [179]).

In the general case not much can be said about the location of the zeros of weighted Chebyshev polynomials. We can say, however, that in the unbounded component of $\mathbf{C} \setminus (\mathcal{S}_w \cup \overline{R_w})$ (cf. (2.11)) most of the zeros are close to $\mathcal{S}_w \cup \overline{R_w}$.

Theorem 3.4. *Suppose that w is an admissible weight and $\{T_n\}$ are the associated Chebyshev polynomials. Then every zero of T_n lies in the convex hull of the set $\mathcal{S}_w \cup \overline{R_w}$, where R_w is defined in* (2.11). *Moreover, if K is any compact subset of the unbounded component of $\mathbf{C} \setminus (\mathcal{S}_w \cup \overline{R_w})$, then there is a number $s = s_K$ depending only on K such that each T_n has at most s zeros in K.*

Nothing similar can be said concerning the bounded components of $\mathbf{C} \setminus (\mathcal{S}_w \cup \overline{R_w})$, see e.g. Example 3.7 below. Also, it is easy to construct examples (consider e.g. Example 3.7, add to Σ some points and define w to be large in these points) showing that the set $\overline{R_w}$ cannot be omitted from Theorem 3.4; for it is possible that, say, $\mathcal{S}_w = \{z \mid |z| = 1\}$, but the number of zeros that T_n, $n = 1, 2, \ldots$, has in $\{z \mid |z| \geq 2\}$ is unbounded with n. Finally, we mention that even if $\overline{R_w} = \emptyset$ and $\mathbf{C} \setminus \mathcal{S}_w$ is connected, it may happen that the T_n's have zeros outside $\mathcal{S}_w$ (take $\Sigma \subset \mathbf{R}$ to be symmetric with respect to 0, $0 \notin \Sigma$ and $w \equiv 1$ on Σ; if n is odd, then $T_n(0) = 0$).

Proof of Theorem 3.4. Let $\operatorname{Con}(\mathcal{S}_w \cup \overline{R_w})$ be the convex hull of the set $\mathcal{S}_w \cup \overline{R_w}$. If $T_n(z) = \prod_{i=1}^n (z - z_i)$, then

$$\|w^n T_n\|_\Sigma = \sup_{z \in \mathcal{S}_w \cup \overline{R_w}} \prod_{i=1}^n |z - z_i| w(z).$$

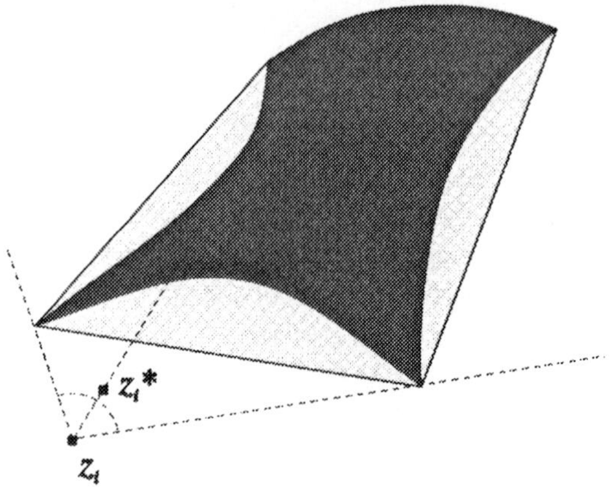

Fig. 3.1

Now if for some i we have $z_i \notin \mathrm{Con}(\mathcal{S}_w \cup \overline{R_w})$, then $\mathcal{S}_w \cup \overline{R_w}$ lies in a cone with vertex at z_i and of opening $< \pi$. But then for the z_i^* given in Figure 3.1 we have for all $z \in \mathcal{S}_w \cup \overline{R_w}$ the inequality

$$|z - z_i^*| < |z - z_i|.$$

Hence for $P_n(z) := (z - z_i^*) \prod_{j \neq i}(z - z_j)$ we get from Theorem 2.7 and the fact that $w > 0$ on $S_w \cup \overline{R}_w$ that

$$\|w^n P_n\|_\Sigma = \|w^n P_n\|_{\mathcal{S}_w \cup \overline{R_w}} < \|w^n T_n\|_{\mathcal{S}_w \cup \overline{R_w}} = \|w^n T_n\|_\Sigma,$$

which is a contradiction. This shows that $z_i \in \mathrm{Con}(\mathcal{S}_w \cup \overline{R_w})$ for all i proving the first statement of the theorem.

To prove the second statement, for a compact set S let us consider the polynomial convex hull $\mathrm{Pc}(S) = \mathbf{C} \setminus \Omega$, where Ω is the unbounded component of the complement $\mathbf{C} \setminus S$ of S. We have to prove that for every compact set K with $K \cap \mathrm{Pc}(\mathcal{S}_w \cup \overline{R_w}) = \emptyset$ the number of zeros of the w-Chebyshev polynomials lying in K is bounded. For this purpose we shall apply the following result.

Lemma 3.5. *If K and S are compact sets and $K \cap \mathrm{Pc}(S) = \emptyset$, then there is a positive integer m and a positive constant $\alpha < 1$ such that if $x_1, \ldots, x_m \in K$ are arbitrary, then there are points $y_1, \ldots, y_m$ so that the rational function*

$$r(z) := \prod_{j=1}^{m} \frac{z - y_j}{z - x_j}$$

has sup norm

$$\|r\|_S \leq \alpha.$$

Proof. We divide the proof of the lemma into three steps.

I. First assume $K = \{z_0\}$, and then without loss of generality we take $z_0 = 0$. We have to prove that for some polynomial $P_m(z) = z^m + \cdots$ we have $|P_m(z)/z^m| < 1$ on S. The substitution $1/z \to \zeta$ carries S onto $S^{-1} := \{\zeta \mid 1/\zeta \in S\}$ and we have to show that

$$|1 + a_1\zeta + \cdots + a_m\zeta^m| < 1, \quad \zeta \in S^{-1},$$

for some positive integer m and complex constants $a_1, \dots, a_m$. But this follows from Runge's theorem ([195, Theorem 13.6]), which ensures that since $0 \notin \mathrm{Pc}(S^{-1})$, the function $1/\zeta$ can be uniformly approximated on $\mathrm{Pc}(S^{-1})$ by polynomials in ζ (after that multiply by ζ).

II. From I and continuity we get that for every $z_0 \in K$ there are constants $\delta(z_0) > 0$, $m(z_0)$, and $0 < \alpha(z_0) < 1$ such that the lemma is true if K is replaced by $K \cap \overline{D_{\delta(z_0)}(z_0)}$, where

$$D_\delta(z_0) := \{z \mid |z - z_0| < \delta\},$$

and m and α are replaced by $m(z_0)$ and $\alpha(z_0)$. By compactness there exists a finite cover $\cup_1^k D_{\delta(z_j)}(z_j)$ of K. Set $D_j := D_{\delta(z_j)}(z_j)$, $M_j := m(z_j)$, $M_0 := \max\{M_1, \dots, M_k\}$, and $\alpha_0 := \max\{\alpha(z_1), \dots, \alpha(z_k)\}$.

III. If we are given m arbitrary points $x_1, \dots, x_m$ from K, then let $m_1, \dots, m_k$ denote the respective number of these that lie in $D_1, \dots, D_k$ (if a point x_j belongs to more than one D_s , we assign it only to the first such D_s); thus $m = m_1 + \cdots + m_k$. Next we group the points in D_1 into subgroups of M_1 elements and apply part II to each subgroup. To those elements (in case M_1 does not divide m_1) which do not belong to any of the $[m_1/M_1]$ subgroups we let the corresponding y_j be simply x_j. Similarly, in D_2 we group the points into $[m_2/M_2]$ groups of M_2 elements and again apply II; etc.

In this fashion we get points $y_1, \dots, y_m$ such that for $z \in S$

$$\left|\prod_1^m \frac{z - y_j}{z - x_j}\right| \le \alpha(z_1)^{[m_1/M_1]} \cdots \alpha(z_k)^{[m_k/M_k]} \le \alpha_0^{(m/M_0)-k},$$

which proves the lemma for $m > M_0 k$. □

We now return to the proof of the second part of Theorem 3.4. Let K be any compact set such that $K \cap \mathrm{Pc}(S) = \emptyset$, where $S := S_w \cup \overline{R_w}$. We can apply Lemma 3.5 and get an m and $\alpha < 1$ with the properties stated there. We claim that no T_n can have m zeros in K, and this will complete the proof.

In fact, if we assume to the contrary that the zeros $x_1, \dots, x_m$ of a fixed T_n do belong to K, then by Lemma 3.5, there are $y_1, \dots, y_m$ such that

$$\left|\prod_{j=1}^m \frac{z - y_j}{z - x_j}\right| \le \alpha < 1$$

for $z \in S$. Thus, for the polynomials

$$T_n^*(z) := T_n(z) \prod_{j=1}^{m} \frac{z - y_j}{z - x_j}$$

we have

$$\|T_n^* w^n\|_{S_w \cup \overline{R_w}} \le \alpha \|T_n w^n\|_{S_w \cup \overline{R_w}} \le \alpha t_n^w .$$

Since this implies

$$\|T_n^* w^n\|_\Sigma \le \alpha t_n^w < t_n^w$$

(see Theorem 2.7), we get a contradiction to the definition of the Chebyshev number t_n^w , and this verifies the claim. □

Concerning the distribution of the zeros of Chebyshev polynomials, in the next section we shall examine in detail the zero distribution of near-extremal polynomials. In particular, those results can be applied to weighted Chebyshev polynomials. Here we only state but do not pause to prove the most widely used case.

Theorem 3.6. *Suppose that w is an admissible weight on Σ, $\{T_n\}$ are the associated Chebyshev polynomials and ν_n, $n = 1, 2, \ldots$, are the normalized counting measures on the zero sets of the T_n's. If $\mathcal{S}_w$ has empty interior and connected complement in* **C**, *then*

$$\lim_{n \to \infty} \nu_n = \mu_w \tag{3.9}$$

in the weak topology of measures.*

In particular, (3.9) *holds if Σ is a subset of the real line.*

This result is no longer true if $\mathcal{S}_w$ has nonempty interior or disconnected complement as the following examples show.

Example 3.7. Let Σ be the unit circle C_1 and $w \equiv 1$ on Σ. Then μ_w is the normalized Lebesgue measure on C_1 and $T_n(z) = z^n$. Thus, (3.9) fails to hold. In this example $\mathcal{S}_w = C_1$ has a disconnected complement in **C**. □

Example 3.8. Let Σ be the unit disk $\overline{D_1} = \{z \mid |z| \le 1\}$, m the normalized (two-dimensional) Lebesgue measure on $\overline{D_1}$ and

$$w(z) = \exp[(1 - |z|^2)/2].$$

It is easy to see that $(1 - |z|^2)/2$ coincides with $U^m(z)$ on $\overline{D_1}$; hence (see Theorem I.3.3) $m = \mu_w$ and so $\mathcal{S}_w = \overline{D_1}$. On the other hand, by circular symmetry, we again have $T_n(z) = z^n$, i.e. (3.9) fails. In this example $\mathcal{S}_w = \overline{D_1}$ has nonempty interior. □

For the n-th root behavior of $|T_n(z)|$ itself we have

Theorem 3.9. *Let w be an admissible weight on Σ and $\{T_n\}$ the associated Chebyshev polynomials. Then*

$$\lim_{n\to\infty} |T_n(z)|^{1/n} = \exp(-U^{\mu_w}(z)) = \exp\left(\int \log|z-t|d\mu_w(t)\right) \tag{3.10}$$

uniformly on compact subsets of $\overline{\mathbf{C}} \setminus \mathrm{Con}(S_w \cup \overline{R_w})$, *where* $\mathrm{Con}(S_w \cup \overline{R_w})$ *denotes the convex hull of the set* $S_w \cup \overline{R_w}$ *(see* (2.11) *for the definition of* R_w*).*

Examples 3.7 and 3.8 show that (3.10) need not be true in $\mathrm{Con}(S_w \cup \overline{R_w})$ or in $S_w \cup \overline{R_w}$.

Since the zeros of T_n lie in $\mathrm{Con}(S_w \cup \overline{R_w})$ (see Theorem 3.4), $T_n(z)^{1/n}$ is well defined in $\mathbf{C}\setminus\mathrm{Con}(S_w \cup \overline{R_w})$. If we take that branch of the n-th root that is positive for $z \to \infty$, $z > 0$, then (3.10) is equivalent to

$$\lim_{n\to\infty} T_n(z)^{1/n} = \exp\left(\int \log(z-t)d\mu_w(t)\right).$$

Proof of Theorem 3.9. We can use Theorem 4.7 from the next section. To apply it all we have to know is that the zeros of every T_n lie in $\mathrm{Con}(S_w \cup \overline{R_w})$, which was proved in Theorem 3.4. □

III.4 Zero Distribution of Polynomials of Asymptotically Minimal Weighted Norm

Let w be an admissible weight on Σ. According to Theorem. I.3.6, for any sequence of monic polynomials $P_n(z) = z^n + \cdots$, $n = 1, 2, \ldots$, we have

$$\liminf_{n\to\infty} (\|w^n P_n\|_\Sigma^*)^{1/n} \geq \exp(-F_w),$$

where $\|f\|_\Sigma^*$ denotes the "sup" norm of f, i.e.

$$\|f\|_\Sigma^* := \inf\{L \mid |f(z)| \leq L \text{ q.e. on } \Sigma\}.$$

We will consider *asymptotically extremal monic polynomials* P_n that satisfy

$$\lim_{n\to\infty} (\|w^n P_n\|_\Sigma^*)^{1/n} = \exp(-F_w). \tag{4.1}$$

We shall show that in several important cases the asymptotic zero distribution of these polynomials coincides with the extremal measure μ_w.

In fact, we prove

Theorem 4.1. *Let w be an admissible weight such that S_w has two-dimensional Lebesgue measure zero, $\{P_n\}$ a sequence of monic polynomials of respective degrees $n = 1, 2, \ldots$ satisfying* (4.1) *and ν_n the normalized counting measure on the zeros of P_n. Then the following two statements are equivalent:*

(a) $\nu_n \to \mu_w$ *as $n \to \infty$ in the weak* topology.*

(b) *For every bounded component R of $\mathbf{C} \setminus \mathcal{S}_w$ and every subsequence $\mathcal{N}$ of the natural numbers there is a $z_0 \in R$ and a subsequence $\mathcal{N}_1 \subset \mathcal{N}$ such that*

$$\lim_{n\to\infty,\, n\in\mathcal{N}_1} |P_n(z_0)|^{1/n} = \exp(-U^{\mu_w}(z_0)). \tag{4.2}$$

Note that, by Theorem 2.1, equation (4.1) implies that

$$\limsup_{n\to\infty,\, n\in\mathcal{N}_1} |P_n(z_0)|^{1/n} \le \exp(-U^{\mu_w}(z_0)).$$

Thus, assertion (b) of Theorem 4.1 roughly says that $|P_n|$ is asymptotically maximal in at least one point of every bounded component of $\mathbf{C} \setminus \mathcal{S}_w$ (and then, of course, it will be maximal everywhere except for points that are accumulation points of the zeros of P_n). Notice that in the unbounded component, $z_0 = \infty$ always has this property and this is the reason why we do not have to impose any condition in Theorem 4.1 concerning the unbounded component of $\mathbf{C} \setminus \mathcal{S}_w$.

A slight modification of the proof (to be given below) of Theorem 4.1 yields

Theorem 4.2. *Let w be an admissible weight such that S_w has empty interior and connected complement in $\mathbf{C}$. Then* (4.1) *implies that $\nu_n \to \mu_w$ in the weak* sense. In particular, this is true if $\Sigma \subseteq \mathbf{R}$.*

Example 4.3. Let $\Sigma = [-1, 1]$ and $w \equiv 1$ on Σ. Then Theorem 4.2 together with Example I.3.5 show the following. If $\{P_n\}$ is any sequence of monic polynomials with $\deg(P_n) = n$ satisfying

$$\lim_{n\to\infty} \|P_n\|_{[-1,1]}^{1/n} = \frac{1}{2},$$

then the corresponding normalized zero measures ν_n converge in the weak* topology to the arcsine distribution $dx/\pi\sqrt{1-x^2}$, $x \in (-1, 1)$. □

The above theorems are best possible; for example, Theorem 4.2 is false if $\mathcal{S}_w$ has nonempty interior or disconnected complement (see Examples 3.7 and 3.8 in Section II.3).

The proof of Theorem 4.1 requires two lemmas.

Lemma 4.4. *Let w be an admissible weight, $\{P_n\}$ a sequence of monic polynomials of respective degrees n satisfying* (4.1), *and ν_n the associated zero counting measure. Then $\nu_n(K) \to 0$ as $n \to \infty$ for every closed subset K of the unbounded component Ω of $\overline{\mathbf{C}} \setminus \mathcal{S}_w$.*

Proof. Set

$$g_n(z) := \frac{1}{n}\log|P_n(z)| - \int \log|z-t|\,d\mu_w(t) + \frac{1}{n}\sum_{z_{n,k}\in\Omega} g_\Omega(z, z_{n,k}),$$

where $\{z_{n,k}\}_{k=1}^n$ are the zeros of P_n and $g_\Omega(z,\zeta)$ is the Green function with pole at ζ for the unbounded component Ω of $\overline{\mathbf{C}} \setminus \mathcal{S}_w$. Then $g_n(z)$ is harmonic in Ω (even at ∞). Now (4.1), Theorems I.1.3, I.4.4 and the properties of Green functions (cf. Section II.4, especially (4.3)(iii) in that section) imply for quasi-every $z' \in \partial\Omega$ that

$$\limsup_{z\to z',\, z\in\Omega} g_n(z) \le \varepsilon_n, \quad \varepsilon_n \to 0.$$

Hence, by the generalized minimum principle (Theorem I.2.4), we conclude that $g_n(z) \le \varepsilon_n$ in Ω, and on letting $z \to \infty$ we get from the symmetry of Green functions

$$\frac{1}{n}\sum_{z_{n,k}\in\Omega} g_\Omega(\infty, z_{n,k}) = \frac{1}{n}\sum_{z_{n,k}\in\Omega} g_\Omega(z_{n,k}, \infty) \le \varepsilon_n.$$

We can write this inequality in the form

$$0 \le \int g_\Omega(z,\infty)d\nu_n^\Omega(z) \le \varepsilon_n, \tag{4.3}$$

where

$$\nu_n^\Omega := \frac{1}{n}\sum_{z_{n,k}\in\Omega} \delta_{z_{n,k}}$$

is the restriction of ν_n to Ω. Thus

$$\int g_\Omega(z,\infty)d\nu_n^\Omega(z) \to 0 \quad \text{as} \quad n \to \infty. \tag{4.4}$$

If K is any closed subset of Ω, then g_Ω is bounded from below on K by a positive constant. Thus (4.4) implies that $\nu_n(K) \to 0$ as $n \to \infty$. □

Next we prove

Lemma 4.5. *With the assumptions of Lemma* 4.4, *condition* (b) *of Theorem* 4.1 *implies that* $\nu_n(K) \to 0$ *as* $n \to \infty$ *for every compact subset* K *of every bounded component* R *of* $\overline{\mathbf{C}} \setminus S_w$.

Proof. If R is a bounded component of $\mathbf{C}\setminus S_w$, then working with the z_0 guaranteed by (b) instead of with ∞, the proof is almost identical to that of Lemma 4.4. Namely, in the same manner we get that for every $\mathcal{N} \subseteq \mathbf{N}$ there is an $\mathcal{N}_1 \subseteq \mathcal{N}$ and a $z_0 \in R$ such that

$$\int g_R(z,z_0)d\nu_n^R(z) \to 0 \quad \text{as} \quad n \to \infty, \ n \in \mathcal{N}_1,$$

where ν_n^R is the restriction of ν_n to R. Hence $\nu_n(K) \to 0$ for $n \to \infty$, $n \in \mathcal{N}_1$, if $K \subseteq R$ is compact. Since this is true for every $\mathcal{N}$, the sequence $\{\nu_n(K)\}_{n=1}^\infty$ must tend to zero. □

Proof of Theorem 4.1. (b)⇒(a) Let σ be a weak* limit point of the measures ν_n on $\overline{\mathbf{C}}$; that is, $\nu_n \to \sigma$ in the weak* sense as $n \to \infty$, $n \in \mathcal{N}$ for some $\mathcal{N}$. It is enough to show that $\sigma = \mu_w$.

In view of Lemma 4.4, σ is a compactly supported unit Borel measure and, from Lemmas 4.4 and 4.5, we get that $\mathrm{supp}(\sigma) \subseteq \mathcal{S}_w$. Furthermore, (4.1) and Theorem I.1.3(e) imply that for quasi-every $z \in \mathcal{S}_w$

$$\int \log|z-t|\,d\nu_n(t) - \int \log|z-t|\,d\mu_w(t) \le \varepsilon_n, \quad \varepsilon_n \to 0. \tag{4.5}$$

We can apply the principle of domination (Theorem II.3.2) to conclude that (4.5) holds for every z.

Now let $\varepsilon > 0$. If $\mathrm{dist}(z, \mathcal{S}_w) > 2\varepsilon$, then (4.5) yields

$$\int \log|z-t|\,d\tilde{\nu}_n(t) - \int \log|z-t|\,d\mu_w(t) \le \varepsilon_n + o(1), \quad \varepsilon_n \to 0,$$

where $\tilde{\nu}_n$ is the normalized counting measure associated with those zeros of P_n that are closer than ε to $\mathcal{S}_w$. Since we also have $\tilde{\nu}_n \to \sigma$ as $n \to \infty$, $n \in \mathcal{N}$, we deduce that for every $z \notin \mathcal{S}_w$

$$-\int \log\frac{1}{|z-t|}\,d\sigma(t) + \int \log\frac{1}{|z-t|}\,d\mu_w(t) \le 0. \tag{4.6}$$

The left-hand side of (4.6) is a harmonic function on $\overline{\mathbf{C}}\setminus\mathcal{S}_w$ that, by assumption (b) and by the argument used in deriving (4.6), has at least one zero in every component of $\overline{\mathbf{C}} \setminus \mathcal{S}_w$ (in the unbounded component this zero occurs at $z = \infty$); hence by (4.6) and the maximum principle it is identically zero on $\overline{\mathbf{C}}\setminus\mathcal{S}_w$. Thus U^σ and U^{μ_w} coincide almost everywhere with respect to two-dimensional Lebesgue measure (recall, it is assumed that $\mathcal{S}_w$ has zero Lebesgue measure), and so $\sigma = \mu_w$ according to Corollary II.2.2, which proves assertion (a).

(a)⇒(b) In view of (4.1) and Theorem 2.1 we have

$$\limsup_{n\to\infty} |P_n(z_0)|^{1/n} \le \exp(-U^{\mu_w}(z_0)), \quad z_0 \in \mathbf{C}. \tag{4.7}$$

Let R be a bounded component of $\mathbf{C} \setminus \mathcal{S}_w$ and for $\varepsilon > 0$ set

$$R_\varepsilon := \{z \mid z \in R, \ \ \mathrm{dist}(z, \mathcal{S}_w) > \varepsilon\},$$

$$\nu_n^{(1)} := \frac{1}{n}\sum_{z_{n,k}\in R_\varepsilon} \delta_{z_{n,k}}, \qquad \nu_n^{(2)} := \nu_n - \nu_n^{(1)},$$

where $\{z_{n,k}\}_{k=1}^n$ are the zeros of P_n and δ_z denotes the unit mass at z. From (a) we get that $\nu_n^{(2)} \to \mu_w$ in the weak* sense. Therefore,

$$\int \log|z-t|\,d\nu_n^{(2)}(t) \to \int \log|z-t|\,d\mu_w(t), \quad z \in R_\varepsilon.$$

We shall show that if $\varepsilon > 0$ is so small that the set R_ε has positive two-dimensional Lebesgue measure $m_2(R_\varepsilon) > 0$ and $\mathcal{N} \subseteq \mathbf{N}$ is an arbitrary subsequence of the natural numbers, then for some $z_0 \in R_\varepsilon$

$$\limsup_{n\to\infty,\, n\in\mathcal{N}} \int \log|z_0 - t|d\nu_n^{(1)}(t) \geq 0.$$

This will prove (4.2) because of the preceding limit relation and (4.7).

Let

$$q_n(z) := \prod_{z_{n,k}\in R_\varepsilon} (z - z_{n,k}).$$

Then $\deg(q_n) =: d_n = o(n)$, and the preceding inequality is equivalent to

$$\limsup_{n\to\infty, n\in\mathcal{N}} |q_n(z_0)|^{1/n} \geq 1. \tag{4.8}$$

For $0 < a < 1$, consider the lemniscate set

$$E_{n,a} := \{z \mid |q_n(z)| \leq a^n\}.$$

By formula (3.7) it has capacity a^{n/d_n}. Hence we can choose a sequence $a_n \to 1-0$ and an $\mathcal{N}_1 \subseteq \mathcal{N}$ such that

$$\sum_{n\in\mathcal{N}_1} [\text{cap}(E_{n,a_n})]^2 < m_2(R_\varepsilon)/10.$$

But then by Lemma 4.6 below

$$\sum_{n\in\mathcal{N}_1} m_2(E_{n,a_n}) < m_2(R_\varepsilon),$$

and so there is a $z_0 \in R_\varepsilon$ that is not contained in any of the sets E_{n,a_n}, $n \in \mathcal{N}_1$. For such a point z_0 inequality (4.8) holds. □

In the preceding proof we used the following lemma.

Lemma 4.6. *Let E be a compact subset of* **C** *and $m_2(E)$ its Lebesgue measure on the plane. Then*

$$\text{cap}(E) \geq \sqrt{\frac{m_2(E)}{\pi e}}.$$

Proof. Let $\lambda := m_2(E) > 0$, $r_0 := \sqrt{\lambda/\pi}$ and

$$D_{r_0}(x) := \{z \mid |z - x| < r_0\}.$$

Now for fixed x, we have

$$\int_E \log\frac{r_0}{|x-t|}dm_2(t) = \int_{E\setminus D_{r_0}(x)} \log\frac{r_0}{|x-t|}dm_2(t)$$

$$+ \int_{E\cap D_{r_0}(x)} \log\frac{r_0}{|x-t|}dm_2(t) \leq \int_{E\cap D_{r_0}(x)} \log\frac{r_0}{|x-t|}dm_2(t),$$

and so, from the definition of r_0, we get

$$\int_E \log \frac{1}{|x-t|} dm_2(t) \le \int_{D_{r_0}(x)} \log \frac{1}{|x-t|} dm_2(t). \tag{4.9}$$

Moreover,

$$\begin{aligned}\int_{D_{r_0}(x)} \log \frac{1}{|x-t|} dm_2(t) &= \int_0^{r_0} r \int_0^{2\pi} \log \frac{1}{|re^{i\varphi}|} d\varphi\, dr \\ &= \pi r_0^2 \log \frac{1}{r_0} + \frac{\pi}{2} r_0^2.\end{aligned}$$

Thus

$$\int_E \log \frac{1}{|x-t|} dm_2(t) \le \lambda \left(\frac{1}{2} \log \frac{\pi e}{\lambda} \right). \tag{4.10}$$

If we divide this inequality by λ^2 and integrate on E with respect to m_2 we get that the logarithmic energy of the unit mass $m_2|_E/\lambda$ concentrated on E is at most

$$\frac{1}{2} \log \frac{\pi e}{m_2(E)}.$$

Hence, the energy V of E (cf. Section I.1) is also bounded by this number and consequently the inequality of the lemma follows from the definition of the logarithmic capacity (see (I.1.5)). □

Now we continue with the

Proof of Theorem 4.2. Following the proof of Lemma 4.4 and the proof of the implication (b)⇒(a) of Theorem 4.1 we get that, for every weak* limit σ of $\{\nu_n\}$, the potential U^σ coincides with U^{μ_w} outside S_w. Since S_w has empty interior and connected complement, it coincides with the boundary of $\Omega := \overline{\mathbb{C}} \setminus S_w$. Now we can invoke Carleson's unicity result Theorem II.4.13 and conclude from $U^\sigma = U^{\mu_w}$ in Ω that $\sigma = \mu_w$. □

When S_w has nonempty interior little can be said about the zero distribution of monic polynomials P_n satisfying (4.1). Indeed, recall Example 3.8 from the preceding section; it is easy to see that for $0 \le a \le \exp(-1/2)$ the polynomials $\{z^n - a^n\}$ satisfy (4.1) and the limit distribution of their zeros is the uniform distribution on the circle $\{z \mid |z| = a\}$, which is quite different from μ_w. We also see from this example (by blending terms from different sequences) that the zeros may not have a limit distribution. Nevertheless, there are restrictions on the weak* limit points of $\{\nu_n\}$ as is shown in the next theorem.

Theorem 4.7. *Let w be an admissible weight, Ω the unbounded component of $\overline{\mathbb{C}} \setminus S_w$, and $\{P_n\}$ a sequence of monic polynomials satisfying* (4.1). *If σ is a weak* limit point of the normalized counting measures $\{\nu_n\}$ associated with the zero sets of the P_n's, then*

(i) $U^\sigma = U^{\mu_w}$ *in Ω; in particular,* $\operatorname{supp}(\sigma) \subseteq \overline{\mathbb{C}} \setminus \Omega$;

(ii) $\widehat{\nu_n} \to \widehat{\mu_w}$ *in the weak* sense, where* $\widehat{\nu}$ *denotes the balayage of the measure* ν *out of* $\mathbf{C} \setminus \overline{\Omega}$ *onto* $\partial\Omega$*; furthermore,* $\widehat{\sigma} = \widehat{\mu_w}$*;*

(iii) *for every entire function* h,

$$\int h \, d\sigma = \int h \, d\mu_w; \tag{4.11}$$

in particular, if the zeros of the P_n*'s are bounded, then*

$$\lim_{n\to\infty} \int h \, d\nu_n = \int h \, d\mu_w; \tag{4.12}$$

(iv) *if* $z \in \Omega$ *is not a limit point of the zeros of the* P_n*'s, then*

$$\lim_{n\to\infty} |P_n(z)|^{1/n} = \exp(-U^{\mu_w}(z)). \tag{4.13}$$

For the concept of balayage see Section II.4.

Conclusion (iii) expresses a "very weak* limit", since weak* would be the same statement but with continuous h. Of course, (iv) holds uniformly away from the zeros of the P_n's, and it shows (cf. Theorem 2.1) that asymptotic minimality on Σ (cf. (4.1)) automatically implies asymptotic maximality outside Σ (or S_w).

We also remark that, in general, nothing more can be said about the zeros of polynomials satisfying (4.1) than the statement in Theorem 4.7. As an example consider the classical case: $\Sigma = \{z \mid |z| = 1\}$, $w \equiv 1$ on Σ. In this case μ_w is the normalized Lebesgue measure on the unit circumference, and by looking at the potentials of the measures in question it is easy to verify the following: if $\{P_n\}$ is any sequence of monic polynomials with bounded zeros and of corresponding degrees $n = 1, 2, \ldots$, such that for the normalized counting measure ν_n on the zero set of P_n we have property (ii), i.e. $\widehat{\nu_n} \to \widehat{\mu_w} = \mu_w$ as $n \to \infty$ in weak* sense, then $\{P_n\}$ also satisfies (4.1) (with $F_w = 0$). In particular, the property expressed in part (i) of Theorem 4.7 is the only restraint for weak* limit points of the measures $\{\nu_n\}$.

Proof of Theorem 4.7. (i) was established in the proof of Theorem 4.1 (see Lemma 4.4 and the proof of Theorem 4.1(b)$\Rightarrow$ (a)).

(ii) Let σ^* be a weak* limit point of the measures $\{\widehat{\nu_n}\}$ and let

$$\lim_{n\to\infty,\, n\in\mathcal{N}} \widehat{\nu_n} = \sigma^*.$$

It is enough to show that $\sigma^* = \widehat{\mu_w}$. From part (i) of the theorem we know that $\mathrm{supp}(\sigma^*) \subseteq \partial\Omega$. By selecting a subsequence from $\mathcal{N}$ we can also suppose that the sequence $\{\nu_n\}_{n\in\mathcal{N}}$ converges to some measure σ in the weak* topology.

By the lower envelope theorem (Theorem I.6.9)

$$\liminf_{n\to\infty,\, n\in\mathcal{N}} U^{\widehat{\nu_n}}(z) = U^{\sigma^*}(z)$$

and

$$\liminf_{n\to\infty,\, n\in\mathcal{N}} U^{\nu_n}(z) = U^{\sigma}(z)$$

hold for quasi-every $z \in \Omega$. If we compare these with the fact that the equalities

$$U^{\widehat{\nu_n}}(z) = U^{\nu_n}(z), \qquad U^{\widehat{\mu_w}}(z) = U^{\mu_w}(z)$$

also hold for quasi-every z, we can conclude from part (i) that

$$U^{\sigma^*}(z) = U^{\widehat{\mu_w}}(z) \tag{4.14}$$

for q.e. $z \in \Omega$. But the measures σ^* and $\widehat{\mu_w}$ are supported on $\partial\Omega$; hence the potentials in (4.14) are continuous in Ω and from this we deduce (4.14) for every $z \in \Omega$. Now (ii) follows from the unicity result Theorem II.4.13.

Assertion (iii) follows from Theorem II.4.7(c) and part (ii).

(iv) follows from (i) and property (4.4); namely the latter implies that for fixed z and large r

$$\lim_{n\to\infty} \left| \prod_{|z_{n,k}|>r} (z - z_{n,k}) \right|^{1/n} = 1.$$

□

For the unweighted case, the following consequence of Theorem 4.7 is often useful.

Corollary 4.8. *Let E be a compact set of positive capacity and connected complement. If $\{P_n\}$ is a sequence of monic polynomials of respective degrees n and P_n has at most $o(n)$ zeros on each compact subset of the interior of E, then the condition*

$$\limsup_{n\to\infty} \|P_n\|_E^{1/n} \le \mathrm{cap}(E) \tag{4.15}$$

implies that $\nu_n \xrightarrow{*} \mu_E$ as $n \to \infty$, where ν_n is the normalized counting measure associated with the zeros of P_n and μ_E is the equilibrium distribution for E.

Note that (4.15) together with (I.3.7) imply

$$\lim_{n\to\infty} \|P_n\|_E^{1/n} = \mathrm{cap}(E).$$

Proof of Corollary 4.8. Let $\Omega := \mathbf{C} \setminus E$ and consider any weak* limit measure σ of $\{\nu_n\}$. Then by Theorem 4.7(i) with ($w \equiv 1$ on E) we have $\mathrm{supp}(\sigma) \subseteq E$ and $U^{\sigma}(z) = U^{\mu_E}(z)$ for all $z \in \Omega$. From Corollary I.4.5 we know that μ_E is supported on $\partial\Omega$ and the same is true for σ because of the $o(n)$ condition and the fact that $\mathrm{supp}(\sigma) \subseteq E$. Hence by the Carleson unicity theorem (Theorem II.4.13) we deduce that $\sigma = \mu_E$. As σ was an arbitrary limit measure, the corollary follows. □

III.5 The Function of Leja and Siciak

In this section we introduce and investigate a function — due to F. Leja and J. Siciak [204] — that gives the smallest upper bound for polynomials majorized by a weight on a set Σ. It will turn out that this function is closely related to the equilibrium potential.

Let $w = \exp(-Q)$ be a quasi-admissible weight on an infinite set Σ, i.e. for the moment we do not assume that the set

$$\Sigma_0 := \{z \mid w(z) > 0\} \tag{5.1}$$

is of positive capacity. Let $\xi^{(n)} = \{\xi_0, \ldots, \xi_n\} = \left\{\xi_0^{(n)}, \ldots, \xi_n^{(n)}\right\}$ be an $(n+1)$-point Fekete set corresponding to w and Σ and set

$$L^{(i)}(z, \xi^{(n)}) := \prod_{k=0,\, k\neq i}^{n} \frac{z - \xi_k}{\xi_i - \xi_k}, \quad i = 0, 1, \ldots, n,$$

which are the basic polynomials of Lagrange interpolation,

$$\Phi^{(i)}(z, \xi^{(n)}) := L^{(i)}(z, \xi^{(n)}) e^{nQ(\xi_i)}, \quad i = 0, 1, \ldots, n,$$

and

$$\Phi_n(z) := \max_i |\Phi^{(i)}(z, \xi^{(n)})|.$$

The sequence $\{\Phi_n(z)^{1/n}\}$ converges to a function $\Phi(z)$ called the *Leja-Siciak function* for every $z \in \mathbf{C}$, as we now show. First we verify the estimate

$$|\Phi^{(i)}(z, \xi^{(n)})| \leq e^{nQ(z)}, \quad z \in \Sigma. \tag{5.2}$$

In fact, in the opposite case there would be an i and a z such that

$$\left(\prod_{k\neq i} |z - \xi_k|\right) e^{-nQ(z)} > \left(\prod_{k\neq i} |\xi_i - \xi_k|\right) e^{-nQ(\xi_i)},$$

which is impossible by the definition of Fekete points, since then the weighted Vandermonde expression (appearing in (1.1)) corresponding to $\{\xi_0, \ldots, \xi_{i-1}, z, \xi_{i+1}, \ldots, \xi_n\}$ would be bigger than that of $\xi^{(n)}$.

Let us now apply the Lagrange interpolation formula to the polynomial $[\Phi^{(i)}]^k$ of degree nk in the form

$$\Phi^{(i)}(z, \xi^{(n)})^k = \sum_{l=0}^{m} \Phi^{(i)}\left(\xi_l^{(m)}, \xi^{(n)}\right)^k L^{(l)}(z, \xi^{(m)}),$$

where k is an arbitrary natural number, and m is any number of the form $m = kn + r$, $0 \leq r < n$. From this formula and from (5.2) for $z = \xi_l^{(m)}$ we get

$$\left|\Phi^{(i)}(z, \xi^{(n)})\right|^k \leq \sum_{l=0}^{m} \left|\Phi^{(l)}(z, \xi^{(m)})\right| e^{-ra},$$

where a is a lower bound for Q, and so

$$\left(\Phi_n(z)^{1/n}\right)^{nk/m} \leq (m+1)^{1/m}\left(\Phi_m(z)^{1/m}\right)e^{-ra/m}.$$

If $m \to \infty$, then $nk/m \to 1$, and it follows that

$$\Phi_n^{1/n} \leq \liminf_{m\to\infty} \Phi_m^{1/m}, \tag{5.3}$$

which implies

$$\limsup_{m\to\infty} \Phi_m^{1/m} \leq \liminf_{m\to\infty} \Phi_m^{1/m},$$

and this proves the convergence claimed above.

It may appear that the function Φ depends upon the particular choice of Fekete points. But, as a consequence of Theorem 5.1 it can be seen that it depends only on w.

In [204] J. Siciak introduced several other sequences that have the same limit $\Phi(z)$. For example, if

$$\overline{\Phi}_n(z) := \inf_{\zeta^{(n)}\subset\Sigma} \max_i |\Phi^{(i)}(z, \zeta^{(n)})|,$$

where $\Phi^{(i)}(z, \zeta^{(n)})$ is defined as above, then it easily follows from the Lagrange interpolation formula that $\{\overline{\Phi}_n^{1/n}\}_{n=1}^{\infty}$ has the same limit as the sequence $\{\Phi_n^{1/n}\}_{n=1}^{\infty}$.

We shall now prove

Theorem 5.1. *Let w be a quasi-admissible weight. Then the function $\Phi(z)$ is the least upper bound of all the functions $|P_n(z)|^{1/n}$, where P_n denotes an arbitrary polynomial of degree n, $n = 1, 2, \ldots$, such that*

$$|P_n(z)| \leq \exp(nQ(z)), \quad z \in \Sigma. \tag{5.4}$$

It is obvious that we get the same result if the degrees of P_n are assumed to be at most n.

Corollary 5.2. *If $\Sigma \neq \mathbf{C}$, then $\Phi(z)$ is finite at a point $z \in \mathbf{C}\setminus\Sigma$ if and only if it is finite at every such point, which, in turn, is equivalent to* $\mathrm{cap}(\Sigma_0) > 0$, *where Σ_0 is the set in* (5.1)*; i.e. to the admissibility of w.*

Proof of Theorem 5.1. Exactly as in the verification of (5.3) one can show that $|P_n(z)|^{1/n} \leq \Phi(z)$ for all polynomials in question. Thus the supremum mentioned in the theorem is at most as large as $\Phi(z)$. On the other hand, (5.2) and the definition of Φ imply that we must have equality. □

Proof of Corollary 5.2. If w is admissible, then it follows from Theorems 2.1 and 5.1 that Φ is finite everywhere.

Now assume $\mathrm{cap}(\Sigma_0) = 0$. We want to show that for $z_0 \notin \Sigma$ we have $\Phi(z_0) = \infty$. Without loss of generality we may assume $z_0 \neq 0$ and $0 \notin \Sigma$. Consider the weight $w_\varepsilon := w + \chi_\varepsilon$ (cf. the proof of Theorem 1.3), where

$$\chi_\varepsilon(z) := \begin{cases} 1 & \text{if } |z| \le \varepsilon \\ 0 & \text{otherwise,} \end{cases}$$

with an $\varepsilon > 0$ for which $|z_0| > \varepsilon$ and the disk $\overline{D}_\varepsilon = \{z \mid |z| \le \varepsilon\}$ does not intersect Σ. For this weight the support S_w of the extremal measure is the circumference $\{z \mid |z| = \varepsilon\}$, the equilibrium potential is $\log(1/|z|)$ for $|z| \ge \varepsilon$ and $F_{w_\varepsilon} = \log(1/\varepsilon)$. If H_n are Fekete polynomials corresponding to w_ε and $P_n(z) = H_n(z)/\|w_\varepsilon^n H_n\|_\Sigma$, then it follows from Corollary 1.10 and from what we have just mentioned, that

$$\lim_{n\to\infty} |P_n(z_0)|^{1/n} = |z_0|/\varepsilon.$$

Clearly, P_n satisfies (5.4), and since $\varepsilon > 0$ is arbitrary, we get $\Phi(z_0) = \infty$. □

In Theorem 2.9 we verified that the supremum of all functions $|P_n(z)|^{1/n}$, where P_n are polynomials of degree n satisfying

$$|P_n(z)| \le e^{nQ(z)}, \quad n = 1, 2, \ldots,$$

quasi-everywhere on Σ, is equal to

$$\exp(-U^{\mu_w}(z) + F_w) \tag{5.5}$$

quasi-everywhere, where U^{μ_w} denotes the equilibrium potential corresponding to w. In other words, if we take the supremum

$$\sup_{\deg(P_n)\le n} |P_n(z)|^{1/n}$$

for $\|P_n w^n\|_\Sigma \le 1$, then we get $\Phi(z)$, while taking the same supremum but for $\|P_n w^n\|_\Sigma^* \le 1$ gives (5.5) quasi-everywhere. Thus, one expects that Φ coincides with (5.5) quasi-everywhere, and below we show that in fact, this is the case. From this point of view, the exceptional set where $|P_n(z)|w(z)^n > 1$ in the case when $\|P_n w^n\|_\Sigma^* \le 1$ (this set has zero capacity and may not be negligible with respect to the supremum norm), can be eliminated:

Corollary 5.3. *Let w be admissible. Then, for quasi-every z,*

$$\Phi(z) = \exp(-U^{\mu_w}(z) + F_w). \tag{5.6}$$

Furthermore, the left-hand side in (5.6) *is never larger than the right-hand side.*

A simple modification of the proof of Theorem 2.9 and the remark made after it show that (5.6) holds outside Σ and it holds at a point $z \in \Sigma$ if and only if the potential is continuous at z and

$$U^{\mu_w}(z) - F_w \ge -Q(z). \tag{5.7}$$

Note that by Theorem I.4.4 the potential U^{μ_w} is continuous quasi-everywhere, and by Theorem I.1.3(d) the relation (5.7) holds quasi-everywhere on Σ.

Proof of Corollary 5.3. We know from Theorems 2.1 and 5.1 that (5.6) with equality replaced by $\leq$ holds everywhere, and this is the last statement of the theorem.

Consider now the polynomials $P_n(z) := \Phi_n^w(z)/\|w^n\Phi_n^w\|_\Sigma$ where Φ_n^w are Fekete polynomials corresponding to w (cf. Section III.1), and let ν_n be the normalized counting measure associated with the zeros of Φ_n^w. According to Theorem 1.3 the sequence $\{\nu_n\}$ converges to μ_w in the weak* topology of measures and so from the lower envelope theorem (Lemma I.6.9) we get that

$$\limsup_{n\to\infty} |\Phi_n^w(z)|^{1/n} = \exp(-U^{\mu_w}(z))$$

quasi-everywhere. Since

$$\lim_{n\to\infty} \|w^n\Phi_n^w\|_\Sigma^{1/n} = \exp(-F_w)$$

(see Theorem 1.9), it follows that

$$\limsup_{n\to\infty} |P_n(z)|^{1/n} = \exp(-U^{\mu_w}(z) + F_w)$$

quasi-everywhere, which, together with the already proved second part verifies the first statement in the theorem, for the P_n clearly satisfy (5.4). □

III.6 Where Does the L^p Norm of a Weighted Polynomial Live?

We have seen in Section III.2 that the supremum norms of weighted polynomials essentially live on $\mathcal{S}_w$, and $\mathcal{S}_w$ is the smallest compact set with this property. In this section we show that concerning $L^p(\sigma)$ norms, where σ is a measure on Σ, a similar role is played by the sets $\mathcal{S}_w^*$ defined in (1.4). The measure σ must satisfy some natural conditions. The following assumptions are convenient to use and are sufficient in applications. First of all, to ensure σ-integrability of weighted polynomials, we assume

$$\int_{|z|>1} \frac{d\sigma(z)}{|z|^K} < \infty \tag{6.1}$$

for some $K > 0$. We also need a certain denseness of σ around points of Σ; namely that for every $z \in \Sigma$

$$\sigma(D_r(z)) \geq c_z r^L, \quad 0 \leq r \leq 1, \tag{6.2}$$

for some constants $L > 0$ independent of z and $c_z > 0$ which, however, may depend on z. Here, as usual,

$$D_r(z) := \{z' \mid |z' - z| < r\}.$$

For such σ's we prove

Theorem 6.1. *Suppose that Σ is regular in the sense that each component of $\mathbf{C} \setminus \Sigma$ is regular with respect to the Dirichlet problem, w is continuous on Σ and assume that $w^{1-\eta}$ is admissible for some $\eta > 0$. Suppose further that σ is a locally finite Borel measure on Σ with properties* (6.1) – (6.2). *Then if $0 < p < \infty$ and $\mathcal{N}$ is any neighborhood of the set*

$$\mathcal{S}_w^* = \{z \in \Sigma \mid U^{\mu_w}(z) - F_w \leq -Q(z)\}, \tag{6.3}$$

then the $L^p(\sigma)$ norm of weighted polynomials $P_n w^n$, $\deg(P_n) \leq n$, lives on $\mathcal{N}$ in the sense that there exist two positive constants D and d independent of n and P_n such that

$$\int |P_n w^n|^p d\sigma \leq \left(1 + De^{-dn}\right) \int_{\mathcal{N}} |P_n w^n|^p d\sigma. \tag{6.4}$$

Furthermore, $\mathcal{S}_w^*$ is the smallest compact set with this property; namely, if S is any compact set such that, for every neighborhood $\mathcal{N}$ of S, (6.4) *holds for all polynomials with some constants D, $d > 0$, then $\mathcal{S}_w^* \subseteq S$.*

Inequality (6.4) says that only an exponentially small fraction of the $L^p(\sigma)$ norm comes from the integral over $\Sigma \setminus \mathcal{N}$. Note also that $\mathcal{S}_w^*$ is compact, so $\mathcal{N} \setminus \mathcal{S}_w^*$ can have as small a σ-measure as we like by appropriately choosing $\mathcal{N}$.

The admissibility of $w^{1-\eta}$ is needed only to ensure σ-integrability of weighted polynomials. If σ is a finite measure, i.e. $\sigma(\Sigma) < \infty$, then this assumption can be dropped and the theorem holds for continuous admissible weights w. We also mention that the continuity of w is essential, without it the conclusion may be false — see Example 6.3 after the proofs.

Sometimes we only need that the $L^p(\sigma)$ norm lives on a set in the n-th root sense. The next result shows that under some natural assumptions on σ the smallest such set is $\mathcal{S}_w$.

Theorem 6.2. *Suppose that the assumptions of Theorem* 6.1 *are satisfied except that now we replace* (6.2) *by the condition*

$$\sigma(D_r(z) \cap \mathcal{S}_w) \geq c_z r^L, \quad 0 \leq r \leq 1, \tag{6.5}$$

for all $z \in \mathcal{S}_w$. Then the $L^p(\sigma)$ norm of weighted polynomials $P_n w^n$, $\deg(P_n) \leq n$, live on $\mathcal{S}_w$ in the sense that

$$\lim_{n\to\infty} \left(\sup_{\deg(P_n) \leq n} \frac{\|P_n w^n\|_{L^p(\sigma)}}{\|P_n w^n\|_{L^p(\sigma|_{\mathcal{S}_w})}} \right)^{1/n} = 1. \tag{6.6}$$

Furthermore, if S is any compact set and

$$\lim_{n\to\infty} \left(\sup_{\deg(P_n) \leq n} \frac{\|P_n w^n\|_{L^p(\sigma)}}{\|P_n w^n\|_{L^p(\sigma|_S)}} \right)^{1/n} = 1, \tag{6.7}$$

then $\mathcal{S}_w \subseteq S$.

Very often we do not need to know $\mathcal{S}_w$ explicitly for verifying (6.5). As a typical example consider the case when $\Sigma \subseteq \mathbf{R}$ consists of a finite number of intervals, Q is convex on each of them and σ is given by a positive continuous density function on Σ (with respect to linear Lebesgue measure). Then $\mathcal{S}_w$ consists of a finite number of intervals (see Theorem IV.1.10(d)); hence (6.5) is automatically satisfied.

Proof of Theorem 6.1. First we prove (6.4). Under the stated assumptions we apply Theorem I.5.1 to conclude the continuity of U^{μ_w}. It also easily follows from Theorem I.1.3(d), the regularity of Σ and the continuity of U^{μ_w} that

$$U^{\mu_w}(z) \geq -Q(z) + F_w \tag{6.8}$$

holds for all $z \in \Sigma$.

Let P_n be an arbitrary polynomial of degree at most n normalized as

$$\|P_n w^n\|_\Sigma = 1. \tag{6.9}$$

The idea of the proof of (6.4) is simple: we use inequality (2.2) of Theorem 2.1 to conclude from (6.8) that $P_n w^n$ is exponentially small outside $\mathcal{N}$ which, together with (6.1), will yield exponentially small $L^p(\sigma)$-integral outside $\mathcal{N}$. On the other hand, we shall show by making use of (6.2) that the total $L^p(\sigma)$-integral is not exponentially small, and these facts easily yield (6.4).

From (6.9) and Theorem 2.1 we obtain for $z \in \Sigma$

$$|P_n(z)w^n(z)| \leq \exp(n(-U^{\mu_w}(z) + F_w - Q(z))). \tag{6.10}$$

Since we have assumed that $w^{1-\eta}$ is admissible for some $\eta > 0$, and this is equivalent to the admissibility of w and the relation

$$\lim_{|z|\to\infty,\, z\in\Sigma} \left[Q(z)(1-\eta) - \log|z|\right] = \infty,$$

we can choose an $R \geq 1$ such that on the right of (6.10) the exponent is smaller than $-(\eta/2)n\log|z| - 1$ for $|z| \geq R$. Thus, for $n > 2K/(\eta/2)p$ we get from (6.1)

$$\int_{|z|>R} |P_n(z)w^n(z)|^p d\sigma(z) \leq \text{const.}\, e^{-np\eta/4}. \tag{6.11}$$

On the other hand, by the choice of the sets $\mathcal{S}_w^*$ and $\mathcal{N}$ and by the continuity of the exponent on the right of (6.10) we can see that there is a $\theta > 0$ such that this exponent is at most $-\theta n$ for $z \notin \mathcal{N}$. Thus,

$$\int_{z\notin\mathcal{N},\, |z|\leq R} |P_n(z)w^n(z)|^p d\sigma(z) \leq \text{const.}\, e^{-np\theta}. \tag{6.12}$$

In view of (6.11) and (6.12) the estimate (6.4) follows once we verify that the $L^p(\sigma)$ norm of $P_n w^n$ cannot be exponentially small (see (6.18) below).

To this end let us consider the sets

$$\Sigma^{(m)} = \{z \in \Sigma \,|\, c_z \geq 1/m\},$$

where the constant c_z is the one appearing in (6.2). If we choose c_z to be the *largest* possible value for which (6.2) holds, then it is easy to see that $\Sigma^{(m)}$ is closed, so we can assume without loss of generality that $\Sigma^{(m)}$ is closed. With χ_S denoting the characteristic function for S we set

$$w_m := w \cdot \chi_{\Sigma^{(m)}} \cdot \chi_{S_w} = w_{|\Sigma^{(m)} \cap S_w},$$

where, for convenience, we identify restrictions $w_{|S}$ with the products $w \cdot \chi_S$. Obviously, w_m is admissible for large m and $w_m \nearrow w_{|S_w}$. But $\mu_{w_{|S_w}} = \mu_w$ and the potential U^{μ_w} is continuous; hence we can invoke Theorem I.6.5 according to which $F_m := F_{w_m}$ converges to F_w, and with $\mu_m := \mu_{w_m}$ the potentials U^{μ_m} uniformly converge to U^{μ_w}. Thus, given $\varepsilon > 0$ we can choose m so that

$$|F_m - F_w| < \varepsilon \quad \text{and} \quad |U^{\mu_m}(z) - U^{\mu_w}(z)| < \varepsilon, \quad z \in \mathbf{C}. \tag{6.13}$$

Set

$$M_{m,n} := \|P_n w_m^n\|.$$

We claim that $M_{m,n}$ is not much smaller than $1 = \|P_n w^n\|_\Sigma$. In fact, by Theorem 2.1, we have for every z

$$|P_n(z)w^n(z)| \leq M_{m,n} \exp(n(-U^{\mu_m}(z) + F_m - Q(z))). \tag{6.14}$$

Inequalities (6.8) and (6.13) show that the right-hand side here is at most $M_{m,n}e^{2n\varepsilon}$, while the left-hand side becomes 1 for some z; hence

$$M_{m,n} \geq e^{-2n\varepsilon}. \tag{6.15}$$

Let z_0 be a point satisfying

$$|P_n(z_0)|w_m^n(z_0) = M_{m,n}.$$

Then by (6.8) and (6.13) in the estimate

$$|P_n(z)| \leq |P_n(z_0)| \exp(n(-U^{\mu_m}(z) + F_m - Q(z_0))),$$

which follows from (6.14), the right-hand side is at most as large as

$$|P_n(z_0)| \exp(n(-U^{\mu_w}(z) + U^{\mu_w}(z_0))) \cdot e^{2\varepsilon n}.$$

From these and the uniform continuity of U^{μ_w} we deduce the existence of a $\delta > 0$ such that for $|z - z_0| \leq \delta$

$$|P_n(z)| \leq |P_n(z_0)|e^{3n\varepsilon}, \tag{6.16}$$

and for $|z - z_0| \leq \delta$, $z \in \Sigma$,

$$|Q(z) - Q(z_0)| \leq \varepsilon.$$

Note that the constants do not depend on P_n or z_0. From (6.16) and Cauchy's formula for the derivative of an analytic function we get

$$|P_n'(z)| \le |P_n(z_0)| e^{3n\varepsilon}\frac{2}{\delta}, \qquad |z-z_0| < \frac{\delta}{2},$$

by which

$$|P_n(z)| > \tfrac{1}{2}|P_n(z_0)| \quad \text{if} \quad |z-z_0| < \tfrac{\delta}{4}e^{-3n\varepsilon}.$$

From this and the relation

$$w(z)/w(z_0) \ge e^{-\varepsilon} \quad \text{if} \quad |z-z_0| < \delta$$

(cf. the choice of δ above) we finally deduce (use also (6.15))

$$|P_n(z)w^n(z)| > \tfrac{1}{2}e^{-3\varepsilon n} \quad \text{if} \quad |z-z_0| < \tfrac{\delta}{4}e^{-3n\varepsilon}, \quad z \in \Sigma. \tag{6.17}$$

Invoking (6.2) and the fact that $z_0 \in \Sigma^{(m)}$, and so $c_{z_0} \ge 1/m$, we conclude that

$$\int |P_n w^n|^p d\sigma \ge \left(\frac{1}{2}e^{-3n\varepsilon}\right)^p \left(\frac{\delta}{4}e^{-3n\varepsilon}\right)^L \cdot \frac{1}{m},$$

and for $n \to \infty$ this implies

$$\liminf_{n\to\infty} \| P_n w^n \|_{L^p(\sigma)}^{1/n} \ge \exp\left(-3\varepsilon\left(1+\frac{L}{p}\right)\right).$$

Since here $\varepsilon > 0$ is arbitrary, we finally get

$$\liminf_{n\to\infty} \| P_n w^n \|_{L^p(\sigma)}^{1/n} \ge 1, \tag{6.18}$$

which was to be proved. Thus, (6.4) is established.

Now we turn to the second half of the theorem, the proof of which is similar to that of Theorem 2.3.

First we verify that $\mathcal{S}_w \subseteq S$. Suppose to the contrary that $\mathcal{S}_w \setminus S$ is not empty, and let $z_0 \in \mathcal{S}_w$ and $\delta > 0$ be such that $D_{2\delta}(z_0) \cap S = \emptyset$. We set $w_0 := w|_{\Sigma \setminus D_\delta(z_0)}$, $\mu_0 = \mu_{w_0}$ and $F_0 = F_{w_0}$. We remark that it is impossible to have

$$U^{\mu_0}(z) - F_0 \ge -Q(z) \tag{6.19}$$

for all $z \in \mathcal{S}_w \cap D_\delta(z_0)$, for then the estimate (2.2) of Theorem 2.1 would yield that the sup norm of weighted polynomials $P_n w^n$, $\deg(P_n) \le n$ live on $\Sigma \setminus D_\delta(z_0)$ which is not the case (see Theorem 2.3). Thus, (6.19) fails at some $z \in \mathcal{S}_w \cap D_\delta(z_0)$, and relabeling this z as z_0 we assume without loss of generality that

$$U^{\mu_0}(z_0) - F_0 < -Q(z_0) - 3\varepsilon$$

for some $\varepsilon > 0$. By continuity (note that $z_0 \notin \mathcal{S}_{w_0}$), we have for some $0 < \delta_1 \le \delta$,

$$U^{\mu_0}(z) - F_0 < -Q(z_0) - 2\varepsilon,$$

for $|z - z_0| \le \delta_1$. Let now Φ_n be the n-th Fekete polynomial associated with w_0. We get from Theorems 1.8 and 1.9 that the polynomials

$$P_n(z) = \Phi_n(z)/\|\Phi_n w^n\| \tag{6.20}$$

satisfy the estimates

$$\int_{\Sigma\setminus D_\delta(z_0)} |P_n w^n|^p d\sigma \le \text{const.} \tag{6.21}$$

(see also the beginning of the proof, in particular (6.11)) and

$$|P_n(z)w^n(z)| \ge \exp(n(-U^{\mu_0}(z) + F_{w_0} - Q(z) - \varepsilon)) \ge e^{n\varepsilon} \tag{6.22}$$

for $|z - z_0| < \delta_1$, $z \in \Sigma$, and all large n. Since the σ-measure of $D_{\delta_1}(z_0) \cap \Sigma$ is positive, we see from (6.21) and (6.22) that (6.4) does not hold for $\mathcal{N} = \Sigma\setminus D_\delta(z_0)$, and this contradiction proves that $z_0 \in \mathcal{S}_w \setminus S$ is indeed impossible.

The preceding proof actually shows that in the case when $\mathcal{S}_w \setminus S$ is not empty, then even (6.7) is impossible, i.e. the above argument proves the second half of Theorem 2.3, as well.

Finally, it is left to show that each $z_0 \in \mathcal{S}_w^* \setminus \mathcal{S}_w$ also belongs to S. Suppose again that to the contrary $z_0 \in (\mathcal{S}_w^* \setminus \mathcal{S}_w) \setminus S$, and $D_{2\delta}(z_0) \cap S = \emptyset$. If now Φ_n are the Fekete polynomials associated with $w_{|\mathcal{S}_w}$ (so that all their zeros lie in $\mathcal{S}_w$), then we get for the polynomials P_n in (6.20) as before (use Theorems 1.8 and 1.9)

$$\begin{aligned} &\liminf_{n\to\infty}\left(\int_{D_\delta(z_0)} |P_n w^n|^p d\sigma\right)^{1/pn} \\ &\ge \min_{z\in D_\delta(z_0)} \exp(-U^{\mu_w}(z) + F_w - Q(z)) \ge 1 - \theta_\delta \end{aligned} \tag{6.23}$$

with $\theta_\delta \to 0$ as $\delta \to 0$, where we used the continuity of U^{μ_w} and Q at z_0 and the relation

$$-U^{\mu_w}(z_0) + F_w - Q(z_0) = 0,$$

which comes from $z_0 \in \mathcal{S}_w^*$ and (6.8). Inequality (6.23) shows that

$$\liminf_{n\to\infty}\left(\int_{D_\delta(z_0)} |P_n w^n|^p d\sigma\right)^{1/pn} \ge 1,$$

and since (6.21) holds in our case, as well, we arrive again at a contradiction with (6.4) if $\mathcal{N} = \Sigma \setminus D_\delta(z_0)$. □

Proof of Theorem 6.2. Again let P_n be a polynomial of degree at most n normalized so that (6.9) holds. The proof given for (6.4) above shows (cf. (6.11) – (6.12)) that $\|P_n w^n\|_{L^p(\sigma)}$ is bounded from above:

$$\|P_n w^n\|_{L^p(\sigma)} \le \text{const.}, \tag{6.24}$$

and at the same time the inequality in (6.17) holds. But in the present case

$$\sigma(\mathcal{S}_w \cap D_r(z_0)) \geq \frac{1}{m} r^L, \quad r = \frac{\delta}{4} e^{-3n\varepsilon},$$

and using this instead of (6.2) we get the following analogue of (6.18) in the preceding proof:

$$\liminf_{n\to\infty} \| P_n w^n \|_{L^p(\sigma \,|\, \mathcal{S}_w)}^{1/n} \geq 1.$$

This and (6.24) prove (6.6).

The second part of Theorem 6.2 was established in the preceding proof (cf. the paragraph after (6.22)). □

The next example shows that the continuity of w in Theorems 6.1 and 6.2 is essential. As for condition (6.2), we only mention that it cannot be essentially relaxed either.

Example 6.3. Let $\Sigma = \mathbf{C}$,

$$w(z) = \begin{cases} 1 & \text{if } -1 \leq z \leq 1 \\ 0 & \text{if } |z| < 2, \; z \notin [-1, 1] \\ \left| z + \sqrt{z^2 - 1} \right|^{-2} & \text{for } |z| \geq 2, \end{cases}$$

and let σ be the planar Lebesgue measure. Then all conditions of Theorem 6.1 are satisfied except the continuity condition on w. Since for the equilibrium potential

$$U^{\mu_K}(z) = \log \frac{1}{\left| z + \sqrt{z^2 - 1} \right|} + \log 2$$

of $K = [-1, 1]$ we have

$$U^{\mu_K}(z) = -Q(z) + \log 2$$

on $[-1, 1]$ and

$$U^{\mu_K}(z) > -Q(z) + \log 2$$

everywhere else, we get from Theorem I.3.3 that $\mu_w = \mu_K$ and $\mathcal{S}_w = \mathcal{S}_w^* = [-1, 1]$. But obviously, if $\mathcal{N} = D_2(0) = \{z \,|\, |z| < 2\}$, then

$$\int_{\mathcal{N}} |P_n w^n|^p d\sigma = 0$$

for all polynomials, so (6.4) is not true. □

III.7 Notes and Historical References

Section III.1

For an interval $[\alpha, \beta]$ the location of the (unweighted) Fekete points is known explicitly. In fact, e.g. for $[-1, 1]$ the n-th Fekete points are unique and they are the zeros of the polynomial $(1-x^2)P_{n-2}^{(1,1)}(x)$, where $P_{n-2}^{(1,1)}(x)$ is the Jacobi polynomial with parameters $(1, 1)$ of degree $(n-2)$. This result is due to Stieltjes; see [216, Section 6.7]. Furthermore, this same result shows that for $w(x) = (1-x)^a(1+x)^b$ on $[-1, 1]$ with $a, b > 0$, the n-th weighted Fekete points are the zeros of the Jacobi polynomial $P_n^{(\alpha,\beta)}$ with parameters $\alpha = 2a(n-1) - 1$, $\beta = 2b(n-1) - 1$.

Lagrange interpolation in Fekete points for a compact set Σ (with $w \equiv 1$) yields a sequence of polynomials converging maximally (in the sense of Walsh) to a given function $f(z)$ analytic on Σ (cf. [229, Section 7.8]).

In the unweighted case, G. Szegő [215] verified the equality of the transfinite diameter and the logarithmic capacity. Note that, for the unweighted case, the second part of Theorem 1.3 immediately yields that the equilibrium measure μ_Σ for a compact set Σ has its support on the outer boundary $\partial_\infty \Sigma$ of Σ, since in this case Fekete points must all lie on $\partial_\infty \Sigma$ (cf. Corollary I.4.5).

The proof of Theorem 1.3 is taken from [219, Lemma 2.2].

Theorem 1.7 is due to J. Siciak [203].

The definitive formulation of Evans' theorem (Theorem 1.11) is the following (see Deny [34]): If μ is a measure with compact support, then the set where $U^\mu(z) = \infty$ is a G_δ-set of zero capacity. Conversely, if E is a bounded G_δ set of zero capacity, then there is a measure μ of compact support such that $U^\mu(z) = \infty$ precisely for $z \in E$.

Section III.2

Theorem 2.1 is taken from Mhaskar-Saff [161], [163] and was preceded by an earlier result of Siciak [203, Lemma 2.1]; see Corollary 5.3 of this chapter. Notice that on letting $z \to \infty$ in inequality (2.2) of Theorem 2.1 we obtain

$$\text{“}\sup_{z \in \mathcal{S}_w}\text{”}[w(z)]^n \, |P_n(z)| \geq \exp(-nF_w)$$

for monic polynomials $P_n(z) = z^n + \cdots$, which was proved earlier in Theorem I.3.6.

The first part of Corollary 2.6 is due to H.N. Mhaskar and E.B. Saff (cf. [163, Corollary 4.2]). The second part and Theorem 2.3 were proved by V. Totik ([220, Lemma 5.2]).

Theorem 2.9 should be compared with Corollary 5.3 in Section III.5 which concerns properties of the Leja-Siciak function Φ.

Section III.3

Theorem 3.1 is essentially due to H.N. Mhaskar and E.B. Saff [163, Theorem 4.3] who used an argument of H. Stahl [211] to handle the set of irregular points of Σ.

Our presentation utilizing Fekete points automatically handles the irregular case. Note that the sequence $\{t_n^w\}$ has the property of logarithmic subadditivity:

$$t_{m+n}^w \le t_m^w t_n^w$$

and hence (cf. Tsuji [222, Section III.5])

$$\lim_{n\to\infty} \left(t_n^w\right)^{1/n} = \inf_{k\ge 1}\left\{\left(t_k^w\right)^{1/k}\right\}.$$

The proof of the first part of Theorem 3.4 follows an argument of Fejér [43] for the unweighted case.

The fundamental Lemma 3.5 is taken from the book of Stahl and Totik [212, Lemma 1.3.2], but essentially goes back to H. Widom [230]. However, its proof is new.

For the case when $\Sigma \subseteq R$, Theorem 3.6 was proved by Mhaskar and Saff [161].

Section III.4

G. Szegő [214] in his investigation of the zeros of partial sums of Taylor series was probably the first one to study asymptotic zero distributions. His work was inspired by that of R. Jentzsch (cf. [86]).

Corollary 4.8 is due to H.-P. Blatt, E. B. Saff, and M. Simkani [17] who used it to prove the following result concerning the zeros of the polynomials $p_n^*(f; z)$ of best uniform approximation to a function f on a compact set E : Suppose f is continuous on E, analytic in the interior E^0 of E, but not analytic in any open set containing E, where the complement of E is connected and regular. Assume further that f does not vanish identically on any component of E^0. Then there exists a subsequence Λ of positive integers such that the normalized counting measures ν_n associated with the zeros of $p_n^*(f; z)$ satisfy $\nu_n \xrightarrow{*} \mu_E$ as $n \to \infty,\ n \in \Lambda$. In particular, every boundary point of E is a limit point of the set of zeros of the sequence of best approximants $\{p_n^*(f; z)\}_1^\infty$.

For sequences of polynomials that are not necessarily monic, the following theorem of Mhaskar and Saff [162] (who generalized an earlier result due to R. Grothmann) is useful. Let w be an admissible weight on Σ and $\{p_n\}$ be any sequence of polynomials with $\deg p_n \le n$. Let Ω denote the unbounded component of $\mathbf{C} \setminus S_w$ and assume that the following two conditions hold for a subsequence of integers Λ :

(i)

$$\limsup_{n\to\infty, n\in\Lambda} \left\|w^n p_n\right\|_{\partial\Omega}^{1/n} \le 1;$$

(ii) there is a point $z_0 \in \Omega$ such that

$$\liminf_{n\to\infty, n\in\Lambda} \left\{\frac{1}{n}\log|p_n(z_0)| + U^{\mu_w}(z_0) - F_w\right\} \ge 0.$$

Then if ν_n denotes the normalized zero counting measure associated with p_n, every weak* limit measure ν of $\{\nu_n\}_{n\in\Lambda}$ is supported on $\mathbf{C} \setminus \Omega$ (the polynomial convex hull of $\mathcal{S}_w$) and, for balayage onto $\partial\Omega$, we have $\widehat{\nu} = \widehat{\mu}_w$.

The proof of Lemma 4.4 is taken from Mhaskar-Saff [162, Lemma 4.1] but also easily follows from Lemma 3.5.

Lemma 4.6 follows the presentation in Tsuji [222, Theorem III.10]. The constant e in the lemma can, in fact, be deleted; see Goluzin [61, Section VII.2].

Theorem 4.7 is similar to a result of Mhaskar-Saff [162, Theorem 2.3].

Section III.5

Theorem 5.1 is due to J. Siciak [203, Theorem 2.2]. F. Leja and his students, especially J. Siciak and J. Górski systematically used the Φ function for approximating Green functions and conformal mappings. See the references concerning their works in the Bibliography.

Section III.6

For special weights on the real line, Theorem 6.1 is well known; see e.g. [138], [143], [169].

A result similar to Theorem 6.2 was proved by Stahl and Totik in [212, Theorem 6.4.1]. This reference also contains several examples concerning the sharpness of the result.

Chapter IV. Determination of the Extremal Measure

In this chapter we shall discuss methods for determining the extremal measure μ_w for the energy problem associated with $w = \exp(-Q)$ on a closed set Σ.

We start by establishing some extremal properties of the support $\mathcal{S}_w$ of μ_w. Determining this set is one of the most important aspects of the energy problem. In fact, by knowing $\mathcal{S}_w$, the extremal potential can be obtained by simultaneously solving Dirichlet problems in connected components of $\mathbf{C} \setminus \mathcal{S}_w$, and then we can launch the recovery machinery of Chapter II to capture μ_w. In this chapter we show that this program can be carried out in many important cases. In doing so, we need results that allow us to determine the support $\mathcal{S}_w$. The points of $\mathcal{S}_w$ can be characterized as points around which weighted polynomials $w^n P_n$ can attain their maximum modulus over Σ. This is useful in establishing if individual points belong to the support or not. The most important property of the support $\mathcal{S}_w$ is that it maximizes the so-called F-functional

$$F(K) := \log \operatorname{cap}(K) - \int Q \, d\omega_K,$$

where ω_K denotes the equilibrium measure of the set K. In several important cases $\mathcal{S}_w$ is essentially the only compact set K for which the F-functional attains its maximum, which allows us to transform the problem of determining $\mathcal{S}_w$ to the problem of determining the maximizing set K for $F(K)$. Of course the determination of the maximizing set for the F-functional can still be quite complicated, but it turns out that sometimes we know in advance some properties of $\mathcal{S}_w$ that allow us to consider the maximum only for a special class of the compact sets K. For example, in the case when Q is defined on a segment and is convex there, $\mathcal{S}_w$ has to be an interval, and the maximizing problem for the F-functional becomes a simple maximum problem in two variables (the endpoints of the unknown interval $\mathcal{S}_w$). This approach leads to some integral equations for the endpoints that can sometimes be explicitly solved. We shall find concrete solutions for Freud, Jacobi and Laguerre weights.

The first method for determining the extremal measure itself is based on the recovery theorems of Chapter II. One of them, namely Theorem II.1.4, determines a measure μ supported on a curve γ via boundary values of analytic functions with real part equal to the potential U^μ outside γ. Since the potential U^{μ_w} is the solution (modulo a constant) of a Dirichlet problem with boundary function $\log w = -Q$, the problem of determining μ_w becomes intimately connected to

Dirichlet problems and analytic conjugation. By applying a conformal mapping to the unit disk, we transform the problem to the determination of the trigonometric conjugate of the image of Q under this conformal mapping. Therefore, the classical theory of trigonometric series can be applied. We shall use this approach to establish smoothness properties of μ_w from those of Q. The relation of the smoothness of μ_w (like being in a Lipshitz class or its Radon–Nikodym derivative being in L^p) to that of Q is roughly the same as the relation of the smoothness of the trigonometric conjugate to that of the original function.

On the interval $[-1, 1]$ an alternative method for finding the extremal measure is to directly solve the integral equation

$$\int_{-1}^{1} \log \frac{1}{|x-t|} g(t)dt = -Q(x) + C, \qquad x \in (-1, 1),$$

using the singular integral

$$L[Q'](t) = \frac{2}{\pi^2} \mathrm{PV} \int_0^1 \frac{\sqrt{1-t^2}\, s Q'(s)}{\sqrt{1-s^2}(s^2-t^2)} ds.$$

With it the solution is given in the form

$$L[Q'](t) + \frac{B}{\pi\sqrt{1-t^2}}.$$

The main difficulty with this approach is to recognize if the obtained solution yields a positive measure, or not. We shall give conditions under which the positivity is automatically satisfied.

In several cases it is of primary importance to know how μ_{w^λ} behaves as λ is changed. We shall establish very precise lower and upper bounds for μ_{w^λ} in terms of μ_w for λ lying close to 1. In doing so we shall need to derive an inequality for measures from those on their potentials, which will be done using Besicovich's covering technique.

Later in the chapter we shall explicitly determine the extremal measures for Freud, Jacobi and Laguerre weights, as well as for certain radially symmetric ones. Finally, we discuss three problems from physics where external fields arise.

IV.1 The Support $\mathcal{S}_w$ of the Extremal Measure

The support $\mathcal{S}_w$ of the extremal measure is one of the most important quantities in determining the extremal measure corresponding to a weight w. In fact, suppose the support is a nice set, say it is bounded by a finite number of smooth Jordan curves and w is a continuous function. Then inside $\mathcal{S}_w$ the potential U^{μ_w} coincides with $-Q$ plus a constant and in the "holes" of $\mathcal{S}_w$ the potential U^{μ_w} is the solution of certain Dirichlet problems (see Section I.3). Thus, $\mathcal{S}_w$ gives a way of determining the equilibrium potential. Now inside $\mathcal{S}_w$ the extremal measure μ_w can be obtained by taking $(-1/2\pi)$-times the Laplacian of U^{μ_w}, i.e. that of $-Q$ (see Theorem II.1.3). Knowing μ_w inside $\mathcal{S}_w$ we can subtract from U^{μ_w} the potential of the part of μ_w lying inside $\mathcal{S}_w$ and if the difference is denoted by $U^{\mu_w^*}$, then the part of μ_w that is supported on the boundary of $\mathcal{S}_w$ can be obtained by taking $(-1/2\pi)$-times the sum of the directional partial derivatives of $U^{\mu_w^*}$ along the normal and along its opposite (see Theorem II.1.5). Thus, for a given w everything is computable — at least in principle — once $\mathcal{S}_w$ is known.

In Section V.1 we shall discuss a simple method by which $\mathcal{S}_w$ can be numerically determined. We would like to emphasize, however, that from a computational point of view the actual determination of $\mathcal{S}_w$ is usually an extremely hard problem. Therefore, knowing properties of the support $\mathcal{S}_w$ can be useful, and in this section we list some of them.

First of all let us mention that $\mathcal{S}_w$, being the support of a measure of finite logarithmic energy, is of positive capacity at every point of $\mathcal{S}_w$ (i.e., the intersection of $\mathcal{S}_w$ with any neighborhood of any point of $\mathcal{S}_w$ is of positive capacity), and every set with this property coincides with some $\mathcal{S}_w$ for some admissible w:

Theorem 1.1. *If S is a compact subset of $\mathbf{C}$ that is of positive capacity at every point of S, then there is an admissible weight w such that $\mathcal{S}_w = S$.*

Proof. By Theorem I.3.8 it is enough to produce a μ such that U^μ is continuous everywhere and $\operatorname{supp}(\mu) = S$. For this purpose we apply Corollary I.6.11 to suitably chosen compact subsets S_k of S having positive capacity to produce continuous potentials $U^{\nu_{S_k}}$. It suffices to choose the S_k's so that they have diameter less than $1/k$, $k = 1, 2, \ldots$, and cover every point of S infinitely many times. Then all we have to do is to form a suitable linear combination $\mu = \Sigma c_k \nu_{S_k}^*$, where the c_k's are positive numbers decreasing so fast that the corresponding series $\Sigma c_k U^{\nu_{S_k}^*}$ of potentials converges uniformly on compact subsets of $\mathbf{C}$. Note that the fact that S is of positive capacity at every one of its points is used to ensure the existence of the sets S_k. □

Next we characterize the points in $\mathcal{S}_w$. To this end we introduce the following

Definition 1.2. We say that the function g attains its *essential maximum modulus* on the set S in the subset $S_1 \subset S$ if $\|g\|_{S\setminus S_1}^* < \|g\|_S^*$, where $\|\cdot\|_S^*$ denotes the "essential supremum norm"; that is, $\|\cdot\|_S^*$ means supremum disregarding sets of zero capacity.

This amounts to saying that for some constant a, $|g| \le a$ quasi-everywhere on $S \setminus S_1$ but $|g| > a$ on a subset of S_1 of positive capacity.

Theorem 1.3. *Let w be an admissible weight with support Σ. Then $z \in \Sigma$ belongs to the support $\mathcal{S}_w$ of the extremal measure μ_w if and only if for every neighborhood B of z there exists a weighted polynomial $w^n P_n$, $\deg P_n \le n$, taking its essential maximum modulus on Σ in $B \cap \Sigma$.*

Proof. Let $z \in \mathcal{S}_w$ and let B be any neighborhood of z. Applying Theorem III.2.3 to the set $\Sigma \setminus B$, it follows that there is a weighted polynomial $w^n P_n$ with

$$\|w^n P_n\|^*_{\Sigma\setminus B} < \|w^n P_n\|^*_{\Sigma},$$

which shows that $w^n P_n$ takes its essential maximum modulus on Σ in $\Sigma \cap B$.

Conversely, if $w^n P_n$ takes its essential maximum modulus on Σ in $\Sigma \cap B$, then by (III.2.5), $B \cap \mathcal{S}_w \neq \emptyset$ and, of course, since this is true for every neighborhood B of z, then $z \in \mathcal{S}_w$. □

Corollary 1.4. *If w is a continuous admissible weight on Σ, and Σ is of positive capacity at each of its points, then a point z belongs to $\mathcal{S}_w$ if and only if for every neighborhood B of z there is a weighted polynomial $w^n P_n$,* $\deg P_n \leq n$, *such that $w^n P_n$ attains its maximum modulus only in B.*

Indeed, with the assumptions of Corollary 1.4, we have $\overline{w} = w$, where $\overline{w}$ is the weight introduced in (III.2.6) and the corollary follows from Theorem 1.3 and the properties of $\overline{w}$ discussed after (III.2.6); see also Corollary III.2.6.

As we have seen, $\mathcal{S}_w$ can be any compact set satisfying the mild condition of Theorem 1.1, so in general its complement $\mathbf{C} \setminus \mathcal{S}_w$ is multiply connected. Let Ω be the unbounded component of $\mathbf{C} \setminus \mathcal{S}_w$. Then we recall from Section I.4 that the boundary $\partial\Omega$ of Ω is called the outer boundary of $\mathcal{S}_w$, and $\mathbf{C} \setminus \Omega$ is called the polynomial convex hull of $\mathcal{S}_w$ and is denoted by $\mathrm{Pc}(\mathcal{S}_w)$. $\mathrm{Pc}(\mathcal{S}_w)$ is just the union of $\mathcal{S}_w$ and the bounded components of its complement, i.e. the union of $\mathcal{S}_w$ and its "holes". Clearly, the outer boundary of $\mathcal{S}_w$ and that of $\mathrm{Pc}(\mathcal{S}_w)$ are the same. Our next aim is to characterize $\mathrm{Pc}(\mathcal{S}_w)$, or equivalently its outer boundary, as the smallest solution of a certain maximum problem.

Let K be a compact subset of Σ of positive capacity, and define

$$F(K) := \log \mathrm{cap}(K) - \int Q \, d\omega_K \tag{1.1}$$

where ω_K denotes the equilibrium measure associated with the set K. This so-called F-functional of Mhaskar and Saff is one of the most powerful tools in finding $\mathcal{S}_w$ and μ_w. Since ω_K is supported on the outer boundary of K (see Corollary I.4.5) and $\mathrm{cap}(K) = \mathrm{cap}(\mathrm{Pc}(K))$, we have the equality $F(K) = F(\mathrm{Pc}(K))$.

Theorem 1.5. *Let w be an admissible weight on Σ. Then the following hold.*

(a) *For every compact set $K \subset \Sigma$ of positive capacity, $F(K) \leq F(\mathcal{S}_w)$.*

(b) $F(\mathcal{S}_w) = F(\mathrm{Pc}(\mathcal{S}_w)) = -F_w$, *where F_w is the modified Robin constant of Section* I.1, (I.1.14).

(c) *If for some compact set $K \subset \Sigma$ of positive capacity the equality $F(K) = F(\mathcal{S}_w)$ holds, then* $\mathrm{Pc}(\mathcal{S}_w) \subset \mathrm{Pc}(K)$, *i.e.* $\mathrm{Pc}(\mathcal{S}_w)$ *is the smallest polynomially convex set maximizing the F-functional.*

(d) *In particular, if Σ has empty interior and connected complement (e.g. $\Sigma \subset \mathbf{R}$), then $\mathcal{S}_w$ is the smallest compact set of positive capacity maximizing the F-functional.*

In many applications a simple convexity argument shows that for certain weights w the set $\mathcal{S}_w$ is convex. If we also have $\Sigma \subset \mathbf{R}$, then of course $\mathcal{S}_w$ is an interval. For intervals on the real line the F-functional is easily computable, and so, in view of part (d) of the theorem, the determination of $\mathcal{S}_w$ becomes a computable maximum problem. We shall apply this procedure in Section IV.5.

Proof of Theorem 1.5. Let $K \subset \Sigma$ be of positive capacity and, as usual, let μ_w be the extremal measure corresponding to w. The inequality

$$U^{\mu_w}(z) \geq -Q(z) + F_w$$

holds quasi-everywhere on K (see Theorem I.1.3(d)); hence it holds ω_K-almost everywhere because ω_K has finite logarithmic energy. Integrating this inequality with respect to ω_K we get

$$\iint \log \frac{1}{|z-t|} d\mu_w(t) d\omega_K(z) \geq -\int Q(z) d\omega_K(z) + F_w. \tag{1.2}$$

Changing the order of integration on the left-hand side and making use of the fact that the potential

$$U^{\omega_K}(z) := \int \log \frac{1}{|z-t|} d\omega_K(t)$$

is smaller than or equal to $\log(1/\text{cap}(K))$ for every $z \in \mathbf{C}$ (see (I.1.4) or Corollary II.3.4), we obtain

$$F(K) \leq -F_w. \tag{1.3}$$

If $K = \mathcal{S}_w$, then, since $U^{\omega_K}(z) = \log(1/\text{cap}(K))$ for quasi-every $z \in K$ (see (I.1.9)), the previous argument together with Theorem I.1.3(e) yields equality in (1.3), and this proves assertions (a) and (b).

If for some K we have $F(K) = -F_w$, then in the previous argument we cannot increase the left-hand side of (1.2) in deriving (1.3), i.e. we must have

$$U^{\omega_K}(z) = \log(1/\text{cap}(K))$$

μ_w-almost everywhere. This already implies $\text{Pc}(\mathcal{S}_w) \subset \text{Pc}(K)$, for in the opposite case we would have a $z_0 \in \mathcal{S}_w \setminus \text{Pc}(K)$ and a neighborhood $D_r(z_0)$ of it that is disjoint from $\text{Pc}(K)$. By the maximum principle, U^{ω_K} is strictly smaller than $\log(1/\text{cap}(K))$ in $D_r(z_0)$ (note that U^{ω_K} is harmonic outside $\text{Pc}(K)$), and $\mu_w(D_r(z_0)) > 0$ which contradicts what we said before, and this proves assertion (c).

Finally, assertion (d) follows from the fact that, if Σ has empty interior and connected complement, then so does $\mathcal{S}_w$ which implies $\text{Pc}(\mathcal{S}_w) = \mathcal{S}_w$. □

Next we discuss how $\mathcal{S}_w$ changes under certain operations on w. In addition to $\mathcal{S}_w$ we need a closely related set $\mathcal{S}_w^*$ defined by

$$\mathcal{S}_w^* := \{z \in \Sigma \mid U^{\mu_w}(z) \leq -Q(z) + F_w\}.$$

This is a closed set because U^{μ_w} and Q are lower semi-continuous, and according to Theorem I.1.3(e) we have $\mathcal{S}_w \subset \mathcal{S}_w^*$. We have already used $\mathcal{S}_w^*$ several times in Chapter III.

Let us measure the distance between two closed sets A and B by the Hausdorff distance

$$\mathrm{dist}\,(A,B) := \max\left(\sup_{z\in A}\inf_{t\in B}|z-t|,\ \sup_{z\in B}\inf_{t\in A}|z-t|\right).$$

With these notations we can state

Theorem 1.6. *Let v and w be admissible weights on Σ. Then the following propositions hold.*

(a) $\mathcal{S}_{\max\{v,w\}} \subset \mathcal{S}_v \cup \mathcal{S}_w$.

(b) *If $z \in \mathcal{S}_v \cap \mathcal{S}_w$, then $z \in \mathcal{S}_{\min\{v,w\}}$ provided in some neighborhood of z one of the functions v or w majorizes the other one.*

(c) *If $\mathcal{S}_w = \mathcal{S}_v$, then for $0 \le \alpha \le 1$ we have $\mathcal{S}_{w^\alpha v^{1-\alpha}} = \mathcal{S}_w$ and $\mu_{w^\alpha v^{1-\alpha}} = \alpha\mu_w + (1-\alpha)\mu_v$.*

(d) *If G is open, $\tilde{\Sigma} := \overline{G \cap \Sigma}$ and $\tilde{w} = w|_{\tilde{\Sigma}}$ is admissible on $\tilde{\Sigma}$, then $\mathcal{S}_w \cap G \subset \mathcal{S}_{\tilde{w}}$.*

(e) *If $\tilde{\Sigma}$ is a closed subset of $\mathcal{S}_w$ of positive capacity, then for $\tilde{w} = w|_{\tilde{\Sigma}}$ we have $\mu_{\tilde{w}} = \widehat{\mu_w}$, where $\widehat{\mu_w}$ denotes the balayage of μ_w onto $\tilde{\Sigma}$.*

(f) *$\mathcal{S}_{w^\lambda}$ is a right continuous (with respect to the Hausdorff metric) decreasing function of the $\lambda \in \mathbf{R}$ values for which w^λ is admissible. Furthermore, if Σ has empty interior and connected complement (in particular if $\Sigma \subset \mathbf{R}$), then*

$$\mathcal{S}^*_{w^{\lambda_0}} \text{“}\subset\text{”} \bigcap_{\lambda<\lambda_0} \mathcal{S}_{w^\lambda} \subset \mathcal{S}^*_{w^{\lambda_0}}, \tag{1.4}$$

where “$\subset$” means inclusion except for a set of zero capacity.

(g) *$\lambda \to \mathcal{S}_{w^\lambda}$ is discontinuous at most at countably many λ. It is continuous at λ_0 if $\mathcal{S}_{w^{\lambda_0}} = \mathcal{S}^*_{w^{\lambda_0}}$, and conversely, if it is continuous at λ_0 and Σ has empty interior and connected complement, then $\mathcal{S}_{w^{\lambda_0}}$ “$=$” $\mathcal{S}^*_{w^{\lambda_0}}$.*

(h) *If $w_1, w_2, \ldots$ are admissible weights converging to an admissible w in a monotone fashion, then $\mathcal{S}_w$ is contained in the closure of the set*

$$\limsup_{n\to\infty} \mathcal{S}_{w_n} := \bigcap_{m=1}^{\infty}\bigcup_{n=m}^{\infty} \mathcal{S}_{w_n}.$$

We shall only give a partial proof for the theorem here; parts (f) and (g) will be proved in Section IV.4 along with related results.

Property (e) can be especially useful, for it allows one to go directly from μ_w to $\mu_{w|_{\tilde{\Sigma}}}$. Recall also that the balayage of a measure onto a compact set K out of $\mathbf{C} \setminus K$ was introduced in Section II.4. We also mention that in (e) the equality $\mathcal{S}_{\tilde{w}} = \tilde{\Sigma}$ need not be true for $\tilde{\Sigma}$ may contain isolated points while $\mathcal{S}_{\tilde{w}}$ cannot.

We illustrate Theorem 1.6 by some examples.

Example 1.7. The inclusion in (a) can be a proper one. Let $\Sigma = \mathbf{C}$ and define v and w as

$$v(z) := \begin{cases} 1 & \text{if } |z| \leq 1 \text{ and } \operatorname{Re} z \leq 0 \\ 0 & \text{otherwise,} \end{cases}$$

and

$$w(z) := v(-z).$$

The extremal measures for the weights v, w, and $\max\{v, w\}$ are just the classical equilibrium distributions for the compact sets where these weights are supported. Thus (cf. Corollary I.4.5) $\mathcal{S}_{\max\{v,w\}}$ coincides with the unit circumference, while $\mathcal{S}_v$ and $\mathcal{S}_w$ are, respectively, the left and right half of it together with the segment $[-i, i]$ of the imaginary axis, i.e. in this case the inclusion in (a) is proper. □

Example 1.8. Assertion (b) is not true without the restriction "provided in some neighborhood of z one of the functions v or w majorizes the other one", i.e. in general the dual of (a): $\mathcal{S}_{\min\{v,w\}} \supset \mathcal{S}_v \cap \mathcal{S}_w$ is false. In fact, let

$$v(z) := \begin{cases} m & \text{if } -2 \leq z \leq -1 \text{ or } a \leq z \leq 2 \\ 1 & \text{if } -1 \leq z \leq a \\ 0 & \text{otherwise,} \end{cases}$$

with the agreement that on the right-hand side the appropriate $\leq$ signs should be replaced by the $<$ sign (depending on the value of m) to make v upper semi-continuous, and let

$$w(z) := v(-z),$$

where the constants $0 \leq a \leq 1$ and $m > 0$ will be chosen in a moment. Actually here we can take $a = 0$ and m large enough. In fact, it easily follows from the Bernstein-Walsh lemma (III.2.4) and Theorem 1.3 that for large m we have $\mathcal{S}_v = [-2, -1] \cup [a, 2]$, $\mathcal{S}_w = [-2, -a] \cup [1, 2]$ and $\mathcal{S}_{\min\{v,w\}} = [-2, -1] \cup [1, 2]$, i.e. for $a = 0$ the point 0 belongs to $\mathcal{S}_v \cap \mathcal{S}_w$ but does not belong to $\mathcal{S}_{\min\{v,w\}}$. □

Example 1.9. Here is an example of a w such that w^λ is admissible for all $\lambda > 0$ and $\lambda \to \mathcal{S}_{w^\lambda}$ is discontinuous at $\lambda = 1$ (cf. (f)). Consider Example 1.8 with $a = 1$ and $m = \exp(-U^{\omega_K}(0))/\operatorname{cap}(K)$, where $K = [-2, -1] \cup [1, 2]$ and U^{ω_K} is the equilibrium potential corresponding to the set K. It easily follows from the Bernstein-Walsh lemma and Theorem 1.3 that $0 \in \mathcal{S}_{w^\lambda}$ for every $1 > \lambda > 0$ but $(-1, 1) \cap \mathcal{S}_w = \emptyset$. Thus $\operatorname{dist}(\mathcal{S}_w, \mathcal{S}_{w^\lambda}) \geq 1$ for all $1 > \lambda > 0$. □

In (1.4) the first inclusion " $\subset$ " cannot be replaced by $\subset$, because we can increase the value of w at any given point z_0 so that $z_0 \in \mathcal{S}^*_{w^{\lambda_0}}$ will be true for the modified w, and of course w remains admissible and $\mathcal{S}_w$ etc. does not change. However, the proof easily implies that if $\mathcal{S}^*_{w^{\lambda_0}}$ is of positive capacity at z_0, then $z_0 \in \cap_{\lambda<\lambda_0} \mathcal{S}_{w^\lambda}$. In particular, if $\mathcal{S}^*_{w^{\lambda_0}}$ is of positive capacity at every one of its points, then $S^*_{w^{\lambda_0}} = \cap_{\lambda<\lambda_0} S_{w^\lambda}$. In any case we can see that the "jumps" of $\lambda \to S_{w^\lambda}$ are closely related to the differences $S^*_w \setminus S_w$ (cf. Example 1.9 above).

Proof of Theorem 1.6. As we have already mentioned, parts (f) and (g) will be proved in Section IV.4 (see Theorem 4.1).

(a) Let $z_0 \in S_{\max\{v,w\}}$. Then, by Theorem 1.3, for every neighborhood D of z_0 there exists an n and a polynomial P_n of degree at most n such that $P_n(\max\{v, w\})^n$ takes its essential maximum modulus on Σ in D, which means (after multiplying P_n by a constant if necessary) that outside D the function $|P_n|(\max\{v, w\})^n$ is quasi-everywhere ≤ 1 but there is a set $A \subset D \cap \Sigma$ of positive capacity where $|P_n|(\max\{v, w\})^n > 1$. Now $A = A_1 \cup A_2$, where A_1 and A_2 are the subsets of A where $v \geq w$ and $w > v$, respectively, and so at least one of them, say A_2, is of positive capacity. But then $|P_n w^n|$ is ≤ 1 quasi-everywhere outside D and > 1 on $A_2 \subset D$, $\operatorname{cap}(A_2) > 0$, which means by Theorem 1.3 that $D \cap S_w \neq \emptyset$. Thus, for any neighborhood D of z_0 we have $D \cap (S_v \cup S_w) \neq \emptyset$, i.e. $z_0 \in S_v \cup S_w$ as we claimed.

(b) If $z_0 \in S_v \cap S_w$ and in a neighborhood D of z_0 we have, say, $w \leq v$ for $z \in D \cap \Sigma$, then whatever other neighborhood D^* of z_0 is given, there is an n and a polynomial P_n, $\deg P_n \leq n$ such that $P_n w^n$ takes its essential maximum modulus in $D \cap D^*$ in the sense of Definition 1.2. But in $D \cap D^*$ the weight $\min\{v, w\}$ coincides with w; therefore $P_n(\min\{v, w\})^n$ also takes its essential maximum modulus in $D \cap D^*$, showing (cf. Theorem 1.3) that $D \cap D^* \cap S_{\min\{v,w\}} \neq \emptyset$. Since this is true for all D^*, we get $z_0 \in S_{\min\{v,w\}}$.

Statement (c) immediately follows from Theorem I.1.3(f) and Theorem I.3.3, and (d) is an obvious consequence of Theorem 1.3.

(e) By Theorem I.1.3 and the concept of the balayage measure

$$U^{\widehat{\mu_w}}(z) + Q(z)$$

coincides quasi-everywhere with a constant function on $\tilde{\Sigma}$. Furthermore, since its potential is bounded on $\tilde{\Sigma}$, the balayage measure $\widehat{\mu_w}$ has finite logarithmic energy (see the discussion in Section II.4). Hence, we can apply Theorem I.3.3 to conclude $\mu_{\tilde{w}} = \widehat{\mu_w}$.

(h) This assertion follows from the fact that the assumptions imply the weak*-convergence of the measures μ_{w_n} to μ_w (see Theorem I.6.2). □

Finally we list some simple but useful geometric properties of S_w. These help to locate the points in S_w.

Theorem 1.10. *Let w be an admissible weight on Σ.*

(a) *If Q is superharmonic in the interior* Int (Σ) *of Σ, then $S_w \subset \partial\Sigma$.*

(b) *If $\Sigma \subset \mathbf{R}$, $I \subset \Sigma$ is an interval and Q is convex on I, then $I \cap S_w$ is an interval.*

(c) *If* $\Sigma \subset [0, \infty)$, $I \subset \Sigma$ *is an interval and* $xQ'(x)$ *increases on* I, *then* $I \cap \mathcal{S}_w$ *is an interval.*

(d) *In particular, if* $\Sigma \subset \mathbf{R}$ *is the union of* k *intervals and* w *satisfies either of the assumptions of* (b) *or* (c) *on each of them, then* $\mathcal{S}_w$ *is the union of intervals at most one of which lies in any component of* Σ. *In this case, if* K *is the union of finitely many (nondegenerate) closed intervals and* $F(K) = F_w$, *then* $K = \mathcal{S}_w$.

(e) *Symmetries of* w *such as axial or circular symmetry are inherited by* $\mathcal{S}_w$.

(f) *If* $\Sigma \subseteq \mathbf{R}$ *and* w *are symmetric with respect to the origin, then* $d\mu_w(t) = d\mu_v(t^2)/2$ *and* $F_w = F_v/2$, *where* v *is the weight on* $\Sigma^2 := \{x^2 \mid x \in \Sigma\}$ *defined by* $v(x) = w(\sqrt{x})^2$.

(g) *If* Σ *is compact and* $Q = -U^\sigma$ *where* σ *is a probability measure with compact support in* $\mathbf{C} \setminus \Sigma$, *then* μ_w *is the balayage of* σ *onto* Σ. *In particular, if* σ *is supported in the unbounded component of* $\mathbf{C} \setminus \Sigma$, *then* $\mathcal{S}_w$ *coincides with the outer boundary of* Σ *minus those points of this outer boundary in a neighborhood of which* Σ *has zero capacity.*

We remark that assertion (d) is particularly useful in practice since to find $\mathcal{S}_w$ (under the stated assumptions) all we have to do is to maximize the F-functional (cf. Theorem 1.5) for sets consisting of at most k intervals.

Proof. (a) Recall from Theorem I.1.3 that

$$\int \log \frac{1}{|z-t|} d\mu_w(t) + Q(z) \geq F_w \tag{1.5}$$

holds quasi-everywhere on Σ, and a similar inequality but with $\geq$ replaced by $\leq$ is true for every $z \in \mathcal{S}_w$. If Q is superharmonic in Int(Σ), then so is the left-hand side of (1.5) and this easily implies that (1.5) must hold everywhere in Int(Σ) (use superharmonicity and Lemma I.2.1). Now (a) follows from the fact that a nonconstant superharmonic function cannot take its minimum at an interior point (the other possibility, namely that the left-hand side of (1.5) is constant on a component B of Int(Σ) implies that the potential

$$U^{\mu_w} := \int \log \frac{1}{|z-t|} d\mu_w(t)$$

is also constant on B, and this again yields that $\mathcal{S}_w = \operatorname{supp}(\mu_w)$ does not intersect B, see Theorem II.2.1).

(b) Assume to the contrary, that $\mathcal{S}_w \cap I$ is not an interval. Then there are two points p_1 and p_2 of $\mathcal{S}_w \cap I$ such that in between them there is no further point of $\mathcal{S}_w \cap I$. But then on $[p_1, p_2]$ the left-hand side of (1.5) is strictly convex (here we use $\mathcal{S}_w \cap (p_1, p_2) = \emptyset$ and the concavity of the log function on any interval not containing the origin), and at p_1 and at p_2 we have (1.5) with $\geq$ replaced by $\leq$, which means that on (p_1, p_2) the strict converse of (1.5) holds, contradicting (1.5). This contradiction proves (b).

(c) Since for positive t the function $x/(x-t)$ is a decreasing function of x on both intervals $(-\infty, t)$ and (t, ∞), it follows from the assumptions that the function

$$x\left(\int \log \frac{1}{|x-t|} d\mu_w(t) + Q(x)\right)'$$

is increasing on every subinterval of I not intersecting $\mathcal{S}_w$. Therefore on every such interval the left-hand side of (1.5) is either monotone, or it is monotone decreasing to a point and after that monotone increasing, and from these observations we get a contradiction exactly as in the proof of (b) if we assume that $I \cap \mathcal{S}_w$ is not an interval.

(d) The first statement is an immediate consequence of statements (b) and (c). To prove the second one we note first of all that $F(K) = F(\mathcal{S}_w)$ implies $\mathcal{S}_w \subset K$ (see Theorem 1.5(d) above). If this inclusion is proper, then on some subinterval $I \subset K \setminus \mathcal{S}_w$ strict inequality holds in (1.5) (see the proof of parts (b) and (c)). Thus, taking into account that $\omega_K(I) > 0$, where ω_K is the equilibrium measure associated with K, we have strict inequality in (1.2) (see also its derivation), and by the proof of Theorem 1.5 we have then $F(K) < -F_w = F(\mathcal{S}_w)$, i.e. in this case $F(K) = F(\mathcal{S}_w)$ is impossible.

(e) This assertion easily follows from the unicity of the extremal measure μ_w (see Theorem I.1.3).

(f) Let $w(x) = \exp(-Q(x))$, $v(x) = \exp(-q(x))$. Then $Q(x) = q(x^2)/2$. By Theorem I.1.3 we have

$$U^{\mu_v}(y) \geq -q(y) + F_v$$

for quasi-every $y \in \Sigma^2$, with equality for quasi-every $y \in \mathrm{supp}(\mu_v)$. By setting $d\mu(t) = d\mu_v(t^2)/2$ and $y = x^2$ this is the same as

$$\int_{t \geq 0} \log \frac{1}{|x^2 - t^2|} 2d\mu(t) \geq -2Q(x) + F_v$$

for quasi-every $x \in \Sigma$, with equality for quasi-every $x \in \mathrm{supp}(\mu)$. But μ is even, so the left-hand side is exactly

$$\int \log \frac{1}{|x-t|} 2d\mu(t) = 2U^{\mu}(x);$$

furthermore, μ has total mass 1. Hence part (f) of the theorem follows by invoking Theorem I.3.3.

(g) To prove (g) all we have to remark is that if σ^* is the balayage of σ onto Σ, then σ^* has finite logarithmic energy and $U^{\sigma^*} - U^{\sigma}$ is constant quasi-everywhere on Σ. Thus we can apply Theorem I.3.3 to conclude $\mu_w = \sigma^*$. □

In what follows we show how Theorem 1.10 can be used to determine the extremal support in some cases.

Theorem 1.11. *Suppose that $w = \exp(-Q)$ is an admissible weight on the real interval $\Sigma = [A, B]$ ($A = -\infty$, $B = \infty$ are allowed). Assume that either*

- *Q is convex on (A, B) or*
- *$[A, B) \subset [0, \infty)$ and $xQ'(x)$ increases on (A, B),*

so that by Theorem 1.10(b),(c) the support $\mathcal{S}_w$ is a finite closed interval, say $\mathcal{S}_w = [a, b]$. Then the endpoints a, b satisfy the following conditions:

(i) *if $b < B$, then*

$$\frac{1}{\pi}\int_a^b Q'(x)\sqrt{\frac{x-a}{b-x}}\,dx = 1; \tag{1.6}$$

(ii) *if $a > A$, then*

$$\frac{1}{\pi}\int_a^b Q'(x)\sqrt{\frac{b-x}{x-a}}\,dx = -1. \tag{1.7}$$

In the limit cases we have

(iii) *if $b = B$, then*

$$\frac{1}{\pi}\int_a^B Q'(x)\sqrt{\frac{x-a}{B-x}}\,dx \le 1; \tag{1.8}$$

(iv) *if $a = A$, then*

$$\frac{1}{\pi}\int_A^b Q'(x)\sqrt{\frac{b-x}{x-A}}\,dx \ge -1. \tag{1.9}$$

Proof. We shall only prove (i) and (iii); the arguments for the left endpoints are similar.

Let us start with (i) in the case when Q is convex. From Theorem 1.5, we know that the F-functional (cf. (1.1)) for w satisfies

$$F(\mathcal{S}_w) = F([a, b]) = \max_{\alpha,\beta} F([\alpha, \beta]),$$

where the maximum is taken over all nondegenerate intervals $[\alpha, \beta] \subset [A, B]$. Now from Example I.3.5 we have

$$\operatorname{cap}([\alpha, \beta]) = \frac{\beta-\alpha}{4}, \qquad d\omega_{[\alpha,\beta]} = \frac{1}{\pi}\frac{dx}{\sqrt{(x-\alpha)(\beta-x)}} \quad \text{on } [\alpha, \beta],$$

and so

$$F([\alpha, \beta]) = \log\left(\frac{\beta-\alpha}{4}\right) - \frac{1}{\pi}\int_\alpha^\beta \frac{Q(x)dx}{\sqrt{(x-\alpha)(\beta-x)}}.$$

On making the change of variable

$$x = \frac{\beta+\alpha}{2} + \left(\frac{\beta-\alpha}{2}\right)\cos\theta, \quad 0 \le \theta \le \pi,$$

we can rewrite $F([\alpha, \beta])$ as

$$F([\alpha,\beta]) = \log\left(\frac{\beta-\alpha}{4}\right) - \frac{1}{\pi}\int_0^\pi Q\left(\frac{\beta+\alpha}{2} + \left(\frac{\beta-\alpha}{2}\right)\cos\theta\right)d\theta. \quad (1.10)$$

Thus, taking right derivative $\partial F/\partial_+\beta$ of F with respect to β ($< B$) we get

$$\frac{\partial F}{\partial_+\beta}([\alpha,\beta]) = \frac{1}{\beta-\alpha} - \frac{1}{2\pi}\int_0^\pi Q'\left(\frac{\beta+\alpha}{2} + \left(\frac{\beta-\alpha}{2}\right)\cos\theta\right)(1+\cos\theta)d\theta$$

$$= \frac{1}{\beta-\alpha} - \frac{1}{(\beta-\alpha)\pi}\int_\alpha^\beta Q'(x)\sqrt{\frac{x-\alpha}{\beta-x}}\,dx, \quad (1.11)$$

where we can differentiate under the integral sign because

$$\frac{Q(x+h)-Q(x)}{h} \searrow Q'(x)$$

as $h \searrow 0$ ($Q'(x)$ denoting here the right derivative), and we have only to appeal to the monotone convergence theorem.

On taking $\alpha = a$ and $\beta = b$, it follows that, if $b < B$,

$$\frac{\partial}{\partial_+\beta}F([a,b]) = \frac{1}{b-a} - \frac{1}{(b-a)\pi}\int_a^b Q'(x)\sqrt{\frac{x-a}{b-x}}\,dx.$$

A similar argument yields the same formula for the left derivative $(\partial/\partial_-\beta)F([a,b])$, where we use the fact that the right and left derivatives of Q are equal a.e. Thus $\partial F/\partial\beta$ exists for $\beta = b$ and since $F([a,b])$ attains its maximum when $\beta = b$, it follows that (1.6) holds.

Next consider (i) in the case when $[A, B) \subset [0,\infty)$ and $xQ'(x)$ increases. Then Q' is bounded on every closed subinterval of (A, B), hence the preceding proof works word for word provided $A < \alpha < \beta < B$, if we appeal in it to the bounded convergence theorem instead of the monotone convergence one. If, however, $\alpha = A$, then $Q(A)$ must be finite (otherwise A could not belong to the support of the extremal measure). This fact combined with $xQ'(x) \nearrow$ yields that $Q'(x)(x-A)$ tends to 0 as $x \to A+0$ when $A > 0$, while for $A = 0$ it follows that $xQ'(x) \geq 0$. In any case we get that $Q'(x)(x-A)$ is bounded on some interval $[A, b+\varepsilon]$, $\varepsilon > 0$, so the preceding proof remains valid by appealing to the dominated convergence theorem when forming the right derivative under the integral sign in (1.11).

Finally, we prove (iii). We have just verified that (1.11) holds for $\beta < B$. As $\beta \nearrow B$, the right-hand side has the limit

$$T := \frac{1}{B-\alpha} - \frac{1}{2\pi}\int_0^\pi Q'\left(\frac{B+\alpha}{2} + \left(\frac{B-\alpha}{2}\right)\cos\theta\right)(1+\cos\theta)d\theta$$

$$= \frac{1}{B-\alpha} - \frac{1}{(B-\alpha)\pi}\int_\alpha^B Q'(x)\sqrt{\frac{x-\alpha}{B-x}}\,dx.$$

Thus, $\partial F([\alpha,\beta])/\partial_+\beta \to T$ as $\beta \to B-0$, and this implies that the left derivative of $F([\alpha,\beta])$ at $\beta = B$ exists and equals T. Since $\beta = B$ maximizes the F-functional, we must have $T \le 0$, and this is exactly the statement in (iii). □

Corollary 1.12. *If $w = \exp(-Q)$ is an admissible weight on $\Sigma = \mathbf{R}$, Q is even, Q' exists on $(0,\infty)$, and the function $xQ'(x)$ is positive and increasing on $(0,\infty)$, then $\mathcal{S}_w = [-a,a]$, where the endpoint a satisfies the equation*

$$\frac{2}{\pi}\int_0^1 \frac{atQ'(at)}{\sqrt{1-t^2}}\,dt = 1. \tag{1.12}$$

Proof. Let $v(x) = w(\sqrt{x})^2$, $x \in [0,\infty)$, so that

$$q(x) := \log(1/v(x)) = 2Q(\sqrt{x}).$$

Then from Theorem 1.10(f), we have

$$\mathcal{S}_w = \{x \mid x^2 \in \mathcal{S}_v\}. \tag{1.13}$$

Moreover, since $xq'(x) = \sqrt{x}Q'(\sqrt{x})$ is positive and increasing on $(0,\infty)$, $\mathcal{S}_v$ is a single interval and, in view of Theorem 1.11, its endpoints satisfy (i) and (ii) of that theorem. But for the left endpoint of $\mathcal{S}_v$ equation (1.7) cannot hold because the integral in (1.7) is positive. Thus $\mathcal{S}_v$ is of the form $\mathcal{S}_v = [0,a^2]$. Furthermore, for the weight $v(x)$, equation (1.6) becomes

$$\frac{1}{\pi}\int_0^{a^2} \frac{q'(x)\sqrt{x}}{\sqrt{a^2-x}}\,dx = \frac{1}{\pi}\int_0^{a^2} \frac{Q'(\sqrt{x})}{\sqrt{a^2-x}}\,dx = 1,$$

or, equivalently,

$$\frac{2}{\pi}\int_0^1 \frac{atQ'(at)}{\sqrt{1-t^2}}\,dt = 1.$$

Since, from (1.13), $\mathcal{S}_w = [-a,a]$, the proof is complete. □

As a straightforward application of the preceding corollary we obtain the following result, which, for weighted polynomials on $\mathbf{R}$, explicitly gives a finite interval on which its supremum norm lives.

Corollary 1.13. *Suppose that $q(x)$ is continuous and even on $\mathbf{R}$, $q'(x)$ exists on $(0,\infty)$ with $xq'(x)$ positive and increasing to infinity on $(0,\infty)$. Then for each integer $n \ge 1$, the equation*

$$\frac{2}{\pi}\int_0^1 \frac{atq'(at)}{\sqrt{1-t^2}}\,dt = n, \quad a > 0, \tag{1.14}$$

has a unique solution $a = a_n$ and, for every polynomial p_n with $\deg p_n \le n$,

$$\|e^{-q(x)}p_n(x)\|_{\mathbf{R}} = \|e^{-q(x)}p_n(x)\|_{[-a_n,a_n]}\,. \tag{1.15}$$

The number a_n is called the *Mhaskar-Rakhmanov-Saff* (MRS) *number* for q.

Proof. Let $w_n(x) := \exp(-q(x)/n)$. Then for each fixed n, the weight w_n satisfies the conditions of Corollary 1.12. Thus $\mathcal{S}_{w_n} = [-a_n, a_n]$, where (cf. (1.12))

$$\frac{2}{\pi}\int_0^1 \frac{a_n t q'(a_n t)/n}{\sqrt{1-t^2}}\,dt = 1;$$

that is, a_n satisfies (1.14). Moreover, the left-hand side of (1.14) is a strictly increasing function of a, and so a_n is the unique solution to (1.14). Finally, since w_n is continuous on $\mathbf{R}$, Corollary III.2.6 asserts that

$$\|e^{-q(x)}p_n(x)\|_{\mathbf{R}} = \|w_n^n p_n\|_{\mathbf{R}} = \|w_n^n p_n\|_{[-a_n,a_n]} = \|e^{-q(x)}p_n(x)\|_{[-a_n,a_n]}$$

for every polynomial p_n with $\deg p_n \le n$. □

Example 1.14 (Freud Weights). The weights $w_\lambda(x) = e^{-|x|^\lambda}$, $\lambda > 0$, on $\mathbf{R}$ clearly satisfy the hypotheses of Corollary 1.12. Hence $\mathcal{S}_{w_\lambda} = [-a, a]$, where

$$\frac{2}{\pi}\int_0^1 \frac{at\lambda(at)^{\lambda-1}}{\sqrt{1-t^2}}\,dt = 1.$$

Since

$$\frac{2}{\pi}\int_0^1 \frac{\lambda t^\lambda\,dt}{\sqrt{1-t^2}} = 2\Gamma\left(\frac{\lambda+1}{2}\right)\Big/\sqrt{\pi}\,\Gamma\left(\frac{\lambda}{2}\right), \tag{1.16}$$

we find

$$a = \gamma_\lambda^{1/\lambda}, \text{ where } \gamma_\lambda := \sqrt{\pi}\,\Gamma\left(\frac{\lambda}{2}\right)\Big/2\Gamma\left(\frac{\lambda+1}{2}\right). \tag{1.17}$$

Thus $\mathcal{S}_{w_\lambda} = [-\gamma_\lambda^{1/\lambda}, \gamma_\lambda^{1/\lambda}]$.

(See Theorem IV.5.1 for the determination of the extremal measure $d\mu_{w_\lambda}$.)

Furthermore, for $q(x) = |x|^\lambda$, the same computation gives the MRS-number $a_n = (n\gamma_\lambda)^{1/\lambda}$, so that by Corollary 1.13,

$$\|e^{-|x|^\lambda}p_n(x)\|_{\mathbf{R}} = \|e^{-|x|^\lambda}p_n(x)\|_{[-(n\gamma_\lambda)^{1/\lambda},\,(n\gamma_\lambda)^{1/\lambda}]}, \tag{1.18}$$

for all polynomials p_n with $\deg p_n \le n$. In particular,

$$\|e^{-x^2}p_n(x)\|_{\mathbf{R}} = \|e^{-x^2}p_n(x)\|_{[-\sqrt{n},\,\sqrt{n}]},$$

$$\|e^{-|x|}p_n(x)\|_{\mathbf{R}} = \|e^{-|x|}p_n(x)\|_{[-\pi n/2,\,\pi n/2]}.$$

□

In the remaining examples of this section we shall appeal to the following identities.

Lemma 1.15. *If* $b_1 > b_2 > 0$, *then*

$$\frac{1}{\pi}\int_0^\pi \log(b_1 \pm b_2\cos\theta)d\theta = \frac{1}{\pi}\int_{-1}^1 \frac{\log(b_1 \pm b_2 t)}{\sqrt{1-t^2}}dt$$

$$= \log\left(b_1 + \sqrt{b_1^2 - b_2^2}\right) - \log 2, \qquad (1.19)$$

$$\frac{1}{\pi}\int_0^\pi \frac{d\theta}{b_1 \pm b_2\cos\theta} = \frac{1}{\sqrt{b_1^2 - b_2^2}}, \qquad (1.20)$$

$$\frac{1}{\pi}\int_0^\pi \frac{\cos\theta\, d\theta}{b_1 \pm b_2\cos\theta} = \pm\frac{\sqrt{b_1^2 - b_2^2} - b_1}{b_2\sqrt{b_1^2 - b_2^2}}. \qquad (1.21)$$

Proof. In Example I.3.5 we showed that

$$\frac{1}{\pi}\int_{-1}^1 \log\frac{1}{|z-t|}\frac{dt}{\sqrt{1-t^2}} = \log 2 - \log|\phi(z)|, \quad z \in \mathbf{C},$$

where $\phi(z) = z + \sqrt{z^2 - 1}$ is the Joukowski transformation mapping of $\overline{\mathbf{C}}\setminus[-1, 1]$ onto the exterior of the unit disk. From this representation (1.19) readily follows. Differentiation of (1.19) with respect to b_1 and b_2 yields (1.20) and (1.21). □

Example 1.16 (Incomplete Polynomials of Lorentz). Let $w(x) = x^{\theta/(1-\theta)}$ on $\Sigma = [0, 1]$, where $0 < \theta < 1$. Then $Q(x) = -(\theta/(1-\theta))\log x$ is convex on $(0, 1)$, so that $\mathcal{S}_w$ is an interval $[a, b]$. Since $w(0) = 0$, clearly a must be positive. Furthermore, since $Q'(x)$ is negative on $(0, 1)$, equation (1.6) of Theorem 1.11 cannot hold. Thus $b = 1$.

To find the left endpoint a one can proceed to solve equation (1.7). But it is actually more convenient to first find a simple formula for the F-functional on $[\alpha, 1]$ and then solve for the zero of its derivative. Indeed, for the given weight, we have (cf. (1.10))

$$F([\alpha, 1]) = \log\left(\frac{1-\alpha}{4}\right) + \frac{\theta}{(1-\theta)\pi}\int_0^\pi \log\left(\frac{1+\alpha}{2} + \left(\frac{1-\alpha}{2}\right)\cos\theta\right)d\theta,$$

which, in view of identity (1.19), can be written as

$$F([\alpha, 1]) = \log\left(\frac{1-\alpha}{4}\right)$$

$$+\frac{\theta}{(1-\theta)}\left[\log\left(\frac{1+\alpha}{2} + \sqrt{\left(\frac{1+\alpha}{2}\right)^2 - \left(\frac{1-\alpha}{2}\right)^2}\right) - \log 2\right]$$

$$= \log\left(\frac{1-\alpha}{4}\right) + \frac{2\theta}{1-\theta}\log(1+\sqrt{\alpha}) - \frac{\theta}{1-\theta}\log 4. \qquad (1.22)$$

On differentiating the last expression with respect to α we obtain

$$\frac{\partial F}{\partial \alpha}([\alpha, 1]) = \frac{-1}{1-\alpha} + \frac{\theta}{1-\theta}\frac{1}{\sqrt{\alpha}+\alpha}.$$

Setting this partial derivative equal to zero yields $a = \theta^2$.

Hence,

$$\mathcal{S}_w = [\theta^2, 1], \text{ for } w(x) = x^{\theta/(1-\theta)} \text{ on } [0, 1]. \tag{1.23}$$

(See Example 5.3 for the determination of μ_w.)

From (1.22) we compute

$$F(\mathcal{S}_w) = F([\theta^2, 1]) = \log(1-\theta) + \frac{1+\theta}{1-\theta}\log(1+\theta) - \frac{1}{1-\theta}\log 4. \tag{1.24}$$

Since $F(\mathcal{S}_w) = -F_w$ (cf. Theorem 1.5(b)), the Chebyshev numbers t_n^w for this weight satisfy (cf. Theorem III.3.1)

$$\lim_{n\to\infty}\left(t_n^w\right)^{1/n} = \frac{(1-\theta)(1+\theta)^{(1+\theta)/(1-\theta)}}{4^{1/(1-\theta)}}. \tag{1.25}$$

□

The weight in the preceding example is (after transforming the interval [0, 1] into [−1, 1]) just a special case of Jacobi weights.

Example 1.17 (Jacobi Weights). Let $w(x) = (1-x)^{\lambda_1}(1+x)^{\lambda_2}$ on $\Sigma = [-1, 1]$, with λ_1, λ_2 positive. Then

$$Q(x) = -\lambda_1 \log(1-x) - \lambda_2 \log(1+x)$$

is convex on $(-1, 1)$ and so $\mathcal{S}_w = [a, b]$ for some a and b. Since $w(1) = w(-1) = 0$, we have $-1 < a < b < 1$ and the formulas (1.6) and (1.7) can be used to compute a, b. Setting $x = (b+a)/2 + (\cos\theta)(b-a)/2$, $0 \le \theta \le \pi$, these formulas become (cf. (1.11))

$$\frac{\lambda_1}{2\pi}\int_0^\pi \frac{(1+\cos\theta)d\theta}{1-\frac{b+a}{2}-\left(\frac{b-a}{2}\right)\cos\theta} - \frac{\lambda_2}{2\pi}\int_0^\pi \frac{(1+\cos\theta)d\theta}{1+\frac{b+a}{2}+\left(\frac{b-a}{2}\right)\cos\theta} = \frac{1}{b-a},$$

$$\frac{\lambda_1}{2\pi}\int_0^\pi \frac{(1-\cos\theta)d\theta}{1-\frac{b+a}{2}-\left(\frac{b-a}{2}\right)\cos\theta} - \frac{\lambda_2}{2\pi}\int_0^\pi \frac{(1-\cos\theta)d\theta}{1+\frac{b+a}{2}+\left(\frac{b-a}{2}\right)\cos\theta} = \frac{-1}{b-a}.$$

On using (1.20) and (1.21) to evaluate the integrals, we obtain

$$\begin{aligned} \lambda_1\left(\sqrt{\frac{1-a}{1-b}}-1\right) + \lambda_2\left(\sqrt{\frac{1+a}{1+b}}-1\right) &= 1, \\ \lambda_1\left(1-\sqrt{\frac{1-b}{1-a}}\right) + \lambda_2\left(1-\sqrt{\frac{1+b}{1+a}}\right) &= -1. \end{aligned} \tag{1.26}$$

To solve this system for a, b it is convenient to introduce the quantities

$$\theta_1 := \frac{\lambda_1}{1+\lambda_1+\lambda_2}, \qquad \theta_2 := \frac{\lambda_2}{1+\lambda_1+\lambda_2}. \tag{1.27}$$

Then the equations (1.26) become

$$\theta_1\sqrt{\frac{1-a}{1-b}} + \theta_2\sqrt{\frac{1+a}{1+b}} = 1,$$

$$\theta_1\sqrt{\frac{1-b}{1-a}} + \theta_2\sqrt{\frac{1+b}{1+a}} = 1,$$

from which we obtain

$$\theta_1 = \frac{1}{2}\sqrt{(1-a)(1-b)}, \qquad \theta_2 = \frac{1}{2}\sqrt{(1+a)(1+b)}.$$

On taking the sum and difference of the squares of θ_1, θ_2, there follows

$$ab = 2\left(\theta_1^2+\theta_2^2\right) - 1, \qquad a+b = 2\left(\theta_2^2-\theta_1^2\right),$$

and on solving this system we get

$$a = \theta_2^2 - \theta_1^2 - \sqrt{\Delta}, \qquad b = \theta_2^2 - \theta_1^2 + \sqrt{\Delta}, \tag{1.28}$$

where

$$\Delta := \left\{1-(\theta_1+\theta_2)^2\right\}\left\{1-(\theta_1-\theta_2)^2\right\}.$$

Thus $\mathcal{S}_w = [\theta_2^2 - \theta_1^2 - \sqrt{\Delta},\ \theta_2^2 - \theta_1^2 + \sqrt{\Delta}]$.

We remark that the above formulas remain valid in the cases when $\lambda_1 = 0$ (for which $b = 1$) and $\lambda_2 = 0$ (for which $a = -1$); compare Example 1.16.

For the determination of the equilibrium measure μ_w, see Example 5.2. □

Example 1.18 (Laguerre Weights). Let $w(x) = x^s e^{-\lambda x}$ on $\Sigma = [0, \infty)$, where $s \geq 0$, $\lambda > 0$. Then $Q(x) = \lambda x - s \log x$ is convex on $(0, \infty)$. Hence the support $\mathcal{S}_w$ is an interval, say $\mathcal{S}_w = [a, b]$. For $s = 0$, we must have $a = 0$ since the contrary assumption would violate (1.7) of Theorem 1.11. On the other hand, for $s > 0$, we have $w(0) = 0$, and so $a > 0$ in this case.

To determine a, b we first compute the F-functional on intervals $[\alpha, \beta] \subset [0, \infty)$. From (1.10) we obtain

$$F([\alpha,\beta]) = \log\left(\frac{\beta-\alpha}{4}\right) - \frac{\lambda}{\pi}\int_0^{\pi}\left(\frac{\beta+\alpha}{2} + \left(\frac{\beta-\alpha}{2}\right)\cos\theta\right)d\theta$$
$$+ \frac{s}{\pi}\int_0^{\pi}\log\left(\frac{\beta+\alpha}{2} + \left(\frac{\beta-\alpha}{2}\right)\cos\theta\right)d\theta,$$

and by evaluating the last integral with the aid of (1.19) we obtain

$$F([\alpha,\beta]) = 2s\log\left(\frac{\sqrt{\beta}+\sqrt{\alpha}}{2}\right) - \lambda\left(\frac{\beta+\alpha}{2}\right) + \log\left(\frac{\beta-\alpha}{4}\right). \tag{1.29}$$

For $s > 0$, the endpoints a, b must satisfy

$$\frac{\partial F}{\partial \alpha}([a,b]) = 0 = \frac{\partial F}{\partial \beta}([a,b]),$$

and, using (1.29), this gives

$$\begin{aligned} \frac{s}{\sqrt{a}(\sqrt{b}+\sqrt{a})} - \frac{\lambda}{2} - \frac{1}{b-a} &= 0, \\ \frac{s}{\sqrt{b}(\sqrt{b}+\sqrt{a})} - \frac{\lambda}{2} + \frac{1}{b-a} &= 0. \end{aligned} \tag{1.30}$$

It is straightforward to determine from this system that

$$\sqrt{ab} = \frac{s}{\lambda} \quad \text{and} \quad a+b = \frac{2}{\lambda}(s+1),$$

which yields

$$a = \frac{1}{\lambda}(s+1-\sqrt{2s+1}), \qquad b = \frac{1}{\lambda}(s+1+\sqrt{2s+1}). \tag{1.31}$$

Notice that these formulas are valid also for $s = 0$. Hence

$$S_w = \left[\frac{1}{\lambda}(s+1-\sqrt{2s+1}),\ \frac{1}{\lambda}(s+1+\sqrt{2s+1})\right]. \tag{1.32}$$

We also compute from (1.29) that

$$F(S_w) = -F_w = \left(\frac{2s+1}{2}\right)\log(2s+1) - (s+1)(\log 2\lambda + 1),$$

so that the Chebyshev numbers t_n^w for this weight satisfy (cf. Theorem III.3.1)

$$\lim_{n\to\infty} (t_n^w)^{1/n} = \frac{(2s+1)^{(2s+1)/2}}{(2\lambda)^{s+1}e^{s+1}}. \tag{1.33}$$

(See Example 5.4 for the determination of the equilibrium distribution μ_w.)

We remark that on replacing s and λ by s/n and λ/n, respectively, Corollary III.2.6 and (1.32) yield that for every polynomial p_n of degree $\le n$,

$$\|x^s e^{-\lambda x} p_n(x)\|_{[0,\infty)} = \|x^s e^{-\lambda x} p_n(x)\|_{[a_n,b_n]}, \tag{1.34}$$

where

$$a_n := \frac{1}{\lambda}\left(s+n-\sqrt{n^2+2sn}\right), \quad b_n := \frac{1}{\lambda}\left(s+n+\sqrt{n^2+2sn}\right).$$

□

IV.2 The Fourier Method and Smoothness Properties of the Extremal Measure μ_w

In this section we discuss a method for determining the equilibrium measure μ_w and its application concerning smoothness properties of μ_w in terms of the smoothness of w on arcs of the support S_w. We call it the Fourier method because its essence is that we transform the problem at hand to a similar one on the unit disk by conformal mapping and apply the results of classical harmonic analysis. The key in the technique is that U^{μ_w} is the solution of certain Dirichlet problems (see Theorem I.4.7), and so under conformal mapping its image will be the solution of the Dirichlet problem on the unit disk with the mapped boundary values, and of course solving the Dirichlet problem in the unit disk is classical Fourier analysis. For understanding the *proofs* of this section one has to be familiar with the results and techniques of classical Fourier analysis even though the statements of the results are self–explanatory.

The first result is

Theorem 2.1. *Suppose that the intersection of S_w with a domain is a simple $C^{1+\delta}$-curve γ for some $\delta > 0$, and let z_0 be any fixed point of γ. If $0 < \alpha < 1$ and $w > 0$ locally satisfies on γ a* Lip α *condition, then the same is true for the function $\mu_w(\gamma_{[z_0,s]})$, $s \in \gamma$.*

Recall that here $\gamma_{[z_0,s]}$ denotes the arc of γ lying between z_0 and s.

In other words, if w is locally in Lip α on γ, then the same is true for μ_w. Of course, the Lip α condition means

$$|w(s_1) - w(s_2)| \le C|s_1 - s_2|^\alpha$$

for all s_1 and s_2 in question (locally, i.e. on every closed subarc). Since γ is assumed to be of class $C^{1+\delta}$ for some $\delta > 0$, and in this concept it is understood that $\gamma' \neq 0$ (cf. Section II.1), this is the same as

$$|w(s_1) - w(s_2)| \le C(\text{length of the arc } \gamma_{[s_1,s_2]})^\alpha,$$

i.e. a Lipschitz condition with respect to arc length.

Proof of Theorem 2.1. Let $\mu = \mu_w$. We keep the notations of Theorem II.1.4 and its proof from the first section of Chapter II. By Theorem II.1.4 it is enough to show that each of H_+ and H_- is in Lip α locally on γ (more precisely, in a left-hand resp. right-hand neighborhood of γ, but since $H_\pm$ turn out to be continuous in the case we are considering — see the proof below — it is enough to consider them on γ) — note that the assumption of Theorem II.1.4 that the potential is bounded is guaranteed in our case by Theorem I.4.3. Consider e.g. H_+, and let $\gamma_{[z_0,z_1]}$ be a subarc of γ. We choose a simply connected domain $D^* \subset \mathcal{D}_+$ of $C^{1+\delta}$-boundary such that $\gamma_{[z_0,z_1]}$ lies on the boundary of D^* and every other boundary point of D^* is contained in $\mathcal{D}_+$ (that we can choose such a D^* follows from the fact that γ was assumed to be of class $C^{1+\delta}$). In particular, then $U^{\mu_w} = \operatorname{Re} H_+$ is

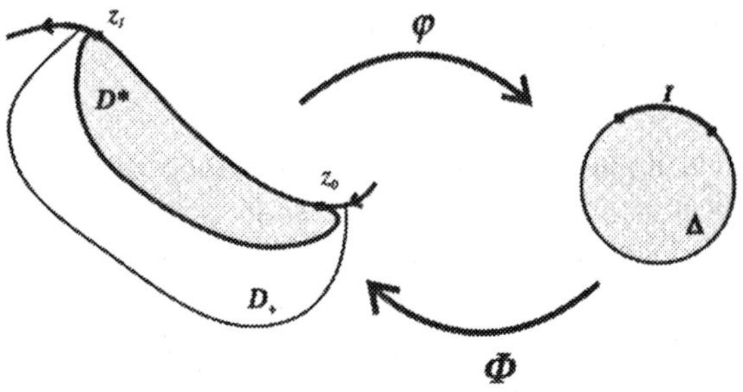

Fig. 2.1

continuous on the part of the boundary of D^* which is disjoint from $\gamma_{[z_0,z_1]}$ (see Figure 2.1) furthermore, by Theorem I.4.8 the potential U^{μ_w} is also continuous on γ. Thus, U^{μ_w} is continuous on the closure $\overline{D^*}$.

Let φ be a conformal map from D^* onto the unit disk Δ, and let Φ be its inverse. Since we have assumed that ∂D^* is of class $C^{1+\delta}$, and δ can be assumed to be smaller than 1, Kellogg's theorem (see [61, Theorem X.1.6]) implies that φ can be continuously extended to γ, furthermore this extension (denoted again by φ) is of class $C^{1+\delta}$.

The potential U^{μ_w} is the solution of the Dirichlet problem on D^* with boundary function equal to $-Q + F_w$ on $\gamma_{[z_0,z_1]}$ and to $U^{\mu_w}\big|_{\partial D^* \setminus \gamma_{[z_0,z_1]}}$ on $\partial D^* \setminus \gamma_{[z_0,z_1]}$. Hence, $U^{\mu_w} \circ \Phi$ is the solution of a Dirichlet problem on Δ corresponding to a bounded boundary function u that agrees modulo a constant with $\log w \circ \Phi$ on the arc $I := \varphi(\gamma_{[z_0,z_1]})$ of the unit circle $C_1(0)$, and obviously $H_+ \circ \Phi$ is the corresponding holomorphic function in Δ. Thus, up to a constant

$$H_+ \circ \Phi(z) = \frac{1}{2\pi} \int_{-\pi}^{\pi} u(e^{it}) \frac{e^{it} + z}{e^{it} - z} dt \tag{2.1}$$

(see [82, Chapter 3]). We shall need that here $u = \log w \circ \Phi + F_w$ is of $\operatorname{Lip} \alpha$ on every subarc of I not containing the endpoints of I, which is immediate from the facts that $w > 0$ is a $\operatorname{Lip} \alpha$ function on γ and Φ is continuously differentiable on I with respect to t by the aforementioned result of Kellogg.

Now it is a standard exercise in singular integrals that then $H_+ \circ \Phi(z)$ is also in $\operatorname{Lip} \alpha$ on any part K of Δ for which $\partial K \cap C_1(0)$ is contained in a closed subarc of I (cf. (2.1)). Going back to H_+ via φ we get that H_+ itself is locally of class $\operatorname{Lip} \alpha$ on γ, which was to prove.

An alternative approach to the $\operatorname{Lip} \alpha$ property of $H_+ \circ \Phi$ is the following. $H_+ \circ \Phi$ is obtained from u as the the result of three operations: taking trigonometric conjugation, addition and taking Poisson-integral. It is well-known that the latter one preserves the $\operatorname{Lip} \alpha$ property locally, so it is enough to verify the same for trigonometric conjugation. Globally this, i.e. the fact that the trigonometric conjugate of a function from the class $\operatorname{Lip} \alpha$, $0 < \alpha < 1$, is again in $\operatorname{Lip} \alpha$, is well-know and is called Privaloff's theorem, see [240, Theorem (13.29)]. To get the local version one only has to represent u as the sum of a Lip α function and a one which vanishes on a prescribed inner subarc I^* of I, which can be achieved by usual cutting techniques. Finally, we only have to mention, that it is obvious from the formula for the trigonometric conjugate that if a function vanishes on I^*, then its trigonometric conjugate is continuously differentiable on any proper subarc of I^*. □

The next result claims the absolute continuity of μ_w with respect to arc measure under mild conditions on w.

Theorem 2.2. *Let γ be as in the preceding theorem but now assume that $w > 0$ is absolutely continuous with respect to the arc measure ds on γ and its (Radon-Nikodym) derivative dw/ds is in $L^p(ds)$ for some $1 < p < \infty$. Then μ_w is absolutely continuous with respect to ds and $d\mu_w/ds$ also belongs to $L^p(ds)$ on any closed subarc of γ.*

A typical example for the situation mentioned in Theorem 2.2 is the real case $\Sigma \subset \mathbf{R}$ when $\mathcal{S}_w$ consists of finitely many intervals. For the derivatives being in L^p globally on the whole γ see Theorem 2.6 below.

Proof of Theorem 2.2. We can follow the proof of Theorem 2.1. If dw/ds is in $L^p(ds)$ on some arc of γ, then (apply Hölder's inequality) w satisfies there a Lip $(1 - 1/p)$ condition; hence by Theorem 2.1, $H_+ \circ \Phi$ satisfies a local Lip $(1 - 1/p)$ condition on I and in some left neighborhood of I.

If we now carefully examine the proof of Theorem 2.1, it turns out that the only thing we have to show is that if u is a continuous function on the unit circle such that on some arc I of this circle $u(e^{it})$ is absolutely continuous with respect to dt and its derivative is locally in L^p, then its Cauchy-Poisson integral (2.1) has the same properties, and it is sufficient to verify the analogous statement for trigonometric conjugation (see the end of the proof of the preceding theorem). But the operation of trigonometric conjugation is bounded in L^p, $1 < p < \infty$; hence the same is true for local L^p spaces over I (see the end of the proof of Theorem 2.1). Furthermore, trigonometric conjugation commutes with differentiation with respect to t, so the result follows. □

As a corollary we show that the conclusion of the first part of Theorem II.1.5 holds for $\mu = \mu_w$ under the conditions on w in Theorem 2.2.

Theorem 2.3. *With the same assumptions as in Theorem 2.2 we have*

$$d\mu_w(s) = -\frac{1}{2\pi}\left(\frac{\partial U^{\mu_w}}{\partial \mathbf{n}_+}(s) + \frac{\partial U^{\mu_w}}{\partial \mathbf{n}_-}(s)\right) ds,$$

where $\partial/\partial \mathbf{n}_\pm$ denote differentiation in the direction of the two normals to γ.

Note that the same formula under more severe assumptions has already been proved in Theorem II.1.5 for general measures (in the present case the potential U^{μ_w} need not be in Lip 1 around γ).

Proof of Theorem 2.3. We only sketch the proof, which follows the proof of the first part of Theorem II.1.5. There are three steps in the argument.

I. First of all we have to show that the normal derivatives $\partial U^{\mu_w}/\partial \mathbf{n}_+(s)$ and $\partial U^{\mu_w}/\partial \mathbf{n}_-(s)$ exist for ds-almost every point on γ, and we have to verify formula (II.1.16). The existence of the derivatives ds-almost everywhere follows from Theorem 2.2 and the second part of Theorem II.1.5 because ds-almost every point of γ is Lebesgue point with respect to ds for any $L^p(ds)$-function (just repeat the classical proof).

The verification of (II.1.16) requires more effort. The exhaustion procedure mentioned in the proof of Theorem II.1.5 can be carried out provided we can find a family of curves γ_τ, $\tau < 1$, lying in R_+ such that γ_τ tends to $\gamma_{[A,B]}$ as τ tends to 1 and with $v = \log 1/|z - z_0|$, $u = U^{\mu_w}$ we have

$$\lim_{\tau\to 1} \int_{\gamma_\tau} v \frac{\partial u}{\partial \mathbf{n}} ds = \int_{\gamma_{[A,B]}} v \frac{\partial u}{\partial \mathbf{n}} ds \tag{2.2}$$

(consider also (2.5) in paragraph II below for estimating $\partial u/\partial \mathbf{n}$ on short curves lying close to A and B).

First we prove this for the special case when R_+ (see the proof of Theorem II.1.5) is the unit disk Δ. In this case, i.e. when we are working on Δ and γ is a subarc I of the unit circle, the Fourier series of $u(e^{it}) = U^{\mu_w}(e^{it})$, which is differentiable a.e. by the choice of R_+, is of the form

$$u(e^{it}) \sim \sum_{k=0}^{\infty} (a_k \cos kt + b_k \sin kt),$$

and we have for any disk $D_\tau(0)$ of radius $\tau < 1$ with center at 0 that on its boundary

$$\frac{\partial u}{\partial \mathbf{n}}(\tau e^{it}) = -\sum_{k=0}^{\infty} (ka_k \cos kt + kb_k \sin kt)\tau^{k-1},$$

which is nothing else than $(-1/\tau)$-times the Poisson integral $P(\tilde{u}'; \tau e^{it})$ of the trigonometric conjugate of the derivative $u'(e^{it})$ of $u(e^{it})$ with respect to t, and of course this latter function is the same as the derivative of the conjugate function. Trigonometric conjugation is a bounded operation in L^p, $1 < p < \infty$, spaces (even locally, see the end of the proof of Theorem 2.1), so $\partial u/\partial \mathbf{n}$ has radial limit

$$\lim_{\tau\to 1-} \frac{\partial u}{\partial \mathbf{n}}(\tau e^{it}) =: -\tilde{u}'(e^{it}) \tag{2.3}$$

for almost all $t \in I$ by Fatou's theorem (see [82, Chapter 3]). It follows from the mean value theorem of calculus that then $\partial u/\partial \mathbf{n}(e^{it})$ also exists for almost all $t \in I$ and we have $\partial u/\partial \mathbf{n}(e^{it}) = -\tilde{u}'(e^{it})$.

Next we use that the Poisson integral represents the boundary function in (local) L^p in the sense

$$\lim_{\tau\to 1-}\int_{I^*}|P(\tilde{u}';\tau e^{it})-\tilde{u}'(e^{it})|^p dt = 0,$$

for every closed subarc I^* of I (repeat the classical proof locally). This and (2.3) easily imply (2.2) (apply Hölder's inequality) with the choice : γ_τ is (part of) the circle $C_\tau(0)$ of radius τ with center at 0 (lying close to I^*).

The existence of (2.2) for the general case can now be shown by applying a conformal mapping φ of R_+ onto the unit disk Δ. In fact, we can choose R_+ to have $C^{1+\delta}$-boundary for some $0<\delta<1$, and then from the fact that under this assumption φ is of class $C^{1+\delta}$ (see Kellogg's theorem mentioned in the proof of Theorem 2.1), it is not difficult to see that certain arcs on the inverse images $\varphi^{-1}(C_\tau(0))$ of the aforementioned circles are suitable as γ_τ.

II. The second problem we encounter if we want to imitate the proof of Theorem II.1.5 is the verification of

$$\lim_{\tau\to 0}\int_{C_r(z_0)} v\frac{\partial u}{\partial \mathbf{n}}ds = 0 \tag{2.4}$$

for any inner point z_0 of the arc $\gamma_{[A,B]}$, where $v\equiv \log 1/r$ on $C_r(z_0)$ and $u=U^{\mu_w}$. The limit relation (2.4) obviously follows if we can show

$$\left|\frac{\partial u}{\partial \mathbf{n}}(z)\right| = O((\operatorname{dist}(z,\gamma))^{-1/p}) \tag{2.5}$$

in a neighborhood of z_0. Actually we can prove more, namely that the derivative of

$$\int \log\frac{1}{z-t}d\mu_w(t),$$

the real part of which is $U^{\mu_w}=u$, satisfies an analogous estimate. This derivative is

$$-\int\frac{1}{z-t}d\mu_w(t), \tag{2.6}$$

and the contribution to this integral of the part of μ_w that is of distance $\geq\theta$ from z_0 with some fixed $\theta>0$ is certainly bounded. Thus, we may disregard this part and we assume that μ_w is supported on γ. We write γ in parametric form $\gamma(t)$, $t\in[-1,1]$ where z_0 corresponds to the parameter value $t=0$. Hölder's inequality implies that if dw/ds is locally in $L^p(ds)$, then w is locally in $\operatorname{Lip}(1-1/p)$; hence we get from Theorem 2.1 that

$$\mu_w(\{\gamma(t)\,|\,c<t<d\})\leq C(d-c)^{1-1/p}$$

for all $-1/2<c<d<1/2$ with a constant C. But this easily yields that if z is sufficiently close to z_0 and ρ is the distance from z to γ, then we can write for (2.6) the estimate

$$\begin{aligned}\left|\int\frac{1}{z-t}d\mu_w(t)\right| &\leq \text{const.}+O(\rho^{-1}\rho^{1-1/p})+\int_\rho^{1/2}O(t^{-2}t^{1-1/p})dt\\ &= O(\rho^{-1/p}),\end{aligned}$$

by which (2.5) is verified.

III. The third problem is to show that with the preceding notations

$$\lim_{r\to 0}\int_{\gamma[C,D]} v\frac{\partial u}{\partial \mathbf{n}} = 0,$$

where $C = C_r$ and $D = D_r$ are the two points where $C_r(z_0)$ intersects γ. However, it was verified in Part I that $\partial u/\partial \mathbf{n}$ is locally in L^p on γ, from which the preceding limit relation immediately follows upon applying Hölder's inequality.

With these three modifications we can now repeat the first part of the proof of Theorem II.1.5 for $\mu = \mu_w$ to get Theorem 2.3. □

From the preceding proof we can see how the Fourier method of the present section works. If we want to determine μ_w on an arc γ of S_w, then, under sufficient smoothness conditions on γ and w, we can apply Theorem 2.1 or Theorem 2.3, so our task is to determine e.g. $\partial U^{\mu_w}/\partial \mathbf{n}_+$. Let D be a simply connected subdomain of $\overline{\mathbf{C}}\setminus S_w$ attached to γ from the left. There are two cases that have to be distinguished because in the unbounded component of $\mathbf{C}\setminus S_w$ the potential is not simply the solution of a Dirichlet problem but the difference of this solution and the Green function (see Theorem I.4.7).

Case I. $\mathbf{n}_+$ *is not the inner normal of the unbounded component* Ω *of* $\overline{\mathbf{C}}\setminus S_w$. Let φ be a conformal mapping from D onto the unit disk Δ with inverse Φ. Then $G := U^{\mu_w}\circ\Phi := g\circ\Phi$ is the solution of Dirichlet's problem in Δ with boundary values equal to a function $u(e^{it})$ that agrees with $U^{\mu_w}\circ\Phi(e^{it}) = g\circ\Phi(e^{it})$, and this in turn agrees with $\log w\circ\Phi(e^{it}) +$ const. for e^{it} belonging to that part of the boundary of Δ which also belongs to $\varphi(\gamma)$. Let $\zeta = \varphi(z)$, $z\in\partial D$, and for clearer notation let $\boldsymbol{\eta}_+$ denote the inner normal in the unit disk. By Kellogg's theorem (see [61, Theorem X.1.6]) Φ is of class $C^{1+\delta}$ if γ is so (without loss of generality $0<\delta<1$ can be assumed). Thus

$$\begin{aligned}\frac{\partial G}{\partial \boldsymbol{\eta}_+}(\zeta_0) &= \lim_{\tau\to 0+}\frac{g(\Phi(\zeta_0+\tau\boldsymbol{\eta}_+(\zeta_0))) - g(z_0)}{\tau}\\ &= \frac{\partial g}{\partial \mathbf{n}_+}(z_0)|\Phi'(\zeta_0)| = \frac{\partial g}{\partial \mathbf{n}_+}(z_0)\frac{1}{|\varphi'(z_0)|},\end{aligned}$$

where we can think of $\varphi'(z_0)$ either as a one-sided derivative or as the limit of the derivative $\varphi'(z)$ as z tends to z_0 from D. Hence, the transformation formula for the normal derivative under conformal mapping is

$$\frac{\partial g}{\partial \mathbf{n}_+}(z_0) = \frac{\partial G}{\partial \boldsymbol{\eta}_+}(\zeta_0)|\varphi'(z_0)|,\quad g = G\circ\varphi,\quad \zeta_0 = \varphi(z_0). \tag{2.7}$$

After (2.3) we saw that on the right $\partial G/\partial\boldsymbol{\eta}_+(\zeta_0)$ is just (under the conditions of Theorem 2.2) $-\tilde{u}'(\zeta_0)$, so finally we get from the formula for the trigonometric conjugate ([240, Section VII.1])

$$\begin{aligned}\frac{\partial U^{\mu_w}}{\partial \mathbf{n}_+}(z_0) &= -|\varphi'(z_0)|\tilde{u}'(\zeta_0) \qquad (2.8)\\ &= |\varphi'(z_0)|\frac{1}{\pi}\int_0^\pi \frac{d}{d\zeta_0}\left(\frac{U^{\mu_w}(\Phi(e^{i(\zeta_0+t)})) - U^{\mu_w}(\Phi(e^{i(\zeta_0-t)}))}{2\tan t/2}\right)dt\\ &= |\varphi'(z_0)|\frac{1}{2\pi}\int_{-\pi}^{\pi}\frac{d}{d\zeta_0}\left(U^{\mu_w}(\Phi(e^{i(\zeta_0+t)}))\right)\cot(t/2)dt,\end{aligned}$$

where $\varphi(z_0) = e^{i\zeta_0}$.

Case II. $\mathbf{n}_+$ *is the inner normal of the unbounded component* Ω *of* $\overline{\mathbf{C}}\setminus S_w$. In this case the potential U^{μ_w} in Ω is of the form $g - g_\Omega$ where g is again the solution of the Dirichlet problem in Ω with boundary condition $\log w +$ const. and g_Ω is the Green function of Ω with pole at infinity (see Theorem I.4.7). The normal derivative of g can be handled as before, while that of g_Ω is directly calculable if we know the conformal mapping from Ω onto the exterior of the unit disk. In fact, if this conformal mapping is Ψ, then $g_\Omega = \log|\Psi|$, and $\partial g_\Omega/\partial \mathbf{n}_+$ turns out to be $|\Psi'|$.

Note that what we have derived is a local result, namely if γ is not as smooth as we have assumed above, the preceding formulae still hold on smooth parts of γ. In particular, if γ is piecewise of class $C^{1+\delta}$ for some $\delta > 0$, then the formula can be applied at any point of γ with the finitely many exceptions where γ' may not exist. Of course, to be able to apply these formulae explicitly, one has to know the conformal mappings involved.

To make things clearer consider the special case when S_w is an interval, say $[-1, 1]$ (other intervals can be handled by linear transformation). Then D can be chosen as $\overline{\mathbf{C}}\setminus[-1, 1]$ for calculating both $\partial U^{\mu_w}/\partial \mathbf{n}_\pm$, and a conformal mapping between D and Δ is $\varphi(z) = z - \sqrt{z^2-1}$ with the branch of the square root that is positive for positive z. The inverse of this is $\Phi(\zeta) = (1/2)(\zeta + \zeta^{-1})$, and so $\Phi(e^{it}) = \cos t$. The present situation falls under the category "Case II" above, so we also need the normal derivative of the Green function of the region $\Omega = \overline{\mathbf{C}}\setminus[-1, 1]$, as well. Since $g_\Omega(z) = \log|z + \sqrt{z^2-1}|$, this derivative can be easily calculated and at $x \in (-1, 1)$ it turns out to be $1/\sqrt{1-x^2}$. Furthermore, in this case the function $u(e^{it})$ agrees modulo a constant with the composition of $\log w$ and $\Phi(e^{it})$. Thus, from Theorems 2.2 and 2.3 and from the preceding formulas we get

Theorem 2.4. *Suppose that* S_w *is the interval* $[-1, 1]$, $w > 0$ *is absolutely continuous and* w' *is in* $L^p[-1, 1]$ *for some* $p > 1$. *Then* μ_w *is locally absolutely continuous with respect to Lebesgue measure on* $(-1, 1)$ *and we have with* $x = \cos\zeta$

$$\mu_w'(x) = \frac{1}{\pi\sqrt{1-x^2}} - \frac{1}{2\pi^2\sqrt{1-x^2}}\int_{-\pi}^{\pi}\frac{w'(\cos t)}{w(\cos t)}\sin t\cot\frac{t-\zeta}{2}\,dt.$$

Furthermore, this derivative is locally in L^p *on* $(-1, 1)$ *provided* $1 < p < \infty$.

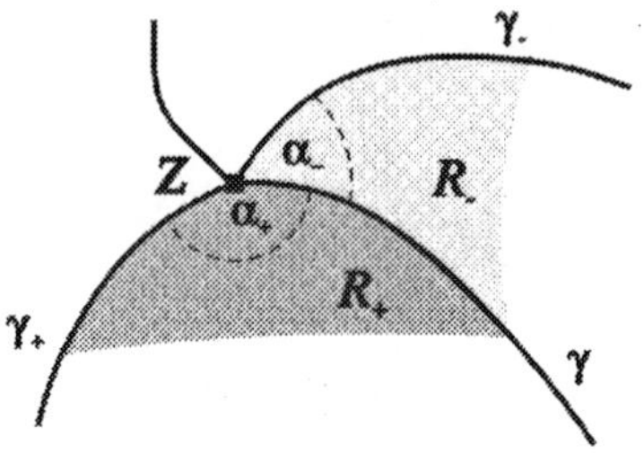

Fig. 2.2

We shall see in Corollary 2.7 that the derivative is globally in $L^p[-1,1]$ for $1 < p < 2$. As for the proof of Theorem 2.4, it was basically given before the formulation of the theorem. The only thing that remained to be mentioned is the fact that by symmetry the normal derivatives $\partial U^{\mu_w}/\partial \mathbf{n}_{\pm}$ agree with each other.

As a corollary of the preceding discussion we prove

Theorem 2.5. *Suppose that $\Sigma \subset \mathbf{R}$ and w is an admissible weight. If $I \subset S_w$ is an interval on which w is positive and locally of class $C^{1+\varepsilon}$ for some $\varepsilon > 0$, then μ_w is absolutely continuous (with respect to linear Lebesgue measure) and its density v_w is continuous inside I.*

Proof. First we prove the theorem in the special case when $I = S_w$. Then without loss of generality we may assume $I = S_w = [-1, 1]$, and the result follows from the representation in Theorem 2.4, because $w'(\cos t)/w(\cos t)$ is locally a Lip ε function, and it is well known that then the same is true for its trigonometric conjugate, which is nothing else than the second integral in the aforementioned representation.

Let overbar denote taking balayage onto I out of $\mathbf{C} \setminus I$. By Theorem 1.6(e) we have

$$\mu_w\big|_I = \overline{\mu_w} = \mu_w\big|_I + \overline{\mu_w\big|_{(\mathbf{R}\setminus I)}}.$$

The special case of the theorem proved above shows that here on the left we have a measure with continuous density, and by Corollary II.4.12 the same is true for the second measure on the right. From these we can conclude the theorem. □

All of the results up to now were of local character. Before closing this section we discuss a global estimate for the derivative of μ_w on the whole γ.

We discuss the situation when the arc γ with one endpoint Z is part of S_w and at Z possibly other arcs of S_w join γ, and we are interested in norm estimates on $d\mu_w/ds$ on the whole γ. The behavior around the other endpoint of γ is not of concern here; if γ satisfies around it the assumptions below, then the result can be applied to this other endpoint, as well.

Thus, suppose that γ lies on the boundary of the connected components R_+ and R_- of $\overline{\mathbf{C}} \setminus S_w$ (see Figure 2.2) and the parts of $\partial R_{\pm} \setminus (\gamma \cup \{Z\})$ lying close to

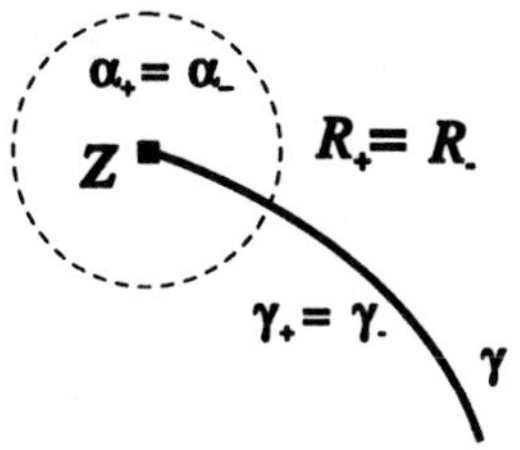

Fig. 2.3

Z are open curves $\gamma_\pm$. Of course, γ_+ and γ_- need not be different and they can coincide with γ, as well (see Figure 2.3). We also suppose that both R_+ and R_- are simply connected in a neighborhood of Z by which we mean that for some neighborhood $D_r(Z)$ of Z the intersections $R_\pm \cap D_r(Z)$ are simply connected. In order to be able to apply the previous results to the curves $\gamma \cup \gamma_\pm \cup \{Z\}$, we assume that each of γ, γ_+ and γ_- can be obtained by intersecting $\mathcal{S}_w$ with a domain (for γ this is obvious for the domain $R_+ \cup R_- \cup \gamma$). We further assume that even the half-closed curves $\gamma \cup \{Z\}$, $\gamma_\pm \cup \{Z\}$ are of class $C^{1+\delta}$ for some $\delta > 0$, and that the angles $\alpha_\pm$ between γ and $\gamma_\pm$ are positive (and of course $\leq 2\pi$). If in a neighborhood of Z the only part of $\mathcal{S}_w$ is $\gamma \cup \{Z\}$ (see Figure 2.3), then we think of γ as a double curve and in this case we set $\gamma_\pm = \gamma$ and $\alpha_\pm = 2\pi$. Finally, let ds_γ and $ds_{\gamma_\pm}$ denote the arc length on γ and $\gamma_\pm$, respectively.

With these notations we are ready to state and prove

Theorem 2.6. *With the assumptions and notations above let*

$$\alpha = \max(\alpha_+, \alpha_-), \qquad 1 < p < \infty,$$

and let γ^* be any closed subarc of $\gamma \cup \{Z\}$ containing Z. If $w > 0$ is absolutely continuous with respect to the arc length ds on $\gamma \cup \gamma_+ \cup \gamma_-$ and on $\gamma \cup \gamma_+ \cup \gamma_-$ we have

$$\frac{dw}{ds}(z)|z - Z|^\Gamma \in L^p(ds),$$

where

$$1 - \frac{1}{p} - \frac{\pi}{\alpha} < \Gamma < 1 - \frac{1}{p},$$

then on γ the measure μ_w is absolutely continuous with respect to ds_γ and on γ^* we have

$$\frac{d\mu_w}{ds_\gamma}(z)|z - Z|^\Gamma \in L^p(ds_{\gamma^*}).$$

In other words, if dw/ds is in $L^p(ds)$ with weight $(\mathrm{dist}(z, Z))^\Gamma$, then $d\mu_w|_\gamma$ is also in the same weighted L^p space on γ^*.

Corollary 2.7. *Suppose that the support S_w is an interval $[a, b]$ on the real line, $1 < p < \infty$, and W is a weight of the form*

$$\prod_{i=0}^{m} |x - x_i|^{\Gamma_i},$$

where $a = x_0 < x_1 < \ldots < x_m = b$,

$$-\frac{1}{p} < \Gamma_i < 1 - \frac{1}{p}$$

if $x_i \in (a, b)$, while

$$\frac{1}{2} - \frac{1}{p} < \Gamma_i < 1 - \frac{1}{p}$$

if $x_i = a$ or b. If $w > 0$ is absolutely continuous on $[a, b]$ and $w' \in L^p(W)$, then μ_w is absolutely continuous (with respect to Lebesgue measure) and $\mu'_w \in L^p(W)$. In particular, $w' \in L^p[a, b]$ implies $\mu'_w \in L^p[a, b]$ provided $1 < p < 2$.

Note that we cannot say that the $L^p(W)$-norm of μ'_w is bounded by the $L^p(W)$-norm of w', for even if w is constant the derivative μ'_w is not zero. The reason for this is the appearance of the Green function in "Case II" above.

Note also that the very last statement of the corollary is not true for $p \geq 2$; consider for example, $w \equiv 1$, for which $\mu'_w(x)$ is the function $\pi^{-1}((x - a)(b - x))^{-1/2}$.

Proof of Theorem 2.6. Clearly, only the behavior around Z is of interest, otherwise we can apply Theorem 2.2. Furthermore, we can apply Theorems 2.2 and 2.3 to any closed part of γ and $\gamma_\pm$; hence the absolute continuity of μ_w with respect to ds follows. To verify the last conclusion it is enough to show (see Theorem 2.3) that e.g. $\partial U^{\mu_w}/\partial \mathbf{n}_+$ is in the weighted L^p-class in question.

Let D be a simply connected subdomain of R_+ containing $\gamma \cup \gamma_+ \cup \{Z\}$ or its intersection with a neighborhood of Z on its boundary. Clearly, without loss of generality we may assume that $\gamma \cup \gamma_+ \cup \{Z\}$ is on the boundary of D and that $\partial D \setminus \{Z\}$ is a Jordan curve of class $C^{1+\delta}$ (otherwise we can take appropriate parts of γ, γ_+ and D). Let φ be a conformal map of D onto the unit disk Δ. Then φ can be extended to the boundary and we can assume that $\varphi(Z) = 1$. Without loss of generality we can also assume $Z = 0$ and that the x-axis is tangent to γ at 0 (see Figure 2.4). Choose a branch of z^{π/α_+} in a neighborhood of 0 cut along γ, and by shrinking D if necessary we can assume that this branch is defined throughout D, and that the image of D under the mapping $\chi(z) = z^{\pi/\alpha_+}$ is a simply connected domain D^*. Obviously 0 stays fixed under this mapping and D^* has $C^{1+\delta}$-boundary because χ changes the angle α_+ at $Z = 0$ of the boundary curve ∂D into the angle π and each of $\gamma \cup \{Z\}$ and $\gamma_+ \cup \{Z\}$ was assumed to be of class $C^{1+\delta}$. φ is the composition $\psi \circ \chi$ of χ with a conformal map φ of D^* onto Δ that carries 0 to 1, and by the Kellogg's theorem ([61, Theorem X.1.6]) used in the proof of Theorem 2.1, ψ is of class $C^{1+\delta}$.

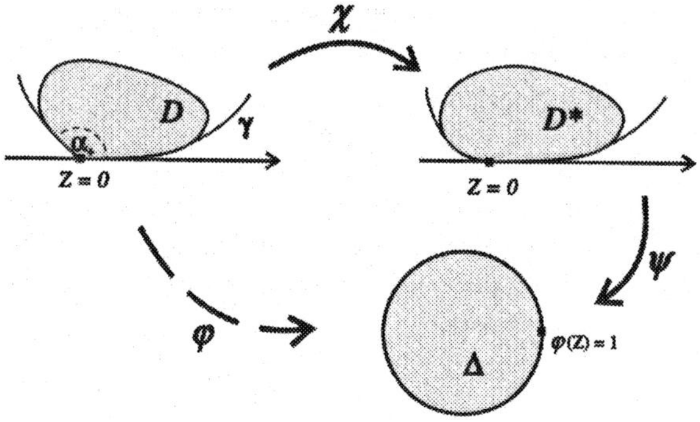

Fig. 2.4

After these preparations we apply formulae (2.7) – (2.8) in "Case I" and the analogous formulae plus an estimate on the normal derivative of the Green function in "Case II".

Consider first "Case I". For $z \in \gamma \cup \gamma_+ \cup \{Z\}$ let $\zeta = \varphi(z) = e^{it}$. Then from what we have said it follows that

$$|1 - \varphi(z)| \sim |z - Z|^{\pi/\alpha_+}, \quad |\varphi'(z)| \sim |z - Z|^{\pi/\alpha_+ - 1}, \quad |z - Z| \sim |t|^{\alpha_+/\pi} \tag{2.9}$$

and

$$ds \sim |t|^{\alpha_+/\pi - 1} dt, \tag{2.10}$$

where $\sim$ indicates that the ratio of the two sides is bounded away from zero and infinity as $z \to Z(= 0)$, $z \in \gamma \cup \gamma_+ \cup \{Z\}$.

Let the arc $\{e^{it} \mid -a \le t \le a\}$ be contained in the image $\varphi(\gamma \cup \gamma_+ \cup \{Z\})$ of $\gamma \cup \gamma_+ \cup \{Z\}$. If $U^{\mu_w}(z) = u(e^{it})$, then $u(e^{it})$ is locally absolutely continuous with respect to dt on $(-a, 0) \cup (0, a)$. Using substitution we can see from the above relations and from the equality $U^{\mu_w}(z) = -Q(z) + F_w$ (see Theorems I.1.3 and I.4.8) that

$$\frac{dw}{ds}(z)|z - Z|^{\Gamma} \in L^p(ds) \quad \text{on} \quad \gamma \cup \gamma_+$$

implies

$$u'(e^{it})t^{(1-\alpha_+/\pi)(1-1/p)+\Gamma\alpha_+/\pi} \in L^p[-a, a],$$

where the prime indicates differentiation with respect to t. On the other hand, by making use of formulae (2.8) and (2.9) – (2.10) and Theorem 2.2 we get that the relation

$$\frac{\partial U^{\mu_w}}{\partial \mathbf{n}_+}(z)|z - Z|^{\Gamma} \in L^p(ds_{\gamma^*})$$

is equivalent to

$$\widetilde{(u')}(e^{it})t^{(1-\alpha_+/\pi)(1-1/p)+\Gamma\alpha_+/\pi} \in L^p[-a/2, a/2].$$

Thus, to verify the theorem it is enough to show that trigonometric conjugation is a bounded operation in the weighted $L^p[-\pi, \pi]$-space with weight t^{Γ^*}, where

$$\Gamma^* = \left(1 - \frac{\alpha_+}{\pi}\right)\left(1 - \frac{1}{p}\right) + \Gamma\frac{\alpha_+}{\pi}$$

(actually we need this statement locally because we want to conclude the above relation on the interval $(-a/2, a/2)$ from an assumption stated for $(-a, a)$, but the global statement implies the local result exactly as in the end of the proof of Theorem 2.1). We have assumed

$$1 - \frac{1}{p} - \frac{\pi}{\alpha} < \Gamma < 1 - \frac{1}{p},$$

which implies

$$1 - \frac{1}{p} - \frac{\pi}{\alpha_+} < \Gamma < 1 - \frac{1}{p},$$

and this gives for Γ^* the inequalities

$$-\frac{1}{p} < \Gamma^* < 1 - \frac{1}{p}.$$

Now the aforementioned boundedness of trigonometric conjugation is contained in Hirschman's multiplier theorem ([81, Theorem 2.1]).

This completes the proof in "Case I". In "Case II" the preceding local proof works as before if we choose D to be a small bounded region. In that case we write the potential U^{μ_w} in the unbounded component Ω of $\overline{\mathbf{C}} \setminus S_w$ in the form $g - g_\Omega$, where g is the solution of the Dirichlet problem in Ω with boundary function of the form $\log w +$ const. and g_Ω is the Green function of Ω with pole at infinity (see Theorem I.4.7). $\partial g/\partial \mathbf{n}_+$ can be shown to be in the weighted $L^p(ds_{\gamma^*})$ space in question exactly as before. As for $\partial g_\Omega/\partial \mathbf{n}_+$, relations analogous to (2.9) show that for the conformal mapping Ψ of Ω onto the exterior of Δ

$$|\Psi'(z)| \sim |z - Z|^{\pi/\alpha_+ - 1}$$

as $z \to Z$; consequently

$$\left|\frac{\partial g_\Omega}{\partial \mathbf{n}_+}(z)\right| \sim |z - Z|^{\pi/\alpha_+ - 1}$$

as $z \to Z$. Thus, for

$$\Gamma > 1 - \frac{1}{p} - \frac{\pi}{\alpha} \geq 1 - \frac{1}{p} - \frac{\pi}{\alpha_+}$$

the function $(\partial g_\Omega/\partial \mathbf{n}_+(z))|z - Z|^\Gamma$ is obviously in $L^p(ds_{\gamma^*})$, and this completes the proof. □

The corollary readily follows if we apply Theorem 2.6 on overlapping intervals of $[a, b]$ containing at most one x_i. Note that the angles $\alpha_\pm$ in this case are π (if $x_i \in (a, b)$) or 2π (if $x_i = a$ or b).

IV.3 The Integral Equation

In this section we shall consider a second method for finding the equilibrium measure. It consists of directly solving the integral equation for the equilibrium potential that we obtain if we apply Theorem I.1.3. Let us begin with the symmetric case.

Theorem 3.1. *Let f be a continuous even function on $[-1, 1]$ such that $f'(x)$ exists a.e. in $[-1, 1]$ and, for some $1 < p < 2$, $f'(x)/\sqrt{1-x^2} \in L^p[-1, 1]$. Then the following assertions hold:*

(a) *The integral equation*

$$\int_{-1}^{1} \log \frac{1}{|x-t|} g(t)\, dt = -f(x) + C_f, \qquad x \in (-1, 1), \tag{3.1}$$

has a solution of the form

$$g(t) = g(f; t) = L[f'](t) + \frac{B_f}{\pi\sqrt{1-t^2}}, \qquad a.e.\ t \in (-1, 1), \tag{3.2}$$

and

$$C_f = \frac{2}{\pi} \int_0^1 \frac{f(t)}{\sqrt{1-t^2}} dt + \log 2, \tag{3.3}$$

where

$$L[f'](t) := \frac{2}{\pi^2} \mathrm{PV} \int_0^1 \frac{\sqrt{1-t^2}\, sf'(s)}{\sqrt{1-s^2}(s^2-t^2)} ds, \text{ for a.e. } t \in (-1, 1), \tag{3.4}$$

PV *denoting principal value, and*

$$B_f := 1 - \frac{1}{\pi} \int_{-1}^{1} \frac{sf'(s)}{\sqrt{1-s^2}} ds. \tag{3.5}$$

Furthermore,

$$\int_{-1}^{1} L[f'](t) dt = \frac{1}{\pi} \int_{-1}^{1} \frac{sf'(s)}{\sqrt{1-s^2}} ds, \tag{3.6}$$

and

$$\int_{-1}^{1} g(t) dt = 1. \tag{3.7}$$

(b) *For some C_1 independent of f,*

$$\left\| L[f'](s)(1-s^2)^{-1/2} \right\|_{L^p[-1,1]} \le C_1 \left\| f'(s)(1-s^2)^{-1/2} \right\|_{L^p[-1,1]}. \tag{3.8}$$

(c) *If the function $sf'(s)$ is increasing in $(0, 1)$, then*

$$L[f'](t) = \frac{2}{\pi^2} \int_0^1 \frac{(1-t^2)^{1/2}(sf'(s) - tf'(t))}{\sqrt{1-s^2}(s^2-t^2)} ds > 0. \tag{3.9}$$

for almost every $t \in (-1, 1)$.

Proof. Consider the equation

$$\mathrm{PV}\int_{-1}^{1} \psi(s)/(t-s)\,ds = f'(t), \qquad t \in (-1, 1), \tag{3.10}$$

with f' continuous in $[-1, 1]$ and vanishing on small intervals around ± 1. If ψ' exists and is bounded in each compact subinterval of $(-1, 1)$ and ψ is integrable in $(-1, 1)$, then integrating (3.10) for t from -1 to x, we obtain for $x \in (-1, 1)$

$$\int_{-1}^{1} \log|x-s|\psi(s)\,ds - \int_{-1}^{1} \log(1+s)\psi(s)\,ds = f(x) - f(-1). \tag{3.11}$$

The integration may be justified as follows: split the principal value integral on the left-hand side of (3.10) into integrals over $\{s \mid |t-s| \geq \varepsilon\}$ and $\{s \mid |t-s| < \varepsilon\}$ for some small $\varepsilon = \varepsilon(t) > 0$; in the first case interchange integrals, and then use the fact that

$$\mathrm{PV}\int_{|t-s|\leq\varepsilon} \frac{\psi(s)}{t-s}\,ds = \int_{|t-s|\leq\varepsilon} \frac{\psi(s)-\psi(t)}{t-s}\,ds = o(1), \quad \varepsilon \to 0,$$

by the boundedness of ψ'. Finally, let $\varepsilon \to 0$. Thus, every sufficiently smooth solution ψ of (3.10) generates a solution of (3.1).

Now let us consider (3.10). If, for example, f''' is continuous in $[-1, 1]$, it is known that for any real number A

$$\psi(t) = \frac{(1+t)^{1/2}}{\pi^2(1-t)^{1/2}}\,\mathrm{PV}\int_{-1}^{1} \frac{(1-s)^{1/2} f'(s)}{(1+s)^{1/2}(s-t)}\,ds + \frac{A}{(1-t^2)^{1/2}} \tag{3.12}$$

is a solution of (3.10); see, for example, [165, p. 249, eqn. (88.1) and p. 251, eqn. (88.8)]. Note that in the case we are considering, the parameters in [165] are $p = q = 1$ and $c_1 = -1, c_2 = 1$. It is easily seen from the continuity of f''' that ψ' exists and is continuous in $(-1, 1)$. Thus, when f''' is continuous in $[-1, 1]$, the function ψ solves the equation (3.11). Assume in addition that f is even, so that f' is odd. Then we see from (3.12) that for $t \in (0, 1)$,

$$\begin{aligned}\psi(t) &= 2\pi^{-2}(1-t^2)^{1/2}\mathrm{PV}\int_{0}^{1} \frac{sf'(s)}{(1-s^2)^{1/2}(s^2-t^2)}\,ds + A(1-t^2)^{-1/2} \qquad (3.13)\\ &= L[f'](t) + A(1-t^2)^{-1/2}.\end{aligned}$$

Using the fact that f' is odd we also obtain via the same method that

$$L[f'](t) = \pi^{-2}(1-t^2)^{1/2}\,\mathrm{PV}\int_{-1}^{1} \frac{f'(s)}{(1-s^2)^{1/2}(s-t)}\,ds. \tag{3.14}$$

Now let us set $A := B_f/\pi$, so that $g(f;t) := \psi(t)$ is given by (3.2). Integrating (3.11) against the equilibrium measure of $[-1, 1]$ we obtain from (I.1.7) that

$$\log\frac{1}{2}\int_{-1}^{1}\psi(s)ds - \frac{1}{\pi}\int_{-1}^{1}\frac{f(x)}{\sqrt{1-x^2}}dx \tag{3.15}$$
$$= \int_{-1}^{1}\log(1+s)\psi(s)ds - f(-1).$$

Here, by (3.2),

$$\int_{-1}^{1}\psi(s)ds = \int_{-1}^{1}L[f'](s)ds + B.$$

From the form (3.14) of $L[f'](t)$, we can get (3.6) as follows. For $z \in \overline{\mathbf{C}} \setminus [-1, 1]$, set

$$H(z) := \int_{-1}^{1}\frac{f'(s)}{(1-s^2)^{1/2}(s-z)}ds.$$

Then H is analytic in $\overline{\mathbf{C}}\setminus[-1, 1]$ and, according to the Sokhotskyi-Plemelj formula (cf. [79, Section 14.1]), for $t \in [-1, 1]$ the PV integral in (3.14) is given by the average $(H^+(t) + H^-(t))/2$, where $H^+(t)$ denotes the (nontangential) limit of $H(z)$ as $z \to t$ from the upper half-plane and $H^-(t)$ is the corresponding limit from the lower half-plane. Hence, from (3.14),

$$L[f'](t) = \frac{(1-t^2)^{1/2}}{2\pi^2}\left[H^+(t) + H^-(t)\right], \quad t \in [-1, 1]. \tag{3.16}$$

Also, with the principal branch of the logarithm, the function

$$\sqrt{1-z^2} = -iz\sqrt{1-z^{-2}} = -iz\exp\left(\tfrac{1}{2}\log(1-z^{-2})\right)$$

is analytic in $\mathbf{C} \setminus [-1, 1]$ and satisfies

$$(\sqrt{1-t^2})^+ = (1-t^2)^{1/2} \geq 0 \quad \text{for } t \in [-1, 1].$$

Thus (3.16) can be written as

$$L[f'](t) = \frac{1}{2\pi^2}\left[\left(\sqrt{1-t^2}H(t)\right)^+ - \left(\sqrt{1-t^2}H(t)\right)^-\right]. \tag{3.17}$$

Next, consider the Laurent expansion of $\sqrt{1-z^2}H(z)$ about $z = \infty$:

$$\sqrt{1-z^2}H(z) = a_0 + \frac{a_1}{z} + \frac{a_2}{z^2} + \cdots.$$

Then $2\pi i a_1$ is given by the integral of $\sqrt{1-z^2}H(z)$ around a large positively oriented circle centered at $z = 0$, and since this circle can be continuously deformed to the interval $[-1, 1]$ without changing the value of the integral, we get in the limit

$$2\pi i a_1 = -\int_{-1}^{1}\left(\sqrt{1-t^2}H(t)\right)^+ dt + \int_{-1}^{1}\left(\sqrt{1-t^2}H(t)\right)^- dt.$$

Hence, by (3.17), we see that

$$2\pi^2 \int_{-1}^{1} L[f'](t)\,dt = -2\pi i a_1. \tag{3.18}$$

Finally, from the representations

$$\sqrt{1-z^2} = -iz + \frac{i}{2z} + \cdots, \quad H(z) = -\sum_{k=0}^{\infty}\left(\int_{-1}^{1} \frac{f'(s)s^k}{(1-s^2)^{1/2}}\,ds\right)\frac{1}{z^{k+1}},$$

which are valid near infinity, we compute that

$$a_1 = i\int_{-1}^{1} \frac{f'(s)s}{(1-s^2)^{1/2}}\,ds,$$

(recall that f' is odd, so the first term in the expansion of $H(z)$ vanishes) which, together with (3.18), yield (3.6).

Next, taking account of (3.5), we obtain

$$\int_{-1}^{1} g(s)ds = \int_{-1}^{1} \psi(s)ds = 1. \tag{3.19}$$

Using (3.15) and (3.19) to substitute for

$$\int_{-1}^{1} \log(1+s)\psi(s)ds - f(-1)$$

in (3.11), we obtain (3.3). Thus we have established (3.1) to (3.7) of (a) in the special case that f''' is continuous in $[-1, 1]$.

Before we can complete the proof of (a) in the general case we have to verify (b). Note that by (3.14),

$$\frac{\pi^2 L[f'](t)}{\sqrt{1-t^2}} = \mathrm{PV}\int_{-1}^{1} \frac{f'(s)}{(1-s^2)^{1/2}(s-t)}ds$$

is the Hilbert transform on $\mathbf{R}$ of the function defined as $f'(t)/\sqrt{1-t^2}$ on $(-1, 1)$ and 0 elsewhere. According to M. Riesz' theorem [56, pp. 128-129] the Hilbert transform is a bounded operator from $L^p(\mathbf{R})$ to $L^p(\mathbf{R})$, so $L[f'](x)$ exists almost everywhere and (3.8) is valid whenever $f'(t)/\sqrt{1-t^2} \in L^p[-1, 1]$. This proves part (b).

We now return to the proof of (a) in the general case. Let f satisfy the hypotheses, and let $F'''(x)$ be continuous in $[-1, 1]$. Since (3.1) is valid for F, we get by Hölder's inequality with $q^{-1} + p^{-1} = 1$

$$\left|\int_{-1}^{1} \log \frac{1}{|x-t|} g(f;t)dt + f(x) - C_f\right|$$

$$= \left|\int_{-1}^{1} \log \frac{1}{|x-t|} (g(f;t) - g(F;t))dt + f(x) - F(x) - C_f + C_F\right|$$

$$\leq \left(\int_{-1}^{1} \left|\log|x-t|\right|^q dt\right)^{1/q} \{\|L[f' - F']\|_{L^p[-1,1]} + C_1|B_f - B_F|\}$$

$$+2\|f - F\|_{[-1,1]}$$

with

$$C_1 := \|\{\pi(1-t^2)^{1/2}\}^{-1}\|_{L^p[-1,1]},$$

where we used (3.2) and (3.3).

In view of (3.8) for arbitrary $\varepsilon > 0$ we can chóose F to be a polynomial satisfying

$$\|f - F\|_{[-1,1]} < \varepsilon$$

and

$$\|L[f' - F'](t)(1-t^2)^{-1/2}\|_{L^p[-1,1]} < \varepsilon.$$

Then it follows from the previous estimate and the definition (3.5) of B_f, B_F that the right-hand side may be made arbitrarily small for $x \in (-1, 1)$. Thus, the left-hand side is 0 and so (3.2) – (3.5) always define a solution of (3.1). By a similar approximation argument we get that (3.6) and (3.7) persist.

Finally we prove part (c). Formally differentiating the equality

$$\frac{1}{\pi}\int_{-1}^{1} \frac{\log|x-t|}{\sqrt{1-t^2}} dt = \log\frac{1}{2}, \qquad x \in [-1, 1]$$

(recall from Example I.3.5 that $\{\pi(1-t^2)^{1/2}\}^{-1}$ is the equilibrium measure of the interval $[-1, 1]$) we obtain

$$\frac{1}{\pi}\text{PV}\int_{-1}^{1} \frac{1}{\sqrt{1-t^2}(s-x)} dt = 0, \qquad s \in (-1, 1). \tag{3.20}$$

The differentiation can be justified by integrating the function defined by the left-hand side of (3.20) and observing that its integral is constant on $[-1, 1]$ because of the preceding identity. Since $(1-t^2)^{-1/2}$ is continuously differentiable in $(-1, 1)$, the argument is the same as that used to justify the passage from (3.10) to (3.11).

From (3.20), we deduce that

$$\frac{1}{\pi}\text{PV}\int_{0}^{1} \frac{s}{\sqrt{1-t^2}(s^2-x^2)} dt = 0, \qquad s \in (-1, 1).$$

Thus, (3.9) follows from (3.4). If $sf'(s)$ is increasing in $(0, 1)$, then $sf'(s) - tf'(t)$ has the same sign as $s - t$, and so $L[f'](t) > 0$, a.e. $t \in (0, 1)$. Moreover, $L[f'](t)$

is clearly even. The existence of the integral in (3.9) as an ordinary Lebesgue integral then follows from the monotone convergence theorem and the positivity of the integrand. □

We shall also need the following theorems which give the equilibrium measure in an important special case.

Theorem 3.2. *Let* $w(x) = \exp(-Q(x))$ *be such that* $S_w = [0, 1]$ *and that* $tQ'(t)$ *is nondecreasing on* $(0, 1)$. *Then the density of the equilibrium measure* $d\mu_w(t) = v(t)dt$ *is given by*

$$v(t) = \frac{1}{\pi^2}\sqrt{\frac{1-t}{t}}\int_0^1 \frac{sQ'(s) - tQ'(t)}{s-t}\frac{1}{\sqrt{s(1-s)}}ds + \frac{D}{\sqrt{t(1-t)}}, \tag{3.21}$$

$t \in (0, 1)$, *where*

$$D = \frac{1}{\pi} - \frac{1}{\pi^2}\int_0^1 \sqrt{\frac{s}{1-s}}Q'(s)ds,$$

and here $D \geq 0$.

Theorem 3.3. *Let us assume that* w *is a symmetric weight for which* $S_w = [-1, 1]$, Q *is differentiable and* $xQ'(x)$ *increases on* $(0, 1)$. *Then the density of the equilibrium measure is given by formula* (3.2) *with* $f(x) = -Q(x)$.

This theorem says that the additional *a priori* knowledge $S_w = [-1, 1]$ automatically implies that the solution of the integral equation (3.1) furnishes the extremal measure.

Theorem 3.3 is an easy consequence of the following proof, so we only verify Theorem 3.2.

Proof of Theorem 3.2. That $D \geq 0$ was the content of Theorem 1.11, (1.8).

Let $f(x) = Q(x^2)/2$, $x \in [-1, 1]$. It was shown in Theorem 3.1 that the integral equation

$$\int_{-1}^1 \log\frac{1}{|x-t|}g(t)dt = -f(x) + C$$

where C is some constant has a solution $g(t)$ of the form

$$g(t) = \frac{2}{\pi^2}\sqrt{1-t^2}\int_0^1 \frac{sf'(s) - tf'(t)}{(1-s^2)^{1/2}(s^2-t^2)}ds + \frac{D_1}{\sqrt{1-t^2}},$$

where

$$D_1 = \frac{1}{\pi} - \frac{1}{\pi^2}\int_{-1}^1 \frac{sf'(s)}{\sqrt{1-s^2}}.$$

Furthermore, g is even and has total integral 1 over $[-1, 1]$. In view of $f(s) = Q(s^2)/2$ we have $D_1 = D$, and the integrand in the expression for g is nonnegative. Hence, g is nonnegative on $[-1, 1]$. If we set $h(t) = g(\sqrt{t})/\sqrt{t}$, $t \in [0, 1]$, then h will have total integral 1 over $[0, 1]$, and by the symmetry of g its potential satisfies

$$\int_0^1 \log\frac{1}{|x-u|}h(u)du = 2\int_0^1 \log\frac{1}{|x-t^2|}g(t)dt = 2\int_{-1}^1 \log\frac{1}{|\sqrt{x}-t|}g(t)dt$$

$$= \quad -2f(\sqrt{x})+2C = -Q(x)+2C$$

for every $x \in [0, 1]$. On applying Theorem I.3.3 we can conclude that $d\mu_w(t) = h(t)dt$. Now if we carry out the substitution $f(s) = Q(s^2)/2$ and $u = s^2$ in the formula for g, we obtain the form of v stated in the theorem. □

IV.4 Behavior of μ_{w^λ}

Frequently one needs to also consider the weight w^λ for some $\lambda > 1$ along with w. In this section we discuss the behavior of $\mathcal{S}_{w^\lambda}$ and μ_{w^λ} as λ increases. First we restate parts (f) and (e) from Theorem 1.6 and give their proofs.

Theorem 4.1. (a) *$\mathcal{S}_{w^\lambda}$ is a right continuous decreasing function of the $\lambda \in \mathbf{R}$ values for which w^λ is admissible. Furthermore, if Σ has empty interior and connected complement, in particular, if $\Sigma \subset \mathbf{R}$, then*

$$\mathcal{S}^*_{w^{\lambda_0}} \text{ “}\subset\text{” } \bigcap_{\lambda<\lambda_0} \mathcal{S}_{w^\lambda} \subset \mathcal{S}^*_{w^{\lambda_0}}, \tag{4.1}$$

where “$\subset$” means inclusion except for a set of zero capacity.

(b) *$\lambda \to \mathcal{S}_{w^\lambda}$ is discontinuous at most at countably many λ. It is continuous at λ_0 if $\mathcal{S}_{w^{\lambda_0}} = \mathcal{S}^*_{w^{\lambda_0}}$, and conversely, if it is continuous at λ_0 and Σ has empty interior and connected complement, then $\mathcal{S}_{w^{\lambda_0}}$ “=” $\mathcal{S}^*_{w^{\lambda_0}}$.*

Example 4.2. Let Σ be the disk $\{z \,|\, 0 \le |z| \le 2\}$, and let $w(z) = 1/|z|$ if $|z| \ge 1$ and $w(z) = 1$ otherwise. It is easy to see that if ω_1 and ω_2 are the equilibrium measures on $C_1(0) = \{z \,|\, |z| = 1\}$ and on $C_2(0) = \{z \,|\, |z| = 2\}$, respectively, then for $\lambda < 1$ we have $\mu_{w^\lambda} = \lambda\omega_1 + (1-\lambda)\omega_2$ (apply Theorem I.3.3), while $\mu_w = \omega_1$. Thus, for $\lambda < 1$ we have $\mathcal{S}_{w^\lambda} = C_1(0) \cup C_2(0)$, while $\mathcal{S}_w = C_1(0)$. Furthermore, $\mathcal{S}^*_w$ is the whole disk Σ, so the left relation in (4.1) fails to hold. The reason for this is that Σ has non–empty interior.

A similar example (restrict w to the union of the cricles $C_1(0)$ and $C_2(0)$) shows that in the assumption that Σ has connected complement cannot be dropped, either. □

Proof of Theorem 4.1 (a). To prove that $\lambda \to \mathcal{S}_{w^\lambda}$ is decreasing, it is enough to verify that $\mathcal{S}_{w^\lambda} \subset \mathcal{S}_w$ for $\lambda > 1$. Let $z_0 \in \mathcal{S}_{w^\lambda}$ and $r > 0$ be arbitrary. By Theorem 1.3 there is a natural number n, a polynomial P_n of degree at most n, a set $A \subset D_r(z_0) \cap \Sigma$ of positive capacity and an $\eta > 0$ such that on A the modulus $|P_n w^{n\lambda}|$ is at least as large as $1+\eta$, and outside $D_r(z_0)$ we have $|P_n w^{n\lambda}| \le 1$ quasi-everywhere. Then w must have a positive lower bound on A, say $w(x) \ge m > 0$, $x \in A$. Furthermore, let M be an upper bound for w. We may assume $m < 1 < M$. For a positive integer l consider the polynomial P_n^l. We clearly have

$$(w(x))^{[ln\lambda]+1}|P_n(x)|^l \geq (1+\eta)^l m, \quad x \in A,$$

while for quasi-every $x \in \Sigma \setminus D_r(z_0)$

$$(w(x))^{[ln\lambda]+1}|P_n(x)|^l \leq M.$$

For large l this means that the weighted polynomial $w^{[ln\lambda]+1}P_n^l$, where $\deg P_n^l \leq [ln\lambda]+1$, takes its essential maximum modulus in the sense of Definition 1.2 in $D_r(z_0)$, and so, by Theorem 1.3, $D_r(z_0) \cap \mathcal{S}_w \neq \emptyset$. Since this is true for every $r > 0$, we get $z_0 \in \mathcal{S}_w$, and $\mathcal{S}_{w^\lambda} \subset \mathcal{S}_w$ is proved.

Next we prove the right continuity of $\lambda \to \mathcal{S}_{w^\lambda}$ in the Hausdorff metric, and it is enough to do so again at $\lambda = 1$. As before, the proof is based on Theorem 1.3. If $z \in \mathcal{S}_w$ and $\varepsilon > 0$, then there is a weighted polynomial $P_n w^n$ taking its essential maximum modulus in $D_\varepsilon(z) \cap \Sigma$. But then obviously $P_n w^{n\lambda}$ will also take its essential maximum modulus in $D_\varepsilon(z) \cap \Sigma$ for some $\lambda = \lambda(z) > 1$. This means that $D_\varepsilon(z) \cap \mathcal{S}_{w^\lambda} \neq \emptyset$, i.e. $\operatorname{dist}\{z, \mathcal{S}_{w^\lambda}\} < \varepsilon$. We can select finitely many points $z_1, \ldots, z_k$ such that the sets $D_\varepsilon(z_i)$, $i = 1, 2, \ldots, k$, cover $\mathcal{S}_w$. If $1 < \lambda < \min\{\lambda(z_i) \mid 1 \leq i \leq k\}$, then every point in $\mathcal{S}_w$ is closer to $\mathcal{S}_{w^\lambda}$ than 2ε. Since we also have $\mathcal{S}_{w^\lambda} \subset \mathcal{S}_w$, this means that the Hausdorff distance between $\mathcal{S}_w$ and $\mathcal{S}_{w^\lambda}$ is at most 2ε, proving the claimed right-continuity.

We prove (4.1) again only for $\lambda_0 = 1$, and while doing so we assume that w^λ is admissible for some $\lambda < 1$, and in what follows we shall consider only such λ's. Since for constants $c > 0$ we have $\mathcal{S}_{cw} = \mathcal{S}_w$ and $\mathcal{S}_{cw}^* = \mathcal{S}_w^*$ (for the latter note that with the notations of Theorem I.1.3, $V_{cw} = V_w - 2\cdot\log c$), we can assume without loss of generality that $w \leq 1$. We start with the verification of

$$\bigcap_{\lambda<1} \mathcal{S}_{w^\lambda} \subset \mathcal{S}_w^*. \tag{4.2}$$

First we prove that for every z

$$\lim_{\lambda \to 1-0} U^{\mu_{w^\lambda}}(z) - F_{w^\lambda} = U^{\mu_w}(z) - F_w. \tag{4.3}$$

Since $w \leq 1$, we have $w^\lambda \geq w$ for $\lambda < 1$, and so Corollary I.4.2 implies that the right-hand side is at most as large as the left-hand one. Now let z be fixed and $\varepsilon > 0$ arbitrary. Then, for some $\lambda < 1$,

$$\lambda(U^{\mu_w}(z) - F_w) < U^{\mu_w}(z) - F_w + \varepsilon$$

is also true, and $\lambda(U^{\mu_w}(z) - F_w) + \log|z|$ is bounded from below around infinity. On applying Theorem I.4.1 we get

$$U^{\mu_{w^\lambda}}(z) - F_{w^\lambda} \leq \lambda(U^{\mu_w}(z) - F_w),$$

which, together with the preceding inequality proves that in (4.3) the left-hand side is at most as large as the right-hand one. Thus, (4.3) is proved.

The limit relation (4.3) easily implies (4.2) because on the support $\mathcal{S}_{w^\lambda}$ we must have

$$U^{\mu_{w^\lambda}}(z) - F_{w^\lambda} \le -\lambda Q(z)$$

(Theorem I.1.3(e)), and letting here $\lambda \to 1-0$ we arrive at $\bigcap_{\lambda<1} \mathcal{S}_{w^\lambda} \subset \mathcal{S}_w^*$, i.e. at (4.2).

We remark that this part of the proof has not used the assumption that Σ has empty interior and connected complement.

The proof of

$$\mathcal{S}_w^* \text{ ``}\subset\text{'' } \bigcap_{\lambda<1} \mathcal{S}_{w^\lambda} \tag{4.4}$$

is based on the following lemma.

Lemma 4.3. *Let* $\lambda < 1$ *and suppose that* $z_0 \in \Sigma \setminus \mathrm{Pc}\,(\mathcal{S}_{w^\lambda})$ *satisfies*

$$\int \log \frac{1}{|z_0 - t|} d\mu_{w^\lambda}(t) \ge -\lambda Q(z_0) + F_{w^\lambda}. \tag{4.5}$$

Then

$$\int \log \frac{1}{|z_0 - t|} d\mu_w(t) > -Q(z_0) + F_w.$$

Recall that $\mathrm{Pc}(\mathcal{S}_{w^\lambda})$ is the polynomial convex hull of the compact set $\mathcal{S}_{w^\lambda}$, defined as the complement in $\mathbf{C}$ of the unbounded component of $\overline{\mathbf{C}} \setminus \mathcal{S}_{w^\lambda}$.

Taking this lemma for granted we can easily prove (4.4). Let $\lambda_1, \lambda_2, \ldots$ be an increasing sequence converging to 1. By Theorem I.1.3(d) inequality (4.5) with $\lambda = \lambda_1, \lambda_2, \ldots$ holds for quasi-every $z_0 \in \mathcal{S}_w^*$. But the lemma says that every such point must belong to every $\mathrm{Pc}(\mathcal{S}_{w^\lambda}) = \mathcal{S}_{w^\lambda}$, $\lambda = \lambda_1, \lambda_2, \ldots$ (note that in case Σ has empty interior and connected complement, so automatically $\mathrm{Pc}(\mathcal{S}_{w^\lambda}) = \mathcal{S}_{w^\lambda}$). Hence in view of the first assertion of part (a) of the theorem, it must belong to every $\mathcal{S}_{w^\lambda}$, $\lambda < 1$.

To complete the proof of part (a) of Theorem 4.1 we need only give the

Proof of Lemma 4.3. The compactness of $\mathcal{S}_{w^\lambda}$ and $z_0 \notin \mathrm{Pc}\,(\mathcal{S}_{w^\lambda})$ imply the existence of a polynomial P such that

$$|P(z_0)| > 1 \quad \text{and} \quad |P(z)| < 1/2 \quad \text{for all} \quad z \in \mathcal{S}_{w^\lambda}. \tag{4.6}$$

Let $r := N/((1/\lambda) - 1)$, where N is the degree of P. The inequalities in (4.6) and Theorem I.1.3(d) imply

$$\log \frac{1}{|P(z)|} + rU^{\mu_w}(z) \ge -rQ(z) + rF_w + \log 2$$

quasi-everywhere on $\mathcal{S}_{w^\lambda}$. We also know from Theorem I.1.3(f) that quasi-everywhere on $\mathcal{S}_{w^\lambda}$

$$U^{\mu_{w^\lambda}}(z) - F_{w^\lambda} = -\lambda Q(z).$$

Since $r/\lambda = N + r$, we get from the preceding two inequalities that quasi-everywhere on $\mathcal{S}_{w^\lambda}$

$$\log \frac{1}{|P(z)|} \quad + \quad rU^{\mu_w}(z) \tag{4.7}$$

$$\geq \quad (N+r)U^{\mu_{w^\lambda}}(z)(N+r) - F_{w^\lambda} + rF_w + \log 2.$$

The measure μ_{w^λ} has finite logarithmic energy, so we can apply the principle of domination Theorem II.3.2 and obtain that (4.7) holds for all $z \in \mathbf{C}$. In particular, with $z = z_0$ we obtain from (4.5) and (4.7) that

$$\log \frac{1}{|P(z_0)|} + r\left(\int \log \frac{1}{|z_0 - t|} d\mu_w(t) + Q(z_0) - F_w\right) \geq \log 2.$$

Finally, this and $\log 1/|P(z_0)| < 0$ (see (4.6)) imply

$$\int \log \frac{1}{|z_0 - t|} d\mu_w(t) + Q(z_0) - F_w \geq r^{-1} \log 2 > 0,$$

and this completes the proof of Lemma 4.3. □

Proof of Theorem 4.1 (b). It is easy to see that $\lambda \to \mathcal{S}_{w^\lambda}$ is continuous at λ_0 if and only if

$$\mathcal{S}_{w^{\lambda_0}} = \bigcap_{\lambda < \lambda_0} \mathcal{S}_{w^\lambda}$$

(use part (a) of Theorem 4.1 and a compactness argument). Thus, if $\mathcal{S}_{w^{\lambda_0}} = \mathcal{S}^*_{w^{\lambda_0}}$, then part (a) yields the continuity in question (recall that $\bigcap_{\lambda<\lambda_0} \mathcal{S}_{w^\lambda} \subset \mathcal{S}^*_{w^{\lambda_0}}$ was proved in part (a) without the assumption that Σ has empty interior and connected complement), and the continuity of $\lambda \to \mathcal{S}_{w^\lambda}$ at λ_0 yields via (4.1) that $\mathcal{S}_{w^{\lambda_0}}$ "=" $\mathcal{S}^*_{w^{\lambda_0}}$.

The statement concerning the cardinality of the points of discontinuity is also easy to verify. In fact, if $\lambda \to \mathcal{S}_{w^\lambda}$ is discontinuous at λ_0, then there is a $\delta = \delta(\lambda_0) > 0$ and a $z = z(\lambda_0)$ such that z belongs to every $\mathcal{S}_{w^\lambda}$, $\lambda < \lambda_0$, but $\text{dist}\{z, \mathcal{S}_{w^{\lambda_0}}\} = \delta$. Since the support $\mathcal{S}_{w^\lambda}$ decreases as λ increases (see part (a) of the theorem), for any natural number m the inequalities $\delta(\lambda_0), \delta(\lambda_1) > 1/m$, $\lambda_0 \neq \lambda_1$, imply $|z(\lambda_0) - z(\lambda_1)| \geq 1/m$. Thus, we can have only countably many points λ with $\delta(\lambda) > 0$, i.e. only countably many points of discontinuity. □

Theorem 4.1 expresses a very general feature of the support of the extremal measure; namely it decreases as we raise the exponent of w. Now if it happens that $\mathcal{S}_{w^\lambda} = \mathcal{S}_w$, then either of the measures μ_{w^λ} or μ_w determines the other one. For clearer notation, let ω_S denote the equilibrium measure of the compact set S.

Lemma 4.4. *If* $\lambda > 1$ *and* $\mathcal{S}_{w^\lambda} = \mathcal{S}_w$, *then*

$$\mu_w = \frac{1}{\lambda}\mu_{w^\lambda} + \left(1 - \frac{1}{\lambda}\right)\omega_{\mathcal{S}_w}. \tag{4.8}$$

Conversely, if

$$\mu_w \geq \left(1 - \frac{1}{\lambda}\right)\omega_{\mathcal{S}_w}$$

for some $\lambda > 1$, *then* $\mathcal{S}_{w^\lambda} = \mathcal{S}_w$, *and* (4.8) *holds.*

Proof. Suppose that $\mathcal{S}_{w^\lambda} = \mathcal{S}_w$. The measure μ on the right in (4.8) is a positive measure of finite logarithmic energy and of total mass 1. By Theorem I.1.3 and Corollary II.3.4

$$U^\mu(z) \le U^{\mu_w}(z) - F_w + \frac{1}{\lambda} F_{w^\lambda} + \left(1 - \frac{1}{\lambda}\right) \log \frac{1}{\operatorname{cap}(\mathcal{S}_w)}$$

at quasi-every point of $\mathcal{S}_{w^\lambda} = \mathcal{S}_w$, hence we get the same inequality for every $z \in \mathbf{C}$ by the principle of domination. The same argument gives the reverse inequality. Thus, the potentials of the two sides of (4.8) coincide everywhere, and we can conclude (4.8) from the unicity theorem Corollary II.2.2.

To prove the converse statement we suppose

$$\mu_w \ge \left(1 - \frac{1}{\lambda}\right) \omega_{\mathcal{S}_w},$$

and consider

$$\nu = \lambda \left(\mu_w - \left(1 - \frac{1}{\lambda}\right) \omega_{\mathcal{S}_w} \right).$$

By the assumption this is a positive unit measure of finite logarithmic energy (observe that $\nu \le \lambda \mu_w$), and by Theorem I.1.3 and Corollary II.3.4 its potential equals $-\lambda Q(z) + F$ with $F = \lambda(F_w + (1 - 1/\lambda) \log \operatorname{cap}(\mathcal{S}_w))$ for quasi-every $z \in \mathcal{S}_w = \operatorname{supp}(\nu)$ and is at least as large as $-\lambda Q(z) + F$ quasi-everywhere on Σ. On invoking Theorem I.3.3 we conclude that $\nu = \mu_{w^\lambda}$. □

Now we consider the general case, namely when we do not assume that the two supports $\mathcal{S}_w$ and $\mathcal{S}_{w^\lambda}$ coincide. In this case we shall get a two sided estimate on the measure μ_{w^λ} which will be very close in spirit to the equality from the preceding theorem. To do this we shall need the following theorem.

Theorem 4.5. *Let μ and ν be two measures of compact support, and let Ω be a domain in which both potentials U^μ and U^ν are finite and satisfy with some constant c the inequality*

$$U^\mu(z) \le U^\nu(z) + c, \qquad z \in \Omega. \tag{4.9}$$

If A is the subset of Ω in which equality holds in (4.9), *then $\nu|_A \le \mu|_A$, i.e. for every Borel subset B of A the inequality $\nu(B) \le \mu(B)$ holds.*

Proof. To prove Theorem 4.5 first we need a definition and a covering lemma.

Let B be a set in the plane and let Γ be a set of open disks such that to each z of B there are disks of Γ with center z and of arbitrarily small positive radii. We then say that Γ covers B in the *Vitali narrow sense.*

Lemma 4.6 (Besicovic's Covering Theorem). *Let μ be a finite Borel measure in the plane and let B be a bounded Borel set such that $\mu(\mathbf{C} \setminus B) = 0$. If Γ is a set of disks that covers B in the Vitali narrow sense, then Γ contains a countable subset $\overline{\Gamma} = \{\Delta_i\}$ of non-overlapping disks such that $\mu(B) = \mu(\cup \Delta_i)$.*

We shall prove the lemma after completing the proof of Theorem 4.5. The latter is now based on the following fact: Let A be the set defined in Theorem 4.5. Then there is a set of disks Γ that covers A in the Vitali narrow sense and is such that for μ, ν as introduced in this theorem, if $\Delta \in \Gamma$, then

$$\nu(\Delta) \leq \mu(\Delta). \tag{4.10}$$

We postpone the proof of this fact and first show how to deduce Theorem 4.5 from it.

Let B be a Borel subset of the set A and let Γ be the set of disks associated with A as guaranteed above. Let $\varepsilon > 0$ be an arbitrary constant and let O be an open set containing B such that $\mu(O) \leq \mu(B) + \varepsilon$. Let Γ_1 consist of all of the disks of Γ which lie in O. We see that Γ_1 covers B in the Vitali narrow sense. Let ν_1 be the measure ν restricted to B. We have, by Lemma 4.6, a countable collection of disks of Γ_1, say $\overline{\Gamma}_1 = \{\Delta_i\}$, which are non-overlapping, and satisfy $\nu_1(\cup\Delta_i) = \nu_1(B)$. We can then write

$$\begin{aligned} \nu(B) &= \nu_1(B) = \nu_1(\cup\Delta_i) = \sum \nu_1(\Delta_i) \leq \sum \nu(\Delta_i) \\ &\leq \sum \mu(\Delta_i) = \mu(\cup\Delta_i) \leq \mu(O) \leq \mu(B) + \varepsilon. \end{aligned}$$

Since this argument can be carried out for every $\varepsilon > 0$, we finally obtain the required inequality $\nu(B) \leq \mu(B)$.

To prove the above claim concerning the existence of a covering in the Vitali narrow sense we first we recall from Theorem II.1.2 that if μ is a finite measure of compact support on the plane, then for any z_0 and $r > 0$ the mean value

$$L(U^\mu; z_0, r) = \frac{1}{2\pi} \int_{-\pi}^{\pi} U^\mu(z_0 + re^{i\theta}) d\theta$$

exists as a finite number, and $L(U^\mu; z_0, r)$ is a nonincreasing function of r that is absolutely continuous on any closed subinterval of $(0, \infty)$. Furthermore,

$$\lim_{r \to 0} L(U^\mu; z_0, r) = U^\mu(z_0). \tag{4.11}$$

If r is a value for which $\frac{d}{dr} L(U^\mu; z_0, r)$ exists, then $\Delta(z_0, r) := \{z \,\big|\, |z - z_0| < r\}$ is called a *regular disk*, and for a regular disk

$$\mu(\Delta(z_0, r)) = -r \frac{d}{dr} L(U^\mu; z_0, r). \tag{4.12}$$

Now let A be the set from Theorem 4.5, and let z_0 be any point in A. We want to prove that there is a sequence $r_n = r_n(z_0)$, $r_n > 0$, $\lim_{n\to\infty} r_n = 0$, such that

$$\nu(\Delta(z_0, r_n)) \leq \mu(\Delta(z_0, r_n)).$$

The collection

$$\Gamma = \{\Delta(z_0, r_n(z_0)) \mid z_0 \in A, \; n = 1, 2, \ldots\}$$

then covers A in the Vitali narrow sense and has the desired further property (4.10).

Let $U_1(z_0) = U^\mu(z_0)$ and $U_2(z_0) = U^\nu(z_0)$, and for $i = 1, 2$ set

$$L_i(z_0, r) := \frac{1}{2\pi}\int_{-\pi}^{\pi} U_i(z_0 + re^{i\theta})d\theta.$$

Now since $U_1(z) \leq U_2(z) + c$ in a neighborhood of z_0, we have $L_1(z_0, r) \leq L_2(z_0, r) + c$ for every small $r > 0$. For $z_0 \in A$ the potentials $U_i(z_0)$, $i = 1, 2$, are finite, and $U_1(z_0) = U_2(z_0) + c$. By subtracting we get

$$L_1(z_0, r) - U_1(z_0) \leq L_2(z_0, r) - U_2(z_0). \tag{4.13}$$

Since $L_1(z, r)$ is absolutely continuous on any $[\varepsilon, R]$ we have

$$\int_\varepsilon^R \frac{d}{ds}L_1(z_0, s)ds = L_1(z_0, R) - L_1(z_0, \varepsilon).$$

By the non-positivity of $\frac{d}{ds}L_1(z_0, s)$, we can let here $\varepsilon \to 0$ and use the monotone convergence theorem to obtain

$$\int_0^R \frac{d}{ds}L_1(z_0, s)ds = L_1(z_0, r) - U_1(z_0).$$

Since all steps hold when the subscript 1 is replaced by 2, we have from (4.13)

$$\int_0^R \frac{d}{ds}L_1(z_0, s)ds \leq \int_0^R \frac{d}{ds}L_2(z_0, s)ds,$$

so that for a set of r of positive measure on $[0, R]$ we have

$$\frac{d}{dr}L_1(z_0, r) \leq \frac{d}{dr}L_2(z_0, r).$$

Since this argument can be repeated for any value of R, however small, we get from (4.12) that there are $r_n = r_n(z_0)$, $r_n > 0$, $r_n \to 0$, such that

$$\mu(\Delta(z_0, r_n) \geq \nu(\Delta(z_0, r_n)),$$

which completes the proof of the claim about the existence of the cover in the Vitali narrow sense with property (4.10). □

In order to continue with the proof of Lemma 4.6, we need the following lemmas.

Lemma 4.7. *Let Γ be a set of disks in the plane such that the center of no one of them is inside another, and let C be a disk whose radius does not exceed the radius of any disk of Γ. Then the number of the disks in Γ intersecting C is at most* 21.

Proof. Let O, O_1, O_2 be the centers of C and two other disks of Γ intersecting C. Without loss of generality we may assume the radius of C to be 1.

We shall find the minimum value of the angle O_1OO_2 in the two cases (a) $OO_1 \le \frac{3}{2}$, $OO_2 \le \frac{3}{2}$ and (b) $OO_1 \ge \frac{3}{2}$, $OO_2 \ge \frac{3}{2}$. In both cases we obviously have $O_1O_2 \ge 1$.

(a) It is easy to see that in this case the angle O_1OO_2 attains its minimum value for $OO_1 = OO_2 = \frac{3}{2}$ and $O_1O_2 = 1$. We have $\cos O_1OO_2 = \frac{7}{9}$ and consequently

$$38^0 < O_1OO_2 < 39^0.$$

From this we conclude that the set Γ cannot contain more than 9 disks intersecting the disk C and having their centers at a distance $\le \frac{3}{2}$ from O.

(b) Assume $OO_2 \le OO_1$. We have

$$O_1O_2 \ge \max\{1,\ OO_1 - 1\}.$$

Obviously if the angle O_1OO_2 has its minimum value, then

$$O_1O_2 = \max\{1,\ OO_1 - 1\}$$

and OO_2 is equal either to $\frac{3}{2}$ or to OO_1.

Consider the following two cases.

(i) $OO_1 \ge 2$. Then $O_1O_2 = OO_1 - 1 = l \ge 1$. We have to examine two triangles

$$OO_2 = \tfrac{3}{2},\quad OO_1 = l+1,\quad O_1O_2 = l$$

and

$$OO_2 = l+1,\quad OO_1 = l+1,\quad O_1O_2 = l.$$

In the first case $\cos O_1OO_2 = \frac{2}{3} + \frac{5}{12(l+1)}$ and it attains its maximum value for $l = 1$; in the second case $\cos O_1OO_2 = 1 - \frac{1}{2}\left(\frac{l}{l+1}\right)^2$, and this too attains its maximum value for $l = 1$. Thus, $\max\cos O_1OO_2 = \frac{7}{8}$ and therefore $\min O_1OO_2 = \arccos\frac{7}{8}$ i.e.

$$28^0 < \min O_1OO_2 < 29^0.$$

(ii) $OO_1 \le 2$. Then the minimum value of the angle O_1OO_2 is attained for $O_1O_2 = 1$, $OO_1 = 2$, and OO_2 equal either to $\frac{3}{2}$ or to 2. As in (i) we have $\cos O_1OO_2 = \frac{7}{8}$ and

$$28^0 < O_1OO_2 < 29^0.$$

Thus the set Γ cannot contain more than 12 disks intersecting C and having their centers at a distance $\ge \frac{3}{2}$ from O.

From (a) and (b) the lemma follows. □

As a consequence we get

Lemma 4.8. *Given a bounded set Γ of disks on the plane the center of no one disk lying inside another disk, then the set Γ can be split into 22 subsets Γ_i, $i = 1, \dots, 22$, such that no pair of disks of the same subset meet.*

Proof. Enumerate the disks of Γ in a sequence $C_1, C_2, \dots$ in order of non-increasing radii. Such an enumeration is possible since, for any $\rho > 0$, Γ contains at most a finite number of disks of radius $> \rho$. Then take C_1 to Γ_1, C_2 to $\Gamma_2, \dots, C_{22}$ to Γ_{22}. The radius of the disk C_{23} is not larger than that of any of the preceding disks, so we can conclude from Lemma 4.7 that it cannot be met by more than 21 of them. Consequently, there exists at least one $i \leq 22$, such that C_i does not meet C_{23}. Take C_{23} to Γ_i. Similarly the disk C_{24} cannot be met by more 21 of the disks $C_1, \dots, C_{23}$, and consequently there exists an $i' \leq 22$ such that no disk of $\Gamma_{i'}$ meets C_{24}. Take C_{24} to $\Gamma_{i'}$ and so on. In this way the required distribution can be carried out. □

Proof of Lemma 4.6. Without loss of generality we assume that the radii of the disks in Γ are bounded. Let r_1 be the upper bound on the radii of the disks of Γ and c_1 a disk of Γ of radius $> r_1/2$, r_2 the upper bound of radii of those disks of Γ whose centers are not inside c_1, and c_2 one of them of radius $> r_2/2$, r_3 the upper bound of radii of those disks of Γ whose centers are not inside c_1, c_2, and c_3 one of them of radius $> r_3/2$ and so on. It is easy to see by the construction that the set C, which is the union of the interior of the disks $c_1, c_2, \dots$ cover the whole of B, so $\mu(C) = \mu(B)$. The disks $c_1, c_2, \dots$ satisfy the conditions of Lemma 4.8, and therefore their set can be split into 22 sets C_i, $i = 1, 2, \dots, 22$ so that disks of the same C_i have no points in common. At least for one i we have $\mu(\cup C_i) \geq \frac{1}{22}\mu(C)$ where $\cup C_i$ denotes the union of the disks in C_i, and therefore we can choose a finite set Γ_0 of disks of C_i, so that

$$\mu(\cup\Gamma_0) > \tfrac{1}{23}\mu(B),$$

$$\mu(B \setminus \cup\Gamma_0) < \tfrac{22}{23}\mu(B).$$

Denote by Γ' the set of those disks of Γ that are outside $\cup\Gamma_0$. The set Γ' covers the set $B \setminus \cup\Gamma_0$ in the Vitali narrow sense. By the preceding argument there is a finite set Γ_1 of disjoint disks of Γ', such that

$$\mu(\cup\Gamma_1) > \tfrac{1}{23}\mu(B \setminus \cup\Gamma_0),$$

$$\mu(B \setminus ((\cup\Gamma_0) \cup (\cup\Gamma_1))) < \tfrac{22}{23}\mu(B \setminus \cup\Gamma_0) < \left(\tfrac{22}{23}\right)^2\mu(B).$$

Similarly, we find finite sets $\Gamma_2, \Gamma_3, \dots$ of disks of Γ, such that the disks in $\overline{\Gamma} := \Gamma_0 \cup \Gamma_1 \cup \Gamma_2 \cup \cdots$ have no points in common and that for any n

$$\mu(B \setminus ((\cup\Gamma_0) \cup (\cup\Gamma_1) \cup \cdots \cup (\cup\Gamma_n))) < \left(\tfrac{22}{23}\right)^{n+1}\mu(B).$$

Thus, we have

$$\mu(B \setminus \overline{\cup\Gamma}) = 0, \qquad \mu(\overline{\cup\Gamma}) = \mu(B),$$

which proves the lemma. □

Now we are in the position to prove

Theorem 4.9. *If* $\lambda > 1$, *then*

$$\mu_w\Big|_{\mathcal{S}_{w^\lambda}} \le \frac{1}{\lambda}\mu_{w^\lambda} + \left(1 - \frac{1}{\lambda}\right)\omega_{\mathcal{S}_{w^\lambda}}$$

and

$$\mu_w\Big|_{\mathcal{S}_{w^\lambda}} \ge \frac{1}{\lambda}\mu_{w^\lambda} + \left(1 - \frac{1}{\lambda}\right)\omega_{\mathcal{S}_w}\Big|_{\mathcal{S}_{w^\lambda}}.$$

Recall that $\omega_{\mathcal{S}}$ denotes the equilibrium measure of the compact set $\mathcal{S}$.

Proof of Theorem 4.9. Consider the two potentials corresponding to the two measures μ_w and μ_{w^λ} with some $\lambda > 1$. It follows from Theorem I.1.3 and Corollary I.4.5 that with $\omega = \omega_{\mathcal{S}_w}$

$$\frac{1}{\lambda}(U^{\mu_{w^\lambda}}(z) - F_{w^\lambda}) + \left(1 - \frac{1}{\lambda}\right)\left(U^{\omega}(z) - \log\frac{1}{\operatorname{cap}(\mathcal{S}_w)}\right) \ge U^{\mu_w}(z) - F_w$$

for quasi-every $z \in \mathcal{S}_w$; hence by the principle of domination we have this inequality everywhere (recall that if a measure μ has finite logarithmic energy, then every set of zero capacity has zero μ–measure). Furthermore, Theorem I.1.3, (I.1.9) and Theorem 4.1(a) imply that the equality sign holds for quasi-every $z \in \mathcal{S}_{w^\lambda}$. Now each of the measures μ_w, μ_{w^λ} and ω have finite logarithmic energy; hence they vanish on sets of zero capacity. Thus by applying Theorem 4.5 we can conclude the second inequality:

$$\mu_w\Big|_{\mathcal{S}_{w^\lambda}} \ge \frac{1}{\lambda}\mu_{w^\lambda} + \left(1 - \frac{1}{\lambda}\right)\omega_{\mathcal{S}_w}\Big|_{\mathcal{S}_{w^\lambda}}.$$

The first one can be shown with the same argument if we notice that with $\overline{\omega} = \omega_{\mathcal{S}_{w^\lambda}}$ we have

$$\frac{1}{\lambda}(U^{\mu_{w^\lambda}}(z) - F_{w^\lambda}) + \left(1 - \frac{1}{\lambda}\right)\left(U^{\overline{\omega}}(z) - \log\frac{1}{\operatorname{cap}(\mathcal{S}_{w^\lambda})}\right) \le U^{\mu_w}(z) - F_w$$

for quasi-every $z \in \mathcal{S}_{w^\lambda}$; hence by the principle of domination we have this inequality everywhere. Furthermore, by Theorem I.1.3, Corollary I.4.5 and Theorem 4.1(a) the equality sign holds for quasi-every $z \in \mathcal{S}_{w^\lambda}$. Thus, the first inequality follows as before from the Theorem 4.5. □

In order to apply the preceding theorem one needs a convenient criterion for concluding that a point x_0 from $\mathcal{S}_w$ belongs to some $\mathcal{S}_{w^\lambda}$, $\lambda > 1$, (recall from Theorem 4.1(a) that these supports are decreasing). We shall only need this result for the case when w is supported on the real line. In what follows let v_w denote the density of μ_w with respect to linear Lebesgue measure (Radon–Nikodym derivative). Whenever we speak of v_w on an interval, then we shall assume that μ_w is absolutely continuous on that interval with respect to Lebesgue measure.

Theorem 4.10. *Let $\Sigma \subseteq \mathbf{R}$ be part of the real line and w an admissible weight on Σ. Suppose x_0 is a point in the (one dimensional) interior of S_w, the density v_w of μ_w is continuous in a neighborhood of x_0, and $v_w(t) > \varepsilon_0$ for $|t - x_0| \le \varepsilon_1$ for some $\varepsilon_0 > 0$ and $\varepsilon_1 > 0$. Then for $\lambda \le 1/(1 - \varepsilon_0\varepsilon_1)$ the point x_0 is in the interior of S_{w^λ}; furthermore, v_{w^λ} is also continuous and positive in a neighborhood of x_0.*

Proof. We begin with the following observation. If for w we consider minimizing the weighted energy (I.1.12) on Σ and also on some closed set $S_w \subseteq \Sigma_1 \subseteq \Sigma$, then we arrive at the same extremal measure μ_w. Seeing that $S_{w^\lambda} \subseteq S_w$ (Theorem 4.1(a)), this shows that in the proof we may assume without loss of generality that $\Sigma = S_w$.

Let ν_0 be the measure the density of which is ε_0 on $[x_0 - \varepsilon_1/2, x_0 + \varepsilon_1/2]$ and 0 otherwise, and consider the positive measure

$$\nu_1 := \frac{1}{1 - \varepsilon_0\varepsilon_1}(\mu_w - \nu_0)$$

of total mass 1, and the weight function

$$w_1(x) := \exp(U^{\nu_1}(x))$$

that it generates. By Theorem I.3.3 the extremal measure corresponding to w_1 coincides with ν_1, and so $x_0 \in S_{w_1}$. Hence (see Theorem 1.3) if $B \subseteq [x_0 - \varepsilon/2, x_0 + \varepsilon/2]$ is a (one dimensional) neighborhood of x_0, then there is a polynomial P_n such that $w_1^n|P_n|$ attains its essential maximum in $B \cap \Sigma$.

The potential of the measure

$$\frac{1}{1 - \varepsilon_0\varepsilon_1}\nu_0$$

is symmetric about x_0, attains its maximum at x_0 and decreases to the right and increases to the left of x_0. But then for the weight

$$w_2(x) = w_1(x)\exp(U^{\nu_0/(1-\varepsilon_0\varepsilon_1)}(x))$$

the weighted polynomial $w_1^n P_n$ can attain its essential maximum only in B. Since this can be done for every small neighborhood B of x_0, it follows that $x_0 \in S_{w_2}$, again by Theorem 1.3. However, the weight function

$$w_2(x) = \exp(U^{\mu_w}(x)/(1 - \varepsilon_0\varepsilon_1))$$

and w^λ with $\lambda = 1/(1 - \varepsilon_0\varepsilon_1)$ differ on S_w only in a multiplicative constant. This together with the relation $\Sigma = S_w$ means (see Theorem I.3.3) that $\mu_{w^\lambda} = \mu_{w_2}$, and so $S_{w^\lambda} = S_{w_2}$ (see also Theorem 4.1(a)). Thus, $x_0 \in S_{w^\lambda}$ as is claimed in the theorem. Furthermore, the same proof can be carried out with x_1 in place of x_0 for every $x_1 \in [x_0 - \varepsilon_1/2, x_0 + \varepsilon_1/2]$; hence x_0 is actually in the interior of S_{w^λ}.

It remains to establish the continuity and positivity of the density function v_{w^λ} at x_0. Let $I = [x_0 - \varepsilon_1/2, x_0 + \varepsilon_1/2]$, and let an overbar denote taking balayage onto I out of $\mathbf{C} \setminus I$. By Theorem 1.6(e) we have

$$\mu_w\Big|_I = \overline{\mu_w}, \qquad \mu_{w^\lambda}\Big|_I = \overline{\mu_{w^\lambda}},$$

and then Theorem 4.4 gives

$$\mu_w\Big|_I = \frac{1}{\lambda}\mu_{w^\lambda}\Big|_I + \left(1 - \frac{1}{\lambda}\right)\omega_I,$$

from which we get the formula

$$\begin{aligned}\mu_{w^\lambda}\big|_I &= \overline{\mu_{w^\lambda}} - \overline{\mu_{w^\lambda}\big|(\mathbf{R}\setminus I)} = \lambda\mu_w\Big|_I - (\lambda-1)\omega_I - \overline{\mu_{w^\lambda}\big|(\mathbf{R}\setminus I)} \\ &= \lambda\mu_w\Big|_I + \lambda\overline{\mu_w\big|(\mathbf{R}\setminus I)} - (\lambda-1)\omega_I - \overline{\mu_{w^\lambda}\big|(\mathbf{R}\setminus I)}.\end{aligned}$$

Now on the right each term beginning with the second one has continuous density in the interior of I (see Corollary II.4.12); furthermore, by the assumption the first term also has continuous density at x_0, and these prove that μ_{w^λ} has continuous density function at x_0.

Finally, the positivity of the density can be seen as follows. If we define ν_0 as the measure the density of which is ε_0 on $[x_0 - 3\varepsilon_1/4, x_0 + 3\varepsilon_1/4]$, then the preceding argument will yield for $\lambda' = 1/(1 - \frac{3}{2}\varepsilon_0\varepsilon_1)$ the conclusion of the theorem. Now Theorem 4.9 implies with w replaced by w^λ and λ replaced by $\lambda'/\lambda > 1$ that in a neighborhood I of x_0 the measure μ_{w^λ} is bounded from below by $(1 - \lambda/\lambda')\omega_{S_{w^\lambda}}$, which has positive density on I. □

IV.5 Exponential and Power-Type Functions

In this section we consider some important concrete weights on the real line. The results here will be used in Chapters VI and VII.

First we determine the equilibrium distribution for the case $\Sigma = \mathbf{R}$, $w(x) = \exp(-c|x|^\lambda)$ with $c, \lambda > 0$. To this end consider the so-called Ullman distribution with density function

$$s_\lambda(t) := \frac{\lambda}{\pi}\int_{|t|}^1 \frac{u^{\lambda-1}}{\sqrt{u^2 - t^2}}du \tag{5.1}$$

on $[-1, 1]$. For its potential we have (writing instead of the measure its density as a parameter in U)

$$\begin{aligned}-U^{s_\lambda}(x) &= \frac{\lambda}{\pi}\int_{-1}^1 \log|x-t| \int_{|t|}^1 \frac{u^{\lambda-1}}{\sqrt{u^2-t^2}}du dt \\ &= \int_0^1 \lambda u^{\lambda-1}\frac{1}{\pi}\int_{-u}^u \frac{\log|x-t|}{\sqrt{u^2-t^2}}dt du.\end{aligned}$$

The expression after $u^{\lambda-1}$ is just the negative of the equilibrium potential of the interval $[-u, u]$; hence it equals $\log u - \log 2$ if $|x| \le u$ and $\log|x + \sqrt{x^2 - u^2}| - \log 2$ if $|x| > u$ (see (I.1.8)). As always, we take that branch of the square root that is positive on the positive part of the real axis. Thus, for $-1 \le x \le 1$ we can continue the above equality as

$$= -\log 2 + \int_{|x|}^{1} \lambda u^{\lambda-1} \log u \, du + \int_0^{|x|} \lambda u^{\lambda-1} \log|x + \sqrt{x^2 - u^2}| \, du$$

$$= -\log 2 - \frac{1}{\lambda} + |x|^\lambda \left(\frac{1}{\lambda} + \int_0^1 \lambda v^{\lambda-1} \log|1 + \sqrt{1 - v^2}| \, dv \right).$$

Integration by parts yields for the last integral the value

$$\int_0^1 \frac{v^{\lambda-1}}{1 + \sqrt{1 - v^2}} \frac{v^2}{\sqrt{1 - v^2}} \, dv = \int_0^1 \frac{v^{\lambda-1}}{\sqrt{1 - v^2}} \, dv - \frac{1}{\lambda},$$

where we used the identity $v^2 = (1 - \sqrt{1 - v^2})(1 + \sqrt{1 - v^2})$. Since

$$\gamma_\lambda := \int_0^1 \frac{v^{\lambda-1}}{\sqrt{1 - v^2}} \, dv = \Gamma\left(\frac{\lambda}{2}\right) \Gamma\left(\frac{1}{2}\right) \Big/ 2\Gamma\left(\frac{\lambda}{2} + \frac{1}{2}\right),$$

we finally have for $x \in [-1, 1]$

$$U^{s_\lambda}(x) = -\gamma_\lambda |x|^\lambda + \log 2 + \frac{1}{\lambda}.$$

Now let $|x| > 1$, $x \in \mathbf{R}$. By symmetry we can assume $x > 1$. Exactly as above we compute

$$U^{s_\lambda}(x) = \log 2 - \int_0^1 \lambda u^{\lambda-1} \log|x + \sqrt{x^2 - u^2}| \, du \tag{5.2}$$

$$= \log 2 - \log|x + \sqrt{x^2 - 1}| - \int_0^1 \frac{u^{\lambda-1}}{x + \sqrt{x^2 - u^2}} \frac{u^2}{\sqrt{x^2 - u^2}} \, du$$

$$= \log 2 - \log|x + \sqrt{x^2 - 1}| - \int_0^1 \frac{x u^{\lambda-1}}{\sqrt{x^2 - u^2}} \, du + \frac{1}{\lambda}$$

$$= -\log|x + \sqrt{x^2 - 1}| - x^\lambda \int_0^{1/x} \frac{v^{\lambda-1}}{\sqrt{1 - v^2}} \, dv + \frac{1}{\lambda} + \log 2.$$

Since

$$\frac{d \log(x + \sqrt{x^2 - u^2})}{dx} = (x^2 - u^2)^{-1/2},$$

from (5.2) we get for the derivative of the potential for $x > 1$

$$\left(U^{s_\lambda}(x)\right)' = -\int_0^1 \frac{\lambda u^{\lambda-1}}{\sqrt{x^2-u^2}}du = -\lambda x^{\lambda-1}\int_0^{1/x} \frac{u^{\lambda-1}}{\sqrt{1-u^2}}du \geq -\lambda\gamma_\lambda x^{\lambda-1}, \tag{5.3}$$

furthermore $U^{s_\lambda}(x)$ coincides with $-\gamma_\lambda x^\lambda + \log 2 + 1/\lambda$ at $x = 1$. Therefore on $\mathbf{R}\setminus[-1,1]$ the potential is above $-\gamma_\lambda|x|^\lambda + \log 2 + 1/\lambda$ while on $[-1,1]$ these two functions coincide. On applying Theorem I.3.3 we can conclude that if $w(x) = \exp(-\gamma_\lambda|x|^\lambda)$, then μ_w is given by the density function s_λ and $F_w = \log 2 + 1/\lambda$.

We can summarize our findings in

Theorem 5.1. *Let $w(x) = \exp(-\gamma_\lambda|x|^\lambda)$, $x \in \mathbf{R}$, where $\lambda > 0$ and*

$$\gamma_\lambda = \Gamma\left(\frac{\lambda}{2}\right)\Gamma\left(\frac{1}{2}\right)\Big/ 2\Gamma\left(\frac{\lambda}{2}+\frac{1}{2}\right).$$

Then $\mathcal{S}_w = [-1,1]$, $F_w = \log 2 + 1/\lambda$, and μ_w is absolutely continuous with respect to Lebesgue measure with density function

$$\frac{\lambda}{\pi}\int_{|t|}^1 \frac{u^{\lambda-1}}{\sqrt{u^2-t^2}}du, \quad t \in [-1,1].$$

Furthermore, on $\mathbf{R}\setminus[-1,1]$ the potential U^{μ_w} is given by

$$-\log|x+\sqrt{x^2-1}| - |x|^\lambda \int_0^{1/|x|} \frac{v^{\lambda-1}}{\sqrt{1-v^2}}dv + \frac{1}{\lambda} + \log 2.$$

On $\mathbf{C}\setminus\mathbf{R}$ the potential is

$$-\log|x+\sqrt{x^2-1}| - \operatorname{Re}\int_0^1 \frac{xu^{\lambda-1}}{\sqrt{x^2-u^2}}du + \frac{1}{\lambda} + \log 2.$$

The last formula is obtained as the first three lines of (5.2) if we write

$$\log|x+\sqrt{x^2-u^2}| = \operatorname{Re}\log(x+\sqrt{x^2-u^2}).$$

Since

$$\log|x+\sqrt{x^2-1}| - \log 2$$

is the Green function of the domain $\mathbf{C}\setminus[-1,1]$ with pole at infinity, we get as a consequence of Theorem I.4.7 that the solution of the Dirichlet problem in $\mathbf{C}\setminus[-1,1]$ with boundary function $|x|^\lambda$, $x \in [-1,1]$, is given by

$$\frac{1}{\gamma_\lambda}\operatorname{Re}\int_0^1 \frac{xu^{\lambda-1}}{\sqrt{x^2-u^2}}du - \frac{1}{\lambda\gamma_\lambda}.$$

Of course, by linear substitution we can find the corresponding quantities for any weight $w(x) = \exp(-c|x|^\lambda)$. For example, $\mathcal{S}_w$ will be

$$[-\gamma_\lambda^{1/\lambda}c^{-1/\lambda}, \gamma_\lambda^{1/\lambda}c^{-1/\lambda}]. \tag{5.4}$$

For later reference let us record here that for $1 < |x| < 2$ the previous formulae (see (5.3)) give

$$(U^{\mu_w}(x))' + \lambda\gamma_\lambda x^{\lambda-1} = \lambda|x|^{\lambda-1}\int_{1/|x|}^{1}\frac{u^{\lambda-1}}{\sqrt{1-u^2}}du \sim ||x|-1|^{1/2}, \tag{5.5}$$

where $\sim$ indicates that the ratio of the two sides lies in between two absolute constants (in the range of the arguments indicated), and so

$$(U^{\mu_w}(x) - F_w) + \gamma_\lambda|x|^\lambda \sim ||x|-1|^{3/2} \tag{5.6}$$

for $|x| \in (1,2)$, and clearly

$$(U^{\mu_w}(x) - F_w) + \gamma_\lambda|x|^\lambda \sim |x|^\lambda \tag{5.7}$$

when $|x| > 2$.

Next we consider Jacobi weights.

Example 5.2 (Jacobi Weights). Let $w(x) = (1-x)^{\lambda_1}(1+x)^{\lambda_2}$ on $\Sigma = [-1,1]$, with λ_1, λ_2 positive. The support $\mathcal{S}_w = [a,b]$ was determined in Example 1.17 (formulas for a, b appear in equations (1.28)). We shall show that

$$d\mu_w(x) = \frac{1}{\pi}\frac{(1+\lambda_1+\lambda_2)}{1-x^2}\sqrt{(x-a)(b-x)}\,dx, \quad a \le x \le b. \tag{5.8}$$

To derive this formula, we shall determine U^{μ_w} and compute its normal derivatives so as to obtain μ_w via Theorem 2.3. Indeed, let

$$\zeta = \varphi(z) := \frac{\sqrt{z-a}+\sqrt{z-b}}{\sqrt{z-a}-\sqrt{z-b}} = \frac{2z-a-b+2\sqrt{(z-a)(z-b)}}{b-a} \tag{5.9}$$

denote the mapping of $\Omega := \overline{\mathbf{C}} \setminus [a,b]$ onto the exterior of the unit disk and set $\zeta_{-1} := \varphi(-1)$, $\zeta_1 := \varphi(1)$. Then, as previously remarked (c.f. Theorem I.4.7),

$$U^{\mu_w}(z) = g(z) - g_\Omega(z,\infty) + F_w = g(z) - \log|\varphi(z)| + F_w, \quad z \in \Omega, \tag{5.10}$$

where g is the solution of the Dirichlet problem in Ω with boundary data

$$-Q(x) = \log w(x) = \lambda_1\log(1-x) + \lambda_2\log(1+x), \quad a \le x \le b.$$

This solution can be seen via inspection to be

$$g(z) = h(z) - \lambda_1\log\left|\frac{\zeta-\zeta_1}{\zeta_1\zeta-1}\right| - \lambda_2\log\left|\frac{\zeta-\zeta_{-1}}{\zeta_{-1}\zeta-1}\right| - (\lambda_1+\lambda_2)\log|\zeta|, \tag{5.11}$$

where

$$h(z) := \lambda_1\log|1-z| + \lambda_2\log|1+z|.$$

(Note that the three terms involving ζ in (5.11) vanish on $[a,b]$ and cancel the singularities of h at $z = \pm 1$ and $z = \infty$.) Thus, from (5.10), we have

$$U^{\mu_w}(z) = h(z) - \lambda_1 \log\left|\frac{\zeta-\zeta_1}{\zeta_1\zeta-1}\right| - \lambda_2 \log\left|\frac{\zeta-\zeta_{-1}}{\zeta_{-1}\zeta-1}\right|$$
$$-(1+\lambda_1+\lambda_2)\log|\zeta| + F_w. \tag{5.12}$$

To evaluate the normal derivative $\partial U^{\mu_w}/\partial \mathbf{n}_+$ we first compute

$$D(z) := \frac{d}{dz}\left\{-\lambda_1 \log\left(\frac{\zeta-\zeta_1}{\zeta_1\zeta-1}\right) - \lambda_2 \log\left(\frac{\zeta-\zeta_{-1}}{\zeta_{-1}\zeta-1}\right) - (1+\lambda_1+\lambda_2)\log\zeta\right\}$$

$$= \frac{d\zeta}{dz}\left\{-\lambda_1\frac{\left(\zeta_1-\dfrac{1}{\zeta_1}\right)}{(\zeta-\zeta_1)\left(\zeta-\dfrac{1}{\zeta_1}\right)} - \lambda_2\frac{\left(\zeta_{-1}-\dfrac{1}{\zeta_{-1}}\right)}{(\zeta-\zeta_{-1})\left(\zeta-\dfrac{1}{\zeta_{-1}}\right)} - (1+\lambda_1+\lambda_2)\frac{1}{\zeta}\right\}. \tag{5.13}$$

As is easily verified from (5.9) we have the identities

$$\frac{d\zeta}{dz} = \frac{\zeta}{\sqrt{(z-a)(z-b)}}, \qquad (\zeta-\zeta_{\pm 1})\left(\zeta-\frac{1}{\zeta_{\pm 1}}\right) = \frac{4(z\mp 1)\zeta}{(b-a)},$$

$$\zeta_1 - \frac{1}{\zeta_1} = \frac{4\sqrt{(1-a)(1-b)}}{b-a}, \quad \zeta_{-1} - \frac{1}{\zeta_{-1}} = \frac{4\sqrt{(1+a)(1+b)}}{a-b}, \tag{5.14}$$

where the square roots are positive. Substituting these expressions into (5.13) we obtain

$$D(z) = \frac{1}{\sqrt{(z-a)(z-b)}}\left[\frac{-\lambda_1\sqrt{(1-a)(1-b)}}{z-1}\right.$$
$$\left. + \lambda_2\frac{\sqrt{(1+a)(1+b)}}{z+1} - (1+\lambda_1+\lambda_2)\right]. \tag{5.15}$$

Next, we observe from the previously obtained equations (1.26) for a, b that the bracketed expression in (5.15) vanishes for $z=a$ and $z=b$. Thus

$$D(z) = \frac{1}{\sqrt{(z-a)(z-b)}}\left[\frac{(1+\lambda_1+\lambda_2)(z-a)(z-b)}{1-z^2}\right]$$
$$= (1+\lambda_1+\lambda_2)\frac{\sqrt{(z-a)(z-b)}}{1-z^2}. \tag{5.16}$$

Finally, since $\partial h/\partial \mathbf{n}_+ = 0$, we obtain from (5.12) and (5.16) that

$$\begin{aligned}\frac{\partial}{\partial \mathbf{n}_+}U^{\mu_w}(x) &= \text{Re } \{iD(z)\}|_{z\to x,\ \text{Im } z>0}\\ &= -\frac{(1+\lambda_1+\lambda_2)}{1-x^2}\sqrt{(x-a)(b-x)}, \quad a<x<b,\end{aligned}$$

and since $\partial U^{\mu_w}/\partial \mathbf{n}_+ = \partial U^{\mu_w}/\partial \mathbf{n}_-$, the formula (5.8) follows from Theorem 2.3. □

Example 5.3 (Incomplete Polynomials). For the weight $w(x) = x^{\theta/(1-\theta)}$, $0<\theta<1$, on $\Sigma = [0, 1]$, we determined in Example 1.16 that $\mathcal{S}_w = [\theta^2, 1]$. We now claim that

$$d\mu_w(x) = \frac{1}{(1-\theta)\pi x}\sqrt{\frac{x-\theta^2}{1-x}}dx, \quad \theta^2 \le x \le 1. \tag{5.17}$$

Indeed, this follows as a limit case of the preceding example where we take $\lambda_1 = 0$ and $\lambda_2 = \theta/(1-\theta)$. It can be easily verified that formula (5.8) remains valid in this case and that $a = 2\theta^2 - 1$, $b = 1$; that is, for $\hat{w}(t) = (1-t)^{\theta/(1-\theta)}$ on $-1 \le t \le 1$, we have

$$d\mu_{\hat{w}}(t) = \frac{(1+\theta/(1-\theta))}{\pi(1+t)}\sqrt{\frac{t+1-2\theta^2}{1-t}}dt, \quad 2\theta^2-1 \le t \le 1.$$

On changing the interval $[-1, 1]$ into $[0, 1]$ via the linear transformation $t = 2x-1$, we get (5.17). □

Finally, we calculate the extremal measure for Laguerre weights.

Example 5.4 (Laguerre Weights). For $w(x) = x^s e^{-\lambda x}$ on $\Sigma = [0, \infty)$, where $s \ge 0$, $\lambda > 0$, we established in Example 1.18 that $\mathcal{S}_w = [a, b]$ with a, b given by the formulas in (1.31). Now we show that

$$d\mu_w(x) = \frac{\lambda}{\pi x}\sqrt{(x-a)(b-x)}, \quad a<x<b. \tag{5.18}$$

Again it is easy to explicitly determine U^{μ_w}. With the same notation as in Example 5.2, we immediately solve the Dirichlet problem in $\Omega = \overline{\mathbf{C}} \setminus [a, b]$ with boundary data

$$-Q(x) = s \log x - \lambda x$$

to arrive at the formula

$$\begin{aligned}U^{\mu_w}(z) &= s\log|z| - \lambda \text{ Re } z - s\log\left|\frac{\zeta-\zeta_0}{\zeta_0\zeta-1}\right|\\ &\quad + \lambda \text{ Re }\sqrt{(z-a)(z-b)} - (1+s)\log|\zeta| + F_w, \end{aligned}\tag{5.19}$$

where $\zeta = \varphi(z)$ and $\zeta_0 := \varphi(0)$. To compute $\partial U^{\mu_w}/\partial \mathbf{n}_+$ we first determine

$$H(z) := \frac{d}{dz}\left\{-s\log\left(\frac{\zeta-\zeta_0}{\zeta_0\zeta-1}\right)-(1+s)\log\zeta+\lambda\sqrt{(z-a)(z-b)}\right\}$$

$$= \frac{d\zeta}{dz}\left\{-s\frac{\zeta_0-\dfrac{1}{\zeta_0}}{(\zeta-\zeta_0)\left(\zeta-\dfrac{1}{\zeta_0}\right)}-(1+s)\frac{1}{\zeta}\right\}+\frac{\lambda}{2}\frac{(2z-a-b)}{\sqrt{(z-a)(z-b)}}. \tag{5.20}$$

From the identities

$$\frac{d\zeta}{dz}=\frac{\zeta}{\sqrt{(z-a)(z-b)}},\quad (\zeta-\zeta_0)\left(\zeta-\frac{1}{\zeta_0}\right)=\frac{4z\zeta}{b-a},\quad \zeta_0-\frac{1}{\zeta_0}=-\frac{4\sqrt{ab}}{b-a},$$

we find

$$H(z)=\frac{1}{\sqrt{(z-a)(z-b)}}\left[\frac{s\sqrt{ab}}{z}-(1+s)+\frac{\lambda}{2}(2z-a-b)\right].$$

Next we observe from the formulas (1.30) of Example 1.18 that the expression in brackets vanishes for $z=a$ and $z=b$. Thus

$$H(z)=\frac{1}{\sqrt{(z-a)(z-b)}}\cdot\left[\frac{\lambda(z-a)(z-b)}{z}\right],$$

and we find from (5.19) and (5.20) that

$$\begin{aligned}\frac{\partial U^{\mu_w}}{\partial \mathbf{n}_+}(x) &= \operatorname{Re}\{iH(z)\}|_{z\to x,\ \operatorname{Im} z>0}\\ &= -\frac{\lambda\sqrt{(x-a)(b-x)}}{x},\quad a<x<b.\end{aligned}$$

Since $\partial U^{\mu_w}/\partial \mathbf{n}_+ = \partial U^{\mu_w}/\partial \mathbf{n}_-$, we get the formula (5.18) from Theorem 2.3.

In the above derivation we tacitly assumed that $s>0$. However, formula (5.18) remains valid in the limiting case when $s=0$. Indeed for $w(x)=e^{-\lambda x}$ on $[0,\infty)$ we obtain $a=0,\ b=2/\lambda$ and

$$d\mu_w(x)=\frac{\lambda}{\pi}\sqrt{\frac{(2/\lambda)-x}{x}}\,dx,\quad 0<x<2/\lambda.$$

This last formula also follows immediately from Theorem 3.2. □

IV.6 Circular Symmetric Weights

We have seen in the preceding sections that explicit determination of μ_w is usually a difficult problem. Now we discuss a special case when the explicit solution is easy to obtain; this is the case when w is circular symmetric satisfying some additional assumptions.

Thus, let $\Sigma = \mathbf{C}$, $w(z) = \exp(-Q(z))$ where $Q(z)$ is a radially symmetric function: $Q(z) = Q(|z|)$. About Q we assume that on $\mathbf{R}_+$ it is differentiable with absolutely continuous derivative bounded below,

$$\lim_{r\to\infty} (Q(r) - \log r) = \infty \tag{6.1}$$

(this is needed for the admissibility of w) and that it satisfies either one of the conditions

(i) $rQ'(r)$ increasing on $(0, \infty)$;
(ii) Q is convex on $(0, \infty)$.

Let $r_0 \geq 0$ be the smallest number for which $Q'(r) > 0$ for all $r > r_0$, and we set R_0 to be the smallest solution of $R_0 Q'(R_0) = 1$. It is easy to see that $r_0 < R_0$ and both of these numbers are finite (cf. (6.1)). With this choice for r_0 and R_0 we shall prove

Theorem 6.1. *Let $w = \exp(-Q)$ be a radially symmetric weight with Q satisfying* (6.1) *and either of the conditions of* (i) *or* (ii), *and assume that Q' is absolutely continuous on $\mathbf{R}_+$. If r_0 and R_0 are defined as above, then S_w is the ring*

$$\{z \mid r_0 \leq |z| \leq R_0\}$$

and μ_w is given by

$$d\mu_w(z) = \frac{1}{2\pi}(rQ'(r))' dr d\varphi, \quad z = re^{i\varphi}. \tag{6.2}$$

F_w is given by $Q(R_0) - \log R_0$, and the equilibrium potential is

$$U^{\mu_w}(z) = \begin{cases} F_w - Q(r_0), & \textit{if } |z| < r_0 \\ F_w - Q(z), & \textit{if } r_0 \leq |z| \leq R_0 \\ \log 1/|z|, & \textit{if } |z| > R_0. \end{cases} \tag{6.3}$$

Proof. Using the formula

$$\frac{1}{2\pi}\int_{-\pi}^{\pi} \log \frac{1}{|z - re^{i\varphi}|} d\varphi = \begin{cases} \log 1/|r|, & \text{if } |z| \leq r \\ \log 1/|z|, & \text{if } |z| > r \end{cases}$$

(see (0.5.5)), one can get by simple computation that the potential of the measure given on the right of (6.2) is the function appearing on the right of (6.3). Since the choice of r_0 and R_0 easily imply that $Q(z) \geq Q(r_0)$ if $|z| < r_0$ and

$$-Q(z) \leq \frac{1}{|z|} + \log R_0 - Q(R_0)$$

if $|z| > R_0$, the theorem follows from Theorem I.3.3 because the measure defined by (6.2) has total mass 1. □

Example 6.2. For $w(z) = |z|^\alpha \exp(-|z|^\beta)$, $\alpha \geq 0$, $\beta > 0$, on $\Sigma = \mathbf{C}$ we have

$$S_w = \left\{z \mid (\alpha/\beta)^{1/\beta} \leq |z| \leq ((1+\alpha)/\beta)^{1/\beta}\right\}$$

and

$$d\mu_w(z) = \frac{1}{2\pi}\beta^2 r^{\beta-1} dr\, d\phi, \qquad z = re^{i\phi}.$$

Indeed, $Q(r) = -\alpha \log r + r^\beta$ satisfies the conditions of Theorem 6.1, and for this Q we find $r_0 = (\alpha/\beta)^{1/\beta}$, $R_0 = ((1+\alpha)/\beta)^{1/\beta}$. □

IV.7 Some Problems from Physics

In this section, we shall discuss some problems from physics that are related to the weighted energy problem in the presence of an external field.

IV.7.1 Contact Problem of Elasticity

Suppose that the lower half-plane $\{z \mid \operatorname{Im} z < 0\}$ consists of an elastic material, and a rigid punch (stamp) with profile $y = Q(x)$ is pressed into the lower half-plane with total force λ (see Figure 7.1). Then there is a region $S = S(Q, \lambda)$ where the punch is in contact with the elastic material, and is above it outside S. On S itself the elastic material exerts a counterforce, and if we assume equilibrium, then this total counterforce has to be equal to λ. The exerted force appears as pressure $p(t)$ on the contact region on the punch. Thus, $p(t) \geq 0$ for $t \in S$, and $p(t) = 0$ for $t \notin S$. The exerted force is equal to the total pressure, i.e.

$$\int_S p(t)\, dt = \lambda.$$

Under some simplifying assumptions it turns out (see [165, Section 115], [58, Sections 2.2 and 6.2]) that the displacement of the surface of the elastic material is given by $\int \log|x-t|\; p(t)\,dt$ (strictly speaking this is true only for small displacements and around the contact region, but we shall not worry about such details). Thus, if $D(\geq 0)$ denotes the vertical displacement of the punch from its initial position where it is resting (without applied force) on the elastic material, then we must have the equations

$$\int \log|x-t|\; p(t)\,dt = Q(x) - D, \qquad x \in S,$$

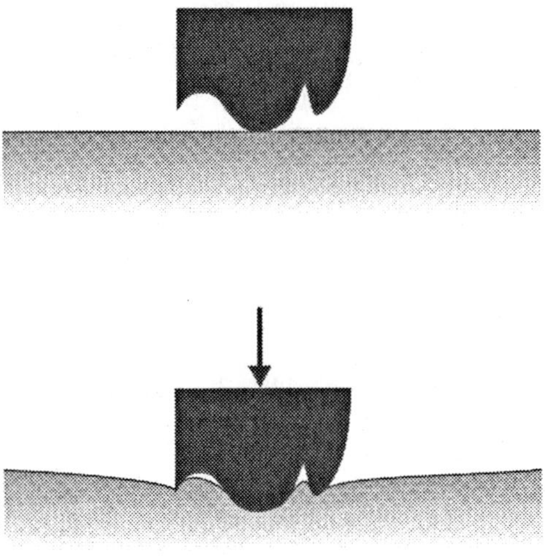

Fig. 7.1

and

$$\int \log|x-t|\, p(t)\, dt < Q(x) - D, \qquad x \notin S.$$

Dividing through by λ and setting $d\mu = p(t)dt/\lambda$, we see that the probability measure μ satisfies the basic equations

$$\int \log \frac{1}{|x-t|}\, d\mu(t) = -\frac{Q(x)}{\lambda} + D/\lambda, \qquad x \in \operatorname{supp}(\mu),$$

and

$$\int \log \frac{1}{|x-t|}\, d\mu(t) \geq -\frac{Q(x)}{\lambda} + D/\lambda, \qquad x \notin \operatorname{supp}(\mu),$$

and hence (see Theorem I.3.3) it is the equilibrium measure for the weighted energy problem with external field Q/λ. The physical meaning of F_w (with $w = \exp(-Q/\lambda)$) is that λF_w is the vertical displacement D of the punch.

This physical interpretation readily explains many of the theorems in this monograph, and is a useful device for making reasonable guesses on the behavior of weighted potentials. For example, it is intuitively clear, that if the shape of a second punch is Q^*, and $Q^*(x) \geq Q(x)$ with equality at some point in S, then to achieve the same vertical displacement for Q^* one needs to exert less force than for Q. Thus, if the force is the same in both cases, then the displacement for Q^* will be bigger than that for Q, and this is exactly the content of formula (I.4.2).

The balayage problem (at least for sets on the real line) onto K is interpreted as follows. If μ is given, then let $Q(x) = -U^\mu(x) - m$ be the profile of the punch, where m is the minimum of $-U^\mu$. Then under the force $\|\mu\|$ the displacement will be $-m$. Suppose now that we freeze the portion of the pressed surface over K,

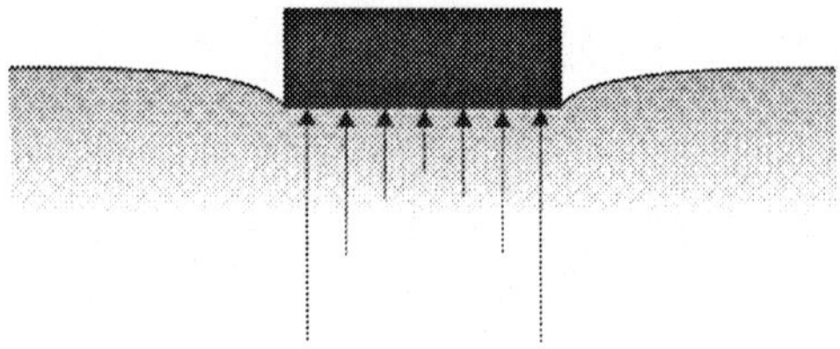

Fig. 7.2

and would like to have a punch for which the shape of the contact region (under total pressing force $\|\mu\|$) is exactly the shape of this frozen part. Then we have to have the frozen part as the stamp, and since force is redistributed to this frozen part, this frozen part will sink deeper into the elastic material than its position originally was. This explains why the constant c in Theorem II.4.7 is positive. Intuitively it is also clear that other parts of the surface of the elastic material will not sink as deep as the part over K, and this is exactly part (b) of that theorem.

As further examples, let us consider some statements concerning the support of the extremal measure, which is the same in our case as the region of contact. It is clear that as we exert a greater force $\lambda_1 > \lambda_2$, then the region of contact increases. This is exactly the statement $\mathcal{S}_{w^{1/\lambda_2}} \subset \mathcal{S}_{w^{1/\lambda_1}}$ given in Theorem 4.1(a). If the shape of the punch Q is convex, then we must have an interval as the contact region, as is asserted in Theorem 1.10(b). Finally, it is clear that the lowest point of the punch must always be in the contact region, but it is easy to come up with a punch over a finite interval $[a, b]$ such that its maximum point is also in the contact region, and yet the contact region is not the whole interval $[a, b]$.

Finally, the elasticity interpretation also tells something about the equilibrium measure, as well. For example, if we increase the contact force on the punch, then the contact pressure also increases, and this is the physical interpretation of the inequality $\mu_w|_{\mathcal{S}_{w^\alpha}} \geq \mu_{w^\alpha}/\alpha$, $\alpha > 1$, which follows from Theorem 4.9. It is also clear that around downward cusps in the contact region the pressure becomes infinite, i.e. at such points the extremal measure has infinite density. If the shape of the punch is a nice differentiable function, then we expect that the elastic material and the punch will separate smoothly, i.e. in such cases at the endpoints of the contact region the pressure becomes zero (see e.g. the equilibrium distribution for the Freud and Jacobi weights in Section IV.5). On the other hand, if at an endpoint in the contact region the punch has a break point, then the separation is not smooth, and pressure builds up at this endpoint. Consider for example the shape of a rectangular punch over an interval $[a, b]$, when the contact pressure is a constant times the equilibrium measure, and hence has a $1/\sqrt{x}$ type singularity; see Figure 7.2.

IV.7.2 Distribution of Energy Levels of Quantum Systems

It is well known in statistical physics that many physical systems can be described by the statistics of the eigenvalues of ensembles of random matrices. Some parts of the theory for the so-called unitary matrix ensembles are connected to weighted energy as we now briefly describe. For all the results below see the literature mentioned in the Notes section at the end of this chapter, in particular [87] and [177].

Let $\mathcal{H}_n$ be the set of all $n \times n$ Hermitian matrices $M = (m_{i,j})_{i,j=1}^n$, and let there be given a probability distribution on $\mathcal{H}_n$ of the form

$$p_n(M)dM = D_n^{-1} \exp(-n\mathrm{Tr}\{V(M)\})dM,$$

where $V(\lambda)$, $\lambda \in \mathbf{R}$, is a real-valued function that increases sufficiently fast at infinity (typically an even polynomial in quantum field theory applications), $\mathrm{Tr}\{H\}$ denotes the trace of the matrix H,

$$dM = \prod_{k=1}^{n} dm_{k,k} \prod_{k<j} d\,\mathrm{Re}\, m_{k,j}\; d\,\mathrm{Im}\, m_{k,j}$$

is the "Lebesgue" measure for the Hermitian matrices, and D_n is just a normalizing constant so that the total integral of $p_n(M)dM$ is one. The fact that V increases sufficiently fast guarantees the existence of D_n.

Every matrix $M \in \mathcal{H}_n$ has n real eigenvalues, which carry the physical information on the system when it is in the state described by M. The joint probability distribution of the eigenvalues of the random matrices $M \in \mathcal{H}_n$ is given by the density

$$q_n(\lambda_1, \ldots, \lambda_n) = d_n^{-1} \prod_{i \neq j} |\lambda_i - \lambda_j| \prod_{j=1}^{n} e^{-nV(\lambda_j)}, \tag{7.1}$$

where d_n is again a normalizing constant. This distribution is basically the square of the weighted Vandermonde expression of Section III.1 for the case $Q(x) = V(x)/2$ (there is an $(n-1)/n$ factor difference that can be neglected). The quantity

$$N_n(\Delta) = \frac{\#\{\text{eigenvalues in } \Delta\}}{n}$$

is of special importance, and is the random variable that equals the normalized number of eigenvalues in the interval Δ. The expected value $EN_n(\Delta)$ of $N_n(\Delta)$ is obtained by integrating (7.1) for all λ_j, $j \geq 2$ over $(-\infty, \infty)$, and for λ_1 over Δ. As $n \to \infty$, the normalized expected number of eigenvalues $EN_n(\Delta)$ has a limit $\mu(\Delta)$, and it turns out that μ is the equilibrium distribution for the external field $Q = V/2$ studied in this monograph. In statistical physics μ is known as the integrated density of states, and its density as the density of states.

The model that we have outlined is known as the *unitary ensemble* associated with V.

The most studied example is when $V(x) = 2x^2$. In this case the random matrix ensemble with probability distribution

$$p_n(M)dM = D_n^{-1}\exp(-2n\sum_{i,j=1}^{n}|m_{i,j}|^2)dM,$$

is called the Gaussian ensemble, and for it the density of states is given by

$$d\mu(t) = \frac{2}{\pi}\sqrt{1-t^2},$$

which is the celebrated Wigner's semi-circle law. This is a special case of general Freud weights $V(x) = 2\gamma_\alpha|x|^\alpha$, where the normalizing constant is

$$\gamma_\alpha = \Gamma\left(\frac{\alpha}{2}\right)\Gamma\left(\frac{1}{2}\right)\Big/2\Gamma\left(\frac{\alpha}{2}+\frac{1}{2}\right).$$

For the general case we have by Theorem 5.1 the formula

$$\frac{\alpha}{\pi}\int_{|t|}^{1}\frac{u^{\alpha-1}}{\sqrt{u^2-t^2}}du, \quad t \in [-1,1],$$

for the density of states. In a similar manner, other weights from Chapter IV yield concrete examples with known density of states.

The distribution (7.1) has the following form:

$$d_n^{-1}\begin{vmatrix} 1 & 1 & \cdots & 1 \\ \lambda_1 & \lambda_2 & \cdots & \lambda_n \\ \lambda_1^2 & \lambda_2^2 & \cdots & \lambda_n^2 \\ \vdots & \vdots & \ddots & \vdots \\ \lambda_1^{n-1} & \lambda_2^{n-1} & \cdots & \lambda_n^{n-1} \end{vmatrix}^2 \prod_{i=1}^{n} e^{-nV(\lambda_i)},$$

and here we can add to any row any linear combination of the other rows. In particular, if $p_j(w^n, x)$ are the orthonormal polynomials (cf. Chapter VII) with respect to $w^{2n}(x)$, $w(x) = \exp(-Q(x))$, $Q(x) = V(x)/2$, defined as

$$\int p_j(w^n, x)p_k(w^n, x)w^{2n}(x)\,dx = \begin{cases} 1 & \text{if } j = k \\ 0 & \text{if } j \neq k, \end{cases}$$

then the preceding expression equals

$$(d_n^*)^{-1}\left|p_{i-1}(w^n, \lambda_j)w^n(\lambda_j)\right|^2_{1\le i,j\le n},$$

where d_n^* is another normalizing constant built from d_n and the leading coefficients of the $p_j(w^n, \cdot)$. By computing the square of the determinants we obtain that the joint probability distribution of the eigenvalues (states) has the form

$$(d_n^*)^{-1} \left|K_n(\lambda_i, \lambda_j)\right|_{1\le i,j\le n},$$

where

$$K_n(t,s) = \sum_{j=0}^{n-1} p_j(w^n, t)w^n(t)p_j(w^n, s)w^n(s)$$

is the so-called reproducing kernel for the weight w^n. It turns out that the normalizing constant d_n^* is $n!$, which follows from the formulae

$$\int K_n(t,\tau)K_n(\tau,s)\,d\tau = K_n(t,s), \qquad \int K_n^2(t,s)\,dt\,ds = n,$$

$$\int K_n(t,t)\,dt = n.$$

These same formulae can be used to find the joint probability distribution of $l \le n$ eigenvalues $\lambda_1, \dots, \lambda_l$, which is given by

$$\frac{1}{n(n-1)\cdots(n-l+1)} \left|K_n(\lambda_j, \lambda_k)\right|_{1\le j,k\le l}.$$

In particular, when $l = 1$ we get

$$EN_n(\Delta) = \int_\Delta \frac{K_n(\lambda,\lambda)}{n}\,d\lambda,$$

where the integrand is known in the theory of orthogonal polynomials as the n-th (weighted) Christoffel function associated with the weight w^n.

These formulae indicate the role of orthogonal polynomials in statistical physics, and also show why orthogonal polynomials with varying weights w^n are important. We shall return to these polynomials in Chapter VII for the case of Freud weights.

IV.7.3 An Electrostatic Problem for an Infinite Wire

In this subsection we would like to illustrate the theory developed in this book through a somewhat artificial example taken from physics. We could have chosen other, more natural examples but the present one illustrates well the methods we can use without complicated calculations.

Suppose a static charge is placed on a straight wire that has identical circular cross sections. If we assume that the length of the wire is large compared to the radius of its cross sections, then the distribution of the charge far from the endpoints of the wire will be very similar to what we would get for an infinitely long wire, which is obviously the uniform distribution of the charge on the surface. Now what happens if we do not neglect the mass of the charged particles and, besides the repellent force between the charged particles we also take into account the effect of Earth's gravitation and the gravitational force between the mass particles?

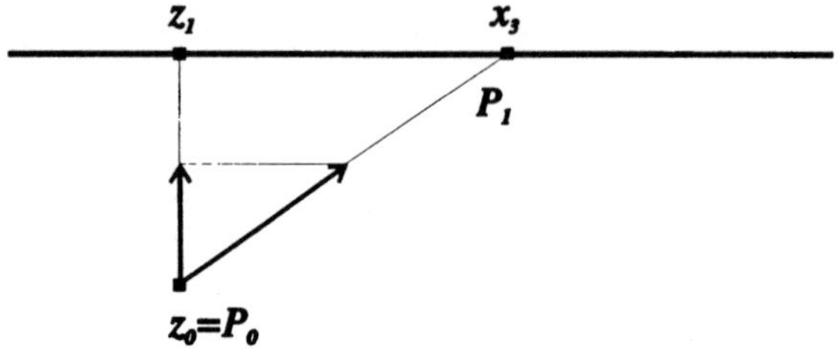

Fig. 7.3

Thus, let us assume that a homogeneous charged mass is placed on (the surface of) a horizontal infinite wire. Let the x_3 coordinate axis in the rectangular coordinate system (x_1, x_2, x_3) be the axis of the wire and let R be the radius of its cross section. We assume that the charge/mass ratio is constant and then the repellent force (coming from the charges) and the attraction force (coming from gravitation) between two charged point masses m_1 and m_2 that are of distance r apart can be combined into a force equal to $\rho m_1 m_2/r^2$, where ρ is a constant (this total force is attractive if $\rho < 0$ and repelling if $\rho > 0$).

Let us identify the (x_1, x_2) plane with $\mathbf{C}$ and let the x_2 direction, i.e. the imaginary direction be the vertical one. The distribution of the mass is obviously homogeneous in the x_3 direction, hence it is given by a distribution $d\mu(z)dx_3$ for some measure μ on the disk $\overline{D}_R = \{z \mid |z| \le R\}$. Our aim is to determine μ. By normalization we can assume without loss of generality that μ has total mass 1.

Let z_0 and z_1 be two points of $\overline{D}_R$. First we determine the force on the particle at z_0 originating from the particles on the line (or infinitesimal cylinder) $\{(z_1, x_3) \mid x_3 \in \mathbf{R}\}$. The direction of this force is the direction $z_0 z_1$ and its magnitude is

$$2\rho \int_0^\infty \frac{|z_1 - z_0|}{(|z_1 - z_0|^2 + x_3^2)^{3/2}} dx_3 = \frac{\rho}{|z_1 - z_0|} \int_{-\infty}^\infty \frac{dt}{(1 + t^2)^{3/2}} = \frac{2\rho}{|z_1 - z_0|},$$

where we used that the $z_0 z_1$ component of the force F^* between the particle elements at $P_0 = z_0$ and $P_1 = (z_1, x_3)$ satisfies (see Figure 7.3)

$$|F^*| = \frac{\rho}{r^2} = \frac{\operatorname{dist}(z_0, z_1)}{(\operatorname{dist}(P_0, P_1))^3}.$$

Thus, the effect of the line $\{(z_1, x_3) \mid x_3 \in \mathbf{R}\}$ on the particle at z_0 is the same as the effect of a particle placed at z_1 provided that on the plane the force between two particles is proportional with the *reciprocal of their distances* with proportionality factor 2ρ. With this observation we have transformed the problem of the determination of μ into a planar one.

Since forces proportional with the reciprocal of the distances between the elements correspond to logarithmic potential fields (see also [90]), our equilibrium problem takes the form of minimizing the energy

$$\iint 2\rho \log \frac{1}{|z-t|} d\mu(z) d\mu(t) + \int g \operatorname{Im}(z) d\mu(z)$$

for all probability measures supported on $\overline{D}_R$, where g is the gravitation constant. Here the term

$$\int g \operatorname{Im}(z) d\mu(z)$$

expresses the potential energy in the gravitation field of Earth, and in the present case this gravitation field is the external field.

Dividing through by 2ρ we finally arrive at the problem of minimizing the weighted energy

$$\iint \log \frac{1}{|z-t|} d\mu(z) d\mu(t) + 2 \int Q d\mu$$

with $\Sigma = \overline{D}_R$ and $Q(z) = g \operatorname{Im}(z)/4\rho$, and then μ_w will be the solution of the problem. Since Q is harmonic, it follows from Theorem 1.10(a) that μ_w is supported on the boundary of $\Sigma = D_R$.

Suppose now that the support S_w is actually equal to this boundary, i.e. $S_w = C_R = \{z \mid |z| = R\}$ (for conditions ensuring this, see the end of this section). Our next aim is to determine the equilibrium potential U^{μ_w}. To this end we need the solutions of the Dirichlet problem in both connected components of $\overline{\mathbf{C}} \setminus C_R$ with boundary function equal to

$$-Q(z) = \frac{-g}{4\rho} \operatorname{Im}(z)$$

(see Theorem I.4.7). In D_R this is obviously given by

$$-\frac{g}{4\rho} \operatorname{Im}(z),$$

while in $\overline{\mathbf{C}} \setminus \overline{D}_R$ the solution is

$$\frac{g \cdot R^2}{4\rho} \operatorname{Im}\left(\frac{1}{z}\right).$$

Hence, by Theorem I.4.7

$$U^{\mu_w}(z) - F_w = \begin{cases} -\dfrac{g}{4\rho} \operatorname{Im}(z) & \text{if } z \in \overline{D}_R \\[2ex] \dfrac{gR^2}{4\rho} \operatorname{Im}\left(\dfrac{1}{z}\right) - \log \dfrac{|z|}{R} & \text{if } z \notin \overline{D}_R, \end{cases} \tag{7.2}$$

where we used that the Green function of $\overline{\mathbf{C}} \setminus \overline{D}_R$ with pole at infinity is $\log |z|/R$.

Let $\mathbf{n}_+$ and $\mathbf{n}_-$ be the normals to C_R in the direction of D_R and $\mathbf{C} \setminus \overline{D}_R$, respectively. By Theorem II.1.5 the equilibrium measure μ_w is given by

$$d\mu_w = -\frac{1}{2\pi}\left(\frac{\partial U^{\mu_w}(s)}{\partial \mathbf{n}_-} + \frac{\partial U^{\mu_w}(s)}{\partial \mathbf{n}_+}\right) ds$$

where ds denotes the arc measure on C_R. From (7.2) we easily obtain for $s = Re^{i\varphi}$

$$\frac{\partial U^{\mu_w}(s)}{\partial \mathbf{n}_+} = \frac{\partial}{\partial r}\left(\frac{g}{4\rho} r \cdot \sin\varphi\right)\bigg|_{r=R} = \frac{g}{4\rho}\sin\varphi$$

and

$$\frac{\partial U^{\mu_w}(s)}{\partial \mathbf{n}_-} = \frac{\partial}{\partial r}\left(-\frac{gR^2}{4\rho}\frac{1}{r}\sin\varphi\right)\bigg|_{r=R} - \frac{\partial}{\partial r}\log\frac{r}{R}\bigg|_{r=R} = \frac{g}{4\rho}\sin\varphi - \frac{1}{R}.$$

Hence, μ_w is given by

$$d\mu_w(s) = \frac{1}{2\pi}\left(\frac{1}{R} - \frac{g}{2\rho}\sin\varphi\right) d\varphi, \quad s = Re^{i\varphi}. \tag{7.3}$$

Since this must be a positive measure, we must have $2\rho \geq Rg$ in the above derivation. On the other hand, it can be shown by direct computation that for $2\rho \geq Rg$ the measure on the right has potential given in (7.2) modulo a constant, and so, by Theorem I.3.3, this measure will be the equilibrium measure, i.e. (7.3) holds.

IV.8 Notes and Historical References

Section IV.1

Theorem 1.3 and its corollary are due to V. Totik [220, Lemma 5.3].

M. v. Golitschek, G. G. Lorentz and Y. Makovoz [60] have introduced the concept of a "minimal essential set" in the case when the weight is continuous on a closed set $\Sigma \subset \mathbf{R}$. Corollary 1.4 shows that in this case, the support $\mathcal{S}_w$ coincides with the minimal essential set.

The F-functional in (1.1) was first introduced by H. N. Mhaskar and E. B. Saff in [161] for $\Sigma \subset \mathbf{R}$ and in [163] for $\Sigma \subset \mathbf{C}$. Theorem 1.5 is due to Mhaskar-Saff [163, Theorem 3.2], [161, Theorem 2.1].

Parts (b),(d) of Theorem 1.10 are due to Mhaskar-Saff [161, Theorem 2.2].

Formulas (1.6) and (1.7) appear in [60], but are implicit in the earlier works of A. A. Gonchar and E. A. Rakhmanov [63], [64]. Formula (1.14) of Corollary 1.13 is given by Mhaskar-Saff in [159] and is also implicit in the cited papers of Gonchar-Rakhmanov.

The support interval for Freud weights (Example 1.14) was determined by Mhaskar-Saff [157] and also by Rakhmanov [189]. The former reference also contains (1.18).

The support interval for incomplete polynomials (Example 1.16) appears in the works of Lorentz [136], Stahl [210], Saff and Varga [198], [199], [200], and v. Golitschek [59], among others.

For Jacobi weights (Example 1.17), formulas for the endpoints of the support interval first appeared in the paper of Moak, Saff and Varga [164], and were also derived by Lachance, Saff and Varga [110], and Saff, Ullman and Varga [197].

The support interval for Laguerre weights (1.18) was first obtained by Mhaskar and Saff [158].

Section IV.2

The Fourier method and the results of this section are taken from Totik [219].

Section IV.3

The discussion of the integral equation (3.1) follows that of D. S. Lubinsky and E. B. Saff [142]. Theorem 4.10 is from Totik [220, Lemma 8.5].

The support $\mathcal{S}_w$ and the equilibrium measure μ_w have been recently studied by a different approach based on inverse spectral method with roots in the Lax–Levermore theory for KdV equations by P. Deift, T. Kriecherbauer and K. T-R. McLaughlin [33]. They consider smooth fields Q on $[-1, 1]$, and using a differential equation for the weighted Fekete polynomials, they get a lower bound for the spacing of weighted Fekete points via a Schur-type comparison theorem. Then by a clever argument they derive from here information on the equilibrium distribution. For example, they prove that if Q is real analytic, then $\mathcal{S}_w$ consists of a finite number of intervals. They also study in detail the case when $Q(x)$ is cx^m for integer m. Their results can be briefly described as follows.

When m is an even integer and c is negative, then this is the (truncated) Freud weight, for which the support $\mathcal{S}_w$ is always an interval. For even m and positive c the support $\mathcal{S}_w$ can consists of at most 3 intervals (at least for certain ranges of c): as we decrease with c from plus infinity the support $\mathcal{S}_w$ increases. For large values of c the support consists of two intervals. Then by decreasing c we reach a critical value where a third interval appears in the middle. Then a range of c is skipped in the analysis, and before we reach another critical value, the support consists again of three intervals. After that critical value the three intervals combine into a single interval. (In a recent paper A. B. J. Kuijlaars and P. Dragnev [109] showed that in the region where the analysis of the paper [33] does not apply, the support is still the union of three intervals.) In a similar manner, if m is odd, then the support always consists of at most two intervals (again a range of c is not included in the analysis). In each case the endpoints of the support, as well as the equilibrium measure are explicitly given in terms of some singular integrals with algebraic kernels.

Section IV.4

The results in this section are taken from Totik [220]. See also Totik and Ullman [221].

For Besicovic's covering theorem see [12]. The proof we gave is the original one.

Theorem 4.5 is due to Ch. J. de La Vallée Poussin and has been rediscovered and generalized many times (see the discussion in Section 2.2 in [221]).

Section IV.5

For Freud weights, the distribution (5.1) was first introduced by J. L. Ullman [224] in his study of orthogonal polynomials on **R**. His work was inspired by that of Nevai and Dehesa [170] who computed the moments for Freud weights.

The distribution function for Jacobi weights (Example 5.2) was first derived by Saff-Ullman-Varga [197] using an electrostatics approach and the Stieltjes inversion formula. For incomplete polynomials (Example 5.3), H. Stahl [210] utilized the method of quadratic trajectories to determine $d\mu_w(x)$. For Laguerre weights (Example 5.4) Fourier methods were used by Mhaskar-Saff [158] to determine the distribution.

Section IV.7

The relation between the contact problem of elasticity with the weighted energy problem was observed and used by A. B. J. Kuijlaars and W. Van Assche [107] (see also [58]).

The theory of random matrices has a vast and fast growing literature. Its basics can be found in the book of M. L. Mehta [152]. The random matrix approach is an alternative way to treat the weighted energy problem, and this approach has been extensively used in the last few years by E. Brézin, A. Figotin, K. Johansson, L. A. Pastur, M. Shcherbina, A. Zee, and many others. It was especially Pastur who achieved deep results via random matrix techniques for orthogonal polynomials, distribution of states, the energy problem and the so-called universality law. We refer the interested readers to the book of Pastur and Figotin [176], and to the papers [13], [24], [27], [37], [38], [87], [177], [178], [236].

Chapter V. Extremal Point Methods

The fact that the weighted equilibrium potential simultaneously solves a certain Dirichlet problem on connected components of $\mathbf{C} \setminus \mathcal{S}_w$ coupled with the fact that the Fekete points are distributed according to the equilibrium distribution leads to a numerical method for determining Dirichlet solutions. However, the determination of the Fekete points is a hard problem, so first we consider an associated sequence a_n that is adaptively generated from earlier points according to the law: a_n is a point where the weighted polynomial expression

$$|(z - a_0)(z - a_1) \cdots (z - a_{n-1})w(z)^n|$$

takes its maximum on Σ. These so-called Leja points are again distributed like the equilibrium distribution, so we can use them in place of weighted Fekete points.

The aforementioned Dirichlet problems are formulated on the connected components of $\mathbf{C} \setminus \mathcal{S}_w$. Therefore, in order to apply the extremal point method for solving a concrete Dirichlet problem with boundary function $-Q$ on a given region R we have to make sure that R is a connected component of $\mathbf{C} \setminus \mathcal{S}_w$, where $w = \exp(-Q)$. This is not generally the case, but (at least for smooth Q) we can do the following: with some appropriate small λ represent λQ as $(q + \lambda Q) - q$, where q is a "strongly convex" (i.e. positive definite) function. We are going to show that, for both external fields $(q + \lambda Q)$ and q defined on ∂R, the supports of the associated equilibrium measures contain ∂R; therefore R appears as a connected component in the complement of these supports. Hence, we can numerically solve the appropriate Dirichlet problems via the extremal point method (with boundary functions $-(q + \lambda Q)$ and $-q$), and the solution for the original problem (with boundary function $-Q$) is obtained by taking linear combinations.

We shall show that this program can be carried out and also that the method is stable from a numerical point of view. In fact, the adaptive feature of Leja points seems likely to ensure even greater stability than what is established here.

We shall also formulate similar algorithms for finding Green functions and conformal maps of simply connected domains.

V.1 Leja Points and Numerical Determination of μ_w

We have seen in Chapters II and IV that exact determination of the extremal measure and its support is usually a hard problem. In Section III.1 we showed that Fekete points are distributed like μ_w; hence these could be used as a discrete substitute for the extremal measure. However, exact determination of an n-Fekete set is equally hard for it is equivalent to an extremal problem in n variables. On the other hand, Fekete points are of great value in practice; for example they are almost ideal nodes for interpolation (cf. Theorem III.1.12). Hence it is worth looking for simple procedures that generate points similar to Fekete points.

In this section we give a very simple method for numerically determining the support $\mathcal{S}_w$ of the extremal measure μ_w and the extremal measure itself corresponding to an admissible weight w. The method originated from F. Leja [118] (unweighted case) and J. Górski [65] (three-dimensional case).

Let $a_0 \in \Sigma$ be an arbitrary point such that $w(a_0) \neq 0$ and inductively define a_n, $n = 1, 2, \ldots$, as a point where the weighted polynomial expression

$$|(z - a_0)(z - a_1) \cdots (z - a_{n-1})w(z)^n|$$

takes its maximum on Σ. Such a point always exists because w is upper semicontinuous and $zw(z) \to 0$ as $|z| \to \infty$. However, a_n need not be unique. We call these points Leja points, although "weighted Leja points" is also a widely used terminology.

Theorem 1.1. *Let w be an admissible weight on Σ. Then the above defined sequence $\{a_n\}_{n=0}^{\infty}$ has limit distribution μ_w.*

Recall, this means that the measures

$$\sigma_n := \frac{1}{n+1} \sum_{i=0}^{n} \delta_{a_i}, \tag{1.1}$$

where δ_a denotes the unit mass placed at a, converge to μ_w in the weak* topology of measures.

Figure 1.1 displays the Leja points $a_0, \ldots, a_{400}$ for the radially symmetrical weight $w(z) = |z|^2 \exp(-|z|^2)$ on $\Sigma = \mathbf{C}$, where we have taken $a_0 = 1.0$. From Example IV.6.2 it follows that $\mathcal{S}_w$ is the annulus $\{z \mid 1 \le |z| \le \sqrt{1.5}\}$ and, indeed, Fig. 1.1 illustrates the fact (from Theorem 1.1) that this annulus is the limiting support of the measures σ_n. Figure 1.2 shows the Leja points $a_0 = 1.0, a_1, \ldots, a_{400}$ for the weight $w(z) = \exp(-|z|^2)$ restricted to the first quadrant. Although $\mathcal{S}_w$ is not explicitly known in this case, Fig. 1.2 suggests that it is a circular-like sector.

Proof of Theorem 1.1. Let

$$s_n := \sum_{0 \le j < k \le n} \log \frac{1}{|a_k - a_j|} + n \sum_{j=0}^{n} Q(a_j)$$

$$= \sum_{k=1}^{n} \left(\sum_{j=0}^{k-1} \log \frac{1}{|a_k - a_j|} + kQ(a_k) + \sum_{j=0}^{k-1} Q(a_j) \right).$$

Since a_k is a point where the minimum of

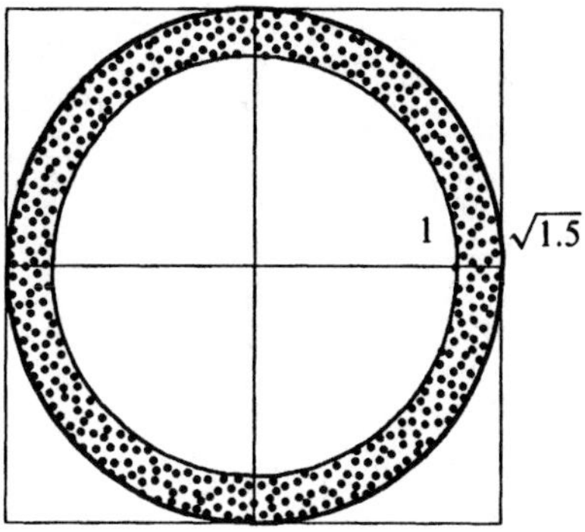

Fig. 1.1 Leja points $a_0 = 1.0, a_1, \ldots, a_{400}$ for $w(z) = |z|^2 \exp(-|z|^2)$ on $\Sigma = \mathbf{C}$

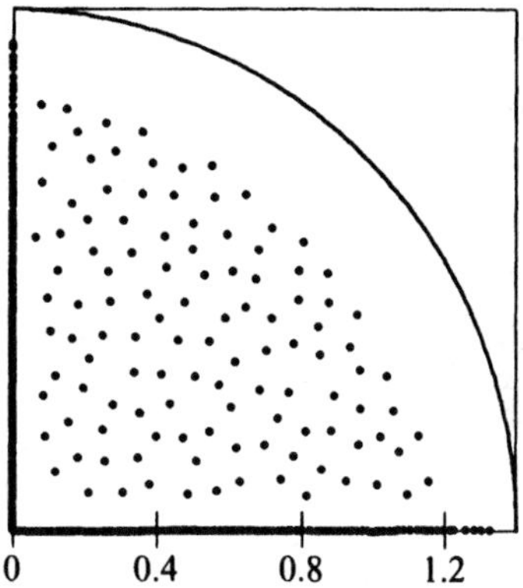

Fig. 1.2 Leja points $a_0 = 1.0, a_1, \ldots, a_{400}$ for $w(z) = \exp(-|z|^2)$ on $\Sigma = \{z = x + iy \mid x \geq 0, y \geq 0\}$

$$\sum_{j=0}^{k-1} \log \frac{1}{|z - a_j|} + kQ(z) + \sum_{j=0}^{k-1} Q(a_j)$$

is attained, we have for any z

$$\begin{aligned} s_n &\leq \sum_{k=1}^{n} \left(\sum_{j=0}^{k-1} \log \frac{1}{|z - a_j|} + kQ(z) + \sum_{j=0}^{k-1} Q(a_j) \right) \\ &= \sum_{k=1}^{n} \sum_{j=0}^{k-1} \left(\log \frac{1}{|z - a_j|} + Q(z) + Q(a_j) \right). \end{aligned} \tag{1.2}$$

Integrating this with respect to μ_w we get

$$s_n = \sum_{k=1}^{n} \sum_{j=0}^{k-1} \left(U^{\mu_w}(a_j) + \int Q d\mu_w + Q(a_j) \right).$$

Now we recall that by Theorem III.2.1 every a_n, $n \geq 1$, belongs to the compact set

$$\mathcal{S}_w^* = \{z \in \Sigma \mid U^{\mu_w}(z) \leq -Q(z) + F_w\} \tag{1.3}$$

(see also the remark after that theorem). Hence we get from the preceding estimate the inequality

$$\begin{aligned} s_n &\leq \sum_{k=2}^{n}\sum_{j=1}^{k-1}\left(F_w + \int Q d\mu_w\right) + \sum_{k=1}^{n}\left(U^{\mu_w}(a_0) + \int Q d\mu_w + Q(a_0)\right) \\ &= \frac{n(n-1)}{2}\left(F_w + \int Q d\mu_w\right) + O(n), \end{aligned} \tag{1.4}$$

where in the last step we used that $Q(a_0)$ is finite.

Let now $\mathcal{N} \subset \mathbf{N}$ be an arbitrary subsequence of the natural numbers. From $\{\sigma_n\}_{n\in\mathcal{N}}$ (cf. (1.1)) we can select a subsequence $\{\sigma_n\}_{n\in\mathcal{N}_1}$ converging in the weak*-topology to a unit measure σ of compact support (note also that all a_n, $n \geq 1$, belong to $\mathcal{S}_w^*$ in (1.3)). Setting

$$\log_M(z) := \begin{cases} \log|z| & \text{if } \log|z| \leq M \\ M & \text{otherwise} \end{cases}$$

we can write

$$\begin{aligned} s_n &= \frac{1}{2}\left\{\sum_{j\neq k}\log\frac{1}{|a_j - a_k|} + 2n\sum_{j=0}^{n} Q(a_j)\right\} \\ &\geq \frac{(n+1)^2}{2}\left\{\iint \log_M \frac{1}{|z-t|} d\sigma_n(z) d\sigma_n(t) + 2\frac{n}{n+1}\int Q d\sigma_n\right\} \\ &\qquad -M\frac{n+1}{2}. \end{aligned}$$

Since Q is lower semi-continuous, we have by Theorem 0.1.4

$$\liminf_{n\to\infty,\, n\in\mathcal{N}_1} \int Q d\sigma_n \geq \int Q d\sigma,$$

and so in view of (1.4)

$$\begin{aligned} I_w(\mu_w) &= F_w + \int Q d\mu_w \geq \liminf_{n\to\infty,\, n\in\mathcal{N}_1} \frac{2}{(n+1)^2} s_n \\ &\geq \lim_{M\to\infty}\liminf_{n\to\infty}\left(\iint \log_M \frac{1}{|z-t|} d\sigma_n(z) d\sigma_n(t) + \frac{2n}{n+1}\int Q d\sigma_n\right) \\ &\geq \lim_{M\to\infty}\left(\iint \log_M \frac{1}{|z-t|} d\sigma(z) d\sigma(t) + 2\int Q d\sigma\right) \\ &= \iint \log\frac{1}{|z-t|} d\sigma(z) d\sigma(t) + 2\int Q d\sigma = I_w(\sigma). \end{aligned}$$

Thus, $I_w(\sigma) \le I_w(\mu_w)$, and by the unicity of the extremal measure μ_w (see Theorem I.1.3) we have $\sigma = \mu_w$. Since this is true for every weak*-limit point of $\{\sigma_n\}_{n=1}^{\infty}$, the whole sequence converges to μ_w. □

We shall also use the fact that the polynomials

$$P_n(z) := (z - a_0)(z - a_1) \cdots (z - a_{n-1}),$$

where the a_i's are the above constructed points, have asymptotically minimal weighted norm.

Theorem 1.2. *With the above assumptions and notations*

$$\lim_{n\to\infty} \|P_n w^n\|_\Sigma^{1/n} = \exp(-F_w).$$

Proof. Let

$$A_k := \|P_k w^k\|_\Sigma^{1/k} = \{w(a_k)^k \prod_{i=0}^{k-1} |a_k - a_i|\}^{1/k}$$

$$= \left\{\prod_{i=0}^{k-1} |a_k - a_i| w(a_k) w(a_i)\right\}^{1/k} \exp\left(\frac{1}{k}\sum_{i=0}^{k-1} Q(a_i)\right).$$

Then

$$\prod_{k=1}^{n} A_k^k = \exp\left(-s_n + \sum_{i=0}^{n}(n-i)Q(a_i)\right). \tag{1.5}$$

As previously remarked, each a_i, $i \ge 1$, belongs to the set $\mathcal{S}_w^*$ in (1.3) and, by Theorem 1.1, the discrete measures $(1/k)\sum_{i=0}^{k-1}\delta_{a_i}$ converge to μ_w in the weak* sense. Therefore we can apply Lemma III.1.5 and we get that

$$\lim_{k\to\infty} \frac{1}{k}\sum_{i=0}^{k-1} Q(a_i) = \int Q d\mu_w.$$

This implies

$$\lim_{n\to\infty} \frac{2}{n(n+1)} \sum_{i=0}^{n}(n-i)Q(a_i) = \int Q d\mu_w.$$

In the proof of Theorem 1.1 we verified that

$$\lim_{n\to\infty} \frac{2}{n(n+1)} s_n = I_w(\mu_w),$$

and so we get from (1.5) that

$$\lim_{n\to\infty} \frac{2}{n(n+1)} \sum_{k=1}^{n} k \log A_k = -I_w(\mu_w) + \int Q d\mu_w = -F_w. \tag{1.6}$$

We also know from Theorem I.3.6 that

$$\log A_k \geq -F_w, \quad k = 1, 2, \ldots . \tag{1.7}$$

From the definition of the points a_k we obtain

$$A_k^k \leq L A_{k-1}^{k-1},$$

where L is an upper bound for $w(z)|z-u|$ for $z \in \Sigma$ and $u \in S_w^*$. Hence

$$\log A_k \leq \log A_{k-1} + \frac{S}{k} \tag{1.8}$$

with $S := \log L + F_w$ (use also (1.7)). Now we show that (1.6) – (1.8) imply

$$\lim_{k\to\infty} \log A_k = -F_w,$$

and this is what we needed to prove.

Assume that $S \geq 1$. Let $0 < \varepsilon < 1$ and suppose that n is an integer such that

$$\log A_n > -F_w + \varepsilon.$$

Then, for $(1 - \varepsilon/(3S))n \leq k \leq n$ we get from repeated application of (1.8) that

$$\log A_k > -F_w + \varepsilon - \frac{S(n-k)}{(1-\varepsilon/3S)n} \geq -F_w + \varepsilon - \frac{\varepsilon/3}{1-\varepsilon/3S} \geq -F_w + \frac{\varepsilon}{2},$$

and so, in view of (1.7), we have

$$\begin{aligned} \frac{2}{n(n+1)} \sum_{k=1}^{n} k \log A_k &\geq -F_w + \frac{\varepsilon}{2} \frac{2}{n(n+1)} \sum_{(1-\varepsilon/(3S))n \leq k \leq n} k \\ &\geq -F_w + \frac{\varepsilon^2(1-\varepsilon/3S)}{6S}. \end{aligned}$$

Thus, in view of (1.6), we see that $\log A_n > -F_w + \varepsilon$ can hold only for finitely many n; that is

$$\limsup_{n\to\infty} \log A_n \leq -F_w.$$

Combination of this fact with (1.7) yields

$$\lim_{n\to\infty} \log A_n = -F_w,$$

which completes the proof. □

From Theorems 1.1 and 1.2 we immediately get

Corollary 1.3. *With the above notations*

$$\lim_{n\to\infty} \frac{1}{n} \log(|P_n(z)|/\|P_n w^n\|_\Sigma) = -U^{\mu_w}(z) + F_w \tag{1.9}$$

locally uniformly in $\mathbf{C} \setminus (S_w^* \cup \{a_0\})$. *In particular, this is true for any* $z \notin \Sigma$.

Note that by Theorem I.4.7 the right-hand side in (1.9) is the solution of the Dirichlet problem in every bounded component of $\mathbf{C} \setminus \mathcal{S}_w$ with boundary function Q and, in the unbounded component R of $\mathbf{C} \setminus \mathcal{S}_w$, it is the sum of this Dirichlet solution and the Green function of R with pole at the infinity. We will use this observation in Sections V.2–V.3 to provide a method for solving Dirichlet problems and a method for finding Green functions, or, equivalently, conformal mappings. In order that these methods yield numerically feasible procedures and that the Leja points be numerically applicable to computing μ_w, c_w, $\mathcal{S}_w$, t_w (cf. (III.3.5)) etc. we must show a certain stability of this extremal point method. Therefore we now discuss a discretized version of the aforementioned results.

Let $\{\varepsilon_n\}_{n=1}^{\infty}$ be a sequence of positive numbers tending to zero, an let us fix a corresponding sequence of discrete subsets $\{S_n\}_{n=1}^{\infty}$ of Σ in such a way that $S_n \subset S_{n+1}$ and, for each n and $z \in \Sigma$, there is a point in S_n whose distance from z is at most ε_n. Let $\hat{a}_0 \in \Sigma$ be arbitrary ($w(\hat{a}_0) \neq 0$), and for each $n = 1, 2, \dots,$ define $\hat{a}_n \in S_n$ as a point where the weighted polynomial expression

$$|(z - \hat{a}_0)(z - \hat{a}_1) \cdots (z - \hat{a}_{n-1})w(z)^n|$$

takes its maximum on S_n.

Theorem 1.4. *Let w be a continuous admissible weight on Σ, and let $\{\varepsilon_n\}$ satisfy*

$$\lim_{n\to\infty} \varepsilon_n^{1/n} = 0. \tag{1.10}$$

Then the sequence $\{\hat{a}_n\}_{n=0}^{\infty}$ defined above has limit distribution μ_w. Furthermore, for the polynomials

$$\hat{P}_n(z) := (z - \hat{a}_0)(z - \hat{a}_1) \cdots (z - \hat{a}_{n-1}), \quad n = 1, 2, \dots,$$

we have

$$\lim_{n\to\infty} \|\hat{P}_n w^n\|_{\Sigma}^{1/n} = \exp(-F_w).$$

For relaxing the condition (1.10) see the discussion at the end of this section. Before giving the proof of this result we state

Corollary 1.5. *With the notation*

$$\hat{A}_n := |(\hat{a}_n - \hat{a}_0)(\hat{a}_n - \hat{a}_1) \cdots (\hat{a}_n - \hat{a}_{n-1})| w(\hat{a}_n)^n$$

and

$$q_n := -\sum_{i=0}^{n-1} Q(\hat{a}_i),$$

we have

$$\lim_{n\to\infty}\left(\prod_{i=1}^{n}\hat{A}_i\exp(q_i)\right)^{2/n(n+1)}=\frac{1}{c_w},$$

$$\lim_{n\to\infty}\hat{A}_n^{1/n}=t_w=\exp(-F_w),$$

$$\lim_{n\to\infty}\frac{1}{n}\log|\hat{P}_n(z)|=-U^{\mu_w}(z),$$

locally uniformly in $\mathbf{C}\setminus\Sigma$*. Furthermore, for every compact set* K *and every* $\varepsilon>0$

$$\frac{1}{n}\#\{\hat{a}_i\,|\,\hat{a}_i\in K,\quad 1\le i\le n\}-\varepsilon\le\mu_w(K)$$

$$\le\frac{1}{n}\#\{\hat{a}_i\,|\,\mathrm{dist}(\hat{a}_i,K)<\varepsilon,\quad 1\le i\le n\}+\varepsilon$$

for large n*, and* $\mathcal{S}_w$ *coincides with the set of points where the sequence* $\{\hat{a}_n\}$ *has positive density.*

This corollary says that all the important quantities related to w can be determined from the sequence $\{\hat{a}_i\}$.

In many cases the set $\mathcal{S}_w^*$ in (1.3) coincides with $\mathcal{S}_w$, and in every such case $\mathcal{S}_w$ is simply the set of limit points of the sequence $\{\hat{a}_n\}$.

Proof of Theorem 1.4. First we verify that

$$\lim_{n\to\infty}\sup_{\deg P_n\le n,\ P_n\not\equiv 0}\left\{\|P_nw^n\|_\Sigma^{1/n}/\|P_nw^n\|_{S_n}^{1/n}\right\}=1. \tag{1.11}$$

Let $P_n\not\equiv 0$ be a polynomial of degree at most n. By Theorem III.2.1, there exists a point $z_0\in\mathcal{S}_w^*$ such that $\|P_nw^n\|_\Sigma=|P_n(z_0)w(z_0)^n|$. Since $\mathcal{S}_w^*$ is compact, it is contained in some disk

$$D_R:=\{z\,|\,|z|\le R\},\quad R\ge 1.$$

Set

$$\omega_n:=\sup\{|Q(z)-Q(z')|\,|\,z\in\mathcal{S}_w^*,\ z'\in\Sigma,\ |z-z'|\le\varepsilon_n\},$$

and note that since Q is a continuous extended real-valued function and Q is finite on $\mathcal{S}_w^*$, we have $\omega_n\to 0$ as $n\to\infty$.

Suppose $\varepsilon_n<1$ and let $\hat{z}_0\in S_n$ satisfy $|z_0-\hat{z}_0|\le\varepsilon_n$. Then

$$|P_n(z_0)w(z_0)^n|\le|P_n(z_0)|w(\hat{z}_0)^ne^{n\omega_n}.$$

Since $|P_n(z_0)-P_n(\hat{z}_0)|\le\|P_n'\|_{D_{2R}}\varepsilon_n\le(n/2R)\|P_n\|_{D_{2R}}\varepsilon_n$, where we have applied Bernstein's inequality on the circle about the origin of radius $2R$, we have

$$|P_n(z_0)w(z_0)^n|\le e^{n\omega_n}|P_n(\hat{z}_0)w(\hat{z}_0)^n|+e^{n\omega_n}w(\hat{z}_0)^n\|P_n\|_{D_{2R}}\frac{n\varepsilon_n}{2R}. \tag{1.12}$$

But by Theorems III.2.1 and I.4.3, there exists a constant $\rho > 1$ such that

$$\|P_n\|_{D_{2R}} \le \rho^n \|P_n w^n\|_\Sigma,$$

where ρ does not depend on P_n. Hence

$$B_n := e^{n\omega_n} w(\hat{z}_0)^n \|P_n\|_{D_{2R}} \frac{n\varepsilon_n}{2R} \le M^n \rho^n e^{n\omega_n} \frac{n\varepsilon_n}{2R} \|P_n w^n\|_\Sigma,$$

where $M := \max\{w(z) \mid z \in D_{2R} \cap \Sigma\}$. Since $\varepsilon_n^{1/n} \to 0$ and $\omega_n \to 0$, we have

$$B_n \le \frac{1}{2}\|P_n w^n\|_\Sigma, \qquad n \ge n_0.$$

Thus from (1.12),

$$\frac{1}{2}\|P_n w^n\|_\Sigma \le e^{n\omega_n}\|P_n w^n\|_{S_n}, \quad n \ge n_0,$$

which yields (1.11).

Let us remark that since $\hat{z}_0 \in D_{2R}$, it follows from (1.11) and Theorem III.2.1 that the sequence $\{\hat{a}_n\}$ is bounded.

We can now repeat the proof of Theorem 1.1 with $\{\hat{a}_n\}$ in place of $\{a_n\}$, the only difference is that, because of (1.11), the analogue of (1.2) is now

$$s_n \le \sum_{k=1}^{n}\left(\sum_{j=0}^{k-1} \log\frac{1}{|z-a_j|} + kQ(z) + \sum_{j=1}^{k-1} Q(a_j) + o(k)\right)$$

uniformly in z; the rest of the proof is the same. Thus the first part of the theorem is proved.

The second part can be verified via the same argument that we used in Theorem 1.2, since in this case

$$\lim_{k\to\infty}\frac{1}{k}\sum_{i=0}^{k-1} Q(\hat{a}_i) = \int Q\,d\mu_w$$

is a consequence of the first part and the continuity of Q (note also that above we verified the boundedness of the sequence $\{\hat{a}_n\}$). □

Corollary 1.5 is an immediate consequence of Theorem 1.4.

The assumption (1.10) is rather strong; even for relatively small n it requires a tremendous amount of computation to determine the numbers $\hat{a}_n$ successively. Moreover, the required precision (say with $\varepsilon_{20} = 20^{-20}$) is simply not attainable on most computers. Fortunately, in practice (1.10) can be relaxed to $\varepsilon_n = O(n^{-\alpha})$ for some α. In fact, the proof of Theorem 1.4 was based on (1.11) and the result holds for every $\{\varepsilon_n\}$ that ensures (1.11). Fortunately under very general assumptions (1.11) will hold for some sequence $\{\varepsilon_n\}$ with $\varepsilon_n = O(n^{-\alpha})$, as the following examples illustrate.

Example 1.6. Let Σ be the real line and $w(x) = \exp(-|x|^\beta)$, $\beta > 1$. It will be shown in Theorem VI.5.5 that there is a constant C_β such that for every n and every polynomial P_n of degree at most n

$$\|P_n'w\|_\Sigma \leq C_\beta n^{1-1/\beta}\|P_n w\|_\Sigma.$$

Using the substitution $x \to n^{1/\beta}x$ we get from this

$$\|P_n'w^n\|_\Sigma \leq C_\beta n\|P_n w^n\|_\Sigma.$$

Since

$$\|P_n(w^n)'\|_S \leq C_\beta' n\|P_n w^n\|_\Sigma \tag{1.13}$$

also holds in a (one-dimensional) neighborhood S of $\mathcal{S}_w$, we have

$$\|(P_n w^n)'\|_S \leq Cn\|P_n w^n\|_\Sigma,$$

which implies (1.11) for every $\{\varepsilon_n\}$ satisfying $\varepsilon_n \leq 1/2Cn$. □

Although, in general, we do not have weighted Markoff-type inequalities of the form (1.13), we can still often ensure (1.11) for $\varepsilon_n = O(n^{-\alpha})$ with some α. Namely if the weight w is continuous and positive on Σ and in every neighborhood U of every point of Σ

$$\|P_n\|_U \leq C_U\|P_n\|_{U\cap S_n}, \quad n \geq n_S, \tag{1.14}$$

for some constant C_U, where S_n are increasing discrete subsets of Σ satisfying $\mathrm{dist}(S_n, \Sigma) \leq \varepsilon_n$, then (1.11) easily follows from the continuity of the weight by a compactness argument. Property (1.14) depends only on the geometry of Σ and does not depend on the weight as the next example illustrates.

Example 1.7. Let Σ be the real line. It easily follows from the classical Markoff inequality

$$\|p_n'\|_{[-1,1]} \leq n^2\|p_n\|_{[-1,1]}, \quad \deg(p_n) \leq n,$$

that (1.14) is true whenever $\varepsilon_n = o(n^{-2})$. Therefore, Theorem 1.4 holds for every positive continuous weight on the real line with (1.10) replaced by $\varepsilon_n = o(n^{-2})$.

Similar statements hold if Σ is an interval, the unit circumference, the union of finitely many smooth Jordan curves or the union of regions bounded by such curves (note that (1.14) is a local property). □

V.2 The Extremal Point Method for Solving Dirichlet Problems

In this section we apply the theory of weighted potentials to solving Dirichlet problems. Actually the procedure simultaneously yields the solution for several Dirichlet problems.

The basic idea behind the method is very simple. We have seen in Theorem I.4.7 that for admissible weight $w(z) = \exp(-Q(z))$ the expression

$$F_w - U^{\mu_w}(z) =: g(z, Q) \tag{2.1}$$

solves the Dirichlet problem in every bounded component of $\mathbf{C} \setminus \mathcal{S}_w$ with boundary values Q and, in the unbounded component of $\mathbf{C} \setminus \mathcal{S}_w$, (2.1) is the sum of the solution of the analogous Dirichlet problem and the Green function of this unbounded component with pole at the infinity. Furthermore, the function on the left of (2.1) can be determined by the extremal point method of Section V.1 (see especially Corollaries 1.3 and 1.5). Thus, if we wish to solve the Dirichlet problems with boundary function Q on the components of $\mathbf{C} \setminus \Sigma$, where Σ is a compact set, then for the bounded components we have a way to do it *provided* for the weight $w(z) = \exp(-Q(z))$ we have $\Sigma = \mathcal{S}_w$. Unfortunately the latter equality may not hold, but we observe that if for some $\lambda > 0$ the equality $\Sigma = \mathcal{S}_{w^\lambda}$ holds, then

$$\frac{1}{\lambda}(F_{w^\lambda} - U^{\mu_{w^\lambda}}(z)) = \frac{1}{\lambda} g(z, \lambda Q) \tag{2.2}$$

also solves the aforementioned Dirichlet problems (on bounded components of $\mathbf{C} \setminus \Sigma$). The assumption $\Sigma = \mathcal{S}_{w^\lambda}$ for some $\lambda > 0$ is still quite strong, but from here it is a small step to generalize the idea as follows: if we can find a function q and a $\lambda > 0$ such that with $v = \exp(-q)$ and $v_\lambda := \exp(-q - \lambda Q)$ the supports $\mathcal{S}_v$ and $\mathcal{S}_{v_\lambda}$ coincide with Σ, then the function

$$\frac{1}{\lambda}(g(z, q + \lambda Q) - g(z, q)) \tag{2.3}$$

solves the Dirichlet problems in every component of $\mathbf{C} \setminus \Sigma$ – even in the unbounded component of $\mathbf{C} \setminus \Sigma$, since the Green function mentioned above cancels by the subtraction in (2.3). Thus, our task has been reduced to finding appropriate q and λ. We will show that this is always possible for smooth weights, and that in the general case all we have to do is to take limit in (2.3).

First we mention that there are no q's that could serve universally for all Σ and Q on Σ, because the support of any extremal measure corresponding to an admissible weight is compact. Therefore we shall choose q depending on Σ. Actually we can choose any function for q that is positive definite on a connected open set containing Σ. For large R the functions $|z|^2/R$, $\log(R^2 + |z|^2)$ or $r(|z|^2/R)$, where r is any increasing strictly convex function on $[0, 1]$, could all serve as q in the considerations below. For definiteness we will pick the second one.

First we formulate the results.

Theorem 2.1. *Let Σ be a compact set of positive capacity contained in the disk $D_R := \{z \mid |z| \le R\}$, $R \ge 1$, and suppose that Q can be extended to a twice continuously differentiable function (as a function of two real variables) to a neighborhood of Σ. Then with*

$$q(z) := \log(22R^2 + |z|^2) \tag{2.4}$$

there is a $\lambda_0 > 0$ such that the function

$$\frac{1}{\lambda}(g(z, q + \lambda Q) - g(z, q))$$

is the solution of the Dirichlet problem with boundary function Q on every component of $\mathbf{C} \setminus \Sigma$ for $0 < \lambda \le \lambda_0$.

More precisely, if Q can be extended as a twice continuously differentiable function Q^* to D_R with

$$\left|\frac{\partial Q^*}{\partial x}\right| + \left|\frac{\partial Q^*}{\partial y}\right| \le M, \quad \left|\frac{\partial^2 Q^*}{\partial x^2}\right| + \left|\frac{\partial^2 Q^*}{\partial y^2}\right| \le M, \tag{2.5}$$

then $\lambda_0 = 1/(10R^2 M)$ is suitable.

Theorem 2.2. *Let Σ be a compact set of positive capacity, and let Q be a lower semi-continuous function on Σ. Then with the function* (2.4)

$$g_Q(z) := \lim_{\lambda \to 0+} \frac{1}{\lambda}(g(z, q + \lambda Q) - g(z, q)) \tag{2.6}$$

coincides with the solution H_Q^U of the Dirichlet problem with boundary function Q on every component U of $\mathbf{C} \setminus \Sigma$. Furthermore, the convergence in (2.6) *is uniform on $\mathbf{C}$ if Q is continuous and Σ is a regular set.*

By a regular set we mean a set for which every component of the complement is regular with respect to the Dirichlet problem.

Recall also that the Perron–Wiener–Brelot solution H_Q^U of the Dirichlet problem was defined in Section I.2. It may happen that on some components U of $\mathbf{C} \setminus \Sigma$ the function H_Q^U is identically infinite, in which case we keep the notion "solution of the Dirichlet problem" for H_Q^U.

We remark, that the convergence in (2.6) is monotone increasing and, of course, the existence of the limit is part of the statement.

If Σ has connected complement, then the function q can be eliminated:

Theorem 2.3. *Let Σ be a compact set of positive capacity and with connected complement, and Q a lower semi-continuous function on Σ. Then*

$$g_Q(z) := \lim_{\lambda \to 0+} \frac{1}{\lambda}(g(z, \lambda Q) - g(z, 0))$$

coincides with the solution H_Q^U of the Dirichlet problem with boundary function Q on $U := \overline{\mathbf{C}} \setminus \Sigma$. Furthermore, the convergence in the definition of g_Q is monotone and uniform on $\mathbf{C}$ if Q is continuous and Σ is a regular set.

Proof of Theorem 2.1. As we have already discussed above, it would be enough to verify that with the assumptions of the theorem and with $v = \exp(-q)$ and $v_\lambda := \exp(-q - \lambda Q)$, the supports $\mathcal{S}_v$ and $\mathcal{S}_{v_\lambda}$ coincide with Σ for $\lambda \le \lambda_0 := 1/10R^2M$. According to the discussion preceding Theorem IV.1.1, $\mathcal{S}_v$ and $\mathcal{S}_{v_\lambda}$ are of positive capacity at every one of their points; therefore $\Sigma = \mathcal{S}_v$ and $\Sigma = \mathcal{S}_{v_\lambda}$ may not be possible. However, this is only a minor difficulty, namely let Σ^* be the set of all $z \in \Sigma$ such that Σ is of positive capacity in every neighborhood of z. It is easy to see that Σ^* is compact, $\Sigma \setminus \Sigma^*$ is of zero capacity and Σ^* is of positive capacity at every one of its points. Since $\Sigma \setminus \Sigma^*$ is of zero capacity, it easily follows from Evans' theorem, Lemma I.2.3, and the definition of Dirichlet solution (Section I.2) that the solutions of Dirichlet problems on the components of $\mathbf{C} \setminus \Sigma$ coincide with the analogous solutions on the components of $\mathbf{C} \setminus \Sigma^*$ (see the proof of Lemma I.2.6; or see [77, Corollary 8.4]); hence it is enough to verify that $\mathcal{S}_v = \mathcal{S}_{v_\lambda} = \Sigma^*$, and this is what we do below. We need the following.

Lemma 2.4. *Let $\Sigma \subseteq D_R$. If θ is a twice continuously differentiable function on D_R such that*

$$\left|\frac{\partial\theta}{\partial x}\right| + \left|\frac{\partial\theta}{\partial y}\right| \le m, \quad \textit{and} \quad \frac{\partial^2\theta}{\partial x^2} + \frac{\partial^2\theta}{\partial y^2} > 0,$$

then for $m \le 1/3R$ the support $\mathcal{S}_w$ of the extremal measure corresponding to $w(z) := \exp(-\theta(z))$, $z \in \Sigma$, coincides with Σ^*.

Proof of Lemma 2.4. Consider the usual rectangular coordinate system with coordinates (x, y, t) and we identify the (x, y) plane with $\mathbf{C}$ by setting $z = x + iy$. Since θ is assumed to be strictly positive definite in D_R, for every $z_0 \in D_R$ the plane given by the equation

$$t = \theta(z_0) + \frac{\partial\theta}{\partial x}(z_0)(x - x_0) + \frac{\partial\theta}{\partial y}(z_0)(y - y_0) =: \theta(z_0) + \mathrm{Re}\,(a_0(z - z_0)),$$

where $a_0 := \frac{\partial\theta}{\partial x}(z_0) - i\frac{\partial\theta}{\partial y}(z_0)$, lies below the surface $t = \theta(z)$, i.e. for any $z_0 \in D_R$

$$\theta(z_0) + \mathrm{Re}\,(a_0(z - z_0)) - \theta(z) < 0, \qquad z \in D_R, \quad z \ne z_0,$$

or equivalently

$$e^{\theta(z_0)}|e^{a_0(z-z_0)}|e^{-\theta(z)} < 1, \qquad z \in D_R, \quad z \ne z_0. \tag{2.7}$$

We have for any $z \in D_R$

$$|na_0(z - z_0)| \le nm2R \le 2n/3,$$

and so elementary estimates on the convergence of the exponential series shows that with

$$h_n(z) := \sum_{k=0}^{n} \frac{(na_0(z - z_0))^k}{k!}$$

the convergence

$$e^{na_0(z-z_0)} - h_n(z) \to 0, \quad n \to \infty,$$

is uniform in $z \in D_R$. But this together with (2.7) implies that in case of $z_0 \in \Sigma^*$ the weighted polynomials $H_n(z) := (e^{n\theta(z_0)}h_n(z))e^{-n\theta(z)}$ take their essential maximum modulus (cf. Theorem IV.1.3) on D_R in any fixed neighborhood of z_0 for all large n (note that $h_n(z_0) = 1$), which means according to Theorem IV.1.3 that z_0 belongs to $\mathcal{S}_w$. Since $z_0 \in \Sigma^*$ was arbitrary, we have $\mathcal{S}_w = \Sigma^*$ as we claimed. □

Now we return to the proof of Theorem 2.1, and first of all we mention that for the function (2.4) we have in D_R

$$\left|\frac{\partial q}{\partial x}\right| + \left|\frac{\partial q}{\partial y}\right| \le \frac{1}{6R}, \qquad \frac{\partial^2 q}{\partial x^2} + \frac{\partial^2 q}{\partial y^2} > \frac{1}{10R^2}.$$

Hence we can apply the previous lemma to q and for $0 < \lambda \le \lambda_0$ to $q + \lambda Q^*$ in place of θ, where Q^* denotes any extension of Q to D_R that satisfies (2.5), and we obtain $\mathcal{S}_v = \mathcal{S}_{v_\lambda} = \Sigma^*$ as was claimed above. □

Proof of Theorem 2.2. We start the proof by showing that the function

$$\frac{1}{\lambda}(g(z, q + \lambda Q) - g(z, q))$$

is decreasing in $\lambda > 0$ and it is $\le Q(z)$ quasi-everywhere. We shall use the notations of the preceding proof. In fact, if $\lambda' < \lambda$, then by Corollary I.4.2(c) we obtain from the identity

$$q + \lambda' Q = \frac{1}{\lambda}\{\lambda'(q + \lambda Q) + (\lambda - \lambda')q\}$$

the inequality

$$U^{\mu_{v_{\lambda'}}}(z) - F_{v_{\lambda'}} \le \frac{1}{\lambda}\left\{\lambda'\left(U^{\mu_{v_\lambda}}(z) - F_{v_\lambda}\right) + (\lambda - \lambda')\left(U^{\mu_v}(z) - F_v\right)\right\},$$

and after rearranging this we get

$$\frac{1}{\lambda}(g(z, q + \lambda Q) - g(z, q)) \le \frac{1}{\lambda'}\left(g(z, q + \lambda' Q) - g(z, q)\right),$$

which is the claimed monotonicity. By the preceding proof the support of $\mathcal{S}_v$, $v(z) := \exp(-q(z))$, $z \in \Sigma$, coincides with Σ^*, and so we obtain from Theorem I.1.3(f) that $g(z, q) = q$ quasi-everywhere on Σ. On the other hand, Theorem I.1.3(d) says that $g(z, q + \lambda Q) \le q + \lambda Q$ quasi-everywhere, and so

$$\frac{1}{\lambda}(g(z, q + \lambda Q) - g(z, q)) \le Q(z)$$

for quasi-every $z \in \Sigma$ as we claimed.

Since the equilibrium potentials are continuous quasi-everywhere, we immediately get from what we have just proved that if U is a component of $\mathbf{C} \setminus \Sigma$, then

$$\limsup_{z'\to z,\, z'\in U} \frac{1}{\lambda}(g(z, q+\lambda Q) - g(z, q)) \le Q(z)$$

for quasi-every $z \in \partial U$, and so (see the proof of Lemma I.2.6) the function

$$\frac{1}{\lambda}(g(z, q+\lambda Q) - g(z, q))\big|_U$$

is at most as large as the lower solution $\underline{H}_Q^U$ of the Dirichlet problem (see the definition in Section I.2). Hence on U the function g_Q in (2.6), the existence of which has been verified above, is at most as large as the lower solution $\underline{H}_Q^U$ of the Dirichlet problem.

By the lower semi-continuity of Q we can choose functions $Q_n \uparrow Q$ such that each Q_n can be extended to a twice continuously differentiable function on D_R. By Corollary I.4.2(a) the functions $(1/\lambda)(g(z, q+\lambda Q) - g(z, q))$, and also the function $g_Q(z)$, are monotone increasing in Q; therefore $g_Q(z) \ge g_{Q_n}(z)$ for all z and n. On the other hand, Theorem 2.1 implies that for all n the function $g_{Q_n}(z)$ coincides on U with the solution of the Dirichlet problem with boundary function $Q_n\big|_U$, and so for quasi-every $z \in \partial U$ the limit of $g_{Q_n}(z')$ as z' tends to z from inside U is equal to $g_{Q_n}(z)$. Thus, for quasi-every $z \in \partial U$

$$\liminf_{z'\to z,\, z'\in U} g_Q(z') \ge \liminf_{z'\to z,\, z'\in U} g_{Q_n}(z') = g_{Q_n}(z),$$

and if here n tends to infinity, we get

$$\liminf_{z'\to z,\, z'\in U} g_Q(z') \ge Q(z)$$

quasi-everywhere on ∂U. Hence $g_Q\big|_U$ is at least as large as the upper solution $\overline{H}_Q^U$ of the Dirichlet problem with boundary function Q (use also the fact, that g_Q, being the limit of an increasing sequence of harmonic functions in U, is either harmonic on U, or it is identically infinite).

Summarizing, we have shown that on U the function g_Q is at least as large as the upper, and is at most as large as the lower solution of the Dirichlet problem. This means that the upper and lower solutions $\overline{H}_Q^U$ and $\underline{H}_Q^U$ coincide, and g_Q coincides on U with the the solution of the Dirichlet problem.

The last statement concerning uniform convergence follows from Dini's theorem, since then an increasing family of continuous functions converges to a continuous function. □

Proof of Theorem 2.3. Setting $q \equiv 0$, we can imitate the preceding proof since in case Σ has connected complement, $(\partial\Sigma)^*$ is the support of the extremal measure for the weight v identically equal to 1 on Σ, and $S_v = (\partial\Sigma)^*$ was the only property of $v(z) = \exp(-q(z))$ we used in the proof above. □

We close this section with a summary of the algorithm of the extremal point method for solving Dirichlet problems for a smooth function. To emphasize that the function q can be chosen in different ways, here we take $q(z) = |z|^2/R$. Thus, suppose we have to solve the Dirichlet problem with boundary function Q on every component of $\mathbf{C} \setminus \Sigma$, where Σ is a compact set contained in the unit disk D_1, and assume that Q is the restriction to Σ of a twice continuously differentiable function Q^* on D_1 with

$$\left|\frac{\partial Q^*}{\partial x}\right| + \left|\frac{\partial Q^*}{\partial y}\right| \le M, \qquad \left|\frac{\partial^2 Q^*}{\partial x^2}\right| + \left|\frac{\partial^2 Q^*}{\partial y^2}\right| \le M,$$

$M \ge 1$. Set $\lambda_0 = 1/10M$, and for the weights $w_1 := \exp(-|z|^2/10)$ and $w_2 := \exp(-\lambda_0 Q(z) - |z|^2/10)$ construct the sequences $\{\hat{a}_n^{(i)}\}$, $i = 1, 2$, as follows. Let $\{\varepsilon_n\}_{n=1}^{\infty}$ be a sequence of positive numbers satisfying the condition (1.10) or, in case Σ is smooth, the milder conditions discussed at the end of Section V.1, and let us fix a corresponding sequence of discrete subsets $\{S_n\}_{n=1}^{\infty}$ of Σ in such a way that $S_n \subset S_{n+1}$, and for each n and $z \in \Sigma$ there is a point in S_n the distance of which to z is at most ε_n. For $i = 1, 2$, let $\hat{a}_0^{(i)} \in \Sigma$ be arbitrary, and for each $n = 1, 2, \ldots$ define $\hat{a}_n^{(i)} \in S_n$ as a point where the weighted polynomial expression

$$\left|(z - \hat{a}_0^{(i)})(z - \hat{a}_1^{(i)}) \cdots (z - \hat{a}_{n-1}^{(i)}) w_i^n(z)\right|$$

takes its maximum on S_n, i.e. $\hat{a}_n^{(1)}$ is one of the points in S_n for which

$$\max_{z \in S_n} \left\{ \left(\log|z - \hat{a}_0^{(1)}| + \cdots + \log|z - \hat{a}_{n-1}^{(1)}| \right) - n|z|^2/10 \right\}$$

is attained and $\hat{a}_n^{(2)}$ is one of the points in S_n for which

$$\max_{z \in S_n} \left\{ \left(\log|z - \hat{a}_0^{(2)}| + \cdots + \log|z - \hat{a}_{n-1}^{(2)}| \right) - n\lambda_0 Q(z) - n|z|^2/10 \right\}$$

is attained.

Then, with the notations

$$\hat{A}_n(z) := \frac{1}{n}\left(\log\left|z - \hat{a}_0^{(2)}\right| + \cdots + \log\left|z - \hat{a}_{n-1}^{(2)}\right|\right)$$

$$-\frac{1}{n}\left(\log\left|z - \hat{a}_0^{(1)}\right| + \cdots + \log\left|z - \hat{a}_{n-1}^{(1)}\right|\right)$$

$$\hat{A}_n^{(1)} := \frac{1}{n}\left(\log\left|\hat{a}_n^{(1)} - \hat{a}_0^{(1)}\right| + \log\left|\hat{a}_n^{(1)} - \hat{a}_2^{(1)}\right| + \cdots + \log\left|\hat{a}_n^{(1)} - \hat{a}_{n-1}^{(1)}\right|\right)$$

$$-\left|\hat{a}_n^{(1)}\right|^2/10$$

and

$$\hat{A}_n^{(2)} := \frac{1}{n}\left(\log\left|\hat{a}_n^{(2)} - \hat{a}_0^{(2)}\right| + \log\left|\hat{a}_n^{(2)} - \hat{a}_2^{(2)}\right| + \cdots + \log\left|\hat{a}_n^{(2)} - \hat{a}_{n-1}^{(2)}\right|\right)$$

$$-\lambda_0 Q(\hat{a}_n^{(2)}) - \left|\hat{a}_n^{(2)}\right|^2/10,$$

the function

$$\frac{1}{\lambda_0}\left\{\hat{A}_n(z)+\hat{A}_n^{(1)}-\hat{A}_n^{(2)}\right\}$$

converges as $n\to\infty$ to the solution of the Dirichlet problem in question. The convergence is locally uniform in every component of $\mathbf{C}\setminus\Sigma$.

V.3 The Extremal Point Method for Determining Green Functions and Conformal Mappings

In this section we apply the extremal point method of the two previous sections to find the Green function of a region and to determine conformal mappings of simply connected domains.

Let Σ be a compact set of positive capacity with connected complement. In the spirit of what we have done in the preceding section we consider the weight $w\equiv 1$ on Σ, i.e. we actually consider the unweighted case. It follows from Theorem I.4.7 that the expression

$$F_w-U^{\mu_w}(z)=F_w-\int\log\frac{1}{|z-t|}d\mu_w(t)=:g(z,Q) \tag{3.1}$$

is the Green function of $\mathbf{C}\setminus\Sigma$ with pole at the infinity, and the function in (3.1) can be determined by the extremal point method of Sections V.1 and V.2 (see especially Corollaries 1.3 and 1.5 and Theorem 2.3).

If we have to determine the Green function $g_U(z,a)$ with pole at a point $a\in U$ of an arbitrary domain U the complement of which has positive capacity, then by the transformation $u\to 1/(z-a)$ we can transform the problem to the one just discussed. This yields the following procedure for finding g_U: Let $\{\varepsilon_n\}_{n=1}^{\infty}$ be a sequence of positive numbers satisfying the condition (1.10) or, in case Σ is smooth, the milder conditions discussed at the end of Section V.1, and let us fix a corresponding sequence of discrete subsets $\{S_n\}_{n=1}^{\infty}$ of $\Sigma:=\mathbf{C}\setminus\partial U$ in such a way that $S_n\subset S_{n+1}$, and for each n and $z\in\Sigma$ there is a point z^* in S_n with $\operatorname{dist}(z,z^*)\le\varepsilon_n$. Let $\hat{a}_0\in\Sigma$ be arbitrary, and for each $n=1,2,\ldots$ define $\hat{a}_n\in S_n$ as a point where the expression

$$\sum_{i=0}^{n-1}\log\left|\frac{1}{z-a}-\frac{1}{\hat{a}_i-a}\right| \tag{3.2}$$

takes its maximum on S_n. Then

$$\frac{1}{n}\sum_{i=0}^{n-1}\log\left|\frac{1}{z-a}-\frac{1}{\hat{a}_i-a}\right|-\frac{1}{n}\sum_{i=0}^{n-1}\log\left|\frac{1}{\hat{a}_n-a}-\frac{1}{\hat{a}_i-a}\right|$$

$$\equiv\frac{1}{n}\sum_{i=0}^{n-1}\log\left|\frac{(\hat{a}_n-a)(z-\hat{a}_i)}{(z-a)(\hat{a}_n-\hat{a}_i)}\right|$$

converges to $g_U(z,a)$ locally uniformly in $U\setminus\{a\}$.

Note that maximizing (3.2) is the same as maximizing

$$\sum_{i=0}^{n-1} \log|z - \hat{a}_i| - n \log|z - a|,$$

i.e. the Green function problem becomes (or is equivalent to) a weighted potential problem with weight function $w(z) := 1/|z - a|$.

Now suppose $U \neq \mathbf{C}$ is a simply connected domain and we wish to determine the conformal mapping of U onto a disk. As before, without loss of generality, we assume that U contains the infinite point of $\overline{\mathbf{C}}$. Let $\Sigma := \partial U$, and (starting with some $\hat{a}_0 \in \Sigma$) determine $\hat{a}_n$ as a point where the polynomial expression

$$|(z - \hat{a}_0)(z - \hat{a}_1) \cdots (z - \hat{a}_{n-1})|$$

takes its maximum on S_n, where the sets S_n are chosen as before. Since U is simply connected, the function $(\hat{P}_n(z))^{1/n}$, where

$$\hat{P}_n(z) := (z - \hat{a}_0)(z - \hat{a}_1) \cdots (z - \hat{a}_{n-1}),$$

is single-valued in U, and in what follows we take the branch of the n-th root that is positive on the positive real half-line. With these notations we have

Theorem 3.1. *Setting*

$$\hat{A}_n := |(\hat{a}_n - \hat{a}_0)(\hat{a}_n - \hat{a}_1) \cdots (\hat{a}_n - \hat{a}_{n-1})|,$$

the expression

$$(\hat{P}_n(z)/\hat{A}_n)^{1/n}$$

converges to the conformal mapping of U onto the exterior of the unit disk that leaves the infinite point and the direction of the positive real axis invariant. Thus,

$$\frac{1}{(\hat{P}_n(z)/\hat{A}_n)^{1/n}}$$

converges locally uniformly inside U to a conformal mapping of U onto the unit disk.

It is also clear from Theorem 1.4 and Corollary 1.5 that $1/(\hat{P}_n(z))^{1/n}$ converges to a conformal mapping of U onto the disk $D_{1/\mathrm{cap}(\Sigma)}$.

Proof of Theorem 3.1. All we have to mention is that if φ is a conformal mapping of U onto the exterior of the unit disk with $\varphi(\infty) = \infty$, then $|\varphi(z)| = \exp(g_U(z, \infty))$, and so by Theorems 1.4 and I.4.7

$$\lim_{n\to\infty} |(\hat{P}_n(z)/\hat{A}_n)^{1/n}/\varphi(z)| = 1$$

locally uniformly on U. The rest is a standard normal family argument. □

V.4 Notes and Historical References

Section V.1

J. Górski [71] proved a theorem similar to Theorem 1.1 in three dimensions.

The reasoning (1.6)–(1.8) was used by F. Leja in several papers [121]–[124].

Leja points are used as points of interpolation despite the fact that little is known concerning the norm of such interpolation sequences for general compact sets on the complex plane (see e.g. [126]). However, numerical evidence and the philosophy behind their generating idea suggest that they provide good choices for nodes.

Sections V.2–V.3

Methods based on extremal points have been widely used for approximating Dirichlet problems, Green functions and conformal mappings, see the works of F. Leja [121, 120, 122, 124, 125, 119, 118, 112, 113, 114, 115, 116, 117, 123], J. Górski [74, 71, 72, 65, 66, 73, 67, 69, 70, 68], W. Kleiner [100, 92, 91, 96, 93, 94, 95, 97, 98, 99], J. Siciak [204, 201, 203, 202, 205], Ch. Pommerenke [184, 182, 181], K. Menke [153, 154], and H. Kloke [100]. In fact, a major part of the work of the Polish school around F. Leja in the fifties focused on this subject, and this school developed the relevant theory in detail. The procedures of Sections V.2–V.3 are based on the same idea. Some of the results (like Theorems 2.1 and 2.2 where a "strongly convex" function helps in identifying the solution) seem to be new, and in the details the discussion follows the theory developed in Chapter I rather than the works just mentioned. For more on the extremal method and generalizations of it to other kernels or dimensions see the papers mentioned above.

Chapter VI. Weights on the Real Line

This chapter is devoted to a detailed study of some questions that lead naturally to minimal energy problems with external fields.

We start with the approximation problem on closed subsets of the real line by weighted polynomials of the form $w^n P_n$, where w is a given continuous weight function. Note that the exponent n in the weight matches the degree of P_n; therefore this approximation is different from standard weighted approximation, for P_n must balance exponential oscillations in w^n. However, there are questions (like the strong asymptotic behavior of Freud-type orthogonal polynomials to be discussed in the next chapter), which require exactly this kind of approximation. We shall start out with an application of the Weierstrass–Stone theorem that shows that the approximation problem is interesting only on the support $\mathcal{S}_w$ of the extremal measure, for the approximating sequence $w^n P_n$ automatically converges to zero outside $\mathcal{S}_w$. We also get that there is always a set Z such that a function f can be approximated by weighted polynomials of the form $w^n P_n$ if and only if f vanishes on Z. The possibility of approximating any function f that vanishes outside $\mathcal{S}_w$ is intimately connected with the behavior of the extremal measure. In fact, supposing for example that the support $\mathcal{S}_w$ consists of a finite number of intervals and the extremal measure is given by its density $v(t)$, the main theorem of Section VI.1 claims that if v is continuous *and positive* inside $\mathcal{S}_w$, then any $f \in C(\Sigma)$ that vanishes outside $\mathcal{S}_w$ can be arbitrarily well approximated by weighted polynomials $w^n P_n$ on Σ. On the other hand, a single internal zero of v can prevent approximation. Actually we shall prove more, namely that if at a point z_0 the density $v(t)$ behaves like a constant times $|t - z_0|^\lambda$ with $\lambda \neq 0$, then $z_0 \in Z$, i.e. approximation around z_0 is impossible. With a simple transformation the role of the endpoint behavior of v in approximation can also be handled.

The aforementioned approximation problem is closely related to another one on the whole real line by weighted polynomials of the form $W(a_n x) P_n(x)$, where W is a given function (typically an exponential-type weight of the form $\exp(-Q(x))$) and $\{a_n\}$ is an appropriate normalizing sequence. Our results from Chapter IV on the actual determination of the equilibrium distributions allow us to resolve this type of approximation question under fairly general conditions.

After that we shall turn to a related topic, namely to the construction of so-called fast decreasing polynomials that take the value 1 at the origin, and away from that point, they decrease as fast as possible. This "as fast as possible" depends on the distance from the origin, and we shall consider the problem when this decrease is exponential at each point $x \in [-1, 1]$, $x \neq 0$:

$$P_n(0) = 1, \qquad |P_n(x)| \leq Ce^{-n\varphi(x)}. \tag{0.1}$$

We shall obtain sharp conditions on φ that permit such polynomials; a necessary condition being

$$\frac{2}{\pi}\int_0^1 \frac{\varphi(t)}{t^2\sqrt{1-t^2}}dt \leq 1,$$

which turns out to be also sufficient when $\varphi(\sqrt{x})$ is concave. For example, it follows that for $\varphi(x) = x^2$ there are polynomials with the properties (0.1), but for $\varphi(x) = (1+\varepsilon)x^2$ there are no such polynomials no matter how small $\varepsilon > 0$ is. Such fast decreasing polynomials are the building blocks for well localized "partitions of unity" consisting of polynomials and they can also be used as polynomial kernels in convolutions that provide good approximations of the identity kernel (Dirac delta).

The proofs of the aforementioned results all use a very careful discretization technique of the potential. For example, roughly speaking we can say that for $f \equiv 1$ the equality

$$w^n(x)\exp(n(-U^{\mu_w}(z) + F_w)) = 1,$$

that we know holds (quasi-everywhere) on $\mathcal{S}_w$, is the perfect solution of the approximation problem, and the polynomial solution is obtained by discretizing the logarithmic potential U^{μ_w}. This discretization technique and a related one is spelled out in detail in Section VI.4.

Finally we shall consider various weighted norm inequalities for polynomials with so-called Freud weights $w_\lambda(x) = \exp(-|x|^\lambda)$. The analogue of the classical inequalities of Markoff and Nikolskii will be established with constants of precise order. These are based on sharp infinite-finite range inequalities that reduce the norm on infinite intervals to those on explicitly given finite ones, like

$$\|w_\lambda P_n\|_{L^p(\mathbf{R})} \leq C\|w_\lambda P_n\|_{L^p([-cn^{1/\lambda},cn^{1/\lambda}])}, \qquad n = 1, 2, \ldots.$$

These infinite-finite range inequalities are extremely important, for they testify that the norms of weighted polynomials actually live on smaller sets, and hence many classical tools known for finite intervals can be brought into play.

VI.1 The Approximation Problem

Let Σ be a closed subset of the real line and w an admissible weight on Σ. We consider the problem of approximating a continuous function f by weighted polynomials $w^n P_n$, $n = 1, 2, \ldots$ (i.e. we consider the full sequence of degrees). This approximation problem appears in several applications as we shall see later. It must be emphasized that the exponent of the weight w^n changes with n, so this is a different (and in some sense more difficult) type of approximation than what is usually called weighted approximation. In fact, in the present case the polynomial P_n must balance exponential oscillations in w^n. A natural assumption when considering this approximation problem is that w be continuous, and to avoid technical complications we shall also assume that Σ is regular in the sense that for every $x_0 \in \Sigma$

$$\sum_{n=1}^{\infty} \frac{n}{\log(1/\mathrm{cap}(E_n))} = \infty, \tag{1.1}$$

where

$$E_n := \{x \in \Sigma \mid 2^{-n-1} \le |x - x_0| \le 2^{-n}\}$$

(see Wiener's Theorem I.4.6). Then, by Theorems I.5.1 and I.4.4, we have for every $x \in \mathcal{S}_w$ the equality

$$U^{\mu_w}(x) = -Q(x) + F_w, \tag{1.2}$$

and for every $x \in \Sigma$

$$U^{\mu_w}(x) \ge -Q(x) + F_w. \tag{1.3}$$

In particular, this is true if Σ consists of finitely many (finite or infinite) intervals.

We start with a Stone–Weierstrass type theorem.

Theorem 1.1. *Let $\Sigma \subset \mathbf{R}$ be a closed set and w a continuous admissible weight on Σ. Then there exists a closed set $Z = Z(w) \subset \Sigma$ such that a continuous function f on Σ is the uniform limit of weighted polynomials $w^n P_n$, $n = 1, 2, \ldots$, if and only if f vanishes on Z.*

Proof. We shall use the following version of the Stone-Weierstrass theorem ([213, Theorem 5], [194]). Let $C(X)$ denote the family of real-valued continuous functions on a compact Hausdorff space X, and for $\mathcal{A} \subset C(X)$ let $Z(\mathcal{A})$ denote the set of points $x \in X$ such that $f(x) = 0$ for every $f \in \mathcal{A}$, i.e. $Z(\mathcal{A})$ is the common zero set of the family $\mathcal{A}$. Suppose that $\mathcal{A}$ has the following four properties.

(a) If $f, g \in \mathcal{A}$, then $\alpha f + \beta g \in \mathcal{A}$ for all real α and β.
(b) If $f, g \in \mathcal{A}$, then $fg \in \mathcal{A}$.
(c) $\mathcal{A}$ is closed under uniform limits.
(d) If $x_1, x_2 \in X \setminus Z(\mathcal{A})$, then there is an $f \in \mathcal{A}$ such that $f(x_1) \ne f(x_2)$.

Then

$$\mathcal{A} = \{f \in C(X) \mid f \equiv 0 \text{ on } Z(\mathcal{A})\}.$$

This is the version we need to prove the theorem.

Let $X = \Sigma \cap [-a, a]$, where a is so large that outside the interval $[-a, a]$ we have

$$U^{\mu_w}(x) \ge -Q(x) + F_w + 1.$$

Recall now the inequality (see (III.2.2))

$$w^n(x)|P_n(x)| \le \|w^n P_n\|_{S_w} \exp(-n(U^{\mu_w}(x) + Q(x) - F_w)). \tag{1.4}$$

By the choice of a, the left-hand side is at most

$$\|w^n P_n\|_{S_w} e^{-n}$$

when $x \notin X$; therefore the uniform convergence of any sequence $\{w^n P_n\}$ on Σ is equivalent to its uniform convergence on X.

Now let $\mathcal{A}$ be the collection of all continuous functions f on X such that $w^n P_n \to f$ uniformly on X for some polynomials P_n, $\deg P_n \le n$, $n = 1, 2, \ldots$. It is obvious that $\mathcal{A}$ satisfies (a). If $w^n P_n \to f$ and $w^n Q_n \to g$ uniformly on X, then with $R_{2n} := P_n Q_n$, $R_{2n+1} := P_{n+1} Q_n$ the sequence $\{w^n R_n\}$ tends uniformly to fg on X. Therefore $\mathcal{A}$ satisfies (b). From a straightforward diagonal argument it follows that (c) is also satisfied. Finally, to prove (d) we note that for $x_1 \in X \setminus Z(\mathcal{A})$ there is $g \in \mathcal{A}$ with $g(x_1) \neq 0$. Suppose that $w^n P_n$ tends to g uniformly on X. Taking $Q_{n+1}(x) := (x - x_2)P_n(x)$ and $f(x) := w(x)(x - x_2)g(x)$, we find that $w^{n+1} Q_{n+1}$ converges uniformly to f, so $f \in \mathcal{A}$. Since $f(x_1) \neq f(x_2)$, we see that (d) also holds.

Now the theorem follows from the aforementioned Stone–Weierstrass theorem. □

Theorem 1.1 reduces the approximation problem to finding the set Z. Of course, Z may be empty (consider $w \equiv 1$ on a finite interval), and it depends on *global* properties of w. However, as we now show, Z always contains $\Sigma \setminus S_w$, so the approximation problem is interesting only on S_w.

Theorem 1.2. *Let $\Sigma \subset \mathbf{R}$ be a regular closed set, and let w be an admissible weight on Σ. If a sequence $\{w^n P_n\}_{n=1}^{\infty}$ of weighted polynomials converges uniformly on S_w, then $\{w^n(x_0)P_n(x_0)\}$ tends to 0 for every $x_0 \in \Sigma \setminus S_w$.*

Let us note that the mere boundedness of $w^n P_n$ on S_w does not necessarily imply that the sequence $\{w^n(x_0)P_n(x_0)\}$ converges to zero for $x_0 \notin S_w$ (cf. Theorem III.2.1). A counterexample is furnished by the weight w that is 1 on $[-1, 1]$ and equals $(x + \sqrt{x^2 - 1})^{-1}$ on $(1, 2]$, and the Chebyshev polynomials

$$T_n(x) = \cos(n \arccos x) = \frac{1}{2}\left((x + \sqrt{x^2 - 1})^n + (x - \sqrt{x^2 - 1})^n\right)$$

for $x \in [-1, 1]$. In this case μ_w is the arcsine measure $(\pi\sqrt{1 - x^2})^{-1}dx$ on $[-1, 1]$, and it is obvious that $w^n T_n$ is bounded on $S_w = [-1, 1]$, but $w^n(x_0)T_n(x_0) \ge 1/2$ (and $w^n(x_0)T_n(x_0) \to 1/2$) for all $x_0 \in (1, 2]$.

Note, however, that if Q is convex, then $w^n(x_0)P_n(x_0) \to 0$ geometrically fast whenever $x_0 \in \Sigma \setminus S_w$ and the sequence $\{\|w^n P_n\|_{S_w}\}$ is bounded (recall Theorem III.2.1 and use strict convexity of $U^{\mu_w} + Q$ on $\Sigma \setminus S_w$).

Proof of Theorem 1.2. Suppose to the contrary that $w^n P_n$ converges to some f_0 uniformly on S_w, but $w^n(x_0)P_n(x_0)$ does not tend to 0 for some $x_0 \in \Sigma \setminus S_w$.

Let $\mathcal{A}_0$ be the collection of all continuous functions f on S_w that are uniform limits of weighted polynomials $w^n Q_n$, $\deg Q_n \le n$, with the additional property that $Q_n(x_0) = 0$. As in the proof of Theorem 1.1, it is easy to show that $\mathcal{A}_0$ satisfies the assumptions of the Stone–Weierstrass theorem mentioned in that proof. Therefore, there is a set $Z_0 \subset S_w$ such that $f \in \mathcal{A}_0$ if and only if f vanishes on Z_0.

We claim that $f_0 \in \mathcal{A}_0$. Indeed, because $w^{n+1}(x)(x - x_0)P_n(x)$ tends to $w(x)(x - x_0)f_0(x)$ on S_w, we have that $w(x)(x - x_0)f_0(x)$ belongs to $\mathcal{A}_0$, and so $w(x)(x-x_0)f_0(x)$ must vanish on Z_0. Since f_0 vanishes on $\mathcal{S}_w$ exactly where $w(x)(x-x_0)f_0(x)$ does so, it follows that $f_0 \in \mathcal{A}_0$.

Thus, there are polynomials Q_n with $\deg Q_n \le n$ and $Q_n(x_0) = 0$ such that $w^n Q_n$ tends to f_0 on $\mathcal{S}_w$. Then $w^n(P_n - Q_n)$ converges to 0 uniformly on $\mathcal{S}_w$. Because $w^n(x_0)P_n(x_0)$ does not tend to zero, it follows that there is an $\varepsilon > 0$ and an integer n such that $|w^n(P_n - Q_n)| \le \varepsilon$ on $\mathcal{S}_w$ while $|w^n(P_n - Q_n)| > 2\varepsilon$ at x_0. However, this contradicts (1.4) at x_0, for (1.3) shows that the second factor on the right of (1.4) is at most 1. This contradiction proves the theorem. □

Next we turn to conditions guaranteeing approximation. Let $O \subset \Sigma$ be an open subset of the real line. The space of continuous real functions that vanish outside O will be denoted by $C_0(O)$.

Definition 1.3. We say that w has the *approximation property on the open set O* if for every $f \in C_0(O)$ there is a sequence $\{w^n P_n\}_{n=1}^{\infty}$, $\deg P_n \le n$, of weighted polynomials converging uniformly to f on Σ.

Theorem 1.2 implies that one can hope for the approximation property for every continuous function on an open set O only if $O \subseteq \mathcal{S}_w$; that is, O should be part of the interior $\mathrm{Int}(\mathcal{S}_w)$ of $\mathcal{S}_w$. On the other hand, the next result implies that if μ_w has continuous and positive density function on the interior of $\mathcal{S}_w$, then w does have the approximation property on $\mathrm{Int}(\mathcal{S}_w)$.

To formulate the results we introduce the following definition:

Definition 1.4. Let S^w denote the set of points x_0 where the equilibrium measure μ_w has continuous and positive density; that is,

$$d\mu_w(t) = v(t)\,dt$$

in a neighborhood of x_0, and the density function $v = v_w$ is continuous and positive in a neighborhood of x_0. This S^w is called the *restricted support* of μ_w.

Thus, if μ_w has positive and continuous density on $\mathrm{Int}(\mathcal{S}_w)$, then $S^w = \mathrm{Int}(\mathcal{S}_w)$. On the other hand, if at x_0 we have $v(x_0) = 0$, then this x_0 *does not* belong to the restricted support. Note also that S^w is a (possibly empty) open subset of **R**.

Theorem 1.5. *Let w be an admissible weight on $\Sigma \subseteq$ **R**. Then w has the approximation property on the restricted support S^w.*

In the notation of Theorem 1.1 the theorem says that $S^w \cap Z(w) = \emptyset$.

Corollary 1.6. *If μ_w has continuous and positive density on* $\mathrm{Int}(\mathcal{S}_w)$, *then every continuous function that vanishes outside* $\mathrm{Int}(\mathcal{S}_w)$ *can be uniformly approximated on Σ by weighted polynomials of the form $w^n P_n$, where the degree of P_n is at most n.*

Another consequence of Theorem 1.5 is

Theorem 1.7. *Suppose that $\Sigma \subseteq \mathbf{R}$ consists of finitely many disjoint intervals I_j and w is an admissible weight of class $C^{1+\varepsilon}$ for some $\varepsilon > 0$ such that $Q = \log 1/w$ is convex on every I_j. Then w has the approximation property on the interior of the support S_w. The same is true if instead of the convexity we assume $\Sigma \subseteq [0, \infty)$ and that $xQ'(x)$ increases on every subinterval of Σ.*

The proof of these theorems will be given after the discussion of some further results and their consequences. In Theorem 1.7 it is enough to assume the $C^{1+\varepsilon}$ condition on any set containing S_w, and sometimes (as in the case of incomplete polynomials to be discussed below) it is possible to verify this condition without explicitly knowing S_w.

Next we show that Theorem 1.5 is sharp.

Theorem 1.8. *Let w be a continuous admissible weight on the regular set Σ, and let $t_0 \in S_w$ be an interior point of S_w. Suppose the extremal measure μ_w has density v in a neighborhood of t_0 that satisfies*

$$v(t) = L|t - t_0|^{\beta}(1 + o(1)), \qquad t \to t_0, \tag{1.5}$$

for some constants $L > 0$ and $\beta \neq 0$. Then $t_0 \in Z(w)$; in other words, if f is the uniform limit of weighted polynomials $w^n P_n$, then $f(t_0) = 0$.

Actually, the L and β in (1.5) for the left and right neighborhoods could be allowed to be different.

Note that the missing case $\beta = 0$ is just what was covered in Theorem 1.5 (provided v is continuous in a neighborhood of t_0), i.e. in this case t_0 does not belong to $Z(w)$, and the functions to be approximated need not vanish at t_0.

Although the statement in the theorem is the same for $\beta > 0$ and $\beta < 0$, these two cases exhibit completely different characteristics. In fact, a zero in the density ($\beta > 0$) is "very fragile" regarding changes in the weight. Consider for example the weight function $w(x) = \exp(x^2)$ on $[-1, 1]$. It can be shown that the extremal measure has the density

$$v(t) = \frac{2t^2}{\pi\sqrt{1 - t^2}}, \qquad t \in [-1, 1].$$

The origin is an internal zero of the density and v satisfies (1.5) with $\beta = 2$. Therefore, by Theorem 1.8, every approximable function vanishes at 0. If, however, we consider w on any smaller interval, say $w_1 = w$ on $[-1, a]$ for some $0 < a < 1$, then the equilibrium measure μ_{w_1} is obtained from μ_w by forming the balayage of $\mu_w\big|_{(a,1]}$ onto $[-1, a]$ and adding this balayage measure to $\mu_w\big|_{[-1,a]}$ (see Theorem IV.1.6(e)). Hence, μ_{w_1} has continuous and positive density in $(-1, a)$, and every continuous function on $[-1, a]$ is approximable by weighted polynomials. Thus, under the indicated change, the internal zero disappears, and the approximation properties change radically.

On the other hand, a power type singularity ($\beta < 0$ in (1.5)) is robust in the sense that restriction (or extension) of the weight does not affect the type of singularity; hence $t_0 \in Z(w)$ for all such weights. In other words, if $\beta < 0$, then no matter on what set Σ^* we approximate f, $f(t_0)$ must vanish provided Σ^* contains a neighborhood of t_0.

Theorem 1.5 and Corollary 1.6 do not tell if approximation is possible on $\text{Int}(\mathcal{S}_w)$ if the density v vanishes at internal points, and in fact, Theorem 1.8 shows that polynomial type behavior around a zero prevents approximation. However, this does not rule out the possibility of approximation on the whole $\mathcal{S}_w$ even if the density vanishes in certain internal points. Indeed, there is a weight w (see [220, Example 2 in Section 4]) such that $\mathcal{S}_w = [-1, 1]$, μ_w has continuous density in $(-1, 1)$ which vanishes at the origin, and still every continuous f can be uniformly approximated by weighted polynomials of the form $w^n P_n$ on $[-1, 1]$.

Until now we have been considering the approximation property around inner points of the extremal support. The next result clarifies the situation near endpoints of subintervals of $\mathcal{S}_w$.

Theorem 1.9. *Let w be a continuous admissible weight on the regular set Σ, and let $t_0 \in \mathcal{S}_w$ be a point such that for some $\delta > 0$ we have $\Sigma \cap (t_0 - \delta, t_0 + \delta) = [t_0, t_0 + \delta)$. Suppose that the extremal measure μ_w has density v in a right neighborhood of t_0 that satisfies*

$$v(t) = L|t - t_0|^{\beta}(1 + o(1)), \qquad t \to t_0, \tag{1.6}$$

with some constants $L > 0$ and β.

(a) *If $\beta \neq -1/2$, then $t_0 \in Z(w)$. In other words, if f is the uniform limit of weighted polynomials $w^n P_n$, then $f(t_0) = 0$.*
(b) *If $\beta = -1/2$ and v is continuous in a right neighborhood of t_0, then $t_0 \notin Z(w)$, and hence approximable functions need not vanish at t_0.*

Similar statements hold for "right endpoints", i.e. for the case when $\mathcal{S}_w \cap (t_0 - \delta, t_0 + \delta) = (t_0 - \delta, t_0]$.

As applications of the approximation theorems let us consider some concrete examples.

1. *Incomplete polynomials.* Let $\theta > 0$, $\Sigma = [0, 1]$ and $w(x) = x^{\theta/(1-\theta)}$. The weighted polynomials $w^n(x)P_n(x)$ are closely related to the so-called incomplete polynomials of the form

$$P_N(x) = \sum_{k=s_N}^{N} a_k x^k, \tag{1.7}$$

with $s_N/N \to \theta$ (set $N = n/(1-\theta)$), which vanish at the origin with high order. The extremal support is $\mathcal{S}_w = [\theta^2, 1]$ (see Section IV.1, Example 1.16). In this case $Q(x) = -(\theta/(1-\theta)) \log x$ is convex, and we get from Theorem 1.7 that every function $f \in C[0, 1]$ that vanishes outside $(\theta^2, 1)$ is the uniform limit of weighted polynomials $x^{n\theta/(1-\theta)} P_n(x)$.

To see if approximation is also possible for functions that do not vanish at $t_0 = 1$, we need to consider the density of the extremal measure. According to formula (IV.5.17) the density is

$$\frac{1}{(1-\theta)\pi t}\sqrt{\frac{t-\theta^2}{1-t}}, \qquad \theta^2 \le t \le 1,$$

which is of the form $h(t)/\sqrt{1-t}$ for some h continuous around $t_0 = 1$. Therefore, by an application of Theorem 1.9 we conclude that $1 \notin Z$, and so every continuous f on $[0, 1]$ that vanishes on $[0, \theta^2]$ can be uniformly approximated by weighted polynomials $w^n P_n$. Thus, in the terminology of Theorem 1.1, $Z = [0, \theta^2]$.

Since the weights x^α, $0 \le \alpha \le 1$, form a compact subset of $C[\theta^2, 1]$, it also follows as in the proof of Theorem 1.5, that every $f \in C[0, 1]$ that vanishes outside $[\theta^2, 1]$ is the uniform limit of polynomials $x^{[n\theta/(1-\theta)]} P_n(x)$, where $[\cdot]$ denotes integral part, i.e. every such function is the uniform limit of incomplete polynomials. Note that no other function can be the uniform limit of such polynomials (apply Theorem III.2.1 and notice that the exponent in (III.2.2) is negative outside S_w because of the strict convexity of Q).

It is also interesting to see what happens if approximation is sought only on $[\theta^2, 1]$. Since the extremal measure has density that is continuous and positive on $(\theta^2, 1)$ and vanishes like $\sqrt{t-\theta^2}$ in a right neighborhood of θ^2, we get from Theorems 1.5 and 1.9 that in this case $Z = \{\theta^2\}$, i.e. even if we are interested only in approximation on $[\theta^2, 1]$, the function f must vanish at the left endpoint.

2. *Exponential weights.* Let $\Sigma = \mathbf{R}$ and $w(x) = \exp(-c|x|^\alpha)$, $\alpha > 0$, $c > 0$. In this case the extremal support is

$$S_w = [-\gamma_\alpha^{1/\alpha} c^{-1/\alpha}, \gamma_\alpha^{1/\alpha} c^{-1/\alpha}]$$

with

$$\gamma_\alpha = \Gamma(\frac{\alpha}{2})\Gamma(\frac{1}{2})/2\Gamma(\frac{\alpha}{2}+\frac{1}{2}),$$

(see Section IV.5, Theorem IV.5.1). The extremal measure is the Ullman distribution given by the scaling to S_w of the density function

$$\frac{\alpha}{\pi}\int_{|t|}^{1} \frac{u^{\alpha-1}}{\sqrt{u^2-t^2}}du, \qquad t \in (-1, 1),$$

which has a $\sqrt{|t-a|}$ behavior around the endpoints of S_w, while it is positive and continuous on the interior of S_w except for the case $\alpha \le 1$ and $t_0 = 0$. If $\alpha < 1$, then at $t_0 = 0$ the density has a power type singularity $t^{\alpha-1}$, while for $\alpha = 1$ the singularity is of logarithmic type.

Now for $\alpha > 1$ Theorem 1.7 can be applied because $Q(x) = c|x|^\alpha$ is convex. Thus, in this case every $f \in C(\mathbf{R})$ that vanishes outside S_w (and no other one) is the uniform limit of weighted polynomials $e^{-cn|x|^\alpha} P_n(x)$. The result is still valid for $\alpha = 1$ (see Theorem 2.7 in the next section), although we cannot derive this from Theorems 1.7 and 1.5 because $|x|$ is not a $C^{1+\varepsilon}$–function for some $\varepsilon > 0$, and, as we have just seen, the density of the extremal measure has a logarithmic type singularity at $t_0 = 0$. Finally, Theorems 1.5, 1.8, and 1.9 imply that for $0 < \alpha < 1$ functions that vanish outside S_w *and at the origin* are the uniform limits of such weighted polynomials.

If we are interested in the approximation of a continuous f only on $\mathcal{S}_w$, then f still must vanish at the endpoints of $\mathcal{S}_w$ for all $\alpha > 1$, and also at the origin if $\alpha < 1$.

3. *Jacobi weights.* Let $\Sigma = [-1, 1]$ and $w(x) = (1-x)^\alpha(1+x)^\beta$, $\alpha, \beta \geq 0$. Then the support of the extremal measure is (see Section IV.5, Example 1.17)

$$[a, b] := [\theta_2^2 - \theta_1^2 - \sqrt{\Delta}, \theta_2^2 - \theta_1^2 + \sqrt{\Delta}],$$

where $\theta_1 := \alpha/(1+\alpha+\beta)$, $\theta_2 := \beta/(1+\alpha+\beta)$ and

$$\Delta = \{1 - (\theta_1 + \theta_2)^2\}\{1 - (\theta_1 - \theta_2)^2\},$$

and the extremal measure is given (see (IV.5.8)) by the density

$$\frac{1}{\pi}\frac{(1+\alpha+\beta)}{1-t^2}\sqrt{(t-a)(b-t)}, \qquad t \in (a, b).$$

Thus, in this case the set Z determining approximation is the following: $Z = [-1, a] \cup [b, 1]$ if $\alpha, \beta > 0$; $Z = [-1, a]$ if $\alpha = 0$, $\beta > 0$; $Z = [b, 1]$ if $\alpha > 0$, $\beta = 0$ and finally $Z = \emptyset$ if $\alpha = \beta = 0$.

4. *Laguerre weights.* For $w(x) = x^\alpha e^{-\lambda x}$, $\alpha \geq 0$, $\lambda > 0$, $\Sigma = [0, \infty)$ we have (see Section IV.5, Example 1.18)

$$\mathcal{S}_w = [a, b] := [\frac{1}{\lambda}(1 + \alpha - \sqrt{1+2\alpha}), \frac{1}{\lambda}(1 + \alpha + \sqrt{1+2\alpha})],$$

with extremal measure given (see (IV.5.18)) by the density

$$\frac{\lambda}{\pi t}\sqrt{(t-a)(b-t)}, \qquad a < t < b.$$

Thus, in this case $Z = [0, a] \cup [b, \infty)$ if $\alpha > 0$, and $Z = [b, \infty) \doteq [2/\lambda, \infty)$ if $\alpha = 0$.

We now turn to the proofs of the aforestated theorems.

Proof of Theorem 1.5. Let $f \in C_0(S^w)$, and let J_ε denote the set of points $x \in S^w$ that are of distance $\geq \varepsilon$ from the complement $\mathbf{R} \setminus S^w$. First we simplify the problem.

I. Obviously, it is enough to consider f's that are positive in S^w and less than, say, 1.

II. Let J^* be an arbitrary finite interval. Eventually we will choose J^* so that in $\Sigma \setminus J^*$ we have

$$U^{\mu_w}(z) \geq -Q(z) + F_w + 1 \tag{1.8}$$

(cf. Theorem I.1.3 and the definition of the admissibility of w in Section I.1), but in order that we can choose it freely later, it can be arbitrary at this point. Then we claim that it is enough to approximate on J^*, because weighted polynomials $w^n P_n$ that are bounded on $\mathcal{S}_w$ tend to zero outside a fixed compact set (see Theorems

I.1.3 and III.2.1, and note also that J^* is at our disposal at this stage, so we can choose it appropriately to contain $\mathcal{S}_w$ and the compact set in question).

III. It is enough to approximate by the absolute values of weighted polynomials. In fact, if $w^n|P_n|$ uniformly tends to $\sqrt{f}$ on a set, then $w^{2n}|P_n|^2$ uniformly tends to f, and here $|P_n|^2$ is already a real polynomial. This establishes our claim when the degree n is even. For odd degree one can get the statement by approximating f/w with even degree polynomials and then multiplying through by w.

IV. It is enough to show the following: for every $\varepsilon > 0$ and $L > 0$ and for every large n, say $n \geq n_{\varepsilon,L}$, there are a continuous function $g_{L,n}$ and a polynomial Q_n of degree at most n such that the $\{g_{L,n}\}_{n=n_{\varepsilon,L}}^{\infty}$ forms a compact family of continuous functions on J_ε;

$$w^n(x)|Q_n(x)| = \exp(g_{L,n}(x) + R_{L,n}(x)), \qquad x \in J_\varepsilon, \tag{1.9}$$

where the remainder term $R_{L,n}(x)$ satisfies $|R_{L,n}(x)| \leq C_\varepsilon/L$ uniformly in $x \in J_\varepsilon$ with some $C_\varepsilon \geq 1$ depending only on ε, and uniformly in $x \in \Sigma \cap J^*$

$$w^n(x)|Q_n(x)| = e^{o(n)}. \tag{1.10}$$

In fact, suppose this is true, and apply it to w^λ instead of w with some $\lambda > 1$ close to 1. We can do this, because, by Theorem IV.4.10, there is a $\lambda > 1$ such that the set $J_{\varepsilon/2}$ is in the support $\mathcal{S}_{w^\lambda}$ of μ_{w^λ} and it has continuous density there. Furthermore, exactly as U^{μ_w}, the potential $U^{\mu_{w^\lambda}}$ is continuous everywhere; and these are the only properties that we shall use in deriving (1.9) and (1.10) below.

Hence, by choosing $\lambda > 1$ close to 1, we get that there are polynomials $Q_{[n/\lambda]}$ of degree at most $[n/\lambda]$ such that with some $g_{L,n}$ and $R_{L,n}$ as above

$$w^n(x)|Q_{[n/\lambda]}(x)| = \exp(g_{L,n}(x) - (n - \lambda[n/\lambda])Q(x) + R_{L,n}(x)), \qquad x \in J_\varepsilon,$$

and

$$w^n(x)|Q_{[n/\lambda]}(x)| = e^{o(n)}, \qquad x \in \Sigma \cap J^*, \tag{1.11}$$

where now J^* is an interval satisfying (1.8).

Let $1/\lambda < \tau < 1$. Since $0 \leq n - \lambda[n/\lambda] \leq \lambda$, and the family of functions $\{g_{L,n} - sQ \mid n \geq n_{\varepsilon,L},\ 0 \leq s \leq \lambda\}$ (considered on J_ε) is compact, for every large n there are polynomials $S_{[(\tau-1/\lambda)n]}$ of degree at most $[(\tau - 1/\lambda)n]$ such that

$$|S_{[(\tau-1/\lambda)n]}(x) - f(x)\exp(-g_{L,n}(x) + (n - \lambda[n/\lambda])Q(x))|$$
$$\leq \exp(-g_{L,n}(x) + (n - \lambda[n/\lambda])Q(x))/L, \qquad x \in J_{2\varepsilon},$$
$$|S_{[(\tau-1/\lambda)n]}(x)| \leq f(x)\exp(-g_{L,n}(x) + (n - \lambda[n/\lambda])Q(x)), \qquad x \in J_\varepsilon \setminus J_{2\varepsilon},$$

and

$$|S_{[(\tau-1/\lambda)n]}(x)| \leq 1, \qquad x \in J^* \setminus J_\varepsilon. \tag{1.12}$$

Since the disjoint sets $J^* \setminus J_\varepsilon$ and $J_{2\varepsilon}$ consist of finitely many intervals, it follows from Corollary VI.3.6 that there is a $0 < c < 1$ and, for each m, polynomials R_m of degree at most m such that

$$|R_m(x) - 1| \le c^m \qquad \text{for } x \in J_{2\varepsilon}, \tag{1.13}$$

$$|R_m(x)| \le c^m \qquad \text{for } x \in J^* \setminus J_\varepsilon, \tag{1.14}$$

and

$$0 \le R_m(x) \le 1 \qquad \text{for } x \in J_\varepsilon \setminus J_{2\varepsilon}. \tag{1.15}$$

Finally, we set

$$P_n(x) = Q_{[n/\lambda]}(x) S_{[(\tau - 1/\lambda)n]}(x) R_{[(1-\tau)n]}(x),$$

which has degree at most n.

If $\eta > 0$ is given, then first choose $\varepsilon > 0$ so that the maximum of f outside $J_{2\varepsilon}$ is smaller than η; then choose $\lambda > 1$ as above, and finally choose L large enough to have $C_\varepsilon/L < \eta$. Then our estimates show that for sufficiently large n the difference $\big|w^n|P_n| - f\big|$ is at most 3η on the set $J^* \cap \Sigma$. In fact, this estimate is clear on J_ε, and by (1.11), (1.12), and (1.14), the weighted polynomial $w^n P_n$ is exponentially small on $(\Sigma \cap J^*) \setminus J_\varepsilon$ which implies the claim on the rest of $J^* \cap \Sigma$. But by (1.8) and Theorem III.2.1 the same is true on $\Sigma \setminus J^*$ provided n is sufficiently large. This proves that

$$\big|w^n|P_n| - f\big| \le 3\eta$$

for every $x \in \Sigma$ provided n is sufficiently large. This is what we need to prove.

V. Thus, we only have to verify (1.9) and (1.10). Since $w(x)$ and $\exp(U^{\mu_w}(x))$ differ on S_w only in a multiplicative constant, and elsewhere the weight $w(x)$ is smaller than $\exp(U^{\mu_w}(x))$ times this constant, it is enough to show (1.9) and (1.10) with w replaced by $\exp(U^{\mu_w})$.

The rest of the proof consists of a discretization technique for logarithmic potentials. Let I be the smallest interval containing S_w. Without loss of generality we may assume $I = [-1, 1]$. Partition $I = [-1, 1]$ by the points $-1 = t_0 < t_1 < \cdots < t_n = 1$ into n intervals I_j, $j = 0, 1, \ldots, n-1$, with $\mu_w(I_j) = 1/n$ (this is always possible because μ_w is a continuous measure, for the existence of a mass point would contradict the fact that μ_w has finite logarithmic energy). Since v is continuous and positive in S^w, there are two constants c, C (depending on ε) such that if $I_j \cap J_{\varepsilon^2} \neq \emptyset$, then

$$c/n \le |I_j| \le C/n. \tag{1.16}$$

Let ξ_j be the *weight point* of the restriction of μ_w to I_j, i.e. $\xi_j \in I_j$ is the point that satisfies

$$n \int_{I_j} t \, d\mu_w(t) = \xi_j, \tag{1.17}$$

and set

$$Q_n(t) = \prod_j (t - iL/n - \xi_j).$$

We claim that this choice will satisfy (1.9) and (1.10) (with w replaced by $\exp(U^{\mu_w})$).

First consider the partial derivative of $U^{\mu_w}(z)$ at $z = x + iy$ with respect to y:

$$\frac{\partial U^{\mu_w}(z)}{\partial y} = -\int_{-1}^{1} \frac{y}{(x-t)^2 + y^2} d\mu_w(t).$$

The so-called Poisson kernel $-y/((x-t)^2 + y^2)$, is nonnegative for $y < 0$, its integral over $(-\infty, \infty)$ is π, and it converges uniformly to zero outside every interval $(x - \eta, x + \eta)$, $\eta > 0$, as $y \to 0-$. Recall now that in S^w, which is a neighborhood of J_ε, we have $d\mu_w(t) = v(t)dt$ for some continuous function v, and then it is an easy exercise to prove that for $x \in J_\varepsilon$

$$\frac{\partial U^{\mu_w}(z)}{\partial y} = -\int_{-1}^{1} \frac{y}{(x-t)^2 + y^2} d\mu_w(t) \to \pi v(x) \tag{1.18}$$

as $y \to 0-$, and this convergence is uniform in $x \in J_\varepsilon$. This and the mean value theorem imply that

$$U^{\mu_w}(x) - U^{\mu_w}(x - iL/n) = \frac{\pi L v(x)}{n} + o\left(\frac{L}{n}\right) \tag{1.19}$$

uniformly in $x \in J_\varepsilon$.

We know from Theorem I.5.1 that the potential U^{μ_w} is continuous on the whole complex plane (see the assumption (1.1) at the beginning of the present section); hence

$$|U^{\mu_w}(x - iL/n) - U^{\mu_w}(x)| = o(1) \tag{1.20}$$

as $n \to \infty$ uniformly in $x \in \mathbf{R}$.

Let $\mu_n(t) = \mu_w(t - iL/n)$. Then the preceding two estimates give a bound for the difference $U^{\mu_w} - U^{\mu_n}$ on J_ε and on $\mathbf{R}$. Next we estimate for $x \in J_\varepsilon$, $x \in I_{j_0}$

$$\left|\log|Q_n(x)| + nU^{\mu_n}(x)\right| \tag{1.21}$$

$$= \left|\sum_{j=0}^{n-1} n \int_{I_j} \left(\log|x - iL/n - t| - \log|x - iL/n - \xi_j|\right) d\mu_w(t)\right|.$$

Here the integrand is

$$\log\left|1 + \frac{\xi_j - t}{x - iL/n - \xi_j}\right| = \operatorname{Re}\log\left(1 + \frac{\xi_j - t}{x - iL/n - \xi_j}\right). \tag{1.22}$$

Since, for $|I_j| \le 1/L$, the absolute value of

$$\frac{\xi_j - t}{x - iL/n - \xi_j}, \qquad t \in I_j,$$

is at most $1/2$ for large L (check this separately for $|\xi_j - t| \le C/n$ and for the opposite case which can only occur if $I_j \cap J_{\varepsilon^2} = \emptyset$ and hence $|x - \xi_j| > \varepsilon/2$ if n is sufficiently large), it easily follows that then the right hand side in (1.22) can be written in the form

$$(\xi_j - t)\operatorname{Re}\frac{1}{x - iL/n - \xi_j} + O\left(\frac{|\xi_j - t|^2}{|x - iL/n - \xi_j|^2}\right).$$

Now since the integral of the first term on I_j against $d\mu_w(t)$ is zero because of the choice of ξ_j, we have to deal only with the second term. For it we have the upper estimate

$$O\left(\frac{(C/n)^2}{(L/n)^2+(c(j-j_0)/n)^2}\right)$$

if $I_j \cap J_{\varepsilon^2} \neq \emptyset$ and

$$O\left(\frac{|I_j|^2}{\varepsilon^2}\right)$$

otherwise (recall that $x \in J_\varepsilon$).

Now we separate the terms on the right of (1.21) for which $|I_j| \geq 1/L$ (such intervals may occur for example between different subintervals of $\mathcal{S}_w$). For an L there are at most L such terms, and for every such j we must have $I_j \cap J_{\varepsilon^2} = \emptyset$ if n is large (recall (1.16)). Thus, if $g^*_{L,n}(x)$ is the sum of these separated terms, then the $g^*_{L,n}(x)$'s form a compact family on J_ε. For the other terms we can apply the preceding two estimates and so we can continue (1.21) as

$$\left|\log|Q_n(x)| + nU^{\mu_n}(x) - g^*_{L,n}(x)\right| \tag{1.23}$$

$$\leq C_1 \sum_{k=0}^{\infty} \frac{C^2}{L^2+c^2k^2} + C_1 \max_j |I_j| \sum_{I_j \cap J_{\varepsilon^2}=\emptyset} |I_j|\varepsilon^{-2} \leq \frac{C_\varepsilon}{L},$$

if n is sufficiently large.

Now

$$\log|Q_n(x)| + nU^{\mu_w}(x) = (\log|Q_n(x)| + nU^{\mu_n}(x)) + (nU^{\mu_w}(x) - nU^{\mu_n}(x)),$$

and here, by the preceding estimate, the first term is at most C_ε/L in absolute value, while (1.19) shows that the second term is $\pi v(x)L + o(L)$ uniformly in $x \in J_\varepsilon$ as $n \to \infty$. This gives (1.9) (recall that we are working with $\exp(U^{\mu_w})$ instead of w) with $g_{L,n}(x) = g^*_{L,n}(x) + \pi v(x)L$.

The proof of (1.10) is standard: using the monotonicity of the logarithmic function we have, for example, for $x \in I_{j_0}$, $j_0 < j < n-1$, the inequality

$$\log|x - iL/n - \xi_j| \leq n \int_{I_{j+1}} \log|x - iL/n - t| d\mu_w(t),$$

and adding these and the analogous inequalities for $j < j_0$ together one can easily deduce the estimate (see also (1.20))

$$\log|Q_n(x)| + nU^{\mu_n}(x) \leq \left| n \int_{I_{j_0-1}\cup I_{j_0}\cup I_{j_0+1}} \log|x + iL/n - t|\, d\mu_w(t)\right|$$

$$+ \left|\sum_{j=j_0,0,n-1} \log|x - iL/n - \xi_j|\right| \leq 6\log\frac{2Dn}{L}$$

for every $x \in J^*$ with D equal to the diameter of J^*. This and (1.20) prove (1.10). □

Proof of Theorem 1.7. We show that Theorem 1.7 is a consequence of Theorem 1.5.

Let $\Sigma = \cup I_j$ with a finite and disjoint union, and suppose that Q is convex on each of the I_j's; or $\Sigma \subseteq [0, \infty)$ and $xQ'(x)$ increases on every interval of Σ. It follows (see Theorem IV.1.10(d)) that then $\mathcal{S}_w$ consists of finitely many intervals, at most one lying in any of the I_j's.

We recall Theorem IV.2.5, according to which μ_w has continuous density in the interior of $\mathcal{S}_w$.

In order to be able to apply Theorem 1.5 we have to show that the density v_w of the extremal measure μ_w cannot vanish at any interior point of $\mathcal{S}_w$. But this is a consequence of Theorem IV.4.9. In fact, by this theorem we have for any $\lambda > 1$

$$\mu_w\Big|_{\mathcal{S}_{w^\lambda}} \geq \frac{1}{\lambda}\mu_{w^\lambda} + \left(1 - \frac{1}{\lambda}\right)\omega_{\mathcal{S}_w}\Big|_{\mathcal{S}_{w^\lambda}},$$

and this clearly rules out that the density of μ_w vanishes at a point x_0 unless x_0 does not belong to the interior of $\mathcal{S}_{w^\lambda}$. Thus, if $x_0 \in \mathcal{S}_w$ does belong to the interior of some $\mathcal{S}_{w^\lambda}$, $\lambda > 1$, then at x_0 the measure μ_w has positive density. But every interior point x_0 of $\mathcal{S}_w$ must belong to the interior of at least one $\mathcal{S}_{w^\lambda}$. In fact, since every $\mathcal{S}_{w^\lambda}$ consists of intervals at most one of which can lie in any I_j, it is enough to prove that in any neighborhood of any point x_0 of $\mathcal{S}_w$ there is a point x_1 lying in some $\mathcal{S}_{w^\lambda}$. Indeed, then this property and the decreasing character of the supports $\mathcal{S}_{w^\lambda}$ (Theorem IV.4.1) imply our claim concerning every point in $\mathrm{Int}(\mathcal{S}_w)$ lying in the interior of some $\mathcal{S}_{w^\lambda}$, $\lambda > 1$. But if $x_0 \in \mathcal{S}_w$ and B is any neighborhood of x_0, then there is an n and a polynomial P_n of degree at most n such that $w^n|P_n|$ attains its maximum in Σ at some point of $B \cap \Sigma$ and nowhere outside of B (Corollary IV.1.4). By continuity then the same is true of $w^{\lambda n}|P_n|$ for some $\lambda > 1$ sufficiently close to 1, and so again Corollary IV.1.4 implies that $B \cap \mathcal{S}_{w^\lambda} \neq \emptyset$. With these considerations the proof is complete. □

Proof of Theorems 1.8 and 1.9. First we show that Theorem 1.8 for a given β is equivalent to the case $\beta/2 - 1/2$ of Theorem 1.9(a). Indeed, let $t_0 = 0$, and let us first suppose that Theorem 1.9 is true. If the conclusion of Theorem 1.8 is not true, then $0 \notin Z(w)$, and there is a $\delta > 0$ such that $Z(w)$ does not intersect the interval $[-\delta, \delta]$. We also assume $\delta > 0$ to be so small that $I := [-\delta, \delta] \subset \mathcal{S}_w$ and that v satisfies on I the assumption (1.5). Let f be a continuous function on $\mathbf{R}$ with $f(0) = 1$ that vanishes outside I. Then, by Theorem 1.1, there is a sequence of weighted polynomials $w^n P_n$ uniformly converging to f on Σ.

We define

$$w_1(x) := \exp(U^{\mu_w}(x) - F_w).$$

Then $w(x) = w_1(x)$ on $\mathcal{S}_w$, so $\mu_w = \mu_{w_1}$. Furthermore, it follows from Theorem III.2.1 and (1.3) that the sequence $\{w_1^n P_n\}$ is uniformly bounded on $\mathbf{R}$. Now we define

$$w_2(x) := \sqrt{w_1(x)w_1(-x)}, \qquad x \in \Sigma \cup (-\Sigma).$$

Since $w_1^n P_n \to 0$ on $\Sigma \setminus I$, we get from the aforementioned uniform boundedness, that

$$w_2^{2n}(x)P_n(x)P_n(-x) \to f(x)f(-x)$$

uniformly on $\Sigma \cup (-\Sigma)$. But $\Sigma^* := \Sigma \cup (-\Sigma)$ is symmetric and w_2 and $P_n(x)P_n(-x)$ are even on Σ^*; therefore with the substitution $x^2 \to x$ we obtain the weight $w_3(x) := w_2(\sqrt{x})^2$ on $\Sigma^{**} := \{x^2 \mid x \in \Sigma\}$ for which $w_3^n(x)R_n(x) \to g(x)$ uniformly on Σ^{**}, where $R_n(x^2) := P_n(x)P_n(-x)$ and $g(x^2) = f(x)f(-x)$. Now 0 is the left endpoint of Σ^{**} (i.e. the smallest value in Σ^{**}), and by Theorem IV.1.10(f) we have $d\mu_{w_3}(t) = 2d\mu_{w_2}(\sqrt{t})$. If we also note that $d\mu_{w_2}(u) = (d\mu_w(u) + d\mu_w(-u))/2$, it follows from (1.5) that in a right neighborhood of 0 the measure μ_{w_3} has density $v_3(t)$ for which

$$v_3(t) = \frac{v(\sqrt{t}) + v(-\sqrt{t})}{2\sqrt{t}} = Lt^{\beta/2-1/2}(1 + o(1)),$$

and here $\beta/2 - 1/2 \neq -1/2$. But then Theorem 1.9 yields that $0 \in Z(w_3)$, which contradicts that $g(0) = f(0)^2 = 1$ and that g is the uniform limit of the weighted polynomials $w_3^n R_n$. This contradiction proves that $0 \in Z(w)$, as was claimed.

Now suppose that Theorem 1.8 holds. Again let $t_0 = 0$, and suppose to the contrary, that the conclusion of Theorem 1.9 for some $\beta \neq -1/2$ is false. Then $0 \notin Z(w)$, and there is a $\delta > 0$ such that $Z(w)$ does not intersect the interval $[0, \delta]$. We also assume $\delta > 0$ is so small that $I := (0, \delta] \subset S_w$ and that v satisfies on I the assumption (1.6). Let f be any continuous function on Σ with $f(0) \neq 0$ that vanishes outside I. Then, by Theorem 1.1, there is a sequence of weighted polynomials $w^n P_n$ uniformly converging to f on Σ. We set $w_1(x) := w(x^2)^{1/2}$ on $\Sigma^* := \{x \mid x^2 \in \Sigma\}$. Then $w_1^{2n}(x)P_n(x^2) \to f(x^2)$ uniformly on Σ^*. In a similar manner, there is a sequence $w^n(x)Q_n(x)$ of weighted polynomials uniformly converging to $f(x)/\sqrt{w(x)}$ on Σ (recall, that f vanishes outside I). Then $w_1^{2n}(x)Q_n(x^2)$ uniformly converges to $f(x^2)/w_1(x)$, i.e. $w_1^{2n+1}(x)Q_n(x^2)$ converges uniformly to $f(x^2)$ on Σ^*. Thus, we have a full sequence (for both even and odd indices) of weighted polynomials uniformly converging to $f(x^2)$ on Σ^*. But by the discussion above, if we assume (1.6), then μ_{w_1} has density $v_1(t) = tv(t^2) = L|t|^{2\beta+1}(1 + o(1))$ with $2\beta + 1 \neq 0$, so by Theorem 1.8 we have $0 \in Z(w_1)$, which contradicts the fact that the function $f(x^2)$, which is not zero at the origin, is the uniform limit of some weighted polynomials $w_1^n R_n$. This contradiction proves that Theorem 1.9 is true provided Theorem 1.8 is true.

Finally, we show that part (b) of Theorem 1.9 follows with the same kind of substitutions from Theorem 1.5. In fact, assume that $t_0 = 0$ and define $w_1(x) := w(x^2)^{1/2}$ on $\Sigma^* := \{x \mid x^2 \in \Sigma\}$. Then we conclude from $d\mu_w(t) = d\mu_{w_1}(t^2)/2$ that μ_{w_1} has continuous and positive density in a neighborhood of 0. But then, by Theorem 1.5, there is a continuous even function f on Σ^* that does not vanish at 0 which is the uniform limit of weighted polynomials $w_1^{2n} P_{2n}$ on Σ^*. By considering $(P_{2n}(x) + P_{2n}(-x))/2$ we can suppose that P_{2n} is even, say $P_{2n}(x) = R_n(x^2)$, and then the substitution $x^2 \to x$ shows that we have $w^n R_n \to g$ uniformly on Σ, where $g(x) = f(\sqrt{x})$. Since $g(0) \neq 0$, it again follows that $0 \notin Z(w)$ as was claimed.

Thus, so far we have shown that part (b) of Theorem 1.9 is true; and also that for the remaining statements it is enough to prove for each $\beta \neq 0$ either Theorem 1.8 with β or Theorem 1.9 with $\beta/2 - 1/2$. Below we shall prove Theorem 1.9 for $\beta > -1/2$ and Theorem 1.8 for $\beta < 0$, which together will complete the proofs of both Theorems 1.8 and Theorem 1.9. □

The proofs of the aforementioned cases ($\beta > -1/2$ and $\beta < 0$) have distinct character; therefore these two cases will be handled separately.

Proof of Theorem 1.9 for $\beta > -1/2$. Without loss of generality we assume $t_0 = 0$. Also, approximation on Σ implies approximation on S_w; therefore we may assume that $\Sigma = S_w$. This implies, in particular, that Σ is compact, and that we have

$$w(x) = \exp(U^{\mu_w}(x) - F_w) \tag{1.24}$$

for all $x \in \Sigma$ (see (1.2)).

First we formulate the following extremal problem. Define for every n

$$\lambda_n := \inf\{\|w^n P_n\| \mid w^n(0)P_n(0) = 1, \deg(P_n) \leq n\}.$$

Theorem 1.9 follows from the following lower bound on the numbers λ_n:

$$\liminf_{n\to\infty} \lambda_{2n} > 1. \tag{1.25}$$

Indeed, this shows that no continuous function f with $0 \leq f \leq 1$ and $f(0) = 1$ is approximable with arbitrary precision by weighted polynomials; hence, by Theorem 1.1, we must have $0 \in Z(w)$.

Instead of working with polynomials P_n satisfying $w^n(0)P_n(0) = 1$ it is more convenient to work with monic polynomials. We define the weight

$$W(x) := \frac{w(1/x)}{|x|w(0)}, \qquad 1/x \in \Sigma, \tag{1.26}$$

and the extremal error

$$E_n := \inf\{\|W^n Q_n\|_{\Sigma^*} \mid Q_n(x) = x^n + \cdots\}, \tag{1.27}$$

where $\Sigma^* := \{1/x \mid x \in \Sigma\}$ is the image of Σ under the mapping $x \to 1/x$. Note that Σ^* contains a whole interval $[a, \infty)$ on which W behaves like $1/x$, so this W is not admissible in the sense of Section I.1; however, this fact will not cause any trouble below. We claim that $E_{2n} = \lambda_{2n}$. Indeed, if P_{2n} is a polynomial of degree $2n$ satisfying $w^{2n}(0)P_{2n}(0) = 1$, then the monic polynomial Q_{2n} given by

$$Q_{2n}(x) := w^{2n}(0)x^{2n}P_{2n}(1/x)$$

satisfies $W^{2n}(x)Q_{2n}(x) = w^{2n}(1/x)P_{2n}(1/x)$, and this correspondence can be reversed. Now $E_{2n} = \lambda_{2n}$ is immediate from the definition of these quantities.

Thus, Theorem 1.9 will be proven if we can show that

$$\liminf_{n\to\infty} E_n > 1. \tag{1.28}$$

This will be achieved via a de La Vallée Poussin type argument. We will construct monic polynomials $Q_n(x) = \prod_{j=1}^{n}(x - \zeta_j)$, $n \geq n_0$, whose zeros satisfy

$$\zeta_1 < \zeta_2 < \cdots < \zeta_n, \tag{1.29}$$

such that there exist numbers

$$x_1 < \zeta_1 < x_2 < \cdots < x_n < \zeta_n < x_{n+1} \tag{1.30}$$

with the property that

$$|W^n(x_j)Q_n(x_j)| \geq C > 1, \qquad j = 1, \ldots, n+1, \tag{1.31}$$

where $C > 1$ is a constant that does not depend on n and j (actually the x_j's also depend on n, but for simplicity we shall write x_j instead of $x_{j,n}$). From this (1.28) easily follows: if Q_n^* is another monic polynomial, then $R_{n-1} := Q_n - Q_n^*$ is of degree at most $n-1$; furthermore, $W^n(x_j)|Q_n^*(x_j)| < C$ for all j would imply that at each point x_j the polynomial R_{n-1} would have the same sign as Q_n, i.e. it is alternately positive and negative at the points x_j. But then R_{n-1} would have a zero on each of the intervals (x_j, x_{j+1}), $j = 1, 2, \ldots, n$. This accounts for n zeros, which is possible only if $R_{n-1} \equiv 0$, i.e. $Q_n \equiv Q_n^*$. This would, however, contradict (1.31) and the assumption $W^n(x_j)|Q_n^*(x_j)| < C$. Hence, for every Q_n^* we have $W^n(x_j)|Q_n^*(x_j)| \geq C$ for at least one j, and this implies (1.28).

Thus, it remains to construct the points ζ_j and x_j with the properties (1.29)–(1.31).

From (1.24) we obtain for $x \in \Sigma^*$,

$$\begin{aligned}
\log W(x) &= \log w(1/x) - \log w(0) - \log|x| \\
&= U^{\mu_w}(1/x) - U^{\mu_w}(0) - \log|x| \\
&= -\int \log\left|\frac{1}{x} - t\right| d\mu_w(t) + \int \log|t|\, d\mu_w(t) - \log|x| \\
&= -\int \log\left|x - \frac{1}{t}\right| d\mu_w(t).
\end{aligned}$$

From here we find via the transformation $t = 1/s$

$$\log W(x) = \int \log\frac{1}{|x-s|}\, d\nu(s) = U^{\nu}(x), \qquad x \in \Sigma^*, \tag{1.32}$$

where the measure $d\nu(s)$ is the image of $d\mu_w(t)$ under the mapping $t = 1/s$. Since in a right neighborhood of 0 the measure μ_w has density v that satisfies (1.6), for some $a > 0$ on the interval $[a, \infty)$ the measure ν has the form

$$dv(s) = u(s)ds, \qquad s \in [a, \infty), \tag{1.33}$$

with density u satisfying

$$u(s) = L\alpha s^{-1-\alpha}(1 + o(1)), \qquad s \to \infty, \tag{1.34}$$

for some constant L. Here $\alpha := \beta + 1 > 1/2$.

Next we discretize this ν using the technique applied in the proof of Theorem 1.5 above. In fact, let $x_1 < x_2 < \cdots < x_n$ be points such that

$$\nu([x_j, x_{j+1}]) = 1/n, \quad j = 1, \dots, n-1, \qquad \nu([x_n, \infty)) = 1/n. \tag{1.35}$$

These points also depend on n, but for brevity we write x_j instead of $x_{j,n}$. We can also assume $x_1 \in \Sigma_1$, so $x_1 = x_{1,n}$ does not tend to $-\infty$. Then we take $\zeta_1, \dots, \zeta_{n-1}$ to be the corresponding weight points on the intervals $[x_j, x_{j+1}]$, i.e.

$$\zeta_j = n \int_{x_j}^{x_{j+1}} s \, d\nu(s), \qquad j = 1, \dots, n-1. \tag{1.36}$$

We have not yet defined ζ_n and x_{n+1}. These two numbers will be chosen below as $\zeta_n = Ax_n$ and $x_{n+1} = Bx_n$ for some appropriate A and B, but for now let ζ_n be an arbitrary number $> x_n$. Let μ_n be the normalized counting measure on the set $\{\zeta_1, \dots, \zeta_n\}$. Then $\log|Q_n(x)| = -U^{\mu_n}(x)$, so the inequalities (1.31) are equivalent to

$$U^{\nu}(x_j) - U^{\mu_n}(x_j) \ge \frac{C_1}{n}, \qquad j = 1, \dots, n+1, \tag{1.37}$$

where C_1 is a positive constant independent of j and n. We write

$$U^{\nu}(x) - U^{\mu_n}(x) = \sum_{k=1}^{n} \Delta_k(x) \tag{1.38}$$

with

$$\Delta_k(x) := \int_{x_k}^{x_{k+1}} \log\left|\frac{x - \zeta_k}{x - s}\right| d\nu(s), \qquad k = 1, \dots, n-1, \tag{1.39}$$

and

$$\Delta_n(x) := \int_{x_n}^{\infty} \log\left|\frac{x - \zeta_n}{x - s}\right| d\nu(s). \tag{1.40}$$

We have seen in the proof of Theorem 1.5 that the convexity of $\log 1/x$ implies that for $k \le n-1$ and $x \notin (x_k, x_{k+1})$ we have

$$\Delta_k(x) \ge 0;$$

in particular

$$\Delta_k(x_j) \ge 0, \qquad \text{for all } k \le n-1 \text{ and } 1 \le j \le n+1. \tag{1.41}$$

Before going on with the definition of ζ_n and x_{n+1}, it is convenient to state and prove tho following lemma.

Lemma 1.10. *If*

$$L := \frac{1}{\alpha} \lim_{s\to\infty} u(s)s^{1+\alpha}, \tag{1.42}$$

then

$$\lim_{n\to\infty} \frac{x_n^\alpha}{n} = L, \qquad x_n = x_{n,n}, \tag{1.43}$$

and

$$\lim_{n\to\infty} \sup_{t\in[1,\infty)} \left| \frac{nx_n u(x_n t)}{\alpha t^{-1-\alpha}} - 1 \right| = 0. \tag{1.44}$$

Note that the limit in (1.42) exists because of (1.34).

Proof of Lemma 1.10. It is clear that $\lim_{n\to\infty} x_n = \infty$. Let $\varepsilon > 0$. From (1.42) it follows that for n large enough, we have

$$|u(s) - L\alpha s^{-1-\alpha}| < \varepsilon\alpha s^{-1-\alpha}, \qquad s \geq x_n. \tag{1.45}$$

Integrating this inequality from x_n to ∞, we find

$$\left| \int_{x_n}^\infty u(s)\,ds - \int_{x_n}^\infty L\alpha s^{-1-\alpha} ds \right| \leq \varepsilon \int_{x_n}^\infty \alpha s^{-1-\alpha} ds,$$

that is

$$\left| \frac{1}{n} - Lx_n^{-\alpha} \right| \leq \varepsilon x_n^{-\alpha}. \tag{1.46}$$

This proves (1.43). We also find from (1.45) that

$$\left| \frac{u(x_n t)}{\alpha(x_n t)^{-1-\alpha}} - L \right| < \varepsilon, \qquad t \in [1, \infty).$$

Combining this with (1.43) and letting $\varepsilon \to 0$ we obtain (1.44). □

We now return to the proof for $\beta > -1/2$ by giving the definitions of ζ_n and x_{n+1}. We shall choose these points in such a way that $n\Delta_n(x_j) \geq C_1$ will be true for all j and this together with (1.41) imply (1.37), by which the proof will be complete.

We take $\zeta_n = Ax_n$, where the constant $A > 1$ has to be determined in such a way that, for every $x \in \Sigma^* \cap (-\infty, x_n]$, we have $n\Delta_n(x) \geq C_1$ for some positive constant C_1. Choose an R such that $\Sigma^* \cap (-\infty, x_n] \subset [-R, x_n]$. For every x from this interval we have

$$\Delta_n(x) = \int_{x_n}^\infty \log\left(\frac{\zeta_n - x}{s - x}\right) u(s)\,ds.$$

Substitute $s = x_n t$ and take $x = x_n y$, where $-R/x_n \leq y \leq 1$, to obtain

$$n\Delta_n(x_n y) = \int_1^\infty \log\left(\frac{A - y}{t - y}\right) nx_n u(x_n t)\,dt.$$

From (1.44) it follows that

$$\lim_{n\to\infty} n\Delta_n(x_n y) = \int_1^\infty \log\left(\frac{A-y}{t-y}\right)\alpha t^{-1-\alpha}dt, \tag{1.47}$$

and it is easy to see that the convergence is uniform in $y \in [-\delta, 1]$ for every $\delta > 0$. Note that this latter interval contains $[-R/x_n, 1]$ for all sufficiently large n.

We proceed with the calculation of the integral in (1.47) for $y \in [0, 1)$. A change of variables $t \mapsto 1/t$ and an integration by parts leads to

$$\log\left(\frac{A-y}{1-y}\right) - \int_0^1 t^{\alpha-1}(1-ty)^{-1}dt,$$

which can be expanded into a power series as

$$\left(\log A - \frac{1}{\alpha}\right) + \sum_{k=1}^{\infty}\left(\frac{1}{k} - \frac{1}{\alpha+k} - \frac{1}{kA^k}\right)y^k.$$

Now it is easy to show that for $A > \exp(1/\alpha)$ all coefficients in this series are positive. So for such A, the right-hand side of (1.47) is above a positive constant for all $y \in [0, 1)$. But then, by continuity, it will also be positive for all $y \in [-\delta, 1]$ for some $\delta > 0$ as well. Thus, we can deduce from the uniform convergence in (1.47) that

$$\lim_{n\to\infty} n\Delta_n(x) \geq C_1 > 0$$

uniformly for $x \in [-R, x_n]$.

It will also be important below that we choose $A < \alpha/(\alpha - 1)$ in case $\alpha > 1$. This is possible, for $\exp(1/\alpha) < \alpha/(\alpha - 1)$ if $\alpha > 1$, and so for $\alpha > 1$ we can choose A from the interval $(\exp(1/\alpha), \alpha/(\alpha - 1))$.

Let us summarize our findings: there exist constants $A > 1$, $C_1 > 0$, and a positive integer n_1 such that with $\zeta_n = Ax_n$ we have $\Delta_n(x) \geq C_1/n$ for $x \in [-R, x_n]$. In particular,

$$\Delta_n(x_j) \geq \frac{C_1}{n}, \qquad \text{for all } n \geq n_1 \text{ and } j = 1, 2, \ldots, n. \tag{1.48}$$

Moreover, if $\alpha > 1$, we can take $A < \alpha/(\alpha - 1)$.

Next we choose x_{n+1} as $x_{n+1} = Bx_n$ for some $B > A$. All that remains is to find a lower bound for $\Delta_n(x_{n+1})$. From

$$\Delta_n(x_{n+1}) = \int_{x_n}^\infty \log\left|\frac{Bx_n - Ax_n}{Bx_n - s}\right| u(s)\, ds$$

we obtain after the substitution $s = x_n t$

$$n\Delta_n(x_{n+1}) = \int_1^\infty \log\left|\frac{B-A}{B-t}\right| nx_n u(tx_n)\, dt.$$

Then (1.44) implies that

$$\lim_{n\to\infty} n\Delta_n(x_{n+1}) = \int_1^\infty \log\left|\frac{B-A}{B-t}\right| \alpha t^{-\alpha-1}dt = I_1 + I_2. \tag{1.49}$$

Here I_1 is the integral from 1 to B and I_2 is the integral from B to ∞. For I_1 we have

$$I_1 = \int_1^B \log\left(\frac{B-t}{B-A}\right) d(t^{-\alpha} - B^{-\alpha}).$$

After an integration by parts and the change of variable $t \mapsto Bt$ we obtain

$$I_1 = (B^{-\alpha} - 1)\log\left(\frac{B-1}{B-A}\right) + B^{-\alpha}\int_{1/B}^1 \frac{t^{-\alpha}-1}{1-t}dt. \tag{1.50}$$

On applying the change of variable $t \mapsto B/t$, we get for I_2 that

$$\begin{aligned} I_2 &= B^{-\alpha}\int_0^1 \log\left(\frac{1-A/B}{1/t-1}\right)\alpha t^{\alpha-1}dt \\ &= B^{-\alpha}\left(\log(1-\frac{A}{B}) + \int_0^1 (\log t)\alpha t^{\alpha-1}dt - \int_0^1 \log(1-t)\alpha t^{\alpha-1}dt\right), \end{aligned}$$

which yields via integration by parts (cf. the derivation of (1.50)) that

$$I_2 = B^{-\alpha}\left(\log(1-A/B) + \int_0^1 \frac{1-t^\alpha}{1-t}dt - \frac{1}{\alpha}\right).$$

Now we distinguish 3 cases: $\alpha = 1$, $1/2 < \alpha < 1$ and $\alpha > 1$. In the following relations all o and O symbols are for $B \to \infty$.

I. $\alpha = 1$. In this case

$$I_2 = B^{-1}\log(1-A/B) = O(B^{-2}),$$

while (1.50) becomes

$$I_1 = (B^{-1} - 1)\log\left(\frac{B-1}{B-A}\right) + B^{-1}\log B = B^{-1}\log B + O(B^{-1}).$$

Thus, for large B the sum of I_1 and I_2 is positive.

II. $1/2 < \alpha < 1$. In this case we note that

$$\begin{aligned} I_1 &= (B^{-\alpha} - 1)\log\left(\frac{B-1}{B-A}\right) + B^{-\alpha}\int_0^1 \frac{t^{-\alpha}-1}{1-t}dt + O(B^{-1}) \\ &= B^{-\alpha}\int_0^1 \frac{t^{-\alpha}-1}{1-t}dt + O(B^{-1}); \end{aligned}$$

therefore,

$$\begin{aligned} I_1 + I_2 &= B^{-\alpha}\left[\int_0^1 \frac{1-t^\alpha}{1-t}dt + \int_0^1 \frac{t^{-\alpha}-1}{1-t}dt - \frac{1}{\alpha}\right] + O(B^{-1}) \\ &= B^{-\alpha}\left[\int_0^1 \frac{t^{-\alpha}-t^\alpha}{1-t}dt - \frac{1}{\alpha}\right] + O(B^{-1}). \end{aligned}$$

The expression in the square bracket is obviously increasing as α increases; furthermore, for $\alpha = 1/2$ it is zero. Hence, for $\alpha > 1/2$ it is positive, and we conclude again that for sufficiently large B we have $I_1 + I_2 > 0$ (note that this part of the proof would fail for $\alpha \leq 1/2$).

III. $\alpha > 1$. We write

$$\begin{aligned} \int_{1/B}^1 \frac{t^{-\alpha}-1}{1-t}dt &= \int_{1/B}^1 t^{-\alpha}dt + \int_{1/B}^1 \frac{t^{-\alpha+1}-1}{1-t}dt \\ &= B^{\alpha-1}\left(\frac{1}{\alpha-1} + o(1)\right), \end{aligned}$$

so (1.50) gives

$$\begin{aligned} I_1 &= -\log\left(\frac{B-1}{B-A}\right) + B^{-1}\left(\frac{1}{\alpha-1} + o(1)\right) \\ &= B^{-1}\left(1 - A + \frac{1}{\alpha-1} + o(1)\right). \end{aligned}$$

Furthermore, $I_2 = O(B^{-\alpha})$. So

$$I_1 + I_2 = B^{-1}(1 - A + 1/(\alpha-1) + o(1)).$$

Since for $\alpha > 1$ we have chosen A to satisfy $A < \alpha/(\alpha-1)$, it follows that $I_1 + I_2 > 0$ for sufficiently large B.

Thus, in all three cases, $I_1 + I_2 > 0$ for large enough B. Together with (1.49) we have proved that for sufficiently large B and $x_{n+1} = Bx_n$

$$\Delta_n(x_{n+1}) > \frac{C_2}{n}, \qquad \text{for all } n \geq n_2.$$

Taking into account (1.41) and (1.48) we see that

$$\sum_{k=1}^n \Delta_k(x_j) \geq \frac{C_3}{n}$$

for all j with some $C_3 > 0$, and this completes the proof. □

Proof of Theorem 1.8 for $\beta < 0$. We suppose again that $t_0 = 0$. First we claim that with $\lambda = \beta + 1$ and some $B > 0$ we have

$$w(x) = w(0)\exp(-B|x|^\lambda + R(x)), \tag{1.51}$$

where $R(x) = o(|x|^\lambda)$ as $x \to 0$. In fact, by (1.2) we have for $x \in \mathcal{S}_w$,

$$\log w(x) - \log w(0) = U^{\mu_w}(x) - U^{\mu_w}(0) = -\int \log\left|\frac{x-t}{t}\right| d\mu_w(t).$$

Using (1.5) we can find a $\delta > 0$ such that the density v exists on $(-\delta, \delta)$ and satisfies

$$v(t) < 2L|t|^\beta, \qquad t \in (-\delta, \delta). \tag{1.52}$$

Then for $x \in \mathcal{S}_w$,

$$\log w(x) - \log w(0) = -\int_{-\delta}^{\delta} \log\left|\frac{x-t}{t}\right| v(t)\,dt - \int_{\mathcal{S}_w\setminus(-\delta,\delta)} \log\left|\frac{x-t}{t}\right| d\mu_w(t).$$

The second integral is clearly $O(|x|)$ as $x \to 0$. To estimate the first one we put $t = xs$ to get (assuming $x > 0$)

$$-\int_{-\delta}^{\delta} \log\left|\frac{x-t}{t}\right| v(t)\,dt = -\int_{-\delta/x}^{\delta/x} \log\left|\frac{1-s}{s}\right| v(xs)x\,ds,$$

which can be written in the form

$$-x^\lambda \int_{-\delta/x}^{\delta/x} \log\left|\frac{1-s}{s}\right| \left[(x|s|)^{1-\lambda} v(xs)\right] |s|^{\lambda-1} ds.$$

For $s \in \mathbf{R}$, we have $(x|s|)^{1-\lambda}v(xs) \to L$ as $x \to 0$ (because of (1.5)). Furthermore, by (1.52), the integrands are dominated by the integrable function

$$2L\left|\log\left|\frac{1-s}{s}\right|\right| |s|^{\lambda-1}, \qquad s \in \mathbf{R}.$$

Therefore, by Lebesgue's dominated convergence theorem,

$$\lim_{x\to 0}\int_{-\delta/x}^{\delta/x} \log\left|\frac{1-s}{s}\right| \left[(x|s|)^{1-\lambda} v(xs)\right] |s|^{\lambda-1} ds = LI,$$

where

$$I := \int_{-\infty}^{\infty} \log\left|\frac{1-s}{s}\right| |s|^{\lambda-1} ds. \tag{1.53}$$

It remains to prove that $I > 0$. Since both functions $\log 1/|s|$ and $|s|^{\lambda-1}$ peak at the origin and decrease as we move away from it, it is easy to see that there is a $c > 0$ such that for all $M \geq 1$

$$-\int_{-M}^{M} \log|s||s|^{\lambda-1}\,ds + \int_{-M}^{M} \log|s-1||s|^{\lambda-1}\,ds \geq c.$$

Since the limit for $M \to \infty$ of the left-hand side is exactly I, the inequality $I \geq c > 0$ follows.

After this preparation we return to the assertion of the theorem. If it is not true, then $0 \notin Z(w)$, so there is a $\delta > 0$ such that $Z(w) \cap (-\delta, \delta) = \emptyset$. We take δ so small that $[-\delta, \delta] \subset \Sigma$ and $|R(x)| \leq B|x|^\lambda$ on $[-\delta, \delta]$. Let f be a continuous function that is identically one on $[-\delta/2, \delta/2]$ and vanishes outside $[-\delta, \delta]$. Then f vanishes on $Z(w)$, so there is a sequence $w^n P_n$ of weighted polynomials converging to f uniformly on $[-\delta, \delta]$. Then

$$\| \exp(-Bn|x|^\lambda + nR(x))P_n(x) - 1\|_{[-\delta/2,\delta/2]} \to 0. \tag{1.54}$$

In particular, by the choice of δ we have for sufficiently large n

$$\exp(-nD|x|^\lambda)|P_n(x)| \leq 2, \qquad x \in [-\delta/2, \delta/2], \tag{1.55}$$

with $D \geq 2B$. Here D can be replaced by any larger constant; hence we suppose D is so large that the extremal support of the weight function $\exp(-D|x|^\lambda)$, which, by formula (IV.5.4), is given by

$$[-\gamma_\lambda^{1/\lambda} D^{-1/\lambda}, \gamma_\lambda^{1/\lambda} D^{-1/\lambda}],$$

is contained in $[-\delta/2, \delta/2]$. Now it follows from Theorem III.2.1 that

$$\exp(-nD|x|^\lambda)|P_n(x)| \leq 2$$

for all $x \in \mathbf{R}$. Hence the polynomials $Q_n(x) := P_n(n^{-1/\lambda}x)$ satisfy

$$\exp(-D|x|^\lambda)|Q_n(x)| \leq 2, \qquad x \in \mathbf{R}. \tag{1.56}$$

Furthermore, since for every fixed $x \in \mathbf{R}$ we have $nR(n^{-1/\lambda}x) \to 0$, it follows from (1.54) that

$$\exp(-B|x|^\lambda)Q_n(x) \to 1 \tag{1.57}$$

as $n \to \infty$ pointwise on $\mathbf{R}$.

Now we apply Theorem 5.5 and deduce that for $W(x) = \exp(-D|x|^\lambda)$ there is a constant K such that

$$\|WP'\|_{\mathbf{R}} \leq K\|WP\|_{\mathbf{R}}$$

for all polynomials P. Using this repeatedly for Q_n and its derivatives, we obtain from (1.56) that for any $j \geq 0$

$$|Q_n^{(j)}(0)| \leq 2K^j,$$

which immediately implies that for any z in the complex plane and any n

$$|Q_n(z)| = \left| \sum_{j=0}^{\infty} \frac{Q_n^{(j)}(0)}{j!} z^j \right| \leq 2e^{K|z|},$$

i.e. the sequence $\{Q_n(z)\}$ is uniformly bounded on compact subsets of $\mathbf{C}$. Thus, it has a subsequence that converges to an entire function h uniformly on compact subsets of $\mathbf{C}$. On the other hand, we have seen in (1.57) that for every $x \in \mathbf{R}$ the $Q_n(x)$'s converge to $\exp(B|x|^\lambda)$. However, these statements contradict each other, for then we must have $h(x) = \exp(B|x|^\lambda)$, $x \in \mathbf{R}$, and so h cannot be analytic at the origin. This contradiction proves Theorem 1.8 for $\beta < 0$. □

VI.2 Approximation with Varying Weights

In several problems weighted polynomials of the form $W_n P_n$ appear, where $\{W_n\}$ is a sequence of weights, i.e. the weights are not powers of a fixed weight function. In such a case we set $w_n = W_n^{1/n}$ and consider weighted polynomials $w_n^n P_n$ with varying weights w_n. The method of the preceding section yields convergence results in this setting, as well.

Theorem 2.1. *Suppose that* $\{w_n\}$ *is a sequence of weights such that the extremal support* S_{w_n} *is* $[0, 1]$ *for all* n, *and let* O *be an open subset of* $(0, 1)$ *for which the set* $[0, 1] \setminus O$ *is of zero capacity. If the equilibrium measures* μ_{w_n} *are absolutely continuous with respect to Lebesgue measure on* O: $\mu_{w_n}(x) = v_n(x)dx$, *and the densities* v_n *are uniformly equicontinuous and uniformly bounded from below by a positive constant on every compact subset of* O, *then every continuous function that vanishes outside* O *can be uniformly approximated on* $[0, 1]$ *by weighted polynomials* $w_n^n P_n$, $\deg P_n \le n$.

Actually, the sequence $\{w_n^n P_n\}$ can be constructed in such a way that the convergence $w_n^n P_n \to f$ holds uniformly on some larger set $[-\theta, 1+\theta]$, $\theta > 0$ (provided of course the weights are defined there). This easily follows from the proof.

The conclusion is false for every O for which $[-1, 1] \setminus O$ is not of positive capacity, but we shall not prove this. For conditions directly on the w_n themselves that guarantee the assumptions in the theorem, see Theorem 2.3 below.

Proof of Theorem 2.1. We set $\Sigma = J^* = [0, 1]$, $J_\varepsilon = \{x \mid (x - \varepsilon, x + \varepsilon) \subseteq O\}$, and copy the proof of Theorem 1.5. This can be done word for word with one exception, and this is the estimate (1.10). In fact, the heart of the proof of Theorem 1.5 is Theorem IV.4.10, which is valid in the following form with the same proof: *If* v_n *are uniformly equicontinuous on an interval* $[x_0 - \varepsilon_1, x_0 + \varepsilon_1]$ *and* $v_n(t) \ge \varepsilon_0$ *there, then for* $\lambda \le 1/(1 - \varepsilon_0\varepsilon_1/2)$ *the interval* $[x_0 - \varepsilon_1/2, x_0 + \varepsilon_1/2]$ *belongs to the interior of* $S_{w_n^\lambda}$, *and the densities* $v_{w_n^\lambda}$ *are also uniformly equicontinuous there.*

What goes wrong with (1.10)? In the present case we could claim (1.10) only under the assumption that the potentials $U^{\mu_{w_n}}(z)$ are uniformly equicontinuous on $[0, 1]$ as functions of the complex variable z (cf. (1.20)), and this may not be true.

In any case we have (1.10) in the form

$$w_n^n(x)|Q_n(x)| \le C_0 e^{C_0 n}. \tag{2.1}$$

In Section VI.1 the estimate (1.10) is used in conjunction with the estimates (1.13)–(1.15), i.e. with

$$|R_m(x) - 1| \le c^m, \qquad \text{for } x \in J_{2\varepsilon}, \tag{2.2}$$

$$|R_m(x)| \le c^m, \qquad \text{for } x \in J^* \setminus J_\varepsilon, \tag{2.3}$$

and

$$0 \le R_m(x) \le 1, \qquad \text{for } x \in J_\varepsilon \setminus J_{2\varepsilon}, \tag{2.4}$$

where c was *some* positive number less than 1. Now if this c was actually smaller than $\exp(-C_0)$ from (2.1), then the proof in Section VI.1 would be valid in the present case, as well. The proof also shows that any J_η with some fixed but small $\eta > 2\varepsilon$ can stand in (2.2) and (2.4) instead of $J_{2\varepsilon}$; furthermore, in (2.2) we do not actually need geometric convergence, i.e. (2.2) can be replaced by

$$|R_m(x) - 1| = o(1) \qquad \text{for } x \in J_\eta, \tag{2.5}$$

as $m \to \infty$. It is also clear from the proof that (2.4) can be replaced by the uniform boundedness of $R_m(x)$:

$$|R_m(x)| \le 2 \qquad \text{for all } m \text{ and } x \in J^*. \tag{2.6}$$

Thus, it is enough to show that in the present case for *arbitrary* $\eta > 0$ *and* $c > 0$ we can choose an $\varepsilon > 0$ such that (2.3), (2.5), and (2.6) hold for some polynomials R_m of degree at most m whenever m is sufficiently large.

The assumption that $[0, 1] \setminus O$ has zero capacity implies that the capacity of $J^* \setminus J_\varepsilon$ tends to zero; hence our claim follows from the next lemma by setting $S = J_\eta$ and $K = J^* \setminus J_{\eta/2}$ if we apply it to the sets $L = J^* \setminus J_\varepsilon$ with $\varepsilon < \eta/2$, $\varepsilon \to 0$. Thus, the verification of the lemma will complete the proof of Theorem 2.1.

Lemma 2.2. *Let S and K be two disjoint compact subsets of* $[0, 1]$. *Then there is a constant $\delta > 0$ such that for all compact subsets L of K and sufficiently large n there are polynomials P_n of degree at most n such that*

$$|P_n(x)| \le 2, \qquad x \in [0, 1], \tag{2.7}$$

$$|P_n(x) - 1| \le \left(\frac{1}{2}\right)^{\delta n}, \qquad x \in S \tag{2.8}$$

and

$$|P_n(x)| \le (\mathrm{cap}(L))^{\delta n}, \qquad x \in L. \tag{2.9}$$

Proof. Let

$$T_m(z) = \prod_{j=1}^{m} (z - z_j)$$

be the polynomial of degree m that has all its zeros z_j in L and minimizes the norm $\|T_m\|_L$ among all such polynomials. We have discussed these so-called restricted Chebyshev polynomials in Section III.3, where we verified (Theorem III.3.1) that

$$\lim_{m\to\infty} \|T_m\|_L^{1/m} = \mathrm{cap}(L). \tag{2.10}$$

Since all the zeros of T_m belong to $[0, 1]$, we also have

$$|T_m(x)| \le 1, \qquad x \in [0, 1]. \tag{2.11}$$

Now let S_ρ and K_ρ be the set of points in the plane the distance of which to S and K, respectively, is at most ρ, and choose ρ so small that the closures of the sets S_ρ and K_ρ are disjoint. Consider the function $f_m(z)$ which is defined to be 1 on K_ρ and $1/T_m(z)$ on S_ρ. This f_m is analytic on $S_\rho \cup K_\rho$ and we have the bound

$$|f_m(z)| \le \left(\frac{1}{\operatorname{dist}(S_\rho, K_\rho)}\right)^m =: C_1^m.$$

Hence by a classical approximation theorem of Bernstein ([229, Theorem 5, Section 4.5, p. 75]) there is a $\tau < 1$ and there are polynomials R_k of degree at most k such that

$$|R_k(z) - f_m(z)| \le C_1^m \tau^k, \qquad z \in S_{\rho/2} \cup K_{\rho/2}.$$

We set here $k = rm$, where r is so large that $\tau^r < 1/4C_1$ holds. Thus,

$$|R_k(z) - f_m(z)| \le 4^{-m}, \qquad z \in S_{\rho/2} \cup K_{\rho/2}. \tag{2.12}$$

We also get from the Bernstein–Walsh lemma (III.2.4) that with some constant $C_2 \ge 1$

$$|R_k(x)| \le C_2^k C_1^m =: C_3^m, \qquad x \in [0, 1] \tag{2.13}$$

(note that here $C_3 \ge 1$).

We have already used in (1.13)–(1.15) Corollary 3.6 from Section VI.3 according to which if there are two disjoint systems of subintervals of [0, 1], then there are polynomials that take values between 0 and 1 on [0, 1] and geometrically converge to zero and respectively to 1 on the two systems of intervals. Thus, we can choose a $\kappa < 1$ and for all sufficiently large l polynomials Q_l of degree at most l such that

$$|Q_l(x)| \le \kappa^l, \qquad x \in [0, 1] \setminus S_{\rho/2},$$

and

$$|Q_l(x) - 1| < \kappa^l, \qquad x \in S.$$

We set here $l = sm$ with an s such that $\kappa^s < 1/2C_3$ holds, by which we get

$$|Q_l(x)| \le \left(\frac{1}{2C_3}\right)^m, \qquad x \in [0, 1] \setminus S_{\rho/2}, \tag{2.14}$$

and

$$|Q_l(x) - 1| < \left(\frac{1}{2}\right)^m, \qquad x \in S. \tag{2.15}$$

Finally, we set $P_n(x) = T_m(x)R_k(x)Q_l(x)$ which has degree at most $(r + s + 1)m$.

On S we have $f_m(x)T_m(x) = 1$; hence

$$|R_k(x)T_m(x) - 1| = |(R_k(x) - f_m(x))T_m(x)| \le 4^{-m} \cdot 1 \le 2^{-m}$$

by (2.12) and (2.11). If we take into account (2.15), then we can conclude that $|P_n(x) - 1| \le 3 \cdot 2^{-m}$ on S, which proves (2.8).

In the same fashion, on L the product $|R_k(x)Q_l(x)|$ is at most 2^{-m} by (2.13) and (2.14), and we have $|T_m(x)| < (\operatorname{cap}(L))^{m/2}$ for large enough m; hence $\|P_n\|_L \le (\operatorname{cap}(L))^{m/2}$ proving (2.9).

If $x \notin S_{\rho/2}$, then (2.11), (2.13), and (2.14) imply that $|P_n(x)| \le 1 < 2$. Finally, if $x \in S_{\rho/2}$, then the same conclusion follows from (2.15), (2.11), and (2.12), because the latter two imply

$$|T_m(x)R_k(x) - 1| = |T_m(x)|\,|R_k(x) - f_m(x)| \le 4^{-m}.$$

These inequalities yield (2.7), and the proof is complete. □

It is clear how one should modify the proof in order to achieve convergence on some $[-\theta, 1+\theta]$: all one needs to do is to add the sets $[-\theta, 0]$ and $[1, 1+\theta]$ to $L = J^* \setminus J_\varepsilon$. As $\varepsilon, \theta \to 0$ the capacity of the new L will tend to zero, and this is what was needed in the proof. This proves the remark stated immediately after the theorem. □

Next we discuss conditions directly in terms of the weights w_n. We shall always assume that the weights are normalized so that the support $\mathcal{S}_{w_n}$ of the corresponding equilibrium measure is $[0, 1]$.

Theorem 2.3. *Suppose that* $\{w_n\}$, $w_n = \exp(-Q_n)$, *is a sequence of weights such that the extremal support* $\mathcal{S}_{w_n}$ *is* $[0, 1]$ *for all* n, *on every closed subinterval* $[a, b] \subset (0, 1)$ *the functions* $\{Q_n\}$ *are uniformly of class* $C^{1+\varepsilon}$ *for some* $\varepsilon > 0$ *that may depend on* $[a, b]$, *and the functions* $tQ_n'(t)$ *are nondecreasing on* $(0, 1)$ *and there are points* $0 < c < d < 1$ *and an* $\eta > 0$ *such that* $dQ_n'(d) \ge cQ_n'(c) + \eta$ *for all* n. *Then every continuous function that vanishes outside* $(0, 1)$ *can be uniformly approximated on* $[0, 1]$ *by weighted polynomials* $w_n^n P_n$, $\deg P_n \le n$.

Being uniformly in $C^{1+\varepsilon}$ means that the derivatives satisfy uniformly a Lipschitz condition

$$|Q_n'(x) - Q_n'(y)| \le C|x - y|^\varepsilon, \qquad x \in [a, b],\ \ y \in (0, 1),$$

with constants $C = C_{a,b}$ and $\varepsilon = \varepsilon_{a,b} > 0$ independent of x and y. Note that the assumptions require $C^{1+\varepsilon}$ smoothness on Q_n only inside $(0, 1)$.

We can conclude again that $w_n^n P_n \to f$ holds uniformly on some larger set $[-\theta, 1+\theta]$, $\theta > 0$ (provided the weights are defined there).

Corollary 2.4. *Suppose that* $\{w_n\}$, $w_n = \exp(-Q_n)$, *is a sequence of even weights such that the extremal support* $\mathcal{S}_{w_n}$ *is* $[-1, 1]$ *for all* n, *and on* $[0, 1]$ *the functions satisfy the conditions of the preceding theorem. Then every continuous function that vanishes outside* $(-1, 1)$ *and also at the point* 0 *is the uniform limit on* $[-1, 1]$ *of weighted polynomials* $w_n^n P_n$, $\deg P_n \le n$.

What happens around 0 (i.e. what is the situation if the function to be approximated does not vanish at the origin) is quite complicated (see Theorems 2.5 and 2.7 for more details): if $w_n(x)$ for all n is the Freud weight $\exp(-\gamma_\alpha^{1/\alpha}|x|^\alpha)$, with $\alpha > 0$, then clearly all the conditions of the corollary are satisfied for all $\alpha > 0$, but an f that vanishes outside $(-1, 1)$ but not at the origin is approximable by weighted polynomials $w_n^n P_n$ only if $\alpha \ge 1$ (see the discussion below).

Proof of Theorem 2.3. All we have to do is to show that the conditions of Theorem 2.1 are satisfied with $O = (0, 1)$.

Let us recall from Theorem IV.3.2 the representation $d\mu_w(t) = v(t)dt$ with

$$v(t) = \frac{1}{\pi^2}\sqrt{\frac{1-t}{t}} \int_0^1 \frac{sQ'(s) - tQ'(t)}{s-t} \frac{1}{\sqrt{s(1-s)}} ds + \frac{D}{\sqrt{t(1-t)}}, \tag{2.16}$$

and

$$D = \frac{1}{\pi} - \frac{1}{\pi^2} \int_0^1 \sqrt{\frac{s}{1-s}} Q'(s)\, ds,$$

where $D \geq 0$.

Under the conditions of Theorem 2.3 the functions $sQ_n'(s)$ uniformly belong to Lip ε for some ε on every compact subinterval of $[a, b] \subset (0, 1)$, and it is well known from the theory of the singular integrals with Cauchy kernels (see e.g. the Plemelj–Privalov theorem in [165, p. 46]) that then the same is true of the integrals in (2.16) (with Q replaced by Q_n). Hence, the uniform equicontinuity of the densities v_n on compact subsets of $(0, 1)$ has been established.

It remains to show that v_n are uniformly bounded away from zero on every compact subset of $(0, 1)$. Since the second term on the right in (2.16) is nonnegative (because $D \geq 0$), it is enough to show that the first term on the right in (2.16) (again with Q replaced by Q_n) remains above a positive constant (independent of n) on compact subsets of $(0, 1)$.

We set $g(t) = tQ_n'(t)$, and it suffices to show that if g is nondecreasing on $(0, 1)$ and $g(d) \geq g(c) + \eta$, then for all $t \in (0, 1)$

$$\int_0^1 \frac{g(s) - g(t)}{s-t} \frac{1}{\sqrt{s(1-s)}} ds \geq \theta,$$

where $\theta > 0$ depends only on c, d and η. In fact, it is enough to prove this for continuously differentiable g, in which case the claim follows from the fact that

$$\int_0^1 \frac{g(s) - g(t)}{s-t} \frac{1}{\sqrt{s(1-s)}} ds \geq \int_0^1 \frac{g(s) - g(t)}{s-t} ds,$$

and here

$$\begin{aligned}
\int_0^1 \frac{g(s) - g(t)}{s-t} ds &= \int_0^1 \frac{1}{s-t} \int_t^s g'(u)\, du ds \\
&= \int_0^t g'(u) \log \frac{t}{t-u} du + \int_t^1 g'(u) \log \frac{1-t}{u-t} du \\
&\geq \alpha \int_c^d g'(u)\, du = \alpha(g(d) - g(c)) \geq \alpha\eta
\end{aligned}$$

with

$$\alpha = \min\left\{\log\frac{1}{1-c}, \log\frac{1}{d}\right\}.$$

□

Proof of Corollary 2.4. The corollary immediately follows from Theorem 2.1 (applied to the interval $[-1, 1]$ rather than to $[0, 1]$ and to the set $O = (-1, 0) \cup (0, 1)$) if we use the substitution $v_n(x) = w_n(\sqrt{x})^2$, $x \in [0, 1]$. In fact, Theorem IV.1.10(f) describes how the equilibrium measure changes with this substitution and then all we have to do is to apply the preceding proof which shows that the densities of μ_{v_n} are uniformly equicontinuous and uniformly bounded from below on compact subsets of $(0, 1)$. Hence, the same is true for μ_w on $(-1, 0) \cup (0, 1)$, and Theorem 2.1 can be applied. □

Now we apply the theorems obtained so far to solve another approximation problem with varying weights.

Let $W(x) = \exp(-Q(x))$ be a weight function on the real line. We consider the problem of approximating functions by weighted polynomials of the form $W(a_n x)P_n(x)$ for some appropriately chosen normalization constants a_n. If $W(x) = \exp(-c|x|^\alpha)$ is a Freud weight and $a_n = n^{1/\alpha}$, then $W(a_n x)P_n(x) = W(x)^n P_n(x)$; hence in this case the problem is just the one considered in the preceding section. In general, we shall set $w_n(x) = W(a_n x)^{1/n}$, so that $W(a_n x)P_n(x) = w_n(x)^n P_n(x)$, and the approximation problem in question is the one considered in this section with varying weights $\{w_n\}$.

For $W(x) = \exp(-Q(x))$ we shall always assume that Q is even, the derivative of $Q(x)$ exists in $(0, \infty)$, is nonnegative, and $xQ'(x) \nearrow \infty$ as $x \to \infty$. Let us fix n for the moment and consider the weight function $w_n(x) = W(x)^{1/n}$, i.e. for which $Q_n(x) = \log 1/w_n(x) = Q(x)/n$. On applying Corollary IV.1.13 we see that $\|WP_n\|_{\mathbf{R}} = \|WP_n\|_{[-a_n, a_n]}$, where the Mhaskar-Rakhmanov-Saff numbers a_n (that are defined for sufficiently large n) are the solutions of the equations

$$1 = \frac{2}{\pi}\int_0^1 \frac{a_n t Q'(a_n t)/n}{\sqrt{1-t^2}}\,dt.$$

In other words, the supremum norm of weighted polynomials of the form $W(x)R_n(x)$, $\deg R_n \leq n$, lives on $[-a_n, a_n]$. So, by contraction, the norm of weighted polynomials $W(a_n x)P_n(x)$ lives on $[-1, 1]$. It can also be shown (see the end of the proof of Theorem 2.5 below) that if a sequence

$$\{W(a_n x)P_n(x)\}_{n=1}^{\infty}$$

is uniformly bounded on $\mathbf{R}$, then $W(a_n x)P_n(x) \to 0$ as $n \to \infty$ for every $x \notin [-1, 1]$. Hence, if a function f is the uniform limit of such weighted polynomials, then it must vanish outside $(-1, 1)$.

Now we consider the problem: Is it true that every continuous function f that vanishes outside $(-1, 1)$ is the uniform limit of weighted polynomials $W(a_n x)P_n(x)$ (on $\mathbf{R}$, or what amounts to the same, on some interval $[-1-\theta, 1+\theta]$, $\theta > 0$)? If the answer to this problem is yes, then we say that the approximation problem for W of type II (as opposed to the problem we have considered previously) is solvable. We shall see that a special role is played by the point zero, so we start with a result in which the approximation is guaranteed with a restriction at the origin.

Theorem 2.5. *Let $xQ'(x) \nearrow \infty$ as $x \to \infty$ and assume $Q'(x) \geq 0$ for $x > 0$. Suppose that there are $C > 1$ and $\varepsilon > 0$ such that*

$$CQ'(Cx) \geq 2Q'(x)$$

and

$$Q'((1+t)x) \leq Q'(x)(1 + Ct^{\varepsilon})$$

are satisfied for $x \geq x_0$ and $0 < t < 1$. Then every continuous f that vanishes outside $(-1, 1)$ and at the origin is the uniform limit of weighted polynomials of the form $W(a_n x)P_n(x)$, $\deg(P_n) \leq n$.

This settles the approximation problem under a rather weak smoothness assumption on Q provided the condition $f(0) = 0$ is assumed for the function f to be approximated. So it remains to see what happens when f does not vanish at the origin.

It turns out that this type of approximation is closely connected with the problem of S. N. Bernstein that asks whether for every continuous g with the property

$$W(x)g(x) \to 0 \quad \text{as} \quad |x| \to \infty$$

there are polynomials S_n such that

$$\|W(g - S_n)\|_{\mathbf{R}} \to 0 \quad \text{as} \quad n \to \infty.$$

It is known (see e.g. [2, Theorems 3,5]) that in our case (i.e. when $xQ'(x)$ increases to infinity) the necessary and sufficient condition for a positive answer to Bernstein's problem is

$$\int_{-\infty}^{\infty} \frac{\log 1/W(t)}{1+t^2}\, dt = \infty.$$

Theorem 2.6. *If, in the case*

$$\int_{-\infty}^{\infty} \frac{\log 1/W(t)}{1+t^2}\, dt < \infty, \tag{2.17}$$

there are polynomials P_n of degree at most n such that

$$\|W(a_n x)P_n(x) - f(x)\|_{\mathbf{R}} \to 0 \quad \text{as} \quad n \to \infty$$

for some f that does not vanish at the origin, then $W^{-1}(z)$ must be an entire function.

On the other hand, we have

Theorem 2.7. *Let $W = e^{-Q}$ be even, and*

$$\int_{-\infty}^{\infty} \frac{Q(t)}{1+t^2}\, dt = \infty.$$

Further suppose that $Q(x)$ is twice continuously differentiable for large x, $Q'(x) > 0$ for $x > 0$ and $xQ'(x) \nearrow \infty$, and the function

$$T(x) = \frac{(xQ'(x))'}{Q'(x)},$$

lies between two positive constants:

$$0 < A \le T(x) \le B \qquad \textit{for} \quad x \ge x_0, \tag{2.18}$$

and is of slow variation in the sense that

$$\lim_{x\to\infty} \frac{T(\lambda x)}{T(x)} = 1 \qquad \textit{for all} \quad \lambda > 0.$$

Then the approximation problem of type II is solvable for W.

These results can be applied for example to Freud weights $W(x) = \exp(-c|x|^\alpha)$, $c, \alpha > 0$, in which case

$$a_n = \gamma_\alpha^{1/\alpha} c^{-1/\alpha} n^{1/\alpha}, \qquad \gamma_\alpha := \Gamma(\frac{\alpha}{2})\Gamma(\frac{1}{2})/2\Gamma(\frac{\alpha}{2}+\frac{1}{2}),$$

is the number from Section IV.5. We can conclude that for all $\alpha > 0$ every f that vanishes outside $(-1, 1)$ and at the origin is the uniform limit of weighted polynomials $W(a_n x)P_n(x)$. When $f(0) \ne 0$, then this is the case if and only if $\alpha \ge 1$.

Proof of Theorem 2.5. We set

$$Q_n(x) = Q(a_n x)/n, \quad \text{and} \quad w_n(x) = \exp(-Q_n(x)).$$

The hypotheses immediately imply that $Q_n'(x) \sim Q_n'(y)$ if $x \sim y$, $x, y \to \infty$, and that for any fixed $x_0 > 0$ we have $a_n x_0 Q'(a_n x_0) \sim n$, which can be translated as $Q_n'(x_0) \sim 1$ (here $A \sim B$ denotes that the ratio A/B stays away from zero and infinity in the range considered). Taking into account also that $xQ_n'(x)$ is nondecreasing on $(0, 1)$, we can easily conclude from the assumptions of the theorem that the Q_n's uniformly belong to $C^{1+\varepsilon}$ on every compact subinterval of $(0, 1)$; furthermore, there are constants $0 < c < d < 1$ and $\eta > 0$ such that $dQ_n'(d) \ge cQ_n'(c) + \eta$. Hence, Corollary 2.4 can be applied and we obtain the statement of the theorem, at least for concluding uniform convergence on $[-1, 1]$.

However, we know that in Corollary 2.4 the convergence is actually true on a larger set $[-1-\theta, 1+\theta]$, $\theta > 0$, and outside this interval weighted polynomials

$W(a_nx)P_n(x)$ that are bounded on $[-1, 1]$ automatically tend to zero under the given conditions. This can be seen as follows.

For any of the equilibrium measures $\mu = \mu_w$ with even $w(x) = w_n(x) = \exp(-Q(x))$, for which $xQ'(x)$ increases on $[0, \infty)$ and for which $\operatorname{supp}(\mu_w) = [-1, 1]$ we have that

$$x\,(U^\mu(x) + Q(x))' = -2\int_0^1 \frac{x^2}{x^2 - t^2}d\mu(t) + xQ'(x)$$

increases on $[1, \infty)$; furthermore, this expression is nonnegative around $x = 1$ (see Theorem I.1.3(d) and (f) in Section I.1). Since for all $t \in [1/4, 3/4]$ and $x \geq 1 + \theta/2$, $0 < \theta \leq 1$, we have

$$\frac{x^2}{x^2 - t^2} \leq \frac{(1 + \theta/4)^2}{(1 + \theta/4)^2 - t^2} - \frac{\theta}{16},$$

we conclude – see also Theorem I.1.3(f) and the fact that $\mu_w(x)'$ is bounded from below on compact subsets of $(0, 1)$ which was proved in the course of the proof of Theorem 2.3 – that for $x \geq 1 + \theta$ the sum $U^\mu(x) + Q(x) - F_w$ is at least as large as a d_θ for some $d_\theta > 0$ independent of $\mu = \mu_{w_n}$. Now we can make use of Theorem III.2.1 according to which for such x

$$|w^n(x)P_n(x)| \leq M\exp\left(n(-Q(x) - U^\mu(x) + F_w)\right) \leq \exp(-nd_\theta),$$

and here the right-hand side tends to zero. Applying this to $w(x) = w_n(x) = W(a_nx)^{1/n}$ completes the proof. □

Proof of Theorem 2.6. Suppose that $f(0) \neq 0$, say $f(0) = 1$, and there are polynomials P_n with $W(a_nx)P_n(x) - f(x)$ uniformly tending to zero on **R**. Set $f_n(x) = f(x/a_n)$ and $R_n(x) = P_n(x/a_n)$. Then

$$\|f_n - WR_n\|_{\mathbf{R}} \to 0$$

as $n \to \infty$, and hence

$$\|W(W^{-1}f_n - R_n)\|_{\mathbf{R}} \to 0,$$

which implies

$$\|W^2(W^{-1}f_n - R_n)\|_{\mathbf{R}} \to 0.$$

We also have

$$\|W^2W^{-1}\|_{\mathbf{R}\setminus[-\gamma,\gamma]} \to 0 \qquad \text{as } \gamma \to \infty;$$

furthermore, the functions f_n uniformly tend to $f(0) = 1$ on compact subsets of **R** (note that $a_n \to \infty$ as $n \to \infty$). These facts imply

$$\|W^2(W^{-1} - R_n)\|_{\mathbf{R}} \to 0, \qquad \text{as } n \to \infty.$$

But then (2.17) implies that the polynomials $\{R_n\}$ are uniformly bounded on every compact subset of the complex plane (see e.g. [2, Theorems 5,7]). Hence we can select a subsequence from them that converges to an entire function on the whole plane. But $R_n(x) \to W(x)^{-1}$ for every real x, so W^{-1} has to be the entire function in question. □

Proof of Theorem 2.7. First of all let us mention that in view of Theorem 2.5 it is enough to show the following: for every $\varepsilon > 0$ there is a continuous function χ (that may also depend on ε) such that $\chi(0) = 1$, and for all sufficiently large n there are polynomials P_n of degree at most n such that $|W(x)P_n(x) - \chi(x/a_n)| \le \varepsilon$ for all $x \in \mathbf{R}$. In fact, suppose this is true and we want to approximate an f that vanishes outside $(-1, 1)$. Then, in view of Theorem 2.5, for a given $\varepsilon > 0$ we can approximate $f(x) - f(0)\chi(x)$ by a $W(a_n x)R_n(x)$ uniformly on $\mathbf{R}$ with error smaller than ε for all large n. The sum $f(0)P_n(a_n x) + R_n(x)$ multiplied by $W(a_n x)$ will then be closer to f than $(1 + |f(0)|)\varepsilon$.

To accomplish this we need a lemma.

Lemma 2.8. *Under the conditions of Theorem 2.7, there is an even entire function H with nonnegative Maclaurin coefficients such that*

$$W(x)H(x) \to 1 \qquad \text{as} \quad x \to \infty.$$

Proof of Lemma 2.8. The proof follows that of [137, Theorem 5(ii)] (actually, in that paper the same lemma was mentioned under somewhat different conditions).

As before, let $W(x) = \exp(-Q(x))$, and for $x > 0$ let the number q_x be the solution of the equation $q_x Q'(q_x) = 2x$. If we differentiate this equation with respect to x and take into account the definition of the function T, we arrive at the formula

$$\frac{dq_x}{dx} Q'(q_x)T(q_x) = 2,$$

and here $Q'(q_x) = 2x/q_x$. Hence the assumption (2.18) on T implies that for large x we have

$$\frac{1}{Bx} \le \frac{dq_x}{dx}\Big/ q_x \le \frac{1}{Ax},$$

which in turn implies via integration that for large x and $u \ge 1/2$

$$u^{1/B} \le \frac{q_{ux}}{q_x} \le u^{1/A}. \tag{2.19}$$

It is also easy to see that

$$q_n \sim a_n. \tag{2.20}$$

Next we set

$$H(x) = \sum_{n=0}^{\infty} \left(\frac{x}{q_n}\right)^{2n} \frac{1}{\sqrt{\pi n T(q_n)}} e^{Q(q_n)}.$$

We claim that this H satisfies the assertion of the lemma. This will follow from [137, Theorem 5] which asserts that if we define

$$G(x) = \sum_{n=0}^{\infty} \left(\frac{x}{q_n}\right)^{2n} \frac{1}{\sqrt{n}} e^{Q(q_n)},$$

then

$$W(x)G(x)/\sqrt{\pi T(x)} \to 1 \qquad \text{as } x \to \infty. \tag{2.21}$$

(We remark that the present function Q is denoted by $2Q$ in [137]; furthermore, although the assumptions of [137, Theorem 5] require more, the proof for part (ii) of that theorem actually uses only smoothness assumptions that are satisfied in this case.) Note that the two series differ only in the factor $1/\sqrt{\pi T(q_n)}$ which lies between two positive constants.

Let

$$h(x, u) = 2u \log \frac{x}{q_u} + Q(q_u) - \frac{1}{2} \log u.$$

Then the choice of q_u gives that

$$\frac{\partial h(x, u)}{\partial u} = 2 \log \frac{x}{q_u} - \frac{1}{2u};$$

hence the equation

$$\frac{\partial h(x, u)}{\partial u} = 0,$$

can be easily seen to have a unique solution $y = y_x$ for large x that satisfies $x/q_y = e^{1/4y}$. Thus, if $y^* = y_x^*$ is the unique y^* with the property $q_{y^*} = x$, then (2.19) shows that $y_x^*/y_x \to 1$ as $x \to \infty$.

Now let $\Lambda = \Lambda_x = (Ky \log y)^{1/2}$ for some large but fixed K. It was shown in the proof of [137, Theorem 5(ii)] that in the series representing G the sum

$$\left(\sum_{n \le y - \Lambda} + \sum_{n \ge y + \Lambda} \right) e^{h(x,n)} = S(x)$$

is negligible compared to the total sum in the sense that $S(x)/G(x) \to 0$ as $x \to \infty$, i.e. (see (2.21))

$$W(x) \sum_{y - \Lambda < n < y + \Lambda} e^{h(x,n)} = (1 + o(1))\sqrt{\pi T(x)}. \tag{2.22}$$

Hence, the same is true of the series representing H. However, for the remaining indices $y - \Lambda < n < y + \Lambda$ we have

$$\frac{q_n}{x} = \frac{q_n}{q_{y^*}} = \frac{q_n}{q_y} \frac{q_y}{q_{y^*}} \to 1$$

as $x \to \infty$, where we used (2.19) for the first factor on the right, and (2.19) and the relation $y_x^*/y_x \to 1$ established above for the second factor. But this means via the assumption made on T that the two sums

$$\sum_{y - \Lambda < n < y + \Lambda} \frac{1}{\sqrt{\pi T(x)}} e^{h(x,n)}$$

and

$$\sum_{y-\Lambda<n<y+\Lambda} \frac{1}{\sqrt{\pi T(q_n)}} e^{h(x,n)}$$

differ only by a factor that tends to 1 as $x \to \infty$. Thus, the lemma follows from formula (2.22). □

Now let us return to the proof of Theorem 2.7. Let R_m be the m-th partial sum of the Maclaurin expansion of the entire function from the preceding lemma. Since the coefficients of H are nonnegative, we have for any x the inequality $0 \le R_m(x) \le H(x)$; furthermore, $R_m(x) \to H(x)$ uniformly on compact subsets of $\mathbf{R}$. As for the numbers a_n, we can conclude from the assumption (2.18) that there is a positive constant C such that $a_n \le a_{2n} \le Ca_n$ holds for all large n. This and (2.20) easily imply (see (2.18) and (2.19)) that given any $\varepsilon > 0$ there are numbers r and t such that

$$\frac{|H(x) - R_m(x)|}{H(x)} \le \varepsilon \qquad \text{for } |x| \le ra_m$$

and

$$\frac{|R_m(x)|}{H(x)} \le \varepsilon \qquad \text{for } |x| \ge ta_m.$$

Now $W(x)H(x) - 1$ tends to zero at infinity; hence, by the solution to Bernstein's problem, for every $\varepsilon > 0$ there exists a polynomial S such that for all $x \in \mathbf{R}$

$$|W(x)(W^{-1}(x) - H(x) - S(x))| \le \varepsilon.$$

We fix this S. Then

$$|1 - W(x)(H(x) + S(x))| \le \varepsilon,$$

and so in view of the preceding inequalities we obtain for large m that

$$|1 - W(x)(R_m(x) + S(x))| \le (1 + M)\varepsilon \qquad \text{for } |x| \le ra_m, \tag{2.23}$$

$$W(x)|R_m(x) + S(x)| \le (1 + M)\varepsilon \qquad \text{for } |x| \ge ta_m, \tag{2.24}$$

and

$$W(x)|R_m(x) + S(x)| \le M + 2 \tag{2.25}$$

otherwise, where M is an upper bound for WH on $\mathbf{R}$.

The assumptions on Q imply (see also the computation for q_x in the proof of Lemma 2.8) that there is an L such that $2ta_{[n/L]} \le ra_n$ is also true. Now if we set

$$P_n(x) = \frac{1}{l} \sum_{k=1}^{l} (R_{[nL^{-k}]}(x) - S(x)),$$

then we get from (2.23)–(2.25) that

$$|W(x)P_n(x) - \chi(|x|/a_n)| \le (1 + M)\varepsilon + \frac{M + 2}{l},$$

where the function $\chi(x)$ is defined on $[0, \infty)$ as follows: $\chi(L^{-k}) = k/l$ for $k = 1, 2, \ldots, l-1$, $\chi(0) = 1$, $\chi(x) = 0$ for $x \ge 1$ and χ is linear otherwise. Hence the polynomials P_n and the function χ satisfy the requirements from the beginning of the present proof with $(M + 2)\varepsilon$ instead of ε if we choose $l > (M + 2)/\varepsilon$. □

VI.3 Fast Decreasing Polynomials

Fast decreasing polynomials play a significant role in several disciplines of mathematical analysis such as approximation theory, orthogonal polynomials, moment problems, etc. These applications require polynomials P that take the value 1 at the origin and are fast decreasing on $[-1, 1] \setminus \{0\}$ in the sense that

$$P(0) = 1, \quad |P(x)| \leq e^{-\varphi(x)}, \quad x \in [-1, 1], \tag{3.1}$$

where φ is a given even function that typically involves the degree of P. For example, the Markoff-Bernstein-type inequality

$$\|wR'\|_{L^p(-\infty,\infty)} \leq C\|wR\|_{L^p(-\infty,\infty)}, \qquad R \text{ a polynomial,}$$

with an absolute constant C for weights like $w(x) = \exp(-|x|^\beta)$, $0 < \beta < 1$, follows from the existence of polynomials P_n of degree at most n satisfying

$$P_n(0) = 1, \quad |P_n(x)| \leq C\exp(-(n|x|)^\beta), \quad x \in [-1, 1]$$

(see [171]).

The significance of such fast decreasing polynomials lies in the fact that they approximate the "Dirac delta function". Therefore we are going to consider the problem in its generality: find the fastest decreasing polynomials of a given order, or alternatively, find the smallest possible degree for the polynomial P in (3.1). This degree is denoted by n_φ.

The order of n_φ can be estimated by an explicitly computable quantity as follows (see [84]):

Let φ be an even function, right continuous and increasing on $[0, 1]$. Then

$$\frac{1}{6}N_\varphi \leq n_\varphi \leq 12N_\varphi,$$

where $N_\varphi = 0$ if $\varphi(1) \leq 0$ and

$$N_\varphi = 2 \sup_{\varphi^{-1}(0) \leq x < b} \sqrt{\frac{\varphi(x)}{x^2}} + \int_b^{1/2} \frac{\varphi(x)}{x^2} dx + \sup_{1/2 \leq x < 1} \frac{\varphi(x)}{-\log(1-x)} + 1,$$

$b = \min(\varphi^{-1}(1), 1/2)$, otherwise.

Here, for $u \geq 0$,

$$\varphi^{-1}(u) = \sup\{\tau \mid \tau \in [0, 1], \ \varphi(\tau) \leq u\}.$$

If $N_\varphi = \infty$, then the statement of the theorem means that there are no polynomials whatsoever with the stated properties.

As special cases the following hold. Let φ be an even and on $[0, 1]$ increasing function with $\varphi(0) = \varphi(0 + 0) = 0$ and $\varphi(x) \leq C\varphi(x/2)$ for $x \in [0, 1]$. Then there are polynomials P_n of degree at most n satisfying

$$P_n(0) = 1, \quad |P_n(x)| \le D\exp(-dn\varphi(x)), \quad x \in [-1, 1], \quad n = 0, 1, \ldots, \tag{3.2}$$

for some constants $D > 0$ and $d > 0$ if and only if

$$\int_0^1 \frac{\varphi(u)}{u^2} du < \infty.$$

For example,

$$|P(0)| = 1, \quad |P(x)| \le C_1 \exp(-n|x|^\beta)$$

with $\beta > 1$ can be achieved by polynomials of degree $\le Cn$, but for $\varphi(x) = |x|$ we get that the minimal degree n_φ of the polynomials P satisfying

$$|P(0)| = 1, \quad |P(x)| \le C_1 e^{-n|x|}, \quad x \in [-1, 1],$$

satisfies

$$\frac{1}{C} n \log n \le n_\varphi \le Cn \log n.$$

These results do not tell the exact conditions under which inequalities such as (3.2) are possible, i.e. they do not say anything about the constants D and d. In this section we show how the theory of weighted potentials can be applied for constructing fast decreasing polynomials and obtaining sharp results concerning the constants.

We start with a somewhat modified problem. Let Σ be a compact subset of $\mathbf{C}$ that contains the origin, and let us call polynomials P_n, $\deg P_n \le n$, *fast decreasing on* Σ, if they attain the value 1 at $x = 0$ and decrease exponentially away from the origin:

$$P_n(0) = 1, \qquad |P_n(x)| \le e^{-n(\varphi(x)+o(1))}, \qquad x \in \Sigma, \ \deg P_n \le n, \tag{3.3}$$

where φ is continuous on Σ. Here and in what follows, $o(1)$ denotes a quantity that tends to zero uniformly in $x \in \Sigma$ as n tends to infinity. Thus, these polynomials decrease exponentially in n at every point of Σ except 0, and the rate of decrease, which is governed by $e^{-\varphi(x)}$ at a point x, depends on the distance of x from 0. The question is what rates are possible, or in other words, for what functions φ are there polynomials with the above properties? We want φ to be as large, or to grow as fast as possible. An obvious condition on φ is that it satisfies $\varphi(0) \le 0$ (otherwise the two conditions in (3.3) are contradictory).

First let us discuss a general necessary and sufficient condition. Let the weight $w(x)$ be defined on Σ as $w(x) = \exp(\varphi(x))$, i.e. we set $Q(x) = -\varphi(x)$. To simplify matters let us assume that Σ is a regular set in the sense that every connected component of $\overline{\mathbf{C}} \setminus \Sigma$ is a regular open set with respect to the Dirichlet problem. Then we can apply Theorem I.5.1 to deduce that the equilibrium potential U^{μ_w} is continuous on the whole plane, and so the equality

$$U^{\mu_w}(z) - F_w = \varphi(z) \tag{3.4}$$

holds for all $z \in \mathcal{S}_w$ and

$$U^{\mu_w}(z) - F_w \ge \varphi(z) \tag{3.5}$$

holds for all $z \in \Sigma$.

Theorem 3.1. *Let φ be a continuous function on the regular compact set Σ with the property $\varphi(0) = 0$. Then there are polynomials with property* (3.3) *if and only if $U^{\mu_w}(0) = F_w$, where $w(x) = \exp(\varphi(x))$.*

Proof. The necessity is immediate if we consider that (see the inequality (III.2.2) in Section III.2) for every $z \in \mathbf{C}$

$$|w^n(z)P_n(z)| \leq \|w^n P_n\|_\Sigma \exp(n(-U^{\mu_w}(z) + F_w));$$

so if we assume (3.3), then the left-hand side at $z = 0$ is 1, while the first term on the right is $\exp(o(n))$. Thus, we must have $U^{\mu_w}(0) \leq F_w$. But the opposite inequality is immediate from (3.5) because $U^{\mu_w}(0) - F_w \geq \varphi(0) = 0$.

In the proof of the sufficiency part we distinguish two cases according to whether or not $0 \in \mathcal{S}_w$.

Case (i): $0 \notin \mathcal{S}_w$. Let w^* denote the restriction of w to $\mathcal{S}_w$. Then $\mu_w = \mu_{w^*}$. Consider the Fekete polynomials Φ_n^* associated with w^* (see Section III.1). Since $0 \notin \mathcal{S}_w = \Sigma^*$, we can apply Corollaries III.1.10 and III.2.6 to conclude that for the polynomials $P_n(x) = \Phi_n^*(x)/\Phi_n^*(0)$ we have $P_n(0) = 1$ and uniformly for $x \in \Sigma$

$$\begin{aligned}\limsup_{n\to\infty} |P_n(x)|^{1/n} &\leq \limsup_{n\to\infty} \left(\frac{\|w^n \Phi_n^*\|_\Sigma}{|\Phi_n^*(0)|}\right)^{1/n} \frac{1}{w(x)} \\ &= \exp(-F_w + U^{\mu_w}(0))e^{-\varphi(x)}.\end{aligned}$$

Thus, if we assume $U^{\mu_w}(0) = F_w$, then it follows that P_n satisfies (3.3).

Case (ii): $0 \in \mathcal{S}_w$. We have seen in Theorem III.1.3 that the asymptotic distribution of Fekete points coincides with μ_w, and since now we assume that 0 belongs to the support $\mathcal{S}_w$ of μ_w, we can conclude that if $\mathcal{F}_{n+1} = \{z_1^{(n+1)}, \ldots, z_{n+1}^{(n+1)}\} \subseteq \Sigma$ are Fekete points, then the point from $\mathcal{F}_{n+1}$, say $z_1^{(n+1)}$, that is closest to 0 converges to 0 as $n \to \infty$. By the definition of Fekete points, for the polynomials

$$R_n(x) = \prod_{i=2}^{n+1}(x - z_i^{(n+1)})$$

we have

$$|R_n(x)w^n(x)| \leq |R_n(z_1^{(n+1)})w^n(z_1^{(n+1)})|,$$

for all $x \in \Sigma$. On applying Theorem III.2.1 we get for all $z \in \mathbf{C}$

$$|R_n(z)| \leq |R_n(z_1^{(n+1)})w^n(z_1^{(n+1)})| \exp\left(n(F_w - U^{\mu_w}(z))\right).$$

Hence it follows from the continuity of the right-hand side (established before the statement of the theorem) that for

$$P_n(x) = R_n(x + z_1^{(n+1)})/R_n(z_1^{(n+1)})$$

we have $P_n(0) = 1$ and

$$|P_n(x)| \le w^n(z_1^{(n+1)}) \exp\left(n(F_w - U^{\mu_w}(x) + o(1))\right) \le e^{-n(\varphi(x)+o(1))},$$

where we also used (3.5) and the fact that $z_1^{(n+1)}$ tends to zero as n tends to infinity and that $w(0) = 1$. □

After this general theorem let us turn to the most important special case when $\Sigma = [-1, 1]$ and φ is symmetric with respect to the origin. Thus, in this case we are looking for polynomials P_n of degree at most $n = 1, 2, \ldots$ with the properties

$$P_n(0) = 1, \qquad |P_n(x)| \le e^{-n(\varphi(x)+o(1))}, \quad x \in [-1, 1], \quad \deg P_n \le n, \tag{3.6}$$

where the $o(1)$ term is uniform in x. The most important part of the next result is a necessary condition for the existence of such fast decreasing polynomials.

Theorem 3.2. *Let φ be a continuous, even, and on* $[0, 1]$ *increasing function with* $\varphi(0) = 0$. *Set* $w(x) = \exp(\varphi(x))$. *Then the following conditions are equivalent.*

(i) *There are polynomials with property* (3.6).

(ii) *There holds*

$$\frac{2}{\pi}\int_0^1 \frac{U^{\mu_w}(t) - F_w}{t^2\sqrt{1-t^2}}\,dt \le 1.$$

(iii) $U^{\mu_w}(0) = F_w$.

Before giving the proof we discuss some consequences. The theorem immediately implies the following necessary condition for the existence of fast decreasing polynomials: if there are polynomials satisfying (3.6), then

$$\frac{2}{\pi}\int_0^1 \frac{\varphi(t)}{t^2\sqrt{1-t^2}}\,dt \le 1 \tag{3.7}$$

must hold because $\varphi(t) \le U^{\mu_w}(t) - F_w$ for $t \in [-1, 1]$ by (3.5).

The nonsymmetric variant of Theorem 3.2 is: if φ is a continuous increasing function on [0,1] with $\varphi(0) = 0$, then there are polynomials with the property

$$P_n(0) = 1, \qquad |P_n(x)| \le e^{-n(\varphi(x)+o(1))}, \quad x \in [0, 1],$$

if and only if

$$\frac{2}{\pi}\int_0^1 \frac{U^{\mu_w}(t) - F_w}{t^2\sqrt{1-t^2}}\,dt \le 1,$$

where now $w(x)$ is defined as $\exp(\varphi(x^2)/2)$ on $[-1, 1]$. This can be obtained from Theorem 3.2 by the substitution $x \to x^2$. Sometimes we need that the polynomials in (3.6) satisfy some additional requirements, but in most cases the conditions in Theorem 3.2 are sufficient in such a stricter sense, as well. Consider e.g. bell-shaped polynomials, i.e. such that, besides (3.6), they are nonnegative, even and increasing on $[-1, 0]$. We can get such polynomials from those in (3.6) by considering

$$\frac{1}{\gamma_n}\int_{-1}^{x}\left(\frac{1}{2}(P_{[n/2]-1}(t)+P_{[n/2]-1}(-t))\right)^2 t\,dt,$$

where

$$\gamma_n=\int_{-1}^{0}\left(\frac{1}{2}(P_{[n/2]-1}(t)+P_{[n/2]-1}(-t))\right)^2 t\,dt$$

(cf. the proof of Corollary 3.3 below).

Closely related to the existence of polynomials in (3.1) is the following approximation problem for the signum function: for what φ is it possible to find polynomials Q_n of degree at most $n = 1, 2, ...,$ such that

$$|\operatorname{sgn} x - Q_n(x)| \le e^{-n(\varphi(x)+o(1))}, \qquad x\in[-1,1], \tag{3.8}$$

uniformly in x? If we like, we can also request that the polynomials Q_n be odd, monotone increasing on $[-1, 1]$, and satisfy $-1 \le Q_n \le 1$ there. Such polynomials can serve as building blocks for the construction of well localized "partitions of unity" consisting of nonnegative polynomials. In fact, since Q_n imitates the signum function, the construction runs parallel (see e.g. [83, p.156]) with that of a real partition of unity using the signum function (cf. Corollary 3.6 below).

Corollary 3.3. *With the assumptions of Theorem* 3.2 *there are polynomials* Q_n *with the property in* (3.8) *if and only if any of* (i)–(iii) *of Theorem* 3.2 *holds.*

Proof. It is easy to see that if the polynomials $P_n, n = 1, 2, \ldots,$ satisfy (3.6), then the polynomials

$$Q_n(x)=\frac{1}{\gamma_n}\int_{-1}^{x}\left(\frac{1}{2}(P_{[n/2]-1}(t)+P_{[n/2]-1}(-t))\right)^2 dt-1,$$

where

$$\gamma_n=\int_{-1}^{0}\left(\frac{1}{2}(P_{[n/2]-1}(t)+P_{[n/2]-1}(-t))\right)^2 dt$$

satisfy (3.8). Indeed, Markoff's inequality

$$\|R_n'\|_{[\alpha,\beta]}\le\frac{2n^2}{\beta-\alpha}\|R_n\|_{[\alpha,\beta]},\qquad \deg R_n\le n,$$

for the derivative of polynomials yields $\gamma_n \ge n^{-2}/2 = e^{o(n)}$ and, for example, for $-1 \le x < 0$

$$\begin{aligned}\int_{-1}^{x}\left(\frac{1}{2}(P_{[n/2]-1}(t)+P_{[n/2]-1}(-t))\right)^2 dt \quad &\le \quad \int_{-1}^{x} e^{-n(\varphi(t)+o(1))}dt\\ &\le \quad e^{-n(\varphi(x)+o(1))}\end{aligned}$$

by the monotonicity of φ. Conversely, if Q_n satisfies (3.8), then on applying Markoff's inequality on small intervals, it is easy to verify that for some sequence $\{\varepsilon_n\}$ tending to zero the polynomials

$$P_n(x) = Q_n'(x+\varepsilon_n)/Q_n'(\varepsilon_n)$$

satisfy (3.6). For example one can take ε_n as a maximum point for $|Q'(x)|$ on $[-1, 1]$. □

Proof of Theorem 3.2. The equivalence of (i) and (iii) follows from Theorem 3.1. That (ii) implies (iii) is an immediate consequence of the continuity of the equilibrium potential that we established before Theorem 3.1. Thus, it only remains to show that (i) implies (ii).

First of all we verify that for all $\alpha \geq 1$

$$\int_0^1 \frac{\log|1-\alpha t^2|}{t^2\sqrt{1-t^2}}dt = -\pi. \tag{3.9}$$

The integral on the left is the limit of

$$\frac{1}{2}\int_0^{(1-\varepsilon)/\alpha} \frac{\log(1-\alpha u)}{u^{3/2}\sqrt{1-u}}du + \frac{1}{2}\int_{(1+\varepsilon)/\alpha}^1 \frac{\log(\alpha u-1)}{u^{3/2}\sqrt{1-u}}du$$

as $\varepsilon \to 0$. Integrating by parts and letting $\varepsilon \to 0+0$, we arrive at

$$\begin{aligned}\int_0^1 \frac{\log|1-\alpha t^2|}{t^2\sqrt{1-t^2}}dt &= -\alpha \text{ PV}\int_0^1 \sqrt{\frac{1-u}{u}}\frac{1}{1-\alpha u}du \\ &= -\text{PV}\int_0^1 \frac{1-u}{\sqrt{u(1-u)}}\frac{1}{1/\alpha-u}du,\end{aligned}$$

where PV means that the integral is taken in the principal value sense. Now, for $\alpha > 1$, (3.9) follows from

$$\int_0^1 \frac{1}{\sqrt{u(1-u)}}du = \pi$$

and

$$\text{PV}\int_0^1 \frac{1}{\sqrt{u(1-u)}}\frac{1}{1/\alpha-u}du = 0$$

valid for all $1/\alpha \in (0, 1)$ (see (IV.3.20)). Then of course it also follows for $\alpha = 1$, as well.

From (3.9) we can deduce

$$\int_a^1 \frac{\log|1-\alpha t^2|}{t^2\sqrt{1-t^2}}dt \geq -\pi \tag{3.10}$$

for all $\alpha \geq 1$ and $a \in [0, 1]$. In fact, this immediately follows from (3.9) and the facts that $\log|1-\alpha t^2| \geq 0$ for $t \in (a, 1)$ if $\alpha \geq 2/a^2$, while $\log|1-\alpha t^2| \leq 0$ for $t \in (0, a)$ if $1 \leq \alpha \leq 2/a^2$.

Next suppose that for some polynomials P_n of degree at most n we have (3.6). Since φ is assumed to be an even function, (3.6) is also true for $P_n(-x)$, and so by considering $(P_n(x) + P_n(-x))/2$ we can assume without loss of generality that P_n is even. Then P_n can be written in the form

$$P_n(x) = \prod_{i=1}^{k}(1 - \alpha_i x^2), \qquad 2k \le n,$$

and the inequality

$$|1 - \alpha_i x^2| \ge \left|1 - |\alpha_i| x^2\right| \ge \left|1 - \max(1, |\alpha_i|) x^2\right|$$

valid for $x \in [-1, 1]$ shows that without loss of generality we can assume $\alpha_i \ge 1$. On applying (3.10) we get for every $a \in (0, 1)$

$$\frac{1}{\pi}\int_a^1 \frac{\log|P_n(t)|}{t^2\sqrt{1-t^2}} dt \ge -k.$$

This and (3.6) yield

$$\frac{1}{\pi}\int_a^1 \frac{-n(\varphi(t) + o(1))}{t^2\sqrt{1-t^2}} dt \ge -k,$$

and so

$$\frac{2}{\pi}\int_0^1 \frac{\varphi(t)}{t^2\sqrt{1-t^2}} dt \le 1 \tag{3.11}$$

follows from $k/n \le 1/2$ by letting first $n \to \infty$ and then $a \to 0$. (3.11) is almost the required inequality (ii); we only have to replace φ by $U^{\mu_w} - F_w$ in it. But this is trivial because (3.4) and (3.6) imply via the generalized Bernstein–Walsh lemma (Theorem III.2.1) that

$$P_n(0) = 1 \quad \text{and} \quad |P_n(x)| \le \exp(-n(U^{\mu_w}(x) - F_w + o(1))), \quad x \in [-1, 1],$$

also hold, and so we can apply the preceding argument with φ replaced by $U^{\mu_w} - F_w$. □

It is desirable to find conditions for (3.6) directly in terms of φ. We have already seen above that (3.7) is a necessary condition. Next we show that in an important special case (3.7) is also sufficient.

Theorem 3.4. *In addition to the assumptions of Theorem* 3.2 *assume that the function* $\varphi(\sqrt{t})$ *is concave on* $[0, 1]$. *Then there are polynomials satisfying*

$$P_n(0) = 1, \qquad |P_n(x)| \le e^{-n(\varphi(x)+o(1))}, \quad x \in [-1, 1], \quad \deg P_n \le n, \tag{3.12}$$

if and only if

$$\frac{2}{\pi}\int_0^1 \frac{\varphi(t)}{t^2\sqrt{1-t^2}} dt \le 1. \tag{3.13}$$

This result can be applied to any $\varphi(t) = c|t|^\alpha$, $0 < \alpha \le 2$, and we obtain that there are polynomials P_n with

$$P_n(0) = 1, \qquad |P_n(x)| \le e^{-n(c|x|^\alpha + o(1))}, \qquad x \in [-1, 1], \tag{3.14}$$

if and only if $\alpha > 1$ and

$$c \le \left(\frac{2}{\pi}\int_0^1 \frac{t^\alpha}{t^2\sqrt{1-t^2}}dt\right)^{-1} = \sqrt{\pi}\Gamma\left(\frac{\alpha}{2}\right)\Big/\Gamma\left(\frac{\alpha-1}{2}\right).$$

We mention that the latter result is no longer true for $\alpha > 2$. This shows that the assumption in Theorem 3.3 that the function $\varphi(\sqrt{t})$ is concave is crucial.

Proof of Theorem 3.4. We only have to verify the sufficiency of the condition, and, in view of Theorem 3.2, it is enough to show that under the assumptions $U^{\mu_w} - F_w$ and φ coincide on $[-1, 1]$. Actually we shall show that in this case $\mathcal{S}_w$ coincides with $[-1, 1]$, and then it is enough to apply (3.4).

Taking into account the concavity of $\varphi(\sqrt{t})$, condition (3.13) is easily seen to be equivalent to

$$\frac{1}{\pi}\int_0^1 \frac{\varphi'(\sqrt{t})}{t}\sqrt{1-t}\,dt = \frac{2}{\pi}\int_0^1 \frac{\varphi'(t)}{t}\sqrt{1-t^2}\,dt \le 1,$$

and hence, by setting $\psi(t) = 2\varphi(\sqrt{t})$ for $t \in [0, 1]$, we get

$$\frac{1}{\pi}\int_0^1 \frac{\psi'(t)}{\sqrt{t}}\sqrt{1-t}\,dt \le 1. \tag{3.15}$$

Now consider the weight function $v(x) = \exp(\psi(x))$ and the corresponding weighted energy problem on $[0, 1]$, where the function $\psi(t) = 2\varphi(\sqrt{t})$ is concave on $[0, 1]$ by the assumption. According to Theorem IV.1.10(b) the support $\mathcal{S}_v$ is then an interval (note that with the notation of that theorem we have now $Q(t) = -\psi(t)$). Since ψ is increasing on [0,1], this support must contain the point 1 (see Theorem IV.1.3). Thus, $\mathcal{S}_v = [a, 1]$ for some $0 \le a < 1$. We also know (see Theorem IV.1.5) that the support $\mathcal{S}_v$ maximizes the F-functional

$$F(K) = \log\,[\text{cap}(K)] + \int \psi\, d\mu_K$$

among all compact subsets K of $[0, 1]$, where μ_K denotes the equilibrium measure for K. As we have already established, in finding this maximum we can restrict ourselves to sets of the form $K = [a, 1]$, $0 \le a < 1$, for which

$$\begin{aligned} F([a, 1]) &= \log\frac{1-a}{4} + \frac{1}{\pi}\int_a^1 \frac{\psi(t)}{\sqrt{(t-a)(1-t)}}dt \\ &= \log\frac{1-a}{4} + \frac{1}{\pi}\int_0^1 \frac{\psi(t + a(1-t))}{\sqrt{t(1-t)}}dt, \end{aligned}$$

because the logarithmic capacity of $[a, 1]$ is $(1-a)/4$ and the corresponding (unweighted) equilibrium measure is $((t-a)(1-t))^{-1/2}dt/\pi$. Here the right-hand side is a differentiable function of $a > 0$ (note that ψ is absolutely continuous on $(0,1)$) with derivative

$$F'([a,1]) = \frac{-1}{1-a} + \frac{1}{\pi}\int_0^1 \frac{\psi'(t+a(1-t))}{\sqrt{t(1-t)}}(1-t)\,dt.$$

Since ψ is concave on $[0, 1]$, this derivative decreases in a, and by (3.15) its value at $a = 0$ is nonpositive. Hence, the derivative is negative for $0 < a < 1$, and so $F([a, 1])$ takes its maximum at $a = 0$, i.e. $\mathcal{S}_v = [0, 1]$. But then invoking Theorem IV.1.10(f) we can conclude that $\mathcal{S}_w = [-1, 1]$ as we claimed. □

Up to now necessary and sufficient conditions have been established for the existence of fast decreasing polynomials satisfying (3.3) or (3.6). Nothing has yet been said about the $o(1)$ term in the exponent except that it is uniform in x. In the last part of this section this term is dropped (more precisely, it is changed to $O(1/n)$).

Let $\varphi(t) = c|t|^\alpha$, $\alpha \le 2$. We want to construct polynomials P_n of degree at most n with the property

$$P_n(0) = 1, \qquad |P_n(x)| \le Ce^{-nc|x|^\alpha}, \quad x \in [-1,1], \tag{3.16}$$

for some constant C. The necessity part of Theorem 3.2 yields for c the inequality $c \le \sqrt{\pi}\,\Gamma\left(\frac{\alpha}{2}\right)/\Gamma\left(\frac{\alpha-1}{2}\right)$. We are going to show that this inequality is also sufficient.

Theorem 3.5. *Let $\alpha \le 2$. Then there are polynomials P_n with property* (3.16) *if and only if $\alpha > 1$ and*

$$c \le \frac{\sqrt{\pi}\,\Gamma\left(\frac{\alpha}{2}\right)}{\Gamma\left(\frac{\alpha-1}{2}\right)}.$$

Proof. As we have already remarked, the necessity of the condition follows from Theorem 3.4 (see the discussion right after it). Hence it is enough to prove that if $1 < \alpha \le 2$ and $c = \sqrt{\pi}\,\Gamma\left(\frac{\alpha}{2}\right)/\Gamma\left(\frac{\alpha-1}{2}\right)$, then there are polynomials with the property (3.16).

Let us consider the energy problem on $\Sigma = [-1, 1]$ with weight $w(x) = \exp(c|x|^\alpha)$ (note the positive sign in the exponent which makes this weight essentially different from the Freud weights $\exp(-c|x|^\alpha)$). First we show that the corresponding extremal measure is given by the density function

$$v(x) = |x|^{\alpha-1}\frac{\alpha}{\pi}\left(\int_0^{|x|}\frac{u^{2-\alpha}}{(1-u^2)^{3/2}}\,du + d_\alpha\right), \tag{3.17}$$

where the constant d_α is given by

$$d_\alpha = \int_0^1 \frac{1-u^{2-\alpha}}{(1-u^2)^{3/2}}\,du.$$

To derive this form consider the function

$$f(x) = \alpha \frac{1}{\pi\sqrt{1-x^2}} - (\alpha - 1)\frac{\alpha}{\pi}\int_{|x|}^{1} \frac{u^{\alpha-1}}{\sqrt{u^2-x^2}}\,du$$

built up from the Chebyshev and Ullman distributions (see (IV.5.1)), and recall (see Section IV.5) that the Ullman distribution corresponds to the energy problem with respect to the weight $\exp(-\gamma_\alpha |x|^\alpha)$ with

$$\gamma_\alpha = \frac{\Gamma\left(\frac{\alpha}{2}\right)\Gamma\left(\frac{1}{2}\right)}{2\Gamma\left(\frac{\alpha+1}{2}\right)}.$$

(Note the negative sign appearing in this weight.) The function f has total integral 1 over $[-1, 1]$ and, by Theorem IV.5.1, its logarithmic potential for $x \in [-1, 1]$ is of the form const. $+ (\alpha - 1)\gamma_\alpha |x|^\alpha$. But using that $\Gamma(t + 1) = t\Gamma(t)$ and $\Gamma(1/2) = \sqrt{\pi}$, we easily obtain that

$$(\alpha - 1)\gamma_\alpha = (\alpha - 1)\frac{\Gamma\left(\frac{\alpha}{2}\right)\Gamma\left(\frac{1}{2}\right)}{2\Gamma\left(\frac{\alpha+1}{2}\right)} = \frac{\sqrt{\pi}\,\Gamma\left(\frac{\alpha}{2}\right)}{\Gamma\left(\frac{\alpha-1}{2}\right)} = c,$$

and so the potential is of the form const. $+ c|x|^\alpha$. Thus, if we can show that the function f is nonnegative, then we can invoke Theorem I.3.3 to conclude that f is in fact the density of the equilibrium measure in question (actually, the same conclusion can be derived from the principle of domination without referring to the nonnegativity of f, but we shall need the following consideration anyway).

Clearly, it is enough to consider positive values of x. If we write the integral appearing in the definition of f in the form

$$x^{\alpha-1}\int_{1}^{1/x} \frac{u^{\alpha-1}}{\sqrt{u^2-1}}\,du,$$

then it follows that f satisfies the differential equation

$$f'(x) = \frac{\alpha}{\pi}\frac{x}{(1-x^2)^{3/2}} + \frac{\alpha-1}{x} f(x)$$

with initial condition $f(0) = 0$. We can solve this linear equation and get with some constant d that

$$f(x) = x^{\alpha-1}\frac{\alpha}{\pi}\left(\int_0^x \frac{u^{2-\alpha}}{(1-u^2)^{3/2}}\,du + d\right).$$

The value of d can be obtained from the condition that f has integral 1 over $[-1, 1]$. This means that we must have

$$\frac{d}{\pi} + \int_0^1 x^{\alpha-1}\frac{\alpha}{\pi}\int_0^x \frac{u^{2-\alpha}}{(1-u^2)^{3/2}}\,du\,dx = \frac{1}{2},$$

which easily yields the value d_α for d. Since d_α is nonnegative, the nonnegativity of f follows from the preceding expression for $f(x)$, and the same expression verifies (3.17).

When $\alpha = 2$, then $d_\alpha = 0$ and (3.17) takes the form

$$v(x) = \frac{2}{\pi}\frac{x^2}{(1-x^2)^{1/2}},$$

while for $1 < \alpha < 2$ the constant $d_\alpha > 0$ and in this case the density v has order $\sim |x|^{\alpha-1}$ as x approaches 0. Thus, with $\delta = \alpha - 1$ if $1 < \alpha < 2$ and $\delta = 2$ if $\alpha = 2$, the density v of the extremal measure satisfies $v(t) \sim |t|^\delta$ as $t \to 0$ and $v(t) \sim (1-t^2)^{-1/2}$ as $t \to \pm 1$, and otherwise v is continuous and positive. This is all we need of v.

Now let $\mu(t) = v(t)dt$ be the extremal measure. By (3.4) and the fact $S_w = [-1, 1]$ that was proved during the proof of Theorem 3.4, we have $U^\mu(x) = c|x|^\alpha + F_w$ for every $x \in [-1, 1]$. If we can construct polynomials

$$R_n(x) = \prod_{j=0}^{n-1}(x - \xi_j)$$

such that

$$-\log|R_n(x)| - nU^\mu(x) \geq C \tag{3.18}$$

for all $x \in [-1, 1]$, and

$$-\log|R_n(0)| - nU^\mu(0) \leq C, \tag{3.19}$$

then $P_n(x) = R_n(x)/R_n(0)$ will satisfy (3.16). This is where we use the discretization technique of Section VI.1.

Let n be an even number (when n is odd, use $n-1$ in place of n below). Let us divide $[-1, 1]$ by the points $-1 = t_0 < t_1 < \ldots < t_n = 1$ into n intervals I_j, $j = 0, 1, \ldots, n-1$, with $\mu(I_j) = 1/n$, and let ξ_j be the weight point of the restriction of μ to I_j. Set

$$R_n(t) = \prod_{j=0}^{n-1}(t - \xi_j).$$

We claim that these polynomials satisfy (3.18) and (3.19).

We write

$$-\log|R_n(x)| - nU^\mu(x) = \sum_{j=0}^{n-1} n\int_{I_j}\log\left|\frac{x-t}{x-\xi_j}\right| v(t)dt =: \sum_{j=0}^{n-1} L_j(x). \tag{3.20}$$

The proof of (3.19) is very simple: since $\xi_{n/2-1} < t_{n/2} = 0 < \xi_{n/2}$, and the function $\log|0-t|$ is concave on every I_j, we get that every term $L_j(0)$ in (3.20) is at most 0. This proves (3.19).

It is left to prove (3.18). Let $x \in I_{j_0}$. The individual terms in (3.20) are clearly bounded from below when $j = j_0, j_0 \pm 1$. For other j's the integrands are bounded in absolute value by an absolute constant independent of n, x, and $j \neq j_0, j_0 \pm 1$; hence the integrals themselves are also uniformly bounded, for the integral of v on each I_j equals $1/n$.

As we have done in Section VI.1, we write for $x \in I_{j_0}$, and $j \neq j_0, j_0 \pm 1$ the integrand in $L_j(x)$ as

$$\log\left|1 + \frac{\xi_j - t}{x - \xi_j}\right| = \frac{\xi_j - t}{x - \xi_j} + O\left(\left|\frac{\xi_j - t}{x - \xi_j}\right|^2\right),$$

which holds because one can easily verify that

$$\frac{\xi_j - t}{x - \xi_j} \geq -q > -1, \qquad t \in I_j,\ j \neq j_0, j_0 \pm 1,$$

with an absolute constant $0 < q < 1$. Thus, we have

$$L_j(x) = n \int_{I_j} O\left(\left|\frac{\xi_j - t}{x - \xi_j}\right|^2\right) v(t)dt = O\left(\frac{|I_j|^2}{(\xi_j - \xi_{j_0})^2}\right), \tag{3.21}$$

because the integrals

$$\int_{I_j} \frac{\xi_j - t}{x - \xi_j} v(t)dt$$

vanish by the choice of the points ξ_j.

We have to distinguish two cases according as x is closer to one of the endpoints or it is closer to 0.

Case I. x is close to an endpoint. Let us suppose for example that $x \in [-1, -1/2]$. We have to estimate

$$S_1(x) := \sum_{j=0}^{j_0-2} |L_j(x)|$$

and

$$S_2(x) := \sum_{j=j_0+2}^{n-1} |L_j(x)|.$$

We shall do this for the first sum, the second one being similar because, in view of (3.21), the part of S_2 corresponding to the indices for which $\xi_j > -1/4$ is less than

$$C \sum_j |I_j|^2 \leq C \sum_j |I_j| \leq C. \tag{3.22}$$

For $j \leq j_0 - 2$

$$1 + \xi_j \sim \left(\frac{j+1}{n}\right)^2, \qquad |I_j| \sim \frac{j+1}{n^2}, \qquad \xi_{j_0} - \xi_j \sim \left(\frac{j_0}{n}\right)^2 - \left(\frac{j}{n}\right)^2;$$

hence

$$S_1(x) \le \sum_{j=0}^{j_0-2} \frac{C\left((j+1)/n^2\right)^2}{\left((j_0/n)^2 - (j/n)^2\right)^2} \le \sum_{j=0}^{j_0/2} + \sum_{j=j_0/2}^{j_0-2} =: K_1 + K_2.$$

(Recall that $A \sim B$ means that A/B is bounded above and below by positive constants.) Here

$$K_1 \le C \sum_{j=0}^{j_0/2} \frac{(j+1)^2}{j_0^4} = O(1)$$

and

$$K_2 \le C \sum_{j=j_0/2}^{j_0-2} \frac{j_0^2}{((j-j_0)j_0)^2} = O(1),$$

and these verify that

$$S_1(x) = O(1).$$

Case II. x is close to 0. Now suppose that $x \in [-1/2, 1/2]$, say $x \in [0, 1/2]$. Let $\xi_j^* = \xi_{n/2+j-1}$, $I_j^* = I_{n/2+j-1}$, $L_j^* = L_{n/2+j-1}$. Then $\xi_{-j+1}^* = -\xi_j^*$, $I_{-j+1}^* = -I_j^*$, and for $j > 0$

$$\xi_j^* \sim \left(\frac{j}{n}\right)^{1/(\delta+1)}, \qquad |I_j^*| \sim \frac{1}{n^{1/(\delta+1)} j^{\delta/(\delta+1)}},$$

where δ is the number chosen above, i.e. $\delta = \alpha - 1$ if $1 < \alpha < 2$ and $\delta = 2$ if $\alpha = 2$.

If $x \in I_{j_0}^*$, $j_0 > 0$, then we have to estimate

$$S_1(x) \;:=\; \sum_{j=1}^{j_0-2} |L_j^*(x)|,$$

$$S_2(x) \;:=\; \sum_{j=-n/4}^{0} |L_j^*(x)|$$

and

$$S_3(x) := \sum_{j=j_0+2}^{n/4} |L_j^*(x)|,$$

because the contribution of the rest (like that of $\sum_{-n/2}^{-n/4}$) is easily seen to be bounded (use the argument of (3.22)).

We shall estimate $S_1(x) + S_2(x)$; the sum $S_3(x)$ can be similarly handled. Now (3.21) yields

$$S_1(x) \;\le\; \sum_{j=1}^{j_0-2} \frac{C\left(n^{1/(\delta+1)} j^{\delta/(\delta+1)}\right)^{-2}}{\left((j_0/n)^{1/(\delta+1)} - (j/n)^{1/(\delta+1)}\right)^2} \le \sum_{j=1}^{j_0/2} + \sum_{j=j_0/2}^{j_0-2}$$

$$\le\; C \sum_{j=1}^{j_0/2} \frac{j^{-2\delta/(\delta+1)}}{j_0^{2/(\delta+1)}} + C \sum_{j=j_0/2}^{j_0-2} \frac{j^{-2\delta/(\delta+1)}}{((j_0-j) j_0^{-\delta/(\delta+1)})^2} = O(1)$$

and

$$S_2(x) \leq \sum_{j=-n/4}^{0} \frac{C\left(n^{1/(\delta+1)}(|j|+1)^{\delta/(\delta+1)}\right)^{-2}}{\left((j_0/n)^{1/(\delta+1)}+(|j|/n)^{1/(\delta+1)}\right)^2} = \sum_{j<-j_0} + \sum_{j=-j_0}^{0}$$

$$\leq C\sum_{j<-j_0}\frac{1}{j^2} + C\sum_{j=-j_0}^{0}\frac{(|j|+1)^{-2\delta/(\delta+1)}}{j_0^{2/(\delta+1)}} = O(1).$$

□

We conclude this section with a consequence of Corollary 3.3.

Corollary 3.6. *Let J_1 and J_2 be two disjoint compact sets on the real line contained in the interval J. Then there is a constant c, $0 < c < 1$, and, for each m large, polynomials R_m of degree at most m such that*

$$|R_m(x)| \leq c^m \qquad for\ \ x \in J_1,$$

$$|R_m(x) - 1| \leq c^m \qquad for\ \ x \in J_2,$$

and

$$0 \leq R_m(x) \leq 1 \qquad for\ \ x \in J. \tag{3.23}$$

Proof. Clearly, we can assume the sets J_1 and J_2 to consist of finitely many intervals. When $J = [-1, 1]$ and $J_1 \subset [-1, 0)$, $J_2 \subset (0, 1]$, the theorem follows from Corollary 3.3 (just add to the polynomials appearing in that corollary the value 1 for some φ that is positive on $(0, 1]$ and divide the result by two). By linear transformation this yields for any $a \in J$ and $\varepsilon > 0$ a polynomial $P_{n,a}$ satisfying condition (3.23) that is exponentially small on the part of J lying to the left of $a - \varepsilon$, and approximates exponentially fast the function 1 to the right of a. Then for $a < b$ lying in J the polynomial

$$1 - (1 - P_{[n/2],b+\varepsilon}(x))P_{[n/2],a}(x)$$

(of degree at most n) approximates 1 exponentially fast on $J \setminus [a-\varepsilon, b+\varepsilon]$ and is exponentially small on $[a, b]$. Now all we need to do is to multiply these polynomials together for the intervals $[a, b]$ making up J_1. □

VI.4 Discretizing a Logarithmic Potential

In this section we consider the important problem of how to discretize a logarithmic potential. To be more precise, we consider the question: Given a potential U, how can one find a potential close to U that corresponds to a discrete measure. This problem has already been touched upon in the preceding three sections.

We shall consider the problem in a somewhat different form. Let μ be a positive measure on $[-1, 1]$ of total mass 1, and consider its logarithmic potential

$$U^{\mu}(z) = \int \log \frac{1}{|z-t|} \, d\mu(t).$$

In many problems, one needs polynomials P_n of degree at most n that behave in a sense like $\exp(-nU^{\mu}(z))$. In other words, the potential $U^{n\mu}$ is to be replaced by a discrete potential

$$U^{\nu_n}(z) = \sum_{j=0}^{n-1} \log \frac{1}{|z-\xi_j|},$$

where the ξ_j's are the zeros of P_n. The idea of constructing such polynomials is very simple: the measure $n\mu$ has total mass n, so it is natural to try to imitate the distribution of $n\mu$ by the distribution of the points $\{\xi_j\}_{j=0}^{n-1}$, which suggests dividing the support into $n+1$ equal parts with respect to the measure μ and selecting the points ξ_j as the division points (this can be done in every case when the measure μ is continuous in the sense that it does not contain a mass point, which we are always going to assume). We shall discuss later a finer method that uses the weight points of the division intervals.

Let us begin with a general theorem.

Theorem 4.1. *Let μ be a measure on $[-1, 1]$ with total mass $\|\mu\| = 1$. Suppose that there exist K and $\varepsilon > 0$ such that*

$$\int_{|x-t|\le n^{-K}} \Big|\log|x-t|\Big| \, d\mu(t) \le \varepsilon \frac{\log n}{n}, \qquad x \in [-1, 1], \tag{4.1}$$

where n is a positive integer. Then if

$$-1 = y_{0,n} < y_{1,n} < \ldots < y_{n-1,n} < y_{n,n} = 1$$

are chosen so that

$$\int_{y_{j,n}}^{y_{j+1,n}} d\mu(t) = \frac{1}{n}, \qquad j = 0, 1, 2, \ldots, n-1, \tag{4.2}$$

and if

$$P_n(x) := \prod_{j=1}^{n-1} (x - y_{j,n}), \tag{4.3}$$

then

$$|P_n(x)| \le e^{-nU^{\mu}(x)} n^{K+\varepsilon}, \qquad x \in \mathbf{R}, \tag{4.4}$$

$$|P_n(x)| \ge \frac{1}{4} e^{-nU^{\mu}(x)} |x - y_{n_x,n}|, \qquad x \in [-1, 1], \tag{4.5}$$

where $y_{n_x,n}$ denotes the closest zero of $P_n(x)$ to x. Furthermore,

$$|P_n(x)| \ge \frac{1}{|x|+1} e^{-nU^{\mu}(x)}, \qquad x \notin [-1, 1]. \tag{4.6}$$

Proof. First fix $x \in [y_{1,n}, y_{n-1,n}]$ and choose l such that

$$y_{l,n} \le x < y_{l+1,n}.$$

Because $\log|x-t|$ is decreasing in t for $t \le y_{l,n}$, we have for $j \le l-1$

$$\log|x-t| \le \log|x-y_{j,n}|, \qquad t \in (y_{j,n}, y_{j+1,n}),$$

and

$$\log|x-t| \ge \log|x-y_{j,n}|, \qquad t \in (y_{j-1,n}, y_{j,n}).$$

Then taking into account (4.2) we obtain, for $j \le l-1$,

$$\int_{y_{j,n}}^{y_{j+1,n}} \log|x-t|\,d\mu(t) \le \frac{1}{n}\log|x-y_{j,n}| \le \int_{y_{j-1,n}}^{y_{j,n}} \log|x-t|\,d\mu(t),$$

and so

$$\begin{aligned} \int_{y_{1,n}}^{y_{l,n}} \log|x-t|\,d\mu(t) &\le \frac{1}{n}\sum_{j=1}^{l-1}\log|x-y_{j,n}| \\ &\le \int_{-1}^{y_{l-1,n}} \log|x-t|\,d\mu(t). \end{aligned} \tag{4.7}$$

Similarly, since $\log|x-t|$ is increasing in t for $t \ge y_{l+1,n}$, we have

$$\begin{aligned} \int_{y_{l+1,n}}^{y_{n-1,n}} \log|x-t|\,d\mu(t) &\le \frac{1}{n}\sum_{j=l+2}^{n-1}\log|x-y_{j,n}| \\ &\le \int_{y_{l+2,n}}^{1} \log|x-t|\,d\mu(t). \end{aligned} \tag{4.8}$$

Adding (4.7) and (4.8), we obtain for $x \in [y_{1,n}, y_{n-1,n}]$,

$$\begin{aligned} -U^{\mu}(x) - \left[\int_{-1}^{y_{1,n}} + \int_{y_{l,n}}^{y_{l+1,n}} + \int_{y_{n-1,n}}^{1}\right] \log|x-t|\,d\mu(t) \\ \le \frac{1}{n}\sum_{\substack{j=1 \\ j\ne l,l+1}}^{n-1} \log|x-y_{j,n}| \le -U^{\mu}(x) - \int_{y_{l-1,n}}^{y_{l+2,n}} \log|x-t|\,d\mu(t). \end{aligned} \tag{4.9}$$

Now if, say, $y_{l,n}$ is closer to x than $y_{l+1,n}$, that is, $y_{n_x,n} = y_{l,n}$, then by (4.2),

$$\int_{y_{l,n}}^{y_{l+1,n}} \log|x-t|\,d\mu(t) \le \frac{1}{n}\log|x-y_{l+1,n}|.$$

Also $\log|x-t| \le \log 2$ for $t \in [-1,1]$, so from (4.9) we deduce that

$$- U^{\mu}(x) - \frac{\log 4}{n} \le \frac{1}{n}\{\log|P_n(x)| - \log|x - y_{n_x,n}|\}. \tag{4.10}$$

This yields (4.5) for $x \in [y_{1,n}, y_{n-1,n}]$.

Next we use the inequalities

$$\frac{1}{n}\log|x - y_{l,n}| \le \int_{y_{l-1,n}}^{y_{l,n}} \log|x - t|\, d\mu(t),$$

and

$$\frac{1}{n}\log|x - y_{l+1,n}| \le \int_{y_{l+1,n}}^{y_{l+2,n}} \log|x - t|\, d\mu(t),$$

together with (4.9) to obtain

$$\frac{1}{n}\log|P_n(x)| \le -U^{\mu}(x) - \int_{y_{l,n}}^{y_{l+1,n}} \log|x - t|\, d\mu(t). \tag{4.11}$$

We split the integral on the right-hand side of (4.11) into integrals over $\{t \mid |x-t| \le n^{-K}\}$ and $\{t \mid |x - t| \ge n^{-K}\}$. In view of (4.1), the former integral is bounded in absolute value by $\varepsilon(\log n)/n$. The absolute value of the latter integral is clearly bounded by

$$K(\log n)\int_{y_{l,n}}^{y_{l+1,n}} d\mu(t) = K\frac{\log n}{n}.$$

Thus

$$\frac{1}{n}\log|P_n(x)| \le -U^{\mu}(x) + (K + \varepsilon)\frac{\log n}{n}. \tag{4.12}$$

This yields (4.4), at least when $x \in [y_{1,n}, y_{n-1,n}]$.

We proceed to prove the inequalities by estimating $|P_n(x)|$ when $x \notin [y_{1,n}, y_{n-1,n}]$. Suppose, say, $x > y_{n-1,n}$. Then, as before, we obtain

$$\frac{1}{n}\log|x - y_{n-1,n}| + \int_{y_{1,n}}^{y_{n-1,n}} \log|x - t|\, d\mu(t)$$

$$\le \frac{1}{n}\sum_{j=1}^{n-1}\log|x - y_{j,n}| \le \int_{-1}^{y_{n-1,n}} \log|x - t|\, d\mu(t),$$

and so

$$- U^{\mu}(x) - \frac{\log(x + 1)}{n} + A_n(x) \le \frac{1}{n}\log|P_n(x)| \tag{4.13}$$

$$\le -U^{\mu}(x) - \int_{y_{n-1,n}}^{1} \log|x - t|\, d\mu(t),$$

where

$$A_n(x) := \frac{1}{n}\log|x - y_{n-1,n}| - \int_{y_{n-1,n}}^{1} \log|x - t|\, d\mu(t).$$

Here, if $x \in [y_{n-1,n}, 1]$, then

$$A_n(x) \geq \frac{1}{n} \log |x - y_{n-1,n}|;$$

furthermore, we deduce as before,

$$\left| \int_{y_{n-1,n}}^{1} \log |x - t| \, d\mu(t) \right| \leq (K + \varepsilon) \frac{\log n}{n}.$$

However, if $x \in [1, \infty)$, then $A_n(x) \geq 0$ and

$$- \int_{y_{n-1,n}}^{1} \log |x - t| \, d\mu(t) \leq - \int_{y_{n-1,n}}^{1} \log |1 - t| \, d\mu(t) \leq (K + \varepsilon) \frac{\log n}{n}.$$

Then (4.13) yields (4.4)–(4.6) for $x > y_{n-1,n}$. The proof for $x < y_{1n}$ is similar. □

Next we discuss a method that yields sharper results provided we know some further properties of the measure μ.

Theorem 4.2. *Let μ be a measure of total mass 1 and suppose that* supp(μ) *consists of finitely many intervals J_j, and $d\mu(x) = v(x)dx$ where v is a continuous function inside every J_j except for finitely many points, and v has only finitely many zeros. Assume further that if $A = \{a_i\}$ is the set consisting of the zeros and discontinuities of v and of the endpoints of the intervals J_j, then for each i there is a $\delta_i > -1$ such that $v(t) \sim |t - a_i|^{\delta_i}$ in a neighborhood of a_i. Then there are polynomials*

$$P_n(x) = \prod_{j=0}^{n-1} (x - \xi_j)$$

with all their zeros in the support of μ such that, for some constants c and C,

$$|P_n(x)| \leq C \exp(-nU^\mu(x))$$

and

$$|P_n(x)| \geq c \exp(-nU^\mu(x)) \min(1, |x - \xi_{n_x}|)$$

hold for all $x \in \mathbf{R}$, where ξ_{n_x} is the nearest zero of P_n to x.

The relation $v(t) \sim |t - a_i|^{\delta_i}$ is assumed only from one side if a_i is an endpoint of a subinterval of supp(μ). Furthermore, for interior points a_j, the δ_j can be different for the left and right neighborhoods of a_i.

Proof. We shall follow the proof of Theorem 3.5. First suppose that the support of the measure μ is a single interval I. We have to construct polynomials

$$P_n(x) = \prod_{j=0}^{n-1} (x - \xi_j)$$

such that

$$-\log|P_n(x)| - nU^\mu(x) \geq C \tag{4.14}$$

for all $x \in \mathbf{R}$,

$$-\log|P_n(x)| - nU^\mu(x) + \log|x - \xi_{n_x}| \leq C \tag{4.15}$$

for all x whose distance from I is at most 1, and

$$-\log|P_n(x)| - nU^\mu(x) \leq C \tag{4.16}$$

for all x having distance at least 1 from the support of μ.

Let n be an even number (when n is odd, use $n-1$ in place of n below, and add appropriately one more zero to get exact degree n). Partition $I =: [a, b]$ by the points $a = t_0 < t_1 < \ldots < t_n = b$ into n intervals I_j, $j = 0, 1, \ldots, n-1$, with $\mu(I_j) = 1/n$, and let ξ_j be the *weight point* of the restriction of μ to I_j; i.e.

$$\xi_j = n\int_{I_j} tv(t)\,dt. \tag{4.17}$$

Set

$$P_n(t) = \prod_{j=0}^{n-1}(t - \xi_j).$$

We claim that these P_n satisfy (4.14)–(4.16).

First let $x \in I$, say $x \in I_{j_0}$ for some j_0, and notice that then n_x is either j_0, or $j_0 \pm 1$. Furthermore, it is sufficient to verify (4.15) with n_x replaced by j_0.

We write

$$-\log|P_n(x)| - nU^\mu(x) = \sum_{j=0}^{n-1} n\int_{I_j}\log\left|\frac{x-t}{x-\xi_j}\right| v(t)\,dt =: \sum_{j=0}^{n-1} L_j(x) \tag{4.18}$$

and

$$\begin{aligned}-\log|P_n(x)| \quad &- \quad nU^\mu(x) + \log|x - \xi_{n_0}| \\ &= \sum_{j\neq n_0} n\int_{I_j}\log\left|\frac{x-t}{x-\xi_j}\right| v(t)\,dt - n\int_{I_{n_0}}\log\frac{1}{|x-t|}v(t)\,dt \\ &=: \sum_{j\neq n_0} L_j(x) - \tilde{L}_{n_0}(x).\end{aligned}$$

Based on this second formula, the proof of (4.15) (with n_x replaced by j_0) is simple: for $j \neq j_0$ the function $\log|x-t|$ is concave on I_j, so

$$n\int_{I_j}\log|x-t|v(t)\,dt \leq \log|x - \xi_j|,$$

and hence every such term $L_j(x)$ in (4.18) is at most 0. Furthermore, $\tilde{L}_{n_0}(x)$ is clearly bounded from below (actually it is greater than zero for sufficiently large n). These observations prove (4.15).

Next we prove (4.14). Let $x \in I_{j_0}$. The individual terms in (4.18) are easily seen to be bounded from below when $j = j_0, j_0 \pm 1$. For other j's the integrands are bounded in absolute value by an absolute constant independent of n, x, and $j \neq j_0, j_0 \pm 1$. Hence n times the integrals themselves are also uniformly bounded, because the integral of v on each I_j equals $1/n$.

The assumptions on v imply that the ratio of the length of consecutive intervals I_j, $I_{j\pm 1}$ is less than a fixed constant; therefore there is an $L \geq 1$ such that for $x \in I_{j_0}$ and $t \in I_j$ with $|j - j_0| \geq L$ we have

$$\frac{\xi_j - t}{x - \xi_j} \geq -\frac{1}{2}. \tag{4.19}$$

From the previous discussion of the lower boundedness of individual terms, for $|j - j_0| < L$ we have $L_j(x) \geq -C_1$ for an absolute constant C_1. Hence

$$\sum_{|j-j_0|<L} L_j(x) \geq -2C_1 L. \tag{4.20}$$

For other j's, (4.19) holds, and we can utilize the idea that was already used in Sections VI.1 and VI.3: we write, for $x \in I_{j_0}$ and $|j - j_0| \geq L$, the integrand in $L_j(x)$ as

$$\log\left|1 + \frac{\xi_j - t}{x - \xi_j}\right| = \frac{\xi_j - t}{x - \xi_j} + O\left(\left|\frac{\xi_j - t}{x - \xi_j}\right|^2\right).$$

Thus, we have for such j's

$$\begin{aligned} L_j(x) &= n\int_{I_j} O\left(\left|\frac{\xi_j - t}{x - \xi_j}\right|^2\right) v(t)\,dt \\ &= O\left(\frac{|I_j|^2}{(\xi_j - \xi_{j_0})^2}\right), \end{aligned} \tag{4.21}$$

because the integrals

$$\int_{I_j} \frac{\xi_j - t}{x - \xi_j} v(t)\,dt$$

vanish by the choice of the points ξ_j.

Now let us suppose that the closest point to x in the set A (that consists of the endpoints of I and of the zeros and singularities of the density v) is a_m. Without loss of generality we may assume that $a_m = 0$, and let $\delta := \delta_m > -1$, i.e. we have $v(t) \sim |t|^\delta$ as $t \to 0$ (from one side only if $a_m = 0$ is an endpoint of I).

In what follows we assume that $a_m = 0$ is an interior point of I; for the endpoints of I, the necessary modifications in the proof are obvious. Let I_{m_0} be the interval containing 0 and set $\xi_j^* = \xi_{m_0+j}$, $I_j^* = I_{m_0+j}$, $L_j^* = L_{m_0+j}$. Then for $j \neq 0$

$$\xi_j^* \sim \left(\frac{|j|}{n}\right)^{1/(\delta+1)}, \qquad |I_j^*| \sim \frac{1}{n^{1/(\delta+1)}|j|^{\delta/(\delta+1)}}.$$

There is an M such that all the points in the intervals I_j^* with $|j| \leq n/M$ lie at most half as far from any point of A as from $a_m = 0$. If, say, $x \in I_{j_0^*}^*$, $j_0^* > 0$, then we have to estimate (see (4.20))

$$S_1(x) := \sum_{j=1}^{j_0^*-L} |L_j^*(x)|,$$

$$S_2(x) := \sum_{j=-n/M}^{0} |L_j^*(x)|,$$

$$S_3(x) := \sum_{j=j_0^*+L}^{n/M} |L_j^*(x)|,$$

and

$$S_4(x) := \sum{}' |L_j^*(x)|,$$

where $\sum'$ indicates that the summation has to be taken for all other indices $|j - j_0^*| \geq L$ not appearing in S_1, S_2 or S_3. The choice of M guarantees that S_4 is bounded independently of x and n (use the monotonicity argument of the first theorem in this section). For notational convenience we write below j_0 instead of j_0^*.

We shall estimate $S_1(x) + S_2(x)$; the sum $S_3(x)$ can be similarly handled. Now (4.21) yields

$$S_1(x) \leq \sum_{j=1}^{j_0-L} \frac{C\left(n^{1/(\delta+1)} j^{\delta/(\delta+1)}\right)^{-2}}{\left((j_0/n)^{1/(\delta+1)} - (j/n)^{1/(\delta+1)}\right)^2} \leq \sum_{j=1}^{j_0/2} + \sum_{j=j_0/2}^{j_0-L}$$

$$\leq C\sum_{j=1}^{j_0/2} \frac{j^{-2\delta/(\delta+1)}}{j_0^{2/(\delta+1)}} + C\sum_{j=j_0/2}^{j_0-L} \frac{j^{-2\delta/(\delta+1)}}{((j_0-j)j_0^{-\delta/(\delta+1)})^2} = O(1)$$

and

$$S_2(x) \leq \sum_{j=-n/M}^{0} \frac{C\left(n^{1/(\delta+1)}(|j|+1)^{\delta/(\delta+1)}\right)^{-2}}{\left((j_0/n)^{1/(\delta+1)} + (|j|/n)^{1/(\delta+1)}\right)^2} = \sum_{j<-j_0} + \sum_{j=-j_0}^{0}$$

$$\leq C\sum_{j<-j_0} \frac{1}{j^2} + C\sum_{j=-j_0}^{0} \frac{(|j|+1)^{-2\delta/(\delta+1)}}{j_0^{2/(\delta+1)}} = O(1).$$

These estimates prove (4.14) and (4.15) in the case when the support of the measure μ is one interval and $x \in I$. But then by the principle of domination we have (4.14) in this case for all $x \in \mathbf{R}$.

The same argument (in fact, in a simpler form) can be used to show that (4.15)–(4.16) hold for $x \notin I$, as well. Thus, we have proved the theorem for one interval.

Now suppose that the support consists of the intervals $J_1, \ldots, J_k$. Set $\alpha_j = \mu(J_j)$, and choose numbers n_j such that their sum is n, and for every $j = 1, \ldots, k$ we have $|n_j - [\alpha_j n]| \le 1$. Since the sum of the α_j's is $\|\mu\| = 1$, this is clearly possible. Now we can apply the preceding consideration to every measure $\mu_j = (1/\alpha_j)\mu|_{J_j}$ and to the degrees n_j to get polynomials P_{n_j} of degree n_j with all their zeros lying in J_j that satisfy (4.14)–(4.16) with μ etc. replaced by μ_j etc. But then one can immediately see that the product polynomial

$$P_n(x) = \prod_{j=1}^{k} P_{n_j}(x)$$

satisfies the estimates stated in the theorem for suitable C and c (notice that the potential U^μ is uniformly bounded on compact subsets of $\mathbf{R}$ by the properties we assumed on μ). □

VI.5 Norm Inequalities for Weighted Polynomials with Exponential Weights

We shall discuss three types of weighted norm inequalities for polynomials with weight $w(x) = \exp(-c|x|^\lambda)$. The infinite–finite range inequalities reduce questions for polynomials with exponential weights on infinite intervals to those on finite ones, and have turned out to be extremely useful in the theory of Freud-type orthogonal polynomials. The Markoff inequality provides bounds for the weighted norm of the derivative of a polynomial in terms of the weighted norm of the polynomial itself. Finally, Nikolskii-type inequalities compare different L^p norms of weighted polynomials.

We begin with the simplest form of the infinite–finite range inequality.

Theorem 5.1. *Let* $w_\lambda(x) = \exp(-\gamma_\lambda |x|^\lambda)$, $x \in \mathbf{R}$, *where* $\lambda > 0$ *and*

$$\gamma_\lambda = \Gamma\left(\frac{\lambda}{2}\right)\Gamma\left(\frac{1}{2}\right) \Big/ 2\Gamma\left(\frac{\lambda}{2} + \frac{1}{2}\right).$$

Then for every $0 < p < \infty$ *and* $\delta > 0$, *there exist positive constants* c_δ *and* C_δ *such that for any polynomial* P_n *of degree at most n the estimate*

$$\|P_n w_\lambda\|_{L^p(\mathbf{R}\setminus I_\delta)} \le C_\delta e^{-c_\delta n} \|P_n w_\lambda\|_{L^p(\mathbf{R})}, \quad n = 1, 2, \ldots, \tag{5.1}$$

holds, where

$$I_\delta = [-(1+\delta)n^{1/\lambda}, (1+\delta)n^{1/\lambda}]. \tag{5.2}$$

In other words, for every $\delta > 0$ only an exponentially small fraction of the L^p-norm of $P_n w_\lambda$ lives on

$$(-\infty, -(1+\delta)n^{1/\lambda}) \bigcup ((1+\delta)n^{1/\lambda}, \infty).$$

If the weight is $\exp(-c|x|^\lambda)$, then simple dilation reduces the problem to Theorem 5.1 (cf. Section IV.5, (IV.5.4)).

Sometimes one needs a finer estimate such as the following.

Theorem 5.2. *Let* $w(x) = \exp(-\gamma_\lambda|x|^\lambda)$, $\lambda > 0$, *be as above, and let* $0 < p < \infty$. *Then there are constants* c, C *such that*

$$\|P_n w_\lambda\|_{L^p(\mathbf{R}\setminus J(\rho_n))} \le Ce^{-cn\rho_n^{3/2}} \|P_n w_\lambda\|_{L^p(\mathbf{R})}, \quad n = 1, 2, \ldots, \tag{5.3}$$

where

$$J(\rho_n) = [-(1+\rho_n)n^{1/\lambda}, (1+\rho_n)n^{1/\lambda}].$$

In particular, if $\rho_n n^{2/3} \to \infty$, *then*

$$\|P_n w_\lambda\|_{L^p(\mathbf{R})} = (1+o(1))\|P_n w_\lambda\|_{L^p(J(\rho_n))}.$$

Finally, if in the infinite–finite range inequality we need only a constant factor, then we can get an even smaller range which is actually part of the interval $[-n^{-1/\lambda}, n^{-1/\lambda}]$ where the norm of the weighted polynomials in question live (see Theorem III.2.1).

Theorem 5.3. *Let* $w_\lambda(x) = \exp(-\gamma_\lambda|x|^\lambda)$, $\lambda > 0$, *be as before, and* $0 < p < \infty$. *Then for every* K *there exists a constant* $C = C_K$ *such that for any polynomial* P_n *of degree at most* n *the estimate*

$$\|P_n w_\lambda\|_{L^p(\mathbf{R})} \le C_K \|P_n w_\lambda\|_{L^p(I_K^*)}, \quad n = 1, 2, \ldots,$$

holds, where

$$I_K^* = [-(1-Kn^{-2/3})n^{1/\lambda}, (1-Kn^{-2/3})n^{1/\lambda}].$$

Proof of Theorem 5.1. By setting $x = n^{1/\lambda}y$, $R_n(y) = P_n(x)$, and $w(y) = \exp(-\gamma_\lambda|y|^\lambda)$ we get

$$\|P_n w_\lambda\|_{L^p(\mathbf{R})} = n^{1/p\lambda}\|R_n w^n\|_{L^p(\mathbf{R})},$$

$$\|P_n w_\lambda\|_{L^p(\mathbf{R}\setminus I_\delta)} = n^{1/p\lambda}\|R_n w^n\|_{L^p(\mathbf{R}\setminus[-1-\delta,1+\delta])}.$$

Hence the theorem immediately follows from Theorem III.6.1 since for w we have $\mathcal{S}_w = \mathcal{S}_w^* = [-1, 1]$ (see Section IV.5). □

The proof of Theorem 5.2 is very similar to, but actually much simpler than that of Theorem 5.3; therefore first we verify the latter one and only indicate the necessary changes we have to make to get Theorem 5.2.

For the proof we need a lemma.

Lemma 5.4. *Let g be a bounded continous function on $\overline{\mathbf{C}}$ such that* $\log|g|$ *is subharmonic on* $\overline{\mathbf{C}}\setminus[-1,1]$, *which is the closed plane cut along* $[-1,1]$. *Then we have for real u, $|u|>1$, the inequality*

$$|g(u)| \le \frac{|u|+1}{|u|-1}\int_{-1}^{1}|g(t)|\,dt.$$

Proof. Let us consider the mapping

$$z=\frac{1}{2}\left(w+\frac{1}{w}\right); \qquad w=z-\sqrt{z^2-1},$$

that maps $\mathbf{C}\setminus[-1,1]$ onto the unit disk. The function

$$\log\left|g\left(\frac{1}{2}\left(w+\frac{1}{w}\right)\right)\right|+\log|w^2-1|$$

is subharmonic in the unit disk Δ; hence

$$h(w):=\left|g\left(\frac{1}{2}\left(w+\frac{1}{w}\right)\right)\right||w^2-1|$$

is a continuous function on the closed disk $\overline{\Delta}$, which is subharmonic in Δ. Therefore, the values of h are bounded by the harmonic function that coincides with $h(e^{i\theta})=2|g(\cos\theta)||\sin\theta|$ on the boundary of Δ, and this function is given by the Poisson integral of $h(e^{i\theta})$ (see Theorem 0.4.1 and Corollary 0.4.4). Taking into account the form of the Poisson kernel from (0.4.6), we obtain with $r=u-\sqrt{u^2-1}$ for $u>1$ the inequality

$$|g(u)| \le \frac{1}{1-r^2}\frac{1}{2\pi}\int_{-\pi}^{\pi}2|g(\cos\theta)||\sin\theta|\frac{1-r^2}{(1-r)^2+4r\sin^2(\theta/2)}\,d\theta.$$

Here the integral of $|g(\cos\theta)||\sin\theta|$ with respect to θ equals two times the integral of $g(t)$ with respect to $t\in[-1,1]$; therefore we obtain

$$|g(u)| \le \frac{1}{(1-r)^2}\int_{-1}^{1}|g(t)|\,dt.$$

To complete the proof we only have to mention that

$$1-r \ge \sqrt{\frac{u-1}{u+1}}.$$

The proof for negative u is the same. □

Proof of Theorem 5.3. Let $w=w_\lambda$, and consider the energy problem with weight w. Because of the the normalization of w, the extremal support is $[-1,1]$, and in view of Theorem 5.1 our task is to show that for every polynomial P_n of degree at most n we have

$$\|P_n w^n\|_{L^p[-2,2]} \le C_K \|P_n w^n\|_{L^p[-1+Kn^{-2/3}, 1-Kn^{-2/3}]}.$$

To prove this, let w^* be the restriction of w to

$$\Sigma^* := [-1 + 2Kn^{-2/3}, 1 - 2Kn^{-2/3}],$$

and with $\kappa := (1 - 2Kn^{-2/3})^\lambda$ let the weight $w^{**}(x)$ be $\exp(-\gamma_\lambda \kappa^{-1}|x|^\lambda)$ on $\Sigma^{**} := \Sigma = \mathbf{R}$, and consider the corresponding weighted energy problems. The support of the extremal measure corresponding to w^{**} coincides with Σ^* (see Section IV.5, (IV.5.4)), and by Theorem I.1.3 on that support we have the inequality

$$\kappa(U^{\mu_{w^{**}}}(x) - F_{w^{**}}) + (1-\kappa)\left(U^{\mu_{\Sigma^*}}(x) - \log\frac{1}{\operatorname{cap}(\Sigma^*)}\right) \le U^{\mu_{w^*}}(x) - F_{w^*}$$

(actually with the "=" sign), where, as usual, μ_{Σ^*} denotes the equilibrium measure of the interval Σ^*. Therefore, by the principle of domination, the same inequality holds for all $x \in \mathbf{R}$. For $|x| \in [1 - Kn^{-2/3}, 2]$, formula (IV.5.6) gives with some $c > 0$

$$\kappa\,(U^{\mu_{w^{**}}}(x) - F_{w^{**}}) + \gamma_\lambda |x|^\lambda \ge c\left||x| - (1 - 2Kn^{-2/3})\right|^{3/2}$$

(recall that for the weight w^{**} the extremal support is $[-1+2Kn^{-2/3}, 1-2Kn^{-2/3}]$ and not $[-1, 1]$), and since $1 - \kappa \le Dn^{-2/3}$ with some constant D, we obtain from (I.1.8)

$$(1-\kappa)\left(U^{\mu_{\Sigma^*}}(x) - \log\frac{1}{\operatorname{cap}(\Sigma^*)}\right) \ge -2Dn^{-2/3}\sqrt{|x| - (1 - 2Kn^{-2/3})}.$$

These estimates show that for $|x| \in [1 - Kn^{-2/3}, 2]$ we have

$$-n(U^{\mu_{w^*}}(x) - F_{w^*} + \gamma_\lambda |x|^\lambda) \tag{5.4}$$

$$\le -cn\left||x| - (1 - 2Kn^{-2/3})\right|^{3/2} + 2Dn^{1/3}\sqrt{|x| - (1 - 2Kn^{-2/3})}.$$

The function

$$g(z) := |P_n(z)|^p \exp(np(U^{\mu_{w^*}}(z) - F_{w^*}))$$

satisfies the assumptions of Lemma 5.4 with $[-1, 1]$ replaced by the interval $[-1 + 2Kn^{-2/3}, 1 - 2Kn^{-2/3}]$; hence for real x, $|x| \in [1 - Kn^{-2/3}, 2]$, we get

$$|P_n(x)|^p \le 2\exp(-np(U^{\mu_{w^*}}(x) - F_w))\frac{|x| + 1}{|x| - (1 - 2Kn^{-2/3})} \times$$

$$\times \int_{-1+2Kn^{-2/3}}^{1-2Kn^{-2/3}} |P_n(t)|^p \exp(np(U^{\mu_{w^*}}(t) - F_{w^*}))\,dt,$$

and here the exponential term under the integral sign is just $w(t)^{np}$. Therefore, the last inequality and (5.4) yield with $\xi := |x| - (1 - 2Kn^{-2/3})$, for which $\xi \in [Kn^{-2/3}, 3]$ is true, that

$$\|P_n w^n\|^p_{L^p[-2,2]} \le 12\|P_n w^n\|^p_{L^p[-1+Kn^{-2/3},1-Kn^{-2/3}]} \times \tag{5.5}$$

$$\times \left(1 + \int_{Kn^{-2/3}}^{3} \frac{\exp(-cpn\xi^{3/2} + 2Dpn^{1/3}\xi^{1/2})}{\xi}\, d\xi\right),$$

and the substitution $\zeta := n^{2/3}\xi$ shows that the last integral is finite. This proves the theorem. □

Proof of Theorem 5.2. In a fashion similar (but actually much simpler) to the way we obtained (5.5) we can derive from Lemma 5.4 and (IV.5.6) that

$$\|P_n w^n\|^p_{L^p[-2,2]\setminus J(\rho_n)} \le C\|P_n w^n\|^p_{L^p[-1,1]} \left(1 + \int_{\rho_n}^{3} \frac{\exp(-cpn\xi^{3/2})}{\xi}\, d\xi\right),$$

from which the theorem follows as before with the substitution $\zeta = n^{2/3}\xi$. □

We now turn our attention to Markoff inequalities.

Theorem 5.5. *Let w_λ, $\lambda > 0$, be as before, and let P_n be a polynomial of degree at most n, $n = 1, 2, \ldots$. Then for any $1 \le p \le \infty$*

$$\|w_\lambda P_n'\|_{L^p(\mathbf{R})} \le C_{p,\lambda} M(n,\lambda)\|w_\lambda P_n\|_{L^p(\mathbf{R})}, \tag{5.6}$$

where

$$M(n,\lambda) = \begin{cases} n^{1-1/\lambda} & \text{if } \lambda > 1 \\ \log n & \text{if } \lambda = 1 \\ 1 & \text{if } \lambda < 1. \end{cases} \tag{5.7}$$

The numbers $M(n,\lambda)$ are called Markoff constants. Note that they are bounded when $\lambda < 1$, i.e. in this case the weighted norm of the derivative of a polynomial is always less than a fixed constant times the original norm.

We also mention that the same estimates hold for $0 < p < 1$ and that they are sharp (see the notes at the end of this chapter).

Proof of Theorem 5.5. Let $w(x) = \exp(-\gamma_\lambda |x|^\lambda)$. On making use of the substitution $x = n^{1/\lambda}y$ that was applied at the beginning of the proof of Theorem 5.1, our task is to show that for every polynomial P_n of degree at most n we have

$$\|w^n P_n'\|_{L^p(\mathbf{R})} \le C_p M^*(n,\lambda)\|w^n P_n\|_{L^p(\mathbf{R})}, \tag{5.8}$$

where

$$M^*(n,\lambda) = \begin{cases} n & \text{if } \lambda > 1 \\ n\log n & \text{if } \lambda = 1 \\ n^{1/\lambda} & \text{if } \lambda < 1. \end{cases}$$

First let $1 \le p < \infty$. In view of Theorem 5.3 it suffices to prove the same inequality with the $L^p(\mathbf{R})$ norm $\|w^n P_n'\|_{L^p(\mathbf{R})}$ on the left replaced by

$$\|w^n P_n'\|_{L^p[-1+2n^{-2/3},1-2n^{-2/3}]},$$

and this is what we are going to show.

To do this we consider the weighted energy problem on $\Sigma = \mathbf{R}$ with weight w. The function $\log|P_n(z)| + U^{\mu_w}(z) - F_w$ is bounded from above and subharmonic on the plane cut along $[-1, 1]$, so Lemma 5.4 can be applied to

$$g(z) := |P_n(z)|^p \exp(np(U^{\mu_w}(z) - F_w))$$

(see Sections IV.5, I.4 for the continuity of U^{μ_w} on the boundary of $\overline{\mathbf{C}} \setminus [-1, 1]$).

Thus, for real u, $|u| > 1$, we have

$$g(u) \le \frac{|u|+1}{|u|-1} \int_{-1}^{1} g(t)\, dt. \tag{5.9}$$

We shall need an improvement of that estimate for $|u|$ close to 1. In fact, consider the weight $w^*(x) = \exp(-\gamma_\lambda e^{\lambda/n}|x|^\lambda) = w(e^{1/n}x)$. On applying substitution and Theorem I.3.3 we can easily get that the corresponding extremal support is the interval $[-e^{-1/n}, e^{-1/n}]$, the extremal measure is given by $d\mu_{w^*}(t) = d\mu_w(e^{1/n}t)$, and

$$U^{\mu_{w^*}}(z) - F_{w^*} = U^{\mu_w}(e^{1/n}z) - F_w.$$

By formula (IV.5.5), the potential $U^{\mu_w}(x)$ has a continuous derivative for $|x| \in [1/2, 3]$; therefore the previous identity shows that for $x \in [-2, 2]$ the function

$$g^*(x) := |P_n(x)|^p \exp(np(U^{\mu_{w^*}}(x) - F_{w^*}))$$

is smaller than a constant times $g(x)$ and vice versa. Therefore, if we apply (5.9) to g^* and w^* instead of g and w, then for real x, $|x| \ge 1$, we have

$$\begin{aligned} g(x) \le Cg^*(x) &\le C\frac{|x|+1}{|x|-e^{-1/n}} \int_{-e^{-1/n}}^{e^{-1/n}} g^*(t)\, dt \\ &\le \frac{C}{|x|-1+1/2n} \int_{-1}^{1} g(t)\, dt, \end{aligned} \tag{5.10}$$

and this is the improvement we had in mind.

The function

$$G(z) := \frac{1}{\pi} \int_{\mathbf{R}} \frac{y}{(x-t)^2 + y^2} g(t)\, dt, \qquad z = x + iy,$$

is bounded and harmonic on the upper half plane $\mathbf{C}_+$ and continuous up to its boundary, where it coincides with g. Therefore, the generalized minimum principle for superharmonic functions (Theorem I.2.4) applied to $G - g$ shows that for $z \in \mathbf{C}_+$ we have $g(z) \le G(z)$. The same can be said of points on the lower half-plane, and altogether we get for arbitrary y

$$g(x+iy) \le \frac{1}{\pi} \int_{\mathbf{R}} \frac{|y|}{(x-u)^2 + y^2} g(u)\, du. \tag{5.11}$$

Let

$$\delta = \frac{1}{M^*(n,\lambda)}.$$

By Cauchy's formula we have

$$P_n'(x) = \frac{1}{2\pi\delta}\int_{-\pi}^{\pi} P_n(x+\delta e^{i\varphi})e^{-i\varphi}\,d\varphi,$$

and so

$$|P_n'(x)| \le \frac{1}{2\pi\delta}\int_{-\pi}^{\pi} |P_n(x+\delta e^{i\varphi})|\,d\varphi.$$

This implies via Hölder's inequality that

$$|P_n'(x)|^p \le \frac{1}{2\pi\delta^p}\int_{-\pi}^{\pi} |P_n(x+\delta \varepsilon^{i\varphi})|^p\,d\varphi$$

and

$$w(x)^{np}|P_n'(x)|^p \le \tag{5.12}$$

$$\frac{1}{2\pi\delta^p}\int_{-\pi}^{\pi} \exp(np(U^{\mu_w}(x+\delta e^{i\varphi}) - F_w))|P_n(x+\delta e^{i\varphi})|^p\sigma_n(x,\varphi)\,d\varphi,$$

where

$$\sigma_n(x,\varphi) = \exp(-np(U^{\mu_w}(x+\delta e^{i\varphi}) - F_w + \gamma_\lambda|x|^\lambda)).$$

We claim that this last function is uniformly bounded for $x \in [-1+2n^{-2/3}, 1-2n^{-2/3}]$.

In view of Theorem I.1.3 and the continuity properties of the function $|x|^\lambda$ it is enough to verify that, for $|y_0| \le \delta = 1/M^*(n,\lambda)$,

$$|U^{\mu_w}(x+iy_0) - F_w + \gamma_\lambda|x|^\lambda| \le \frac{C}{n} \tag{5.13}$$

with some constant C. But that is easy: the partial derivative of $U^{\mu_w}(x+iy)$ with respect to y is

$$\frac{\partial U^{\mu_w}(x+iy)}{\partial y} = -\int_{-1}^{1} \frac{y}{(x-t)^2+y^2}d\mu_w(t). \tag{5.14}$$

The form of μ_w given in Theorem IV.5.1 shows that for fixed y this is largest in absolute value if $x = 0$. The same form yields that as $t \to 0$

$$\frac{d\mu_w(t)}{dt} \sim \begin{cases} 1 & \text{if } \lambda > 1 \\ \log 1/|t| & \text{if } \lambda = 1 \\ |t|^{\lambda-1} & \text{if } \lambda < 1. \end{cases}$$

Therefore a simple calculation based on the properties of the Poisson kernel in (5.14) shows that uniformly in x

$$\left|\frac{\partial U^{\mu_w}(x+iy)}{\partial y}\right| \le C \begin{cases} 1 & \text{if } \lambda > 1 \\ \log 1/|y| & \text{if } \lambda = 1 \\ |y|^{\lambda-1} & \text{if } \lambda < 1. \end{cases}$$

Finally,

$$\begin{aligned} \left|U^{\mu_w}(x+iy_0) - F_w + \gamma_\lambda |x|^\lambda\right| &= \left|\int_0^{y_0} \frac{\partial U^{\mu_w}(x+iy)}{\partial y}\,dy\right| \\ &\le C \begin{cases} |y_0| & \text{if } \lambda > 1 \\ |y_0|\log 1/|y_0| & \text{if } \lambda = 1 \\ |y_0|^\lambda & \text{if } \lambda < 1, \end{cases} \end{aligned}$$

which verifies (5.13).

Thus, (5.12) yields

$$\int_{-1+2n^{-2/3}}^{1-2n^{-2/3}} w(x)^{np}|P_n'(x)|^p\,dx \le \frac{C}{\delta^p} \sup_{|y|\le\delta} \int_{-1+n^{-2/3}}^{1-n^{-2/3}} g(x+iy)\,dx.$$

If we put this together with (5.11) we can see that

$$\begin{aligned} &\int_{-1+2n^{-2/3}}^{1-2n^{-2/3}} w(x)^{np}|P_n'(x)|^p\,dx \\ &\le \frac{C}{\delta^p} \sup_{|y|\le\delta} \int_{-1+n^{-2/3}}^{1-n^{-2/3}} \int_{\mathbf{R}} \frac{|y|}{(x-u)^2+y^2} g(u)\,du\,dx. \end{aligned}$$

Let us write the inner integral as the sum of integrals over $[-1, 1]$ and $\mathbf{R}\setminus[-1, 1]$. Since the integral of $|y|/((x-u)^2+y^2)$ with respect to x is at most π, we can continue the preceding inequality as

$$\le \frac{C}{\delta^p}\left(\int_{-1}^{1} g(u)\,du + \sup_{|y|\le\delta} \int_{-1+n^{-2/3}}^{1-n^{-2/3}} \int_{|u|\ge 1} \frac{|y|}{(x-u)^2+y^2} g(u)\,du\,dx\right).$$

Now to estimate $g(u)$ in the last integral we make use of (5.10), which, together with $M^*(n,\lambda) = 1/\delta$, shows that the inequality

$$\int_{-1+2n^{-2/3}}^{1-2n^{-2/3}} w(x)^{np}|P_n'(x)|^p\,dx \le CM^*(n,\lambda)^p \|w^n P_n\|_{L^p(\mathbf{R})}^p$$

holds if we can verify that for $|y| \le 1/n$ and $x \in [-1+n^{-2/3}, 1-2n^{-/3}]$ we have

$$\int_{|u|\ge 1} \int_{-1+n^{-2/3}}^{1-n^{-2/3}} \frac{y}{(x-u)^2+y^2} \frac{1}{|u|-1+1/n}\,dx\,du \le C$$

independently of n. By dropping the y^2 from the denominator this follows in a straightforward manner.

The proof is complete for $1 \le p < \infty$.

When $p = \infty$ we have to estimate $w(x)^n |P_n'(x)|$ for $x \in [-1, 1]$ (see Theorem III.2.1). In this case formula (5.12) takes the form

$$w(x)^n |P_n'(x)| \le \frac{1}{2\pi\delta} \int_{-\pi}^{\pi} e^{n(U^{\mu_w}(x+\delta e^{i\varphi})-F_w)} |P_n(x+\delta\varepsilon^{i\varphi})| \sigma_n(x,\varphi)\, d\varphi$$

with

$$\sigma_n(x,\varphi) = \exp(-n(U^{\mu_w}(x+\delta e^{i\varphi}) - F_w + \gamma_\lambda |x|^\lambda)),$$

and the theorem follows from the boundedness of $\sigma_n(x,\varphi)$ (that has been verified above) and from the maximum principle for subharmonic functions. □

Finally, we turn to Nikolskii–type inequalities for weighted polynomials.

Theorem 5.6. *Let w_λ, $\lambda > 0$, be as in the preceding theorems, and let $0 < p$, $q \le \infty$. Then there is a constant C such that for all polynomials of degree at most n, $n = 1, 2, \ldots$, we have*

$$\|P_n w_\lambda\|_{L^p(\mathbf{R})} \le C N_n(\lambda, p, q) \|P_n w_\lambda\|_{L^q(\mathbf{R})}, \tag{5.15}$$

where

$$N_n(\lambda, p, q) = \begin{cases} n^{(1/\lambda)(1/p-1/q)} & \text{if } p \le q \\ M(n,\lambda)^{(1/q-1/p)} & \text{if } q < p, \end{cases} \tag{5.16}$$

with $M(n,\lambda)$ defined in (5.7).

We remark that (5.15) is sharp regarding the order of the Nikolskii constants $N_n(\lambda, p, q)$ (see the notes at the end of this chapter).

Proof of Theorem 5.6. The case $p \le q$ immediately follows from Theorem 5.1 if we apply Hölder's inequality on the finite interval I_δ with some fixed positive δ.

Let now $q < p$. In this case the result can be easily derived from the Markoff inequality (for the supremum norm) in Theorem 5.5. In fact, let $L = \|w_\lambda P_n\|_{L^\infty(\mathbf{R})}$. By Theorem III.2.1 there is a point $x_0 \in [-n^{1/\lambda}, n^{1/\lambda}]$ with $w_\lambda(x_0)|P_n(x_0)| = L$. Without loss of generality we may assume $x_0 \ge 0$ and $w_\lambda(x_0)P_n(x_0) = L$. For technical reasons first let us also suppose that if $0 < \lambda < 1$, then $x_0 \ge 1$. Theorem 5.5 implies that for $x \in [x_0, x_0+1]$ we have $w_\lambda(x)|P_n'(x)| \le C_1 M(n,\lambda) L$, and since $|w_\lambda(x)'| \le 2\lambda M(n,\lambda) w_\lambda(x)$ is also true for such x's, we can conclude that $|(w_\lambda(x)P_n(x))'| \le CM(n,\lambda)L$ with a constant C. This and $w_\lambda(x_0)P_n(x_0) = L$ imply that for $x_0 \le x \le x_0 + 1/2CM(n,\lambda)$ we have $w_\lambda(x)P_n(x) \ge L/2$, and so

$$\|w_\lambda P_n\|_q \ge \left(\int_{x_0}^{x_0+1/2CM(n,\lambda)} (L/2)^q\, dt \right)^{1/q} \ge \frac{1}{2(2C)^{1/q} M(n,\lambda)^{1/q}} L,$$

and this is exactly the inequality (5.15) for the pair q and $p = \infty$.

But then the same automatically follows for any pair $q < p$. In fact,

$$\|w_\lambda P_n\|_p^p \le \|w_\lambda P_n\|_\infty^{p-q} \|w_\lambda P_n\|_q^q \le \left(CM(n,\lambda)^{1/q} \|w_\lambda P_n\|_q\right)^{p-q} \|w_\lambda P_n\|_q^q,$$

and on taking the p-th root we obtain (5.15).

When $0 < \lambda < 1$ and x_0 happens to be in the interval $[0, 1]$, then we can no longer use the previous estimate for the derivative $w_\lambda(x)'$ because this derivative is not bounded around zero. But in such cases we can consider the polynomial $Q_n(x) = P_n(x-1)$ instead of $P_n(x)$. Since $w(x-1) \sim w(x)$ uniformly in $x \in \mathbf{R}$, these two polynomials have comparable weighted L^p and L^q norms, and $w_\lambda(x_0+1)|Q_n(x_0+1)|$ is comparable to the L^∞ norm $\|w_\lambda Q_n\|_{L^\infty(\mathbf{R})}$. Since here $x_0 + 1 \in [1, 2]$, we can repeat the preceding argument for Q and $x_0 + 1$, and the result follows. □

VI.6 Comparisons of Different Weighted Norms of Polynomials

In this section we apply some of the previous results to compare different weighted norms of polynomials with exponential weights. In turn these results will be applied in Section VI.7 for the asymptotic determination of certain n-widths.

Let $\alpha, \beta > 0$ and consider the set $H(\alpha, \beta)$ of those continuous weights $w(x)$ on $\mathbf{R}$ that satisfy

$$\lim_{|x|\to\infty} \frac{\log(1/w(x))}{x^\beta} = \alpha. \tag{6.1}$$

Typical examples are the exponential weights $\exp(-\alpha|x|^\beta)$. Suppose $w_1 \in H(\alpha_1, \beta_1)$ and $w_2 \in H(\alpha_2, \beta_2)$. We address the question of comparing the supremum norms $\|P_n w_1\|$ and $\|P_n w_2\|$ if P_n is a polynomial of degree at most n. In particular, for fixed $w_1 \in H(\alpha_1, \beta_1)$ and $w_2 \in H(\alpha_2, \beta_2)$ we shall be interested in estimating the quantity

$$h_n(\alpha_1, \beta_1, \alpha_2, \beta_2) := \sup\{\|P_n w_1\| / \|P_n w_2\| \mid \deg P_n \le n, \quad P_n \not\equiv 0\}.$$

We shall see that the order of h_n is independent of the actual choice of w_1 and w_2, so we can use the somewhat misleading notation $h_n(\alpha_1, \beta_1, \alpha_2, \beta_2)$.

Obviously, if $\beta_1 > \beta_2$ or $\beta_1 = \beta_2$ and $\alpha_1 \ge \alpha_2$, then $h_n^{1/n} \to 1$; therefore in what follows we shall assume $\beta_1 < \beta_2$ or $\beta_1 = \beta_2$ and $\alpha_1 < \alpha_2$.

Theorem 6.1. *There holds*

$$\lim_{n\to\infty} \frac{\log h_n(\alpha_1, \beta_1, \alpha_2, \beta_2)}{n \log n} = \frac{1}{\beta_1} - \frac{1}{\beta_2}$$

if $\beta_1 < \beta_2$. On the other hand, if $\beta_1 = \beta_2 = \beta$ and $\alpha_1 < \alpha_2$, then

$$\lim_{n\to\infty} \frac{\log h_n(\alpha_1, \beta_1, \alpha_2, \beta_2)}{n} = \delta + \sqrt{\delta^2 - 1},$$

where

$$\delta = \left(1/B^{-1}\left(\frac{2\gamma_\beta \alpha_1}{\alpha_2}; \frac{\beta}{2}, \frac{1}{2}\right)\right)^{1/2}, \quad \gamma_\beta = \sqrt{\pi}\frac{\Gamma(\frac{\beta}{2})}{2\Gamma(\frac{\beta}{2}+\frac{1}{2})},$$

and $B^{-1}(x;\beta,\alpha)$ *denotes the inverse of the Beta-function*

$$B(x;\beta,\alpha)=\int_0^x u^{\beta-1}(1-u)^{\alpha-1}\,du.$$

For example, if $\beta_1=\beta_2=1$, then $\gamma_\beta=\pi/2$, and

$$\lim_{n\to\infty}\frac{\log h_n(\alpha_1,1,\alpha_2,1)}{n}=\cot\left(\frac{\alpha_1}{\alpha_2}\frac{\pi}{4}\right).$$

On the other hand, if $\beta_1=\beta_2=2$, then $\gamma_\beta=1$ and

$$\lim_{n\to\infty}\frac{\log h_n(\alpha_1,2,\alpha_2,2)}{n}=\sqrt{2\frac{\alpha_2}{\alpha_1}-1}.$$

Proof of Theorem 6.1. For $w\in H(\alpha,\beta)$ let $\overline{w}_{\alpha,\beta}(x)=\exp(-\alpha|x|^\beta)$. Then for every $\varepsilon>0$ there is a constant C_ε for which

$$\frac{1}{C_\varepsilon}\overline{w}_{\alpha+\varepsilon,\beta}(x)\le w(x)\le C_\varepsilon\overline{w}_{\alpha-\varepsilon,\beta}(x),$$

and so, in view of the type of asymptotics we need to prove in the theorem, we can assume that w_j is actually the weight $\exp(-\alpha_j|x|^{\beta_j})$, $j=1,2$.

First consider the case $\beta_1<\beta_2$. We make use of the substitution $x\to n^{1/\beta_2}y$. Then we have to estimate the ratios

$$\|R_n(y)\exp(-\alpha_1 n^{\beta_1/\beta_2}|y|^{\beta_1})\|/\|R_n(y)\exp(-\alpha_2 n|y|^{\beta_2})\| \tag{6.2}$$

for an arbitrary monic polynomial $R_n(y)=y^n+\cdots$. Let

$$\|R_n(y)\exp(-\alpha_2 n y^{\beta_2})\|=d_n.$$

Then applying Theorem II.2.1 to the weight $w(x)=\exp(-\alpha_2|x|^{\beta_2})$ we get

$$|R_n(y)|\le d_n\exp(n(\log(|y|+1)+c)),\qquad y\in\mathbf{C},$$

for some constant c. Thus, (6.2) is at most as large as the maximum of

$$\exp\left(-\alpha_1 n^{\beta_1/\beta_2}|y|^{\beta_1}+n(\log(|y|+1)+c)\right)$$

as a function of y. This maximum is attained around a constant times $n^{1/\beta_1-1/\beta_2}$ and is

$$\exp\left(\left(\frac{1}{\beta_1}-\frac{1}{\beta_2}\right)n\log n+O(n)\right),$$

by which we have verified

$$\limsup_{n\to\infty}\frac{\log h_n(\alpha_1,\beta_1,\alpha_2,\beta_2)}{n\log n}\le\frac{1}{\beta_1}-\frac{1}{\beta_2}. \tag{6.3}$$

On the other hand, if we set $P_n(x)=x^n$ in the definition of the quantity h_n, then from the equality

$$\|x^n e^{-\alpha|x|^\beta}\| = \left(\frac{n}{\alpha\beta}\right)^{n/\beta} e^{-n/\beta}$$

we can easily deduce that

$$\liminf_{n\to\infty} \frac{\log h_n(\alpha_1, \beta_1, \alpha_2, \beta_2)}{n \log n} \geq \frac{1}{\beta_1} - \frac{1}{\beta_2}.$$

Together with (6.3) this settles the case when $\beta_1 \neq \beta_2$.

When $\beta_1 = \beta_2 = \beta$, we again use substitution but this time we need more precise calculations. Let

$$\gamma_\beta = \Gamma(\frac{\beta}{2})\Gamma(\frac{1}{2})/2\Gamma(\frac{\beta}{2} + \frac{1}{2}) = \sqrt{\pi}\Gamma(\frac{\beta}{2})/2\Gamma(\frac{\beta}{2} + \frac{1}{2}),$$

be the quantity obtained in Theorem IV.5.1, and substitute $\alpha_2^{-1/\beta}\gamma_\beta^{1/\beta}n^{1/\beta}y$ for x. Exactly as above, we then have to estimate

$$\|R_n(y)\exp(-\alpha\gamma_\beta n|y|^\beta)\|/\|R_n(y)\exp(-\gamma_\beta n|y|^\beta)\|,$$

where R_n is a polynomial of degree n and $\alpha := \alpha_1/\alpha_2 < 1$. Let $w(x) = \exp(-\gamma_\beta n|x|^\beta)$. Since the equilibrium potential U^{μ_w} associated with w is everywhere continuous (cf. Section IV.5), we can conclude from Theorems III.2.1 and III.2.9 that when n is kept fixed, the supremum for R_n of $|R_n(y)|^{1/n}$ under the condition $\|R_n w^n\| \leq 1$ is at most $\exp(-U^{\mu_w}(y) + F_w)$, and this supremum tends to $\exp(-U^{\mu_w}(y) + F_w)$ when $n \to \infty$. Thus, to determine

$$\lim_{n\to\infty} \sup_{R_n} \left(\|R_n(y)\exp(-\alpha\gamma_\beta n|y|^\beta)\|/\|R_n(y)\exp(-\gamma_\beta n|y|^\beta)\|\right)^{1/n},$$

which is the same as

$$\lim_{n\to\infty} h_n(\alpha_1, \beta_1, \alpha_2, \beta_2)^{1/n},$$

we have to calculate the maximum of $\exp(-U^{\mu_w}(y) + F_w - \alpha\gamma_\beta|y|^\beta)$ on $\mathbf{R}$. Since $S_w = [-1, 1]$ (see Theorem IV.5.1), and the potential U^{μ_w} is even and continuous, this maximum is attained for some $y \geq 1$. The derivative of $-U^{\mu_w}(y)+F_w-\alpha\gamma_\beta y^\beta$ when $y > 1$ equals (see formula (IV.5.3) of Section IV.5)

$$\beta y^{\beta-1} \int_0^{1/y} \frac{t^{\beta-1}}{\sqrt{1-t^2}} dt - \alpha\gamma_\beta\beta y^{\beta-1},$$

and this becomes zero for y satisfying

$$\int_0^{1/y} t^{\beta-1}(1-t^2)^{-1/2} dt = \alpha\gamma_\beta. \tag{6.4}$$

Note that the integral on the left with y replaced by 1 is γ_β, and $\alpha = \alpha_1/\alpha_2 < 1$; hence there is one and only one $y > 1$ satisfying (6.4), and obviously this is where the maximum in question is attained. The substitution $t^2 \to v$ shows that this is given by

$$\left(1/B^{-1}\left(\frac{2\gamma_\beta\alpha_1}{\alpha_2};\frac{\beta}{2},\frac{1}{2}\right)\right)^{1/2}.$$

As for the maximum value itself, we substitute into (IV.5.2) the obtained value of y and get from $F_w = \log 2 + 1/\beta$ (see Theorem IV.5.1)

$$\begin{aligned}
&\exp(-U^{\mu_w}(y) + F_w - \alpha\gamma_\beta|y|^\beta)\\
&= \exp\left(\log|y+\sqrt{y^2-1}| + y^\beta\int_0^{1/y} t^{\beta-1}(1-t^2)^{-1/2}\,dt\right.\\
&\qquad \left. -\frac{1}{\beta} - \log 2 + F_w - \alpha\gamma_\beta y^\beta\right)\\
&= \exp\left(\log|y+\sqrt{y^2-1}|\right) = y + \sqrt{y^2-1}.
\end{aligned}$$

This proves the theorem for the case $\beta_1 = \beta_2 = \beta$. □

Since the form of the limit appearing in the preceding theorem is rather complicated in the case $\beta_1 = \beta_2$, we mention the following which is a corollary of the proof.

Corollary 6.2. *If $\beta_1 = \beta_2 = \beta$ and $\alpha_1 < \alpha_2$, then*

$$\begin{aligned}
\left(\frac{\alpha_2}{\alpha_1}\right)^{1/\beta} &\le \liminf_{n\to\infty} h_n(\alpha_1,\beta_1,\alpha_2,\beta_2)^{1/n} \qquad (6.5)\\
&\le \limsup_{n\to\infty} h_n(\alpha_1,\beta_1,\alpha_2,\beta_2)^{1/n} \le 2\left(\frac{\alpha_2}{\alpha_1}\right)^{1/\beta}.
\end{aligned}$$

Proof. First of all we note that the maximum of the function $\log y - s\gamma_\beta y^\beta$ for $y > 0$ is

$$\frac{1}{\beta}\left(\log\frac{1}{s\beta\gamma_\beta} - 1\right). \qquad (6.6)$$

Setting in the definition of h_n the polynomial z^n, we get the lower estimate in (6.5) from (6.6) if we apply the latter one for $s\gamma_\beta = \alpha_1$ and $s\gamma_\beta = \alpha_2$.

From the symmetry of the weight it follows that μ_w, where w is again $\exp(-\gamma_\beta n|x|^\beta)$, is even and so, for $x > 1$ we have $-U^{\mu_w}(y) \le \log y$. Hence with $\alpha = \alpha_1/\alpha_2$

$$\begin{aligned}
\max_{y\ge 1}\,(-U^{\mu_w}(y) + F_w - \alpha\gamma_\beta y^\beta) &\le \max_{y\ge 1}\,(\log y + F_w - \alpha\gamma_\beta y^\beta)\\
&= \frac{1}{\beta}\left(\log\frac{\alpha_2}{\alpha_1\beta\gamma_\beta} - 1\right) + \log 2 + \frac{1}{\beta}\\
&= \frac{1}{\beta}\log\frac{\alpha_2}{\alpha_1} - \frac{1}{\beta}\log\beta\gamma_\beta + \log 2 \le \frac{1}{\beta}\log\frac{\alpha_2}{\alpha_1} + \log 2
\end{aligned}$$

because

$$\beta\gamma_\beta = \int_0^1 \beta t^{\beta-1}(1-t^2)^{-1/2}\,dt > 1.$$

By the proof of Theorem 6.1 this verifies the upper estimate in (6.5). □

We shall need the variant of Theorem 6.1 when the norms are taken on different sets. Let us again consider the sets $H(\alpha,\beta)$ of weight functions w defined by (6.1) but now we suppose that w is defined on $\mathbf{C}$. If $w_1 \in H(\alpha_1,\beta_1)$, $w_2 \in H(\alpha_2,\beta_2)$, and E_1 and E_2 are two closed sets, then we can similarly define as above

$$h_n(\alpha_1,\beta_1,\alpha_2,\beta_2;E_1,E_2)$$

$$:= \sup\left\{\|P_n w_1\|_{E_1}/\|P_n w_2\|_{E_2} \mid \deg P_n \le n,\ P_n \text{ not identically zero}\right\},$$

where $\|\cdot\|_E$ denotes the supremum norm on E.

We say that $E \subseteq \mathbf{C}$ is circularly connected if its "circular projection"

$$\{r \ge 0 \mid E \text{ intersects } C_r := \{z \mid |z| = r\}\}$$

onto $\mathbf{R}_+$ is connected (i.e. it is an interval). With this notion the first half of Theorem 6.1 holds word for word in this more general setting.

Theorem 6.3. *Let E_1 and E_2 be two unbounded circularly connected closed subsets of $\mathbf{C}$ and $\beta_1 < \beta_2$. Then*

$$\lim_{n\to\infty} \frac{\log h_n(\alpha_1,\beta_1,\alpha_2,\beta_2;E_1,E_2)}{n\log n} = \frac{1}{\beta_1} - \frac{1}{\beta_2}. \tag{6.7}$$

The proof shows that if $\beta_1 = \beta_2$, then

$$\frac{\log h_n(\alpha_1,\beta_1,\alpha_2,\beta_2;E_1,E_2)}{n} \tag{6.8}$$

is bounded away from zero and infinity; however its asymptotic behavior strongly depends on the sets E_1, E_2, e.g. it may not have a limit as $n \to \infty$. In a similar manner, if $\beta_1 > \beta_2$ then (6.8) tends to zero.

Proof of Theorem 6.3. That the right-hand side in (6.7) is an asymptotic lower bound for the sequence on the left-hand side is shown exactly as in the preceding proof by considering $P_n(z) = z^n$.

In the proof of the corresponding estimate in the other direction we follow the proof of Theorem 6.1 and may again assume $w_i(z) = \exp(-\alpha_i|z|^{\beta_i})$, and since now we are dealing with the upper estimate we can obviously assume $E_1 = \mathbf{C}$. Let us use again the substitution $z = n^{1/\beta}y$. Then E_2 is mapped into some set $E_2^{(n)}$ under this substitution and we have to estimate (6.2), but now the norms are taken on different sets: $\mathbf{C}$ and $E_2^{(n)}$. The proof of Theorem 6.1 works word for word if $E_2 = \mathbf{R}$ (in which case $E_2^{(n)} = \mathbf{R}$, as well); hence it is enough to show that

$$\left\{ \| R_n(y) \exp(-\alpha_2 n |y|^{\beta_2}) \|_{\mathbf{R}} / \| R_n(y) \exp(-\alpha_2 n |y|^{\beta_2}) \|_{E_2^{(n)}} \right\}^{1/n} \tag{6.9}$$

is bounded as $n \to \infty$ independently of the choice of $R_n(y) = y^n + \cdots$ (recall that now we want to find an upper estimate and the normalizing factor in (6.7) is $n \log n$).

Consider the weight function $v(x) = \exp(-\alpha_2 x^{\beta_2})$ on $\mathbf{R}$ and let $\mathcal{S}_v$ be the support of the corresponding extremal measure. Choose the number $m \geq 2$ so that $\mathcal{S}_v \subseteq [-m, m]$. Further let

$$E_{2,n} = E_2^{(n)} \cap \{z \mid 1 \leq |z| \leq 2\}.$$

Since $E_2^{(n)}$ is circularly connected, for large n every circle $C_r = \{z \mid |z| = r\}$ with $1 \leq r \leq 2$ intersects $E_{2,n}$; hence, by Lemma I.2.1, $E_{2,n}$ has capacity at least $\operatorname{cap}([1,2]) = 1/4$. On applying Theorem I.3.6 to $\Sigma = E_{2,n}$, $w \equiv 1$, in which (classical) case $\exp(-F_w) = \operatorname{cap}(E_{2,n})$, we get that for any polynomial $S_k(y) = y^k + \cdots$ we have the inequality

$$\| S_k \|_{E_{2,n}} \geq \left(\frac{1}{4} \right)^k . \tag{6.10}$$

Let us now factor R_n as $R_n = S_k S_{n-k}^*$, where S_k, S_{n-k}^* are monic polynomials of degree k and $(n-k)$, respectively for some k such that S_k has all its zeros in $|z| \leq 3m$ while S_{n-k}^* has all its zeros in $|z| > 3m$. Then making use of the fact that for $|z'| > 3m$ we have

$$\max_{z \in -[m,m]} |z - z'| \leq 2 \min_{z \in E_{2,n}} |z - z'|$$

we can write, by Theorem II.2.1,

$$\begin{aligned}
\| R_n(y) e^{-\alpha_2 n |y|^{\beta_2}} \|_{\mathbf{R}} &= \| R_n(y) e^{-\alpha_2 n |y|^{\beta_2}} \|_{[-m,m]} \\
&\leq \| R_n \|_{[-m,m]} \leq (4m)^k \| S_{n-k}^* \|_{[-m,m]} \leq (4m)^k 2^{n-k} \min_{y \in E_{2,n}} |S_{n-k}^*(y)| \\
&\leq (4m)^k 2^{n-k} 4^k \| S_k \|_{E_{2,n}} \min_{y \in E_{2,n}} |S_{n-k}^*(y)| \leq (4m)^{2n} \| R_n \|_{E_{2,n}} \\
&\leq (4m)^{2n} e^{\alpha_2 n 2^{\beta_2}} \| R_n(y) e^{-\alpha n |y|^{\beta_2}} \|_{E_{2,n}},
\end{aligned}$$

which proves the boundedness of the expression in (6.9). The above chain of estimates is self explanatory, except perhaps that at the fourth inequality we used (6.10). □

VI.7 n-Widths for Weighted Entire Functions

The results of the preceding section have applications concerning n-widths of entire functions.

Let X be a normed linear space and $Y \subseteq X$ a subset of X. The Kolmogoroff n-width of Y in X is defined by

$$d_n(Y, X) := \inf_{X_n} \sup_{y \in Y} \inf_{x \in X_n} \|x - y\|,$$

where $\inf_{X_n}$ means taking infimum for all n-dimensional subspaces X_n of X.

We shall need the following lemma which is one of the basic results on n-widths.

Lemma 7.1. *If X_{n+1} is an $(n+1)$-dimensional subspace of X and*

$$Y = \{y \in X_{n+1} \mid \|y\|_X \le M\}$$

is the ball in X_{n+1} with radius M, then

$$d_n(Y, X) \ge M.$$

The proof immediately follows from the fact that if X_n is any n-dimensional subspace of X, then there is always $y \in Y$, $\|y\| = M$, such that its best approximation from X_n is the zero element of X_n, i.e.

$$M = \|y\| = \min_{x \in X_n} \|y - x\|,$$

which in turn is an easy consequence of Borsuk's antipodal theorem. For details see [180].

Consider the weights $w \in H(\alpha, \beta)$ from the preceding section and assume now that w is defined on the whole complex plane. We denote by $B(w)$ the set of all entire functions f satisfying

$$|f(z)w(z)| \le 1, \quad z \in \mathbf{C},$$

and if $E \subseteq \mathbf{C}$ is a closed set, then let $X_E(w)$ be the set of all functions f defined and continuous on E for which the norm

$$\|f\|_{X_E(w)} = \sup_{z \in E} |f(z)w(z)|$$

is finite. With this norm $X_E(w)$ is obviously a Banach space and $B(w)$ is a subset of the unit ball of $X_{\mathbf{C}}(w)$.

In this section we investigate the asymptotic behavior of the n-widths $d_n(B(w_1), X_E(w_2))$, where $w_1 \in H(\alpha_1, \beta_1)$ and $w_2 \in H(\alpha_2, \beta_2)$, under the assumption that E is a circularly connected unbounded closed subset of $\mathbf{C}$. If $\beta_1 > \beta_2$ or $\beta_1 = \beta_2$ but $\alpha_1 > \alpha_2$, then it is easy to see by considering functions of the form $\Sigma c_m z^{n_m}$ that $B(w_1)$ is not a subset of $X_E(w_2)$ and the same can happen in the case $\beta_1 = \beta_2$, $\alpha_1 = \alpha_2$. Hence we may assume $\beta_1 < \beta_2$ or $\beta_1 = \beta_2$ and $\alpha_1 < \alpha_2$.

Theorem 7.2. *Let $E \subseteq \mathbf{C}$ be a circularly connected unbounded closed set, $w_1 \in H(\alpha_1, \beta_1)$, $w_2 \in H(\alpha_2, \beta_2)$ and assume that $0 < \beta_1 < \beta_2$. Then*

$$\lim_{n\to\infty} d_n(B(w_1), X_E(w_2))^{1/n\log n} = \exp\left(\frac{1}{\beta_2} - \frac{1}{\beta_1}\right). \tag{7.1}$$

In the case $\beta_1 = \beta_2$, $\alpha_1 < \alpha_2$, the proof gives that $d_n^{1/n}$ lies between two positive constants less than 1. It is not known if the limit

$$\lim_{n\to\infty} d_n(B(w_1), X_E(w_2))^{1/n}$$

exists for $w_1(z) = \exp(-\alpha_1|z|^\beta)$, $w_2 = \exp(-\alpha_2|z|^\beta)$, $\alpha_1 < \alpha_2$, and any E. It is not difficult to show that if $E = \mathbf{C}$, then this limit exists and equals $(\alpha_1/\alpha_2)^{1/\beta}$. In fact, in this case the proof below can be easily modified to yield

$$\left(\frac{\alpha_1}{\alpha_2}\right)^{n/\beta} \le d_n(B(w_1), X_{\mathbf{C}}(w_2)) \le \left(\frac{\alpha_1}{\alpha_2}\right)^{n/\beta} \frac{1}{1-(\alpha_1/\alpha_2)^{1/\beta}} \tag{7.2}$$

for all n (use the results from Section IV.6 to conclude that if $\Sigma = \mathbf{C}$ and $v(z) = \exp(-\alpha_2|z|^\beta)$, then

$$F_v - U^{\mu_v}(z) = \begin{cases} -\alpha_2|z|^\beta & \text{if } |z| < (\alpha_2\beta)^{-1/\beta} \\ \dfrac{1}{\beta}(1+\log(\alpha_2\beta)) + \log|z| & \text{if } |z| \ge (\alpha_2\beta)^{-1/\beta}, \end{cases}$$

which can be used to prove the inequality

$$\|P_n(z)\exp(-\alpha_1|z|^\beta)\|_{\mathbf{C}} \le \left(\frac{\alpha_2}{\alpha_1}\right)^{n/\beta} \|P_n(z)\exp(-\alpha_2|z|^\beta)\|_{\mathbf{C}} \tag{7.3}$$

by the argument given in the preceding section. If we now use (7.3) instead of (7.4) in the proof below, then we get the first inequality in (7.2). The second one easily follows from (7.7) and (7.8) below.

Proof of Theorem 7.2. We claim that if the constants D_n satisfy

$$\|P_n w_1\|_{\mathbf{C}} \le D_n \|P_n w_2\|_E, \quad \deg(P_n) \le n, \tag{7.4}$$

then

$$d_n(B(w_1), X_E(w_2)) \ge \frac{1}{D_n}. \tag{7.5}$$

In fact, (7.4) means that $B(w_1)$ contains every polynomial of degree at most n with $X_E(w_2)$-norm $\le 1/D_n$. But these polynomials form a ball of radius $1/D_n$ in an $(n+1)$-dimensional subspace of $X_E(w_2)$; hence (7.5) follows from Lemma 7.1.

Now

$$\liminf_{n\to\infty} d_n(B(w_1), X_E(w_2))^{1/n\log n} \ge \exp\left(\frac{1}{\beta_2} - \frac{1}{\beta_1}\right) \tag{7.6}$$

follows from (7.5) and Theorem 6.3.

In the proof of the corresponding upper estimate we can assume without loss of generality that $w_1(z) = \exp(-\alpha_1|z|^{\beta_1})$ and $w_2(z) = \exp(-\alpha_2|z|^{\beta_2})$ (cf. the preceding section).

Let X_n be the space of polynomials of degree at most $n-1$. Then X_n is of dimension n and we can use X_n as a test space to estimate d_n from above.

If

$$f(z) = \sum_{k=0}^{\infty} a_k z^k$$

belongs to $B(w_1)$, then on the circle $C_r := \{z \mid |z| = r\}$ we have

$$|f(z)| \leq \exp(\alpha_1 r^{\beta_1}).$$

Hence from Cauchy's formula we get for the k-th coefficient of f

$$|a_k| \leq \exp(\alpha_1 r^{\beta_1}) r^{-k}.$$

The right-hand side attains its minimum for

$$r = \left(\frac{k}{\alpha_1 \beta_1}\right)^{1/\beta_1}$$

from which

$$|a_k| \leq \left(\frac{\alpha_1 \beta_1}{k}\right)^{k/\beta_1} e^{k/\beta_1} \tag{7.7}$$

follows. Similar computation shows that

$$|z^k w_2(z)| \leq \left(\frac{k}{\alpha_2 \beta_2}\right)^{k/\beta_2} e^{-k/\beta_2}. \tag{7.8}$$

Finally, (7.7) and (7.8) yield

$$\left\| f(z) - \sum_{k=0}^{n-1} a_k z^k \right\|_{X_E(w_2)}$$

$$\leq \sum_{k=n}^{\infty} \left(\frac{\alpha_1 \beta_1}{k}\right)^{k/\beta_1} \left(\frac{k}{\alpha_2 \beta_2}\right)^{k/\beta_2} \exp\left(k\left(\frac{1}{\beta_1} - \frac{1}{\beta_2}\right)\right)$$

$$= \exp\left(n \log n \left(\frac{1}{\beta_2} - \frac{1}{\beta_1}\right) + O(n)\right),$$

from which

$$\limsup_{n\to\infty} d_n(B(w_1), X_E(w_2))^{1/n \log n} \leq \exp\left(\frac{1}{\beta_2} - \frac{1}{\beta_1}\right)$$

immediately follows. This and (7.6) prove the theorem. □

VI.8 Notes and Historical References

Section VI.1

The results of this section were taken from [220] by V. Totik (see also [217]) and [106], [104] and [105] by A. B. J. Kuijlaars. They had been preceded by many special results for individual weights.

The type of approximation that is discussed in this section has evolved from G. G. Lorentz' incomplete polynomials. Lorentz [136] studied polynomials on [0, 1] that vanish at zero with high order. That is, he considered polynomials of the form

$$P_n(x) = \sum_{k=s_n}^{n} a_k x^k, \tag{8.1}$$

and he verified that if $s_n/n \to \theta$ and the P_n's are bounded on [0, 1], then $P_n(x)$ tends to zero uniformly on compact subsets of $[0, \theta^2)$. E. B. Saff and R. S. Varga [200] showed that $[0, \theta^2)$ is the largest set with this property. Although here there is no fixed weight, the resemblance to weighted polynomials $w^n P_n$ with $w(x) = x^{\theta/(1-\theta)}$ is apparent, and in fact, it is easy to transform results concerning incomplete polynomials into analogous ones concerning such weighted polynomials, and vice versa.

In our terminology Lorentz' result means that the support of the extremal measure for the weight $w(x) = x^{\theta/(1-\theta)}$, $\Sigma = [-1, 1]$ is $[\theta^2, 1]$. The corresponding approximation problem, namely that every $f \in C[0, 1]$ that vanishes on $[0, \theta^2)$ is the uniform limit of polynomials of the form (8.1), was independently proved by Saff and Varga [200] and M. v. Golitschek [59].

In [196] Saff generalized the problem to exponential weights of the form $w_\alpha(x) = \exp(-c|x|^\alpha)$, $\alpha > 1$. Saff and Mhaskar proved in [157] that in this case the extremal support is

$$\mathcal{S}_w = [-\gamma_\alpha^{1/\alpha} c^{-1/\alpha}, \gamma_\alpha^{1/\alpha} c^{-1/\alpha}], \qquad \gamma_\alpha = \Gamma(\frac{\alpha}{2})\Gamma(\frac{1}{2})/(2\Gamma(\frac{\alpha}{2}+\frac{1}{2})), \tag{8.2}$$

and they also determined the extremal measure (given by the Ullman distribution, see Section IV.5). In [196] Saff conjectured that every continuous function that vanishes outside (8.2) can be uniformly approximated by weighted polynomials $w_\alpha^n P_n$. This was shown to be true in the special case $\alpha = 2$ in [160] by Mhaskar and Saff, and by D. S. Lubinsky and Saff [143] for all $\alpha > 1$. The missing range $0 < \alpha \le 1$ was settled by Lubinsky and Totik [146] who proved that approximation is still possible if $\alpha = 1$, and for $\alpha < 1$ a necessary and sufficient condition that f be the uniform limit of weighted polynomials $w_\alpha^n P_n$ is that f vanishes outside $\mathcal{S}_w$ and at the origin. It was also proven there (for the case $\alpha > 1$) that even if one considers the approximation problem for f only on the interval $\mathcal{S}_w$, then f still must vanish at the endpoints in order to be the uniform limit of weighted polynomials, i.e. approximation is not possible up to the endpoint for nonvanishing functions (this is a consequence of Theorem 1.9). More precisely, the following exact range for the approximation was established: Suppose that $\mathcal{S}_w = [-1, 1]$ and for $n \geq 1$ we are given closed intervals J_n symmetric about zero, and polynomials P_n of degree $\leq n$ such that

$$\lim_{n\to\infty} \|w_\alpha^n P_n - 1\|_{J_n} = 0.$$

Then there exists a sequence $\{\rho_n\}_{n=1}^\infty$ with

$$\lim_{n\to\infty} \rho_n = \infty, \tag{8.3}$$

such that for infinitely many n

$$J_n \subset [-1 + \rho_n n^{-2/3}, 1 - \rho_n n^{-2/3}].$$

Conversely, if $\{\rho_n\}_{n=1}^\infty$ is a sequence satisfying (8.3), then for every continuous $f \in C[-1, 1]$ there exist polynomials P_n of degree at most n such that

$$\lim_{n\to\infty} \|w_\alpha^n P_n - f\|_{[-1+\rho_n n^{-2/3}, 1-\rho_n n^{-2/3}]} = 0,$$

and

$$\sup_n \|w_\alpha^n P_n\|_{\mathbf{R}} < \infty.$$

In [160] the conjecture was made by Mhaskar and Saff that even for general continuous weights w, approximation by weighted polynomials $w^n P_n$ is possible for an f if and only if f vanishes outside $\mathcal{S}_w$. The necessity of the condition was proved by Totik [220] (it also follows from Theorem 1.2). Its sufficiency is not true; a counterexample was given by Totik in [220] with the additional property that the extremal measure has continuous density. If this is not required, then one can consider $w(x) = \exp(-|x|^\alpha)$, $\alpha < 1$, for which, as we have just discussed, approximation is possible only if the function to be approximated vanishes at 0. In [21] the weaker conjecture was stated that at least for the case when $Q = \log 1/w$ is convex, a necessary and sufficient condition for approximation is the same as before, namely that the function vanishes outside $\mathcal{S}_w$. This conjecture of Borwein and Saff follows from Theorem 1.7 under the smoothness assumption that Q is a $C^{1+\varepsilon}$, $\varepsilon > 0$, function on the support $\mathcal{S}_w$ (note that if Q is convex on an interval I, then it is automatically Lip 1 inside I). The sufficiency of the conjecture for general convex Q's has recently been verified by Totik.

We have already mentioned, that if $w(x) = x^{\theta/(1-\theta)}$, then $\mathcal{S}_w = [\theta^2, 1]$ (see [136]). The generalization to Jacobi weights was done by Saff, Ullman and Varga in [197] (see the discussion after Theorem 1.7). In this case the approximation problem was settled by X. He and X. Li [76].

As a "midway" case between Jacobi weights and Freud-type exponential weights lie the Laguerre weights $w(x) = x^\alpha e^{-\lambda x}$, $\alpha \geq 0$, $\Sigma = [0, \infty)$, for which the extremal support was established by Mhaskar and Saff [158].

In all these cases $Q = \log 1/w$ is convex; hence Theorem 1.7 applies.

The basic Stone-Weierstrass type theorem Theorem 1.1 was observed by Kuijlaars [104]. Although it immediately follows from the Stone-Weierstrass theorem, it has turned out to be extremely useful in localizing conditions regarding approximation. Theorem 1.2 is also from Kuijlaars [104]; a somewhat weaker version appeared in [220].

Theorems 1.5 and 1.7 are from [220] by Totik. Their proofs easily yield the following theorem: suppose that w is an admissible weight of class $C^{1+\varepsilon}$ for some $\varepsilon > 0$. Then w has the approximation property on the union of the interiors of the supports $\mathcal{S}_{w^\lambda}$, $\lambda > 1$.

In [220] Totik raised the problem if internal zeros in the density prevent approximation. In [220] he gave examples that showed that the answer was sometimes yes, and sometimes no. Kuijlaars [105], [106] satisfactorily settled the problem by proving Theorems 1.8 and 1.9. He also found the clever technique used in the proof of Theorems 1.8 and 1.9 with which one can transform the approximation problem in internal points to those at endpoints. These results of Kuijlaars solved several open problems, among others the one of G. G. Lorentz on the impossibility of approximating an f with $f(\theta^2) \neq 0$ on the whole $[\theta^2, 1]$ by incomplete polynomials (see the last paragraph of the discussion concerning incomplete polynomials in Section VI.1).

The approximation problem on an unbounded set by weighted polynomials $w^n P_n$ with weights w for which $w(x)|x| \to \alpha \neq 0$ was investigated by P. Simeonov [209]. The approximation around the point infinity depends on how dense the equilibrium measure (which in this case has noncompact support) is around that point. For example, if the density of the equilibrium measure is of the form $v(t)/t^2$ with a continuous and positive function v around infinity, then approximation is possible around that point. In fact, the case of the point infinity can be reduced to that of an ordinary point by the transformation $x \to x/(1+x^2)$.

I. E. Pritsker [185] and Pritsker and R. S. Varga [186], [187] have investigated the approximation by weighted polynomials $w^n(z)P_n(z)$ over compact subsets E of the complex plane. In that case w is assumed to be continuous on E and analytic in the interior of E.

For related questions concerning approximation by weighted rational functions see the notes to Section VIII.5.

Section VI.2

The main results of this section were taken from [220] by V. Totik.

The types of approximation discussed in Theorems 2.5, 2.6 and 2.7 were investigated in detail by D. S. Lubinsky and E. B. Saff in [143] and especially in [142], where the basic theorems for this new kind of weighted approximation have been established. Theorem 2.5 extends these results to somewhat more general weights. Theorems 2.6 and 2.7 were stated by Lubinsky and Totik in [146] (see also [220]). Actually, Theorem 2.6 is an easy consequence of some general results on entire functions of type zero; see the work of N. I. Akhiezer [2].

Lemma 2.8 is essentially due to Lubinsky [137], and is a basic tool in replacing weights with analytic ones having nonnegative Taylor coefficients; a technique that was exploited in different directions by Lubinsky.

Section VI.3

There are many constructions in the literature for generating fast decreasing polynomials (often called *pin polynomials*). These range from infinite products through entire functions to factors of orthogonal polynomials (cf. [35], [171]), just to mention a few. Since the Chebyshev polynomials increase the fastest outside $[-1, 1]$ among all polynomials of a fixed degree, perhaps the most natural approach is to use them in the construction. This was done by K. G. Ivanov and V. Totik in [84], where the fundamental theorem mentioned in the introduction to Section VI.3 was proven.

The potential theoretic approach and the results of this section are due to Totik [219] and [220]. The case $\alpha = 2$ of Theorem 3.5 is due to Lubinsky and Totik [146].

It is worth comparing Theorem 3.5 with Theorem 1.8 from Section VI.1. For example, for $\alpha = 2$ the former one says that there are weighted polynomials $\exp(nx^2)P_n(x)$ that take the value 1 at the origin, and otherwise are bounded on $[-1, 1]$ by an absolute constant. On the other hand, Theorem 1.8 says that such polynomials cannot uniformly converge on the whole interval $[-1, 1]$.

Theorem 3.5 was extended to by D. Benkő [11] who showed that for a large class of functions φ for which $\varphi(\sqrt{x})$ is concave the necessary condition of Theorem 3.2(ii) is already sufficient for the existence of polynomials with properties

$$P_n(0) = 1, \qquad |P_n(x)| \le Ce^{-n\varphi(x)}, \quad x \in [-1, 1]$$

(cf. Theorem 3.4).

The extension of Theorem 3.5 to $\alpha > 2$ follows from the following general result of A. B. J. Kuijlaars and W. Van Assche [108]:

Let ϕ be a C^{n_0}-function on $[0, 1]$ satisfying

$$\phi(0) = \phi'(0) = \cdots = \phi^{(n_0)}(0) = 0,$$

and

$$\phi^{(n_0)} \text{ is increasing and concave on } [0, 1]$$

for some $n_0 \ge 1$ (when $n_0 = 1$ or $n_0 = 2$ assume also that ϕ is a C^3-function on (0,1]). Then there exist polynomials p_n, $\deg p_n \le n$, satisfying with some constant C,

$$p_n(0) = 1, \qquad |p_n(x)| \le C\exp(-cn\phi(x)), \quad x \in [0, 1],$$

if and only if

$$c \le \left[\frac{1-a_0}{2}\int_{a_0}^{1}\frac{\phi(s)-\phi(a_0)}{s-a_0}\,\frac{ds}{\pi\sqrt{(1-s)(s-a_0)}}\right]^{-1},$$

where $a_0 \in (0, 1)$ is the largest solution of the equation

$$\frac{\displaystyle\int_a^1 \frac{\phi(s)}{s} \frac{ds}{\pi\sqrt{(1-s)(s-a)}}}{\displaystyle\frac{1-a}{2}\int_a^1 \frac{\phi(s)-\phi(a)}{s-a} \frac{ds}{\pi\sqrt{(1-s)(s-a)}}} = -\frac{1}{\sqrt{a}}\log\left(\frac{1-\sqrt{a}}{1+\sqrt{a}}\right).$$

Analogues of the results of this section for fast decreasing rational functions were obtained by A. L. Levin and E. B. Saff in [135]. For rationals of the form $r_n(x) = p_n(x)/p_n(-x)$, $\deg p_n \le n$, they proved, using Green potentials (for the half-plane) with external fields, that if $\varphi \in C[0,1]$, $\varphi(0) = 0$, then such r_n's exist satisfying

$$|r_n(x)| \le D\exp(-dn\varphi(x)), \quad x \in [0,1], \quad n \ge 1,$$

for some constants D, $d > 0$ if and only if

$$\int_0^1 \frac{\varphi(x)}{x} dx < \infty$$

(compare with (3.2)). Moreover, if φ is also increasing and concave on $[0, 1]$, then such r_n's exist satisfying

$$|r_n(x)| \le e^{-n(\varphi(x)+o(1))}, \quad x \in [0,1], \quad n \ge 1,$$

if and only if

$$\frac{2}{\pi^2}\int_0^1 \varphi(x)\frac{dx}{x\sqrt{1-x^2}} \le 1$$

(compare Theorem 3.4). If the last inequality is strict, they further show that with some additional smoothness assumptions that

$$|r_n(x)| \le Ce^{-n\varphi(x)}, \quad x \in [0,1], \quad n \ge 1,$$

can be achieved. Levin and Saff apply these results to obtain rational approximations to sgn x and $|x|$ on $[-1, 1]$ that converge geometrically fast for $x \ne 0$.

Section VI.4

The method of distributing the nodes of discretization by imitating the equilibrium distribution goes back at least to Szegő and was used by E. A. Rakhmanov [189] and H. N. Mhaskar and E. B. Saff [161]. Theorem 4.1 was taken (in a corrected form) from [142] by D. S. Lubinsky and E. B. Saff. The finer technique of Theorem 4.2 is due to V. Totik [220]. A. L. Levin and Lubinsky [129] have a different method, based on numerical quadrature, which can also be used to obtain finer estimates than what appears in Theorem 4.1. See also [145] for discretization of logarithmic potentials.

Section VI.5

The infinite-finite range inequality (5.1) goes back to G. Freud [46, 45, 48], who was the first to realize the importance of such an inequality. Later there was a lot of activity in this area, see e.g. the works of P. Nevai [168, 169, 167], W. C. Bauldry [7], D. S. Lubinsky [139]. See Nevai's [169] work for a detailed account of the subject.

The sharper version Theorem 5.2 was proved by Lubinsky and Saff [142] for weights that are much more general than the ones discussed in Section VI.5. Theorem 5.3 is due to Levin and Lubinsky [129].

Theorem 5.3 is sharp: if $\{\rho_n\}$ is a sequence with $\rho_n n^{2/3} \neq O(1)$, then there is a sequence of polynomials $\{P_n\}$ such that $\|P_n w_\lambda\|_{L^p(\mathbf{R})}$ is not bounded by a constant times

$$\|P_n w_\lambda\|_{L^p([-(1-\rho_n)n^{1/\lambda},(1-\rho_n)n^{1/\lambda}])}.$$

This follows (at least for $\lambda > 1$) from the results of [146] about the uniform approximability of any continuous f on $[-1+\rho_n, 1-\rho_n]$ by weighted polynomials $w^n P_n$ when w is a Freud weight (with exponent bigger than 1).

The Markoff constant in Theorem 5.5 is due to G. Freud [46, 45, 47] for $\lambda > 2$, to Levin and Lubinsky [127, 128] for $\lambda > 1$ and to Nevai and Totik [171] for $\lambda \leq 1$. The present proof for Theorem 5.5 utilizes an idea of G. Halász (personal communication) and essentially differs from earlier ones.

For Markoff-type inequalities for more general exponential weights see [144] and [131].

Theorem 5.6 on Nikolskii inequality was proved by Nevai and Totik in [172]. Partial results had been proven by C. Markett, H. N. Mhaskar, E. B. Saff, R. A. Zalik and others, see e.g. [150, 155, 157, 238].

Sections VI.6 – VI.7

H. N. Mhaskar and C. Micchelli [156] investigated $d_n\ (B(w_1), X_E(w_2))$ for special E's and w_1, w_2.

Chapter VII. Applications Concerning Orthogonal Polynomials

The analysis of the asymptotic behavior of orthogonal polynomials was one of the driving forces for the resurgence of interest in potentials with external fields. The relationship between the two subjects can be seen as follows: consider, for example, orthogonal polynomials with respect to the so-called Freud weights $W(x) = \exp(-|x|^\lambda) =: \exp(-Q(x))$. On applying the substitution $x \to n^{1/\lambda}x$ and the defining properties of orthogonal polynomials one arrives at monic polynomials P_n minimizing the integral

$$\int (|P_n|e^{-nQ})^2 .$$

Writing $|P_n|$ in the form $\exp(-nU^\nu)$ with ν equal to the normalized counting measure on the zeros of P_n, and replacing the L^2 norm by the supremum norm we arrive at the problem of maximizing the minimum of

$$U^\nu + Q,$$

which is equivalent to the energy problem discussed in the first part of this book.

In this chapter we use this approach for finding the zero distribution and the asymptotic behavior of Freud-type orthogonal polynomials on the real line. First we shall discuss zero distribution and nth root asymptotics. These are basically equivalent, and they are easily obtained from the fact that the orthonormal polynomials turn out to be asymptotically minimal in the sense of Chapter III, so results from that chapter can be applied. After that we shall prove strong asymptotics for the leading coefficients, which requires the results concerning the approximation problem treated in Section VI.1. Finally, we shall settle the problem on the possible limit distributions of the zeros of orthogonal polynomials with respect to general measures on an interval.

VII.1 Zero Distribution and n-th Root Asymptotics for Orthogonal Polynomials with Exponential Weights

Let μ be a Borel measure on the real line such that $\text{supp}(\mu)$ consists of infinitely many points and the moments of μ

$$\int x^j d\mu(x), \quad j = 0, 1, 2, \ldots,$$

are all finite. Then we can form the *orthonormal polynomials*

$$p_n(\mu; x) = \gamma_n(\mu)x^n + \cdots$$

corresponding to μ with the stipulation that $\gamma_n(\mu) > 0$, and for all natural numbers n and m

$$\int p_n(\mu; x)p_m(\mu; x)\, d\mu(x) = \begin{cases} 1 & \text{if } n = m \\ 0 & \text{otherwise.} \end{cases}$$

If μ is given by its density function v (i.e. $d\mu(x) = v(x)\, dx$) then we shall also speak of the orthogonal polynomials with respect to v (instead of μ).

The *leading coefficient* $\gamma_n(\mu)$ plays a distinguished role. For example, by expanding an arbitrary monic polynomial $x^n + \cdots$ into $\{p_k(\mu; x)\}_{k=0}^n$, it immediately follows from the orthogonality relations that

$$\frac{1}{\gamma_n(\mu)^2} \le \int (x^n + \cdots)^2\, d\mu(x)$$

with equality only for the so-called *monic orthogonal polynomial*

$$\frac{1}{\gamma_n(\mu)} p_n(\mu; x).$$

Hence, the number $1/\gamma_n(\mu)$ is the extremum value in the extremum problem

$$\frac{1}{\gamma_n(\mu)} = \inf_{P_n(x)=x^n+\cdots} \left(\int P_n^2\, d\mu \right)^{1/2}, \tag{1.1}$$

and the monic orthogonal polynomial $p_n(\mu; x)/\gamma_n(\mu)$ is the unique extremal function.

It is well known (see e.g. [216, Theorem III.3.1]) and easy to see that $p_n(\mu; x)$ has exactly n simple real zeros

$$x_{1,n} < x_{2,n} < \cdots < x_{n,n},$$

and each interval of $\mathbf{R}$ contiguous to $\operatorname{supp}(\mu)$ can contain at most one of these zeros. These zeros are good nodes for numerical quadrature, and they are intimately related to the asymptotic behavior of the orthonormal polynomials and to the generating measure itself. Therefore, the description of the distribution of the zeros is an important part of the theory of orthogonal polynomials. Asymptotic properties of orthogonal polynomials (and hence their zero distribution) largely depend on the order of the leading coefficients $\gamma_n(\mu)$ as $n \to \infty$; hence determining the exact order for $\{\gamma_n(\mu)\}$ is another important problem of the theory. In the present chapter we show how these problems can be resolved for some measures using the theory of weighted potentials and polynomials.

Our main objective is to study orthogonal polynomials with respect to the *square* of the *exponential weight* functions (also called *Freud weights*)

$$v_{\alpha,c}(x) := \exp(-c|x|^\alpha), \quad x \in (-\infty, \infty).$$

The associated orthonormal polynomials will be denoted by $p_n(v_{\alpha,c}; x) = \gamma_n(v_{\alpha,c})x^n + \cdots$, keeping in mind that $v_{\alpha,c}^2(x)dx$ is the measure of orthogonality.

For $\alpha = 2$ and $c = 1/2$ we obtain the classical Hermite polynomials $H_n(x)$:

$$\int_{-\infty}^{\infty} H_n(x)H_m(x)e^{-x^2}\,dx = \begin{cases} \sqrt{\pi}\,2^n n! & \text{if } n = m \\ 0 & \text{otherwise.} \end{cases}$$

For these polynomials virtually everything is known, which is due to the fact that we have a generating function, an explicit three-term recurrence relation and even an explicit differential equation at our disposal when dealing with H_n. For example, $H_n(x) = 2^n x^n + \cdots$ (see [216, Section 5.5]) so that

$$\gamma_n(v_{2,1/2}) = \pi^{-1/4}2^{n/2}(n!)^{-1/2},$$

and for the largest zero $x_{n,n}$ of $H_n(x)$ we have the asymptotic formula ([216, Theorem 6.32])

$$x_{n,n} = \sqrt{2n} + O(n^{-1/6}),$$

which can be derived from the differential equation

$$y'' - 2xy' + 2ny = 0$$

for H_n using Sturm's comparison method. However, if $\alpha \neq 2$ most of the tools mentioned above are not available*, so one has to find different methods in this "totally nonclassical" case.

The above form for the largest zero indicates that the zeros of H_n tend to infinity in the sense that their relative density in every finite interval tends to zero with n, i.e. the number of zeros of H_n in any fixed finite interval is $o(n)$. Thus, one has to modify the concept of the usual zero distribution. We shall be interested in the distribution of zeros in the smallest interval $[x_{1n}, x_{nn}]$ containing them. In other words, we will be considering the so-called *normalized* or *contracted zero distribution* that is obtained by dividing the zeros by x_{nn} (note that $x_{1,n} = -x_{n,n}$), and analyzing the resulting distribution on $[-1, 1]$.

The first result describes the asymptotic behavior of these contracted zero distributions.

Theorem 1.1. *Let* $v_{\alpha,c}(x) = \exp(-c|x|^\alpha)$, $x \in \mathbf{R}$, *where* $\alpha, c > 0$, *and set*

$$\gamma_\alpha := \Gamma\left(\frac{\alpha}{2}\right)\Gamma\left(\frac{1}{2}\right)\Big/2\Gamma\left(\frac{\alpha}{2}+\frac{1}{2}\right).$$

If $x_{1,n} < x_{2,n} < \cdots < x_{n,n}$ *denote the zeros of the orthogonal polynomials corresponding to* $v_{\alpha,c}^2$, *then for* $n \to \infty$ *we have*

* According to Y. Chen and M.E.H. Ismail, for $\alpha \geq 1$, second order differential equations (although complicated) exist for the corresponding orthogonal polynomials.

$$\frac{x_{n,n}}{n^{1/\alpha}} \to c^{-1/\alpha}\gamma_\alpha^{1/\alpha}, \tag{1.2}$$

and the limit distribution of the normalized zeros or, equivalently, the asymptotic distribution of the set

$$\{x_{j,n}/n^{1/\alpha}c^{-1/\alpha}\gamma_\alpha^{1/\alpha}\}_{j=1}^n,$$

is given by the Ullman distribution

$$\frac{\alpha}{\pi}\int_{|t|}^1 \frac{u^{\alpha-1}}{\sqrt{u^2-t^2}}\,du, \quad t \in [-1, 1].$$

Recall that in Section IV.5 the Ullman distribution was shown to be the equilibrium distribution for the weight function $\exp(-\gamma_\alpha|x|^\alpha)$. The next result describes the so-called n-th *root asymptotic behavior* for the leading coefficients and the contracted orthogonal polynomials. The reader is cautioned to keep in mind the distinction between the notation for the constant γ_α and that for the leading coefficients $\gamma_n(\cdot)$.

Theorem 1.2. *With the same assumptions as in Theorem* 1.1, *for the orthonormal polynomials $p_n(x) = \gamma_n(v_{\alpha,c})x^n + \cdots$ with respect to $v_{\alpha,c}^2$, we have*

$$\lim_{n\to\infty} \gamma_n(v_{\alpha,c})^{1/n} n^{1/\alpha} = 2c^{1/\alpha}\gamma_\alpha^{-1/\alpha}e^{1/\alpha}, \tag{1.3}$$

and locally uniformly on $\mathbf{R} \setminus [-1, 1]$

$$\lim_{n\to\infty} |p_n(n^{1/\alpha}c^{-1/\alpha}\gamma_\alpha^{1/\alpha}x)|^{1/n} \tag{1.4}$$

$$= \exp\left(\log|x + \sqrt{x^2-1}| + |x|^\alpha \int_0^{1/|x|} \frac{u^{\alpha-1}}{\sqrt{1-u^2}}\,du\right),$$

while on $\mathbf{C} \setminus [-1, 1]$ *we have locally uniformly*

$$\lim_{n\to\infty} |p_n(n^{1/\alpha}c^{-1/\alpha}\gamma_\alpha^{1/\alpha}z)|^{1/n} \tag{1.5}$$

$$= \exp\left(\log|z + \sqrt{z^2-1}| + \operatorname{Re}\int_0^1 \frac{zu^{\alpha-1}}{\sqrt{z^2-u^2}}\,du\right).$$

Proof of Theorem 1.1. We recall from Section VI.5 the Nikolskii type inequality

$$\|P_n v_{\alpha,c}\|_{L^p(\mathbf{R})} \le CN_n(\alpha, p, q)\|P_n v_{\alpha,c}\|_{L^q(\mathbf{R})} \tag{1.6}$$

with $N_n(\alpha, p, q) = \mathcal{O}(n^\tau)$ for some $\tau > 0$. This implies that

$$\lim_{n\to\infty}\left(\left(\min_{P_n}\|P_n v_{\alpha,c}\|_{L^2(\mathbf{R})}\right)\Big/\left(\min_{P_n}\|P_n v_{\alpha,c}\|_{L^\infty(\mathbf{R})}\right)\right)^{1/n} = 1, \tag{1.7}$$

where the infimum is taken for all polynomials of the form $P_n(x) = x^n + \cdots$. Here

$$\min_{P_n} \|P_n v_{\alpha,c}\|_{L^2(\mathbf{R})} = \gamma_n(v_{\alpha,c})^{-1} \tag{1.8}$$

(see (1.1)), while

$$\min_{P_n} \|P_n v_{\alpha,c}\|_{L^\infty(\mathbf{R})} = \min_{R_n} \|R_n w^n\|_{L^\infty(\mathbf{R})} (n^{1/\alpha} c^{-1/\alpha} \gamma_\alpha^{1/\alpha})^n \tag{1.9}$$

where $w(y) = \exp(-\gamma_\alpha |y|^\alpha)$, and both minima are taken for monic polynomials of degree n. The last formula is obtained by the substitution

$$x = n^{1/\alpha} c^{-1/\alpha} \gamma_\alpha^{1/\alpha} y, \quad R_n(y) = P_n(x)/(n^{1/\alpha} c^{-1/\alpha} \gamma_\alpha^{1/\alpha})^n.$$

Using the fact that if p_n are the orthonormal polynomials corresponding to $v_{\alpha,c}^2$, then the monic polynomials $p_n/\gamma_n(v_{\alpha,c})$ minimize the L^2 norm on the left-hand side of (1.8), it follows from (1.7)–(1.9) via Theorems III.3.1 and IV.5.1 that (1.3) is true and also that the monic polynomials

$$R_n^*(x) = p_n(n^{1/\alpha} c^{-1/\alpha} \gamma_\alpha^{1/\alpha} x) \Big/ \left(\gamma_n(v_{\alpha,c})(n^{1/\alpha} c^{-1/\alpha} \gamma_\alpha^{1/\alpha})^n\right) \tag{1.10}$$

are asymptotically optimal in the sense of Section III.4 in the weighted Chebyshev problem with weight function w. Thus, the statement of the theorem concerning the asymptotic distribution of the zeros follows from Theorems III.4.2 and IV.5.1.

This implies that we must have

$$\liminf_{n\to\infty} x_{n,n}/n^{1/\alpha} \geq c^{-1/\alpha} \gamma_\alpha^{1/\alpha}. \tag{1.11}$$

On the other hand, if we assume that for infinitely many n we have

$$x_{n,n} \geq (1 + 2\delta) n^{1/\alpha} c^{-1/\alpha} \gamma_\alpha^{1/\alpha} \tag{1.12}$$

for some $\delta > 0$, then with

$$I_\delta = [-(1+\delta) n^{1/\alpha} c^{-1/\alpha} \gamma_\alpha^{1/\alpha}, (1+\delta) n^{1/\alpha} c^{-1/\alpha} \gamma_\alpha^{1/\alpha}]$$

we get for the monic polynomials

$$q_n(x) = (x - (1+\delta) n^{1/\alpha} c^{-1/\alpha} \gamma_\alpha^{1/\alpha}) p_n(x)/\gamma_n(v_{\alpha,c})(x - x_{n,n})$$

that for infinitely many n the $L^2(I_\delta)$-norm of $q_n v_{\alpha,c}$ is at most $(2+2\delta)/(2+3\delta)$-times the $L^2(I_\delta)$-norm of $(p_n/\gamma_n(v_{\alpha,c}))v_{\alpha,c}$, which easily implies in view of the infinite-finite range inequality of Theorem VI.5.1 that

$$\|q_n v_{\alpha,c}\|_{L^2(\mathbf{R})} < \left\| \frac{p_n}{\gamma_n(v_{\alpha,c})} v_{\alpha,c} \right\|_{L^2(\mathbf{R})}.$$

This, however, is impossible because the monic orthogonal polynomials minimize the weighted L^2 norms for any fixed degree (see (1.1)). Thus, (1.12) can happen only for finitely many n's and this together with (1.11) proves (1.2). □

Proof of Theorem 1.2. We follow the preceding proof. The limit (1.3) is an easy consequence of the asymptotic minimality of R_n^* (see (1.10)) established above if we apply Theorems III.3.1 and IV.5.1 according to which the weighted Chebyshev constant is $\exp(-\log 2 - 1/\alpha)$. In a similar fashion, (1.4) and (1.5) follow from Theorems III.4.7(iv), IV.5.1 and formulae (1.10), (1.3) (use also the fact from the previous theorem that the contracted zeros of the orthogonal polynomials, which are essentially the zeros of the monic polynomial R_n^*, cannot accumulate at any point of $\mathbf{R}\setminus[-1, 1]$ and they have as limit distribution the Ullman measure, which is the μ_w from Theorem IV.5.1). □

VII.2 Strong Asymptotics

Let $w(x) = w_\alpha(x) = e^{-\gamma_\alpha |x|^\alpha}$, $\alpha > 1$, be an exponential weight on $\mathbf{R}$ normalized so that $\mathcal{S}_w = [-1, 1]$ (see Theorem IV.5.1 – this normalization is made for convenience, any other positive constant can replace γ_α on the right), and consider the orthonormal polynomials with respect to w^2:

$$p_n(w; x) = \gamma_n(w)x^n + \cdots$$

defined by the orthogonality relation

$$\int p_n(w; x)p_m(w; x)w^2(x)\,dx = \delta_{n,m}.$$

Let Π_n denote the set of polynomials of degree n and with leading coefficient one, i.e.

$$\Pi_n = \{x^n + \cdots\}.$$

We have already mentioned in (1.1) that the leading coefficient $\gamma_n(w)$ of the orthonormal polynomial p_n gives the constant in a weighted extremal (minimum) problem, namely

$$\frac{1}{\gamma_n(w)^2} = \inf_{P_n \in \Pi_n} \int P_n^2 w^2, \tag{2.1}$$

and it is one of the most important quantities related to p_n. Indeed, their behavior determines the behavior of the p_n's, which can also be seen from the fact that in the recurrence formula

$$xp_n(w; x) = A_{n+1}p_{n+1}(w; x) + A_n p_{n-1}(w; x)$$

the recurrence coefficients are given by

$$A_n = \gamma_{n-1}(w)/\gamma_n(w).$$

Hence, it is important to know the asymptotic behavior of the recurrence coefficients and the leading coefficients. In the previous section we established the so-called n-th root behavior of the $\gamma_n(w)$'s, which was a relatively easy consequence of general results from the first part of this book. Now we completely describe the asymptotic behavior of these leading coefficients.

Theorem 2.1. *For any* $\alpha > 1$,

$$\lim_{n\to\infty} \gamma_n(w_\alpha)\pi^{1/2}2^{-n}e^{-n/\alpha}n^{(n+1/2)/\alpha} = 1. \tag{2.2}$$

Actually the result is also true for $\alpha \le 1$ ([142]), but we shall be content with the proof of the $\alpha > 1$ case.

Corollary 2.2. *With the same assumption as in Theorem* 2.1,

$$\lim_{n\to\infty} n^{-1/\alpha}A_n = \frac{1}{2}. \tag{2.3}$$

Proof of Theorem 2.1. We know from the computation of Section IV.5 that $\mathcal{S}_w = [-1, 1]$, μ_w is given by the Ullman distribution with density

$$v(t) = \frac{\alpha}{\pi}\int_{|t|}^{1} \frac{u^{\alpha-1}}{\sqrt{u^2-t^2}}\,du \tag{2.4}$$

on $(-1, 1)$, and

$$F_w = \log 2 + 1/\alpha. \tag{2.5}$$

We shall need a formula of S. N. Bernstein (see [1, pp. 250–254] or [142, p. 111]): Let R_{2q} be a polynomial of degree $2q$, positive on $(-1, 1)$ with possibly simple zeros at ± 1. Then for $n \ge q$

$$\left(\inf_{P_n\in\Pi_n}\int_{-1}^{1}\frac{\phi^{1/2}}{R_{2q}}P_n^2\right)^{1/2} = \pi^{1/2}2^{-n}\exp\left(\frac{1}{\pi}\int_{-1}^{1}\frac{\log(\phi^{1/4}/R_{2q}^{1/2})}{\phi^{1/2}}\right), \tag{2.6}$$

where $\phi(x) = 1-x^2$. In what follows we abbreviate the geometric mean appearing on the right as $G[\phi^{1/4}/R_{2q}^{1/2}]$, i.e.

$$G[V] := \exp\left(\frac{1}{\pi}\int_{-1}^{1}\frac{\log V(x)}{\sqrt{1-x^2}}\,dx\right). \tag{2.7}$$

We separate the lim inf and lim sup estimates.

Proof of the upper estimate. Let $\rho_n = 1-n^{-2/3}$, and let us carry out the substitution $x = \rho_n^{1/\alpha}n^{1/\alpha}y$ in the integrals in (2.1), and then restrict the integrals to $[-1, 1]$. We get

$$\frac{1}{\gamma_n(w)^2} \ge n^{(2n+1)/\alpha}\rho_n^{(2n+1)/\alpha}\inf_{P_n\in\Pi_n}\int_{-1}^{1} e^{-\gamma_\alpha\rho_n 2n|x|^\alpha}P_n^2(x)\,dx. \tag{2.8}$$

We are going to show with the method of Section VI.1 that there are polynomials H_n of degree at most n such that if

$$h_n(x) = e^{-\gamma_\alpha\rho_n n|x|^\alpha}|H_n(x)|(1-x^2)^{-1/4}, \tag{2.9}$$

then

$$h_n(x) \ge 1 \qquad \text{for} \quad x \in [-1, 1] \tag{2.10}$$

and

$$\lim_{n\to\infty} G[h_n] = 1. \tag{2.11}$$

Then we will have by (2.8) and (2.10)

$$\frac{1}{\gamma_n(w)^2} \geq n^{(2n+1)/\alpha} \rho_n^{(2n+1)/\alpha} \inf_{P_n\in\Pi_n} \int_{-1}^{1} \frac{\phi^{1/2}}{|H_n|^2} P_n^2,$$

and so by Bernstein's formula (2.6)

$$\frac{1}{\gamma_n(w)^2} \geq n^{(2n+1)/\alpha} \rho_n^{(2n+1)/\alpha} \pi 2^{-2n} \left(G[\phi^{1/4}/|H_n|]\right)^2. \tag{2.12}$$

But here

$$\phi^{1/4}(x)/|H_n(x)| = e^{-\gamma_\alpha \rho_n n|x|^\alpha}/h_n(x)$$

and

$$\gamma_\alpha \frac{1}{\pi} \int_{-1}^{1} \frac{|x|^\alpha}{\sqrt{1-x^2}}\, dx = \frac{1}{\alpha}, \tag{2.13}$$

which, together with (2.11) imply that $\rho_n^{2n/\alpha}$ times the geometric mean on the right hand side of (2.12) has the form

$$(1+o(1))e^{-2n/\alpha} \exp\left((2n/\alpha)((1-\rho_n)+\log\rho_n)\right) = (1+o(1))e^{-2n/\alpha},$$

and this proves that

$$\limsup_{n\to\infty} \gamma_n(w)\pi^{1/2}2^{-n}e^{-n/\alpha}n^{(n+1/2)/\alpha} \leq 1.$$

Thus, all we need is the existence of polynomials H_n with properties (2.10) and (2.11).

We follow the proof of Theorem VI.1.5 in Section VI.1, and we are also going to use here the same notation. Let us divide again the interval $[-1, 1]$ into equal subintervals I_k, $k = 0, \ldots, n-1$, modulo μ_w, i.e. with the property $\mu_w(I_k) = 1/n$ (enumerated from the left), and let ξ_k be the weight point of μ_w on I_k. We set again

$$Q_n(t) = \prod_j (t - iL/n - \xi_j). \tag{2.14}$$

By symmetry, we can restrict our attention to the left endpoint -1. Thus, let $x \in I_{j_0}$ with $0 \leq j_0 \leq n/2$. Notice that the Ullman distribution has continuous density on the whole real line (which of course vanishes outside $[-1, 1]$); hence (VI.1.19) holds uniformly on the whole real line. In the estimate of

$$|\log|Q_n(x)| + nU^{\mu_n}(x)| \tag{2.15}$$

$$= \left|\sum_{j=0}^{n-1} n \int_{I_j} \left(\log|x - iL/n - t| - \log|x - iL/n - \xi_j|\right) d\mu_w(t)\right|$$

we shall however need a somewhat finer analysis around the endpoints than what was done in Section VI.1.

For the Ullman distribution (2.4) it immediately follows that

$$v(t) \sim (1-t^2)^{1/2} \qquad \text{as} \quad t \to \pm 1,$$

and this property alone implies for the I_k's and ξ_k's for $0 \le k \le n/2$

$$1+\xi_k \sim \left(\frac{k+1}{n}\right)^{2/3}, \qquad |I_k| \sim \frac{1}{(k+1)^{1/3} n^{2/3}}, \tag{2.16}$$

and analogous estimates hold for $k \ge n/2$. Hence for $j \ne j_0$

$$\operatorname{dist}(\xi_j, I_{j_0}) \sim \frac{1}{n^{2/3}} \sum_{k \in [j_0, j]} \frac{1}{(k+1)^{1/3}} \sim \frac{|j^{2/3} - j_0^{2/3}|}{n^{2/3}}.$$

Since for $x \in I_{j_0}$, the absolute value of the the j-th, $j \ne j_0, j_0 \pm 1$, term in the sum in (2.15) is at most a fixed constant times

$$\left(\frac{|I_j|}{\operatorname{dist}(\xi_j, I_{j_0})}\right)^2$$

(see the argument after (VI.1.21)), it follows from the preceding estimate that

$$\sum_{j \ne j_0, j_0 \pm 1} n \left| \int_{I_j} \left(\log|x - iL/n - t| - \log|x - iL/n - \xi_j|\right) d\mu_w(t) \right| \tag{2.17}$$

$$\le C \sum_{j \ne j_0, j_0 \pm 1} \left(\frac{1}{(j+1)^{1/3} n^{2/3}}\right)^2 \Bigg/ \left(\frac{|j^{2/3} - j_0^{2/3}|}{n^{2/3}}\right)^2 = O(1).$$

Furthermore, direct calculation based on the fact that

$$\int_0^a \left|\log \frac{u}{a}\right| du \le a, \qquad a > 0, \tag{2.18}$$

shows that for $x \in I_{j_0}$ both of the $j_0 - 1$-th and the $j_0 + 1$-st terms are bounded by a fixed constant.

Thus, it has left to estimate the j_0-th term in (2.15). Its absolute value is obviously bounded by

$$\frac{1}{2} \log \frac{(L/n)^2 + |I_{j_0}|^2}{(L/n)^2} \le \frac{1}{2} \log \left(1 + \frac{1}{L^2} O\left(\left(\frac{n}{j_0+1}\right)^{2/3}\right)\right) \le \frac{1}{2} \log \frac{1}{1-x^2}$$

if L is sufficiently large (recall that $x \in I_{j_0}$; hence

$$\left(\frac{n}{j_0+1}\right)^{2/3} \le C\frac{1}{1-x^2}$$

by (2.16)).

Taking into account (VI.1.19), which, as we have seen in Section VI.1, uniformly holds on **R**, we get that the polynomials Q_n from Section VI.1 satisfy

$$\frac{e^{L\pi v(x)}}{C}\sqrt{1-x^2} \le e^{-\gamma_\alpha n|x|^\alpha}|Q_n(x)|e^{nF_w} \le \frac{Ce^{L\pi v(x)}}{\sqrt{1-x^2}} \tag{2.19}$$

uniformly in $x \in [-1, 1]$ and large n, $n \ge n_L$, with some absolute constant C (recall (2.5)), and

$$e^{-\gamma_\alpha n|x|^\alpha}|Q_n(x)|e^{nF_w} = e^{L\pi v(x)+O_x(1/L)} \tag{2.20}$$

where O_x is uniformly bounded on compact subsets of $(-1, 1)$, i.e. $O_x(1/L)$ denotes a quantity $l_{n,x,L}/L$, where $|l_{n,x,L}| \le C_\varepsilon$ for every $x \in [-1+\varepsilon, 1-\varepsilon]$, n and L with some constant C_ε depending only on $\varepsilon > 0$.

Next we improve the estimate (2.19) for x lying close to ± 1. The proof of (2.17) and (2.19) easily yields that there is a $c > 0$ such that

$$e^{-\gamma_\alpha n|x|^\alpha}|Q_n(x)|e^{nF_w} \ge ce^{L\pi v(x)} \quad \text{for all } n \text{ and } \; 0 < 1-|x| \le cn^{-2/3}. \tag{2.21}$$

In fact, recall that, say, $|I_0| \sim (1+\xi_0) \sim n^{-2/3}$, and so (2.18) implies the boundedness of the $j_0(=0)$-th term provided $0 < 1-|x| \le cn^{-2/3}$ and c is sufficiently small.

For later purposes we record here that the proof also gives the following:

$$\frac{1}{C} \le \exp(nU^{\mu_w}(x))|Q_n(x)| \le C \tag{2.22}$$

uniformly on $\mathbf{R} \setminus [-1, 1]$.

Now let us remove the two zeros from Q_n that have the smallest and the largest real parts, respectively. On the interval $[-1+cn^{-2/3}, 1-cn^{-2/3}]$ in absolute value this introduces a factor lying between $c_1/(1-x^2)^{-1}$ and $(1-x^2)^{-3/2}$ (outside $I_0 \cup I_{n-1}$ this factor is $\sim (1-x^2)^{-1}$, but for x lying close to ξ_0 or ξ_{n-1} this factor is $\sim n/L \le (1-x^2)^{-3/2}$). Hence for the so modified polynomial Q_n^* we get from (2.19), (2.20), and (2.21) that

$$\frac{e^{L\pi v(x)}}{C} \le e^{-\gamma_\alpha n|x|^\alpha}|Q_n^*(x)|e^{nF_w} \le \frac{Ce^{L\pi v(x)}}{(1-x^2)^2} \tag{2.23}$$

holds with some constant C uniformly in $x \in [-1, 1]$ and n, and

$$e^{-\gamma_\alpha n|x|^\alpha}|Q_n^*(x)|e^{nF_w} = \frac{e^{L\pi v(x)+O_x(1/L)}}{1-x^2} \tag{2.24}$$

uniformly on compact subsets of $(-1, 1)$.

After these preparatory steps we now return to (2.10) and (2.11). We apply the preceding estimates for $Q^*_{[n\rho_n]}$, which has degree at most $[n-n^{1/3}]$, to conclude

$$\frac{e^{L\pi v(x)}}{C} \le e^{-\gamma_\alpha[\rho_n n]|x|^\alpha}|Q^*_{[n\rho_n]}(x)|e^{[n\rho_n]F_w} \le \frac{Ce^{L\pi v(x)}}{(1-x^2)^2}$$

on $[-1, 1]$ and

$$e^{-\gamma_\alpha[\rho_n n]|x|^\alpha}|Q^*_{[n\rho_n]}(x)|e^{[n\rho_n]F_w} = \frac{e^{L\pi v(x)+O_x(1/L)}}{1-x^2}$$

uniformly on compact subsets of $(-1, 1)$.

Whatever L is, for large n we can easily eliminate the factor $e^{L\pi v(x)}$ by multiplying through by a polynomial of degree $\le n^{1/6}$ that is close $e^{-L\pi v(x)}$ on $[-1, 1]$. This way we get polynomials $Q^{**}_{[n\rho_n+n^{1/6}]}$ that have degree at most $[n\rho_n + n^{1/6}]$ for which

$$\frac{1}{C} \le e^{-\gamma_\alpha[\rho_n n]|x|^\alpha}|Q^{**}_{[n\rho_n+n^{1/6}]}(x)|e^{[n\rho_n]F_w} \le \frac{C}{(1-x^2)^2} \tag{2.25}$$

on $[-1, 1]$ and

$$e^{-\gamma_\alpha[\rho_n n]|x|^\alpha}|Q^{**}_{[n\rho_n+n^{1/6}]}(x)|e^{[n\rho_n]F_w} = \frac{e^{O_x(1/L)}}{1-x^2} \tag{2.26}$$

uniformly on compact subsets of $(-1, 1)$.

Now by choosing L sufficiently large (actually $L = L(n) \to \infty$ very slowly compared to n), then it is easy to find polynomials $R_{n-[n\rho_n+n^{1/6}]}$ of degree at most $n - [n\rho_n + n^{1/6}] \ge n^{1/6}$ such that with

$$H_n = Q^{**}_{[n\rho_n+n^{1/6}]}R_{n-[n\rho_n+n^{1/6}]}$$

both properties (2.10) and (2.11) are satisfied (use exactly as in Section VI.1 the fact that the family of functions $\{-sQ \mid 0 \le s \le 1\}$ considered on $[-1, 1]$ is compact).

The point is that in the definition of h_n in (2.9) the factor $(1-x^2)^{-1/4}$ appears, which only improves the lower estimate in (2.25), and so (2.10) is easy to obtain. Achieving (2.11) at the same time is a simple approximation procedure if we use the upper estimate in (2.25) and the asymptotic relation (2.26).

For those who have not seen this type of approximation argument some further details are as follows. Fix an $\varepsilon > 0$, and then choose and fix L so large that on $[-1+\varepsilon, 1-\varepsilon]$ we have

$$\frac{1-\varepsilon}{1-x^2} \le e^{-\gamma_\alpha[\rho_n n]|x|^\alpha}|Q^{**}_{[n\rho_n+n^{1/6}]}(x)|e^{[n\rho_n]F_w} \le \frac{1+\varepsilon}{1-x^2},$$

which is possible in view of (2.26). Next consider the function G_ε which coincides with $(1-x^2)^{5/4}/(1-\varepsilon)$ on $[-1+2\varepsilon, 1-2\varepsilon]$, equals the constant C from (2.25) on the intervals $[-1, -1+\varepsilon]$ and $[1-\varepsilon, 1]$, and is linear on the intervals $[-1+\varepsilon, -1+2\varepsilon]$ and $[1-2\varepsilon, 1-\varepsilon]$. Now for large n (so large that all the estimates up to this point hold — recall that, because of the possible choice of large L, they are true only for sufficiently large n) there are polynomials R_n^* of degree at most $n - [n\rho_n + n^{1/6}]$ ($\geq n^{1/6}$) such that

$$G_\varepsilon(x) \leq e^{-\gamma_\alpha(n\rho_n - [\rho_n n])|x|^\alpha} R_n^*(x) \leq (1+\varepsilon)G_\varepsilon(x)$$

is true for every $x \in [-1, 1]$. Then for the polynomial $R_{n-[n\rho_n+n^{1/6}]}(x) = R_n^*(x)e^{[n\rho_n]F_w}$ and $H_n = Q^{**}_{[n\rho_n+n^{1/6}]}R_{n-[n\rho_n+n^{1/6}]}$ we will have (2.10) by the choice of G_ε and our lower estimates. On the other hand, we have the upper bound

$$e^{-\gamma_\alpha \rho_n n|x|^\alpha}|H_n(x)| \leq \frac{C^2(1+\varepsilon)}{(1-x^2)^2}$$

on all of $[-1, 1]$, and the inequality

$$(1-x^2)^{1/4} \leq e^{-\gamma_\alpha \rho_n n|x|^\alpha}|H_n(x)| \leq \frac{(1+\varepsilon)^2}{1-\varepsilon}(1-x^2)^{1/4}$$

on $[-1+2\varepsilon, 1-2\varepsilon]$, so the geometric mean of h_n from (2.9) is at most

$$\log\frac{(1+\varepsilon)^2}{1-\varepsilon} + \frac{2}{\pi}\int_{1-2\varepsilon}^1 \frac{\log C^2(1+\varepsilon) + |\log(1-x^2)^{9/4}|}{\sqrt{1-x^2}}\,dx,$$

which can be as small as we like if $\varepsilon > 0$ is sufficiently small. □

Proof of the lower estimate. The proof of the lower estimate is very similar to the above argument. In fact, let now $\rho_n = 1 + n^{-7/12}$, and let us carry out the substitution $x = \rho_n^{1/\alpha}n^{1/\alpha}y$ in the integrals in (2.1). The result is

$$\frac{1}{\gamma_n(w)^2} = n^{(2n+1)/\alpha}\rho_n^{(2n+1)/\alpha}\inf_{P_n\in\Pi_n}\int_{-\infty}^{\infty} e^{-\gamma_\alpha\rho_n 2n|x|^\alpha}P_n^2(x)\,dx.$$

However, since now we want to prove a lower estimate, we cannot restrict the integral to $[-1, 1]$; rather we need the infinite-finite range inequality that was proved in Theorem VI.5.2: for $\rho_n = 1 + n^{-7/12}$

$$\sup_{\deg P_n\leq n}\left(\int_{-\infty}^{\infty} e^{-\gamma_\alpha 2n|x|^\alpha}P_n^2(x)\,dx \Big/ \int_{-\rho_n^{1/\alpha}}^{\rho_n^{1/\alpha}} e^{-\gamma_\alpha 2n|x|^\alpha}P_n^2(x)\,dx\right) \tag{2.27}$$

$$= 1 + o(1)$$

as $n \to \infty$. This tells us that the part of the integrals of weighted polynomials away from the extremal support is negligible. In the above integral not the weight w^{2n} but $(w^{\rho_n})^{2n}$ appears, and the corresponding extremal measure has support $S_{w^{\rho_n}} = [-\rho_n^{-1/\alpha}, \rho_n^{-1/\alpha}]$. Thus, if we apply (2.27), then we can conclude that by restricting the integrals to $[-1, 1]$ we introduce only a constant that tends to zero, i.e.

$$\frac{1}{\gamma_n(w)^2} = (1+o(1))n^{(2n+1)/\alpha}\rho_n^{(2n+1)/\alpha}\inf_{P_n\in\Pi_n}\int_{-1}^{1} e^{-\gamma_\alpha\rho_n 2n|x|^\alpha}P_n^2(x)\,dx. \tag{2.28}$$

Actually, the infinite-finite range inequality (2.27) has to be applied to the intervals $[-\rho_n^{-1/\alpha}, \rho_n^{-1/\alpha}]$ and $[-1, 1]$ rather than to $[-1, 1]$ and $[-\rho_n^{1/\alpha}, \rho_n^{1/\alpha}]$, which only requires a linear transformation, not introducing any new constant in the ratios in question. In general, this linear transformation $x \to \rho_n^{1/\alpha} y$ introduces in our formulae only a constant that tends to 1 as $n \to \infty$; hence in what follows we shall use it without explicit mention.

Let us now consider the weight $w^{\rho_n^2}$ for which $S_{w^{\rho_n^2}} = [-\rho_n^{-2/\alpha}, \rho_n^{-2/\alpha}]$. By (2.19), (2.20), and (2.22) there are polynomials $Q_{[n/\rho_n]}$ of degree at most $[n/\rho_n]$ such that with $w_n = w^{\rho_n^2}$

$$\begin{aligned} \frac{e^{L\pi v(x)}}{C}(\rho_n^{-4/\alpha} - x^2)^{1/2} &\leq e^{-\gamma_\alpha [\rho_n n]|x|^\alpha} |Q_{[n/\rho_n]}(x)| e^{[n/\rho_n]F_{w_n}} \\ &\leq \frac{Ce^{L\pi v(x)}}{(\rho_n^{-4/\alpha} - x^2)^{1/2}} \end{aligned} \tag{2.29}$$

uniformly in $x \in [-\rho_n^{-2/\alpha}, \rho_n^{-2/\alpha}]$ and n,

$$\frac{1}{C} \leq \exp\left(\frac{n}{\rho_n} U^{\mu_{w_n}}(x)\right) |Q_{[n/\rho_n]}(x)| \leq C \tag{2.30}$$

uniformly in n and $x \notin [-\rho_n^{-2/\alpha}(1 - cn^{-2/3}), \rho_n^{-2/\alpha}(1 - cn^{-2/3})]$, and

$$e^{-\gamma_\alpha \rho_n^2 [n/\rho_n]|x|^\alpha} |Q_{[n/\rho_n]}(x)| e^{[n/\rho_n]F_{w_n}} = e^{L\pi v(x) + O_x(1/L)} \tag{2.31}$$

uniformly on compact subsets of $(-1, 1)$ for sufficiently large $n \geq n_L$ (here we have already used that the densities of the equilibium measures associated with $w^{\rho_n^2}$ uniformly tend to $v(x)$). On applying (IV.5.6) we can conclude from (2.30) that

$$\begin{aligned} &\int_{\rho_n^{-2/\alpha} \leq |x| \leq 1} \frac{\log\left(e^{-\gamma_\alpha \rho_n n|x|^\alpha} |Q_{[n/\rho_n]}(x)| e^{(n/\rho_n)F_{w_n}} (1 - x^2)^{1/4}\right)}{\sqrt{1 - x^2}} dx \\ &= O\left(n(\rho_n - 1)^{3/2}(\rho_n - 1)^{1/2} + (\rho_n - 1)^{1/2} \log \frac{1}{\rho_n - 1}\right) \\ &= O(n^{-1/6}) = o(1) \end{aligned}$$

by the choice of the ρ_n's, which is an estimate that is used in (2.34) below.

From (2.30) and (2.31) it easily follows that we can multiply this $Q_{[n/\rho_n]}$ by a suitable $R_{n-1-[n/\rho_n]}$ of degree at most $n - 1 - [n/\rho_n] \geq \frac{1}{2}n^{5/12}$ to get a H_{n-1} with the following properties: H_{n-1} does not vanish on $(-1, 1)$, if

$$h_n(x) = e^{-\gamma_\alpha \rho_n n|x|^\alpha} |H_{n-1}(x)(1 - x^2)^{1/2}| (1 - x^2)^{-1/4}, \tag{2.32}$$

then

$$h_n(x) \leq 1 \qquad \text{for} \quad x \in [-1, 1], \tag{2.33}$$

and

$$\lim_{n \to \infty} G[h_n] = 1. \tag{2.34}$$

From here the proof is the same as in the case of the upper estimate: set

$$R_{2n}(x) = |H_{n-1}(x)|^2(1-x^2)$$

in Bernstein's formula, use (2.33) and (2.34) instead of (2.10) and (2.11), and reverse the corresponding inequalities. Ultimately we obtain

$$\liminf_{n\to\infty} \gamma_n(w)\pi^{1/2}2^{-n}e^{-n/\alpha}n^{(n+1/2)/\alpha} \geq 1,$$

and the proof is complete. □

When we considered asymptotics for the leading coefficients we essentially solved the problem of finding asymptotics for the value in the minimum problem (2.1). Let us now consider the same problem, but in L^p:

$$E_{n,p}(w) := \inf_{P_n\in\Pi_n} \|wP_n\|_{L^p}.$$

The following generalization of (2.2) holds.

Theorem 2.3. *If $w(x) = \exp(-\gamma_\alpha|x|^\alpha)$, with $\alpha > 1$, then*

$$\lim_{n\to\infty} E_{n,p}(w)\sigma_p^{-1}2^{n-1+1/p}e^{n/\alpha}n^{-(n+1/p)/\alpha} = 1, \tag{2.35}$$

where

$$\sigma_p = \left(\Gamma(1/2)\Gamma((p+1)/2)/\Gamma(p/2+1)\right)^{1/p}.$$

Proof. In L^p, Bernstein's formula takes the following form: Let $1 \leq p \leq \infty$, and R_{2q} a polynomial of degree $2q$ which is positive on $(-1, 1)$ with possibly simple zeros at ± 1. Then for $n \geq q$

$$\left(\inf_{P_n\in\Pi_n} \int_{-1}^1 \frac{\phi^{p/2-1/2}}{R_{2q}^{p/2}}|P_n|^p\right)^{1/p}$$

$$= \sigma_p 2^{-n+1-1/p} \exp\left(\frac{1}{\pi}\int_{-1}^1 \frac{\log(\phi^{1/2-1/2p}/R_{2q}^{1/2})}{\phi^{1/2}}\right), \tag{2.36}$$

i.e.

$$E_{n,p}(\phi^{1/2-1/2p}/R_{2q}^{1/2}) = \sigma_p 2^{-n+1-1/p} G\left[\phi^{1/2-1/2p}/R_{2q}^{1/2}\right].$$

Using this formula instead of (2.6) we can imitate the proof of (2.2) and get (2.35) with minor modifications. For example, in the proof of the upper estimate (which corresponds to the lower estimate on $\gamma_n(w)$ discussed in the preceding proof) we have to use the L^p version of infinite-finite range inequality (2.27) (see Theorem VI.5.2), and change (2.32) to

$$h_n(x) = e^{-\gamma_\alpha \rho_n n|x|^\alpha}|H_{n-1}(x)(1-x^2)^{1/2}|(1-x^2)^{(-1+1/p)/2}$$

(respectively (2.9) to

$$h_n(x) = e^{-\gamma_\alpha \rho_n n|x|^\alpha}|H_n(x)|(1-x^2)^{(-1+1/p)/2} \quad),$$

for which (2.33) and (2.34) can be achieved exactly as before.

In a similar manner, only minor changes are needed in the lower estimate (which corresponds to the upper estimate on $\gamma_n(w)$ treated in the next to last proof). □

VII.3 Weak* Limits of Zeros of Orthogonal Polynomials

In this section we consider two problems for zeros of orthogonal polynomials with respect to general measures, the solution of which require weighted potentials and polynomials. Contrary to the preceding sections here μ, the measure of orthogonality, will be of compact support on **R**.

Let μ be a finite Borel measure with support consisting of infinitely many points in $[0, 1]$, and consider the corresponding orthogonal polynomials $P_n(\mu; x) = x^n + \cdots$ with zeros $z_1(\mu, n), \ldots, z_n(\mu, n)$. We are interested in the limiting distribution of these zeros. In general, this distribution does not exist, so to be more precise, we are going to analyze the distributions that arise from some subsequence $\{z_i(\mu, n_k)\}_{1\le i\le n_k}, n_1 < n_2 < \cdots$. This amounts to examining the weak* limits of the "zero measures"

$$\nu(P_n(\mu)) := \frac{1}{n}\sum_{i=1}^{n} \delta_{z_i(\mu,n)},$$

in the space $\mathcal{M}[0, 1]$ of all unit Borel measures supported on $[0, 1]$.

We need the concept of a carrier of μ. By a *carrier* of μ we mean any Borel set the complement of which has zero μ-measure. The *minimal carrier capacity* c_μ is defined as

$$c_\mu := \inf\{\operatorname{cap}(C) \mid C \text{ is a carrier of } \mu\}.$$

For Borel sets C, by $\operatorname{cap}(C)$ we shall always mean inner logarithmic capacity, that is the supremum of the capacities of compact subsets of C. As a general rule we can say that orthogonal polynomials belonging to a measure μ with $c_\mu = 0$ can behave very pathologically while those with $c_\mu > 0$ are more regular (see [212]).

The leading coefficients $\gamma_n(\mu)$ of the corresponding orthonormal polynomials $p_n(\mu; x) = \gamma_n(\mu)x^n + \cdots$ are closely related to the capacity of the support of μ and to c_μ, namely (see e.g. [212, Chapter I])

$$\frac{1}{\operatorname{cap}(\operatorname{supp}(\mu))} \le \liminf_{n\to\infty} \gamma_n(\mu)^{1/n} \le \limsup_{n\to\infty} \gamma_n(\mu)^{1/n} \le \frac{1}{c_\mu}.$$

Now it can be shown ([212, Chapter II]) that if for a sequence $N \subseteq \mathbf{N}$ the limit of $\{\nu(p_n(\mu))\}_{n\in N}$ exists in the weak*-topology, then the limit of $\{\gamma_n(\mu)^{1/n}\}_{n\in N}$ also exists as $n \to \infty$. In the converse direction one only can assert that if

$$\lim_{n\to\infty,\, n\in N} \gamma_n(\mu)^{1/n}$$

exists *and is equal to either* $1/\text{cap}(\text{supp}(\mu))$ *or to* $1/c_\mu$, then the corresponding sequence $\{\nu(p_n(\mu))\}_{n\in N}$ also converges to the equilibrium distribution of $\text{supp}(\mu)$ or of C, respectively, where C is any carrier of μ with $\text{cap}(C) = c_\mu$ (for the concept of equilibrium distribution of a Borel set see below).

So how can the weak* limits of zeros that arise from one measure μ be characterized? If we equip the set $\mathcal{M}[0, 1]$ of probability Borel measures on $[0, 1]$ with the weak*-topology and

$$\mathcal{M}_\mu := \{\nu \mid \nu \text{ is a weak* limit point of the measures } \nu(p_n(\mu))\},$$

then $\mathcal{M}_\mu$ is a connected closed subset of $\mathcal{M}[0, 1]$, and conversely, every connected and closed subset of $\mathcal{M}[0, 1]$ equals $\mathcal{M}_\mu$ for some μ. However, the situation radically changes if we require $c_\mu > 0$. In this case one loses a lot of freedom in constructing weak* limits of zeros of orthogonal polynomials and the situation becomes much more complex. But, with the help of the theory we have presented for weighted polynomials, we can lay a hand on it.

The equilibrium measure of a Borel set $C \subseteq [0, 1]$ of positive capacity is the unique measure ν in $\mathcal{M}[0, 1]$ satisfying $U^\nu \le \log(\text{cap}(C)^{-1})$ with equality on C except for a set of zero capacity. For its existence and uniqueness see [212, Appendix].

A carrier C of μ is called a *minimal carrier* if $\text{cap}(C) = c_\mu$. It is easy to see that there are always minimal carriers. Below we will encounter potentials U^ν that are bounded above and at some point they take on their supremum. We shall write $\text{MAX}\, U^\nu$ for the *set of maximum points* of U^ν with the agreement that $\text{MAX}\, U^\nu$ is empty if the potential is not bounded above (or if its supremum is not attained). With these concepts we prove

Theorem 3.1. *If* $c_\mu > 0$ *and* C *is a minimal carrier of* μ, *then any weak* limit* ν *of the zero distributions* $\{\nu(P_n(\mu))\}$ *satisfies* $C\,\text{“}\subseteq\text{”}\,\text{MAX}\, U^\nu$ *and* $\text{supp}(\nu) \subseteq \overline{C}$, *where* “ $\subseteq$ ” *means inclusion except for a set of zero capacity. Conversely, if* $C \subseteq [0, 1]$ *is of positive capacity and* $\mathcal{M}_C$ *is the set of probability measures* ν *satisfying* $C\,\text{“}\subseteq\text{”}\,\text{MAX}\, U^\nu$ *and* $\text{supp}(\nu) \subseteq \overline{C}$, *then there is a measure* μ *such that* C *is a minimal carrier of* μ *and* $\mathcal{M}_\mu = \mathcal{M}_C$.

This theorem says that all possible weak* limits can occur among the limit distributions of the zeros of orthogonal polynomials, even in the case $c_\mu > 0$.

Proof of Theorem 3.1. We shall work with the monic polynomials

$$P_n(\mu; x) = \frac{1}{\gamma_n(\mu)} p_n(\mu; x)$$

rather than with the orthonormal ones. Let $c_\mu > 0$ and $\nu \in \mathcal{M}_\mu$. That $\text{supp}(\nu) \subseteq \text{supp}(\mu) \subseteq \overline{C}$ follows from the well known fact that in any interval contiguous to $\text{supp}(\mu)$ each of the polynomials $P_n(\mu)$ can have at most one zero (otherwise there would be a polynomial of degree at most $n-1$ having the same sign as $P_n(\mu)$ on $\text{supp}(\mu)$, contradicting orthogonality). Furthermore, above we have already remarked that if the subsequence $\{\nu(P_n(\mu))\}_{n\in N}$ converges to ν in the weak*-topology, then $\{\gamma_n^{1/n}\}_{n\in N}$ also converges to some number α, and by the lower envelope theorem (Theorem I.6.9) $U^\nu(x)$ coincides quasi-everywhere with the lim inf of the sequence

$$\left\{\frac{1}{n}\log(1/|P_n(\mu,x)|)\right\}_{n\in N}.$$

But by general estimates on orthonormal polynomials we have

$$\limsup_{n\to\infty,\, n\in N} |P_n(\mu,x)|^{1/n} = 1/\alpha$$

quasi-everywhere on C and

$$\limsup_{n\to\infty,\, n\in N} |P_n(\mu,x)|^{1/n} \geq 1/\alpha$$

quasi-everywhere ([212]); thus the potential U^ν satisfies $U^\nu(x) \leq \log\alpha$ quasi-everywhere, and equality holds for quasi every $x \in C$. From the lower semicontinuity of U^ν we finally obtain $U^\nu(x) \leq \log\alpha$ everywhere, and C "$\subseteq$" $\mathrm{MAX}\, U^\nu$ has been verified.

The second half of the theorem requires a more sophisticated argument and this is where we use weighted polynomials.

The weak*-topology on $\mathcal{M}[0,1]$ is metrizable and in what follows $d(\cdot,\cdot)$ will denote a metric that generates this topology. E.g. if $\{f_k\}$ is a dense set in the unit ball of $C[0,1]$, then we can set

$$d(f,g) = \sum_{k=0}^{\infty} \frac{1}{2^k}\left|\int (f-g)f_k\right|.$$

Choose a dense sequence $\{\nu_k\}_{k=1}^\infty$ in $\mathcal{M}_C$. Then there is a $C_0 \subseteq C$ such that $\mathrm{cap}(C_0) = \mathrm{cap}(C)$ and $C_0 \subseteq \mathrm{MAX}\, U^{\nu_k}$ for all k. C_0 is not necessarily compact, but we can choose an increasing sequence $C_1 \subseteq C_2 \subseteq \cdots$ of compact subsets of C_0 such that $\mathrm{cap}(C_n) \to \mathrm{cap}(C)$ as n tends to infinity.

Fix a k. From Lemma I.6.10 we know that there is an increasing sequence $\{\nu_k^i\}_{i=1}^\infty$ of measures (with total mass at most one) such that the support of ν_k^i is contained in the support of ν_k, $U^{\nu_k^i}$ is continuous, $\|\nu - \nu_k^i\| \to 0$, and for all x we have

$$U^{\nu_k^i}(x) \to U^{\nu_k}(x), \quad i \to \infty. \tag{3.1}$$

Since on the compact set C_k the sequence $\{U^{\nu_k^i}\}_{i=1}^\infty$ converges to the constant $\max U^{\nu_k}$, we have uniform convergence in (3.1) on C_k. These facts imply that there is a probability measure σ_k ($= \nu_k^i/\|\nu_k^i\|$ for sufficiently large i) such that

(i) $d(\sigma_k, \nu_k) < 1/k$,
(ii) $\mathrm{supp}(\sigma_k) \subseteq \mathrm{supp}(\nu_k) \subseteq \overline{C}$,
(iii) U^{σ_k} is continuous,
(iv) $\max U^{\sigma_k} - 1/k \leq U^{\sigma_k}(x) \leq \max U^{\sigma_k}$ for all $x \in C_k$.

Here and in what follows "max" denotes maximum on $[0,1]$.

We set

$$w_k(x) = \exp\left(U^{\sigma_k}(x) - \max U^{\sigma_k}\right). \tag{3.2}$$

Then w_k is continuous,

$$\exp\left(-\max U^{\sigma_k}\right) \le w_k(x) \le 1 \tag{3.3}$$

for $x \in [0, 1]$ and

$$e^{-1/k} \le w_k(x) \le 1 \tag{3.4}$$

for $x \in C_k$. By Theorem I.3.3 we have $\mu_{w_k} = \sigma_k$ and $F_{w_k} = \max U^{\sigma_k}$; hence we obtain from Theorem III.2.1 that for any polynomial p_n of degree at most n we have

$$\|p_n w_k^n\|_{[0,1]} = \|p_n w_k^n\|_{\mathrm{supp}(\sigma_k)}. \tag{3.5}$$

Now we select a certain finite subset of supp(σ_k) on which the maximum of every weighted polynomial $|p_n w_k^n|$ is comparable with the norm of $p_n w_k^n$. Let

$$\left\{x_0^{(k,n)}, x_1^{(k,n)}, \dots, x_n^{(k,n)}\right\} = \{x_0, x_1, \dots, x_n\}$$

be an $(n+1)$-point Fekete set for the weight w_k restricted to supp(σ_k) (see Section III.1). By Theorem III.1.12

$$\|p_n w_k^n\|_{\mathrm{supp}(\sigma_k)} \le (n+1)\|p_n w_k^n\|_{\{x_i\}_{0\le i\le n}}$$

for every polynomial p_n of degree at most n, and so the choice $\{x_i\}_{i=0}^n$ is suitable for us. If we also take into account that supp$(\sigma_k) \subseteq \overline{C}$ and that w_k is continuous, simple compactness argument yields instead of the Fekete set above an $(n+1)$-point set $S_{n,k} \subseteq C$ such that for all polynomials p_n of degree at most n we have

$$\|p_n w_k^n\|_{[0,1]} = \|p_n w_k^n\|_{\mathrm{supp}(\sigma_k)} \le (n+2)\|p_n w_k^n\|_{S_{n,k}} \tag{3.6}$$

(use that the set of polynomials p_n, $\deg(p_n) \le n$ with $\|p_n w_k^n\|_{\mathrm{supp}(\sigma_k)} \le 1$ is compact).

Next we need the existence of polynomials $P_{n,k}$ of the form $x^n + \cdots$ such that

$$\lim_{n\to\infty} \|P_{n,k} w_k^n\|_{[0,1]}^{1/n} = \exp(-\max U^{\sigma_k}). \tag{3.7}$$

By Theorem III.1.9 and $F_{w_k} = \max U^{\sigma_k}$ we can take as $P_{n,k}$ the Fekete polynomials associated with w_k.

With a sequence $\{n_k\}$ to be chosen below we set

$$\mu_k := \frac{1}{(n_k+1)^2}\omega_{C_k} + \frac{1}{(n_k+1)^3}\sum_{\tau\in S_{n_k,k}} w_k^{2n_k}(\tau)\delta_\tau =: \mu_k' + \mu_k^* \tag{3.8}$$

where ω_{C_k} is the equilibrium measure of the compact set C_k and $S_{n_k,k}$ is the subset of C appearing in (3.6). Finally, let

$$\mu := \sum_{k=1}^{\infty} \mu_k.$$

We will show that $\mathcal{M}_C = \mathcal{M}_\mu$ and C is a minimal carrier of μ, by which the proof will be complete.

That

$$c_\mu \geq \lim_{k\to\infty} \operatorname{cap}(C_k) = \operatorname{cap}(C)$$

is obvious because $\mu \geq (n_k+1)^{-2}\omega_{C_k}$ for all k and it is easy to show that C_k is a minimal carrier of ω_{C_k}. Then, by the first part of the theorem, we must have $\mathcal{M}_\mu \subseteq \mathcal{M}_C$, therefore we only have to show the converse inclusion. This will be done at the end of the proof after some preparation to be done below.

Set

$$T_{n_k}(x) := P_{m_k,k}(x) \prod_{t\in\cup S_{n_r,r},\, r<k} (x-t), \tag{3.9}$$

where the $P_{m_k,k}$ are the polynomials from (3.7), $S_{n_r,r}$ are the sets from (3.6), and where

$$m_k = n_k - \left|\bigcup_{r=1}^{k-1} S_{n_r,r}\right| \geq n_k - \sum_{r=1}^{k-1}(n_r+1).$$

Thus, $\deg(T_{n_k}) = n_k$ and T_{n_k} has leading coefficient 1. Clearly T_{n_k} vanishes on the support of μ_r^* for $1 \leq r < k$, and since on the support of μ_r^* for $1 \leq r \leq k$, which are all contained in C_k, we have by (3.3)–(3.4) and (3.7) the inequality $1 \geq w_k(x) \geq e^{-1/k}$, we get

$$\begin{aligned}
\int T_{n_k}^2\, d\left(\sum_{r=1}^{k} \mu_r'\right) &\leq \|T_{n_k}\|_{C_k}^2 \leq e^{2n_k/k}\|T_{n_k} w_k^{m_k}\|_{[0,1]}^2 \\
&\leq e^{2n_k/k}\|P_{m_k,k} w_k^{m_k}\|_{[0,1]}^2 \\
&\leq e^{2n_k/k}\exp\left(2m_k\left(-\max U^{\sigma_k} + \frac{1}{k}\right)\right) \\
&\leq e^{5n_k/k}\exp(2n_k(-\max U^{\sigma_k}))
\end{aligned}$$

provided n_k is large enough compared to n_{k-1}. By the definition of μ_k^* we have

$$\int T_{n_k}^2\, d\mu_k^* \leq \|T_{n_k} w_k^{n_k}\|_{[0,1]}^2$$

for which the preceding estimate can be applied. Finally,

$$\begin{aligned}
\int T_{n_k}^2\, d\left(\sum_{r=k+1}^{\infty} \mu_k\right) &\leq \sum_{r=k+1}^{\infty} \|\mu_r\| \leq 2\sum_{r=k+1}^{\infty} (n_r+1)^{-2} \\
&\leq \frac{2}{n_{k+1}} \leq \exp(2n_k(-\max U^{\sigma_k}))
\end{aligned}$$

provided n_{k+1} is large compared to n_k.

Until now we have not said anything about the sequence $\{n_k\}$. Let us choose it increasing so fast that all the estimates following (3.9) hold. From them we get

$$\int T_{n_k}^2 \, d\mu \le e^{6n_k/k} \exp(2n_k(-\max U^{\sigma_k})).$$

Now we make use of the basic fact that the monic orthogonal polynomials $P_{n_k}(\mu)$ minimize the $L^2(\mu)$-norm among all monic polynomials of degree n_k. Thus, together with the preceding estimate we also have

$$\int P_{n_k}^2(\mu) \, d\mu \le e^{6n_k/k} \exp(2n_k(-\max U^{\sigma_k})).$$

But here the left-hand side is at least as large as

$$\int P_{n_k}^2(\mu) d\mu_k \ge (n_k+1)^{-3} \| P_{n_k}(\mu) w_k^{n_k} \|_{S_{n_k,k}}^2 \ge (n_k+2)^{-5} \| P_{n_k}(\mu) w_k^{n_k} \|_{[0,1]}^2,$$

where we used property (3.6) of the sets $S_{n_k,k}$. Thus, we can conclude from the last two estimates

$$\| P_{n_k}(\mu) w_k^{n_k} \|_{[0,1]} \le (n_k+2)^3 e^{3n_k/k} \exp(n_k(-\max U^{\sigma_k})). \tag{3.10}$$

Finally we can complete the proof from (3.10) along the lines of Section III.4 (zero distribution of polynomials of asymptotically minimal weighted norm). Let $\sigma \in \mathcal{M}_C$ be arbitrary. By the choice of the measures ν_k and σ_k, there is a sequence $N \subseteq \mathbf{N}$ such that σ_k converges in the weak* topology to σ as k tends to infinity through N. We get from (3.10) that for all $x \in [0, 1]$

$$\frac{1}{n_k} \log \frac{1}{|P_{n_k}(\mu, x)|} \ge U^{\sigma_k}(x) - \frac{3}{k} - \frac{\log(n_k+2)^3}{n_k}, \qquad k \in N. \tag{3.11}$$

By the principle of domination (Theorem II.3.2), this then holds for all $x \in \mathbf{C}$ (recall that the potential U^{σ_k} is continuous; hence σ_k has finite logarithmic energy). Now if κ is a weak* limit point of the measures $\nu(P_{n_k}(\mu))$, $k \in N$, then going to infinity through a suitable subsequence of N we get from (3.11) for $x \notin [0, 1]$ (note that all the zeros of P_{n_k} are contained in $[0, 1]$)

$$U^\kappa(x) \ge U^\sigma(x).$$

At infinity the difference of the left and right-hand sides vanishes, so by the minimum principle for harmonic functions, $U^\kappa(x) = U^\sigma(x)$ holds true outside $[0, 1]$. Thus, these two potentials coincide except for a set of two-dimensional Lebesgue measure zero; hence $\kappa = \sigma$ follows from Corollary II.2.2. □

VII.4 Notes and Historical References

The theory of orthogonal polynomials is so vast that it is impossible to give even a sketch of the results. Here we restrict ourselves to those few subjects that have been discussed in this chapter.

Section VII.1

A thorough account of Freud type orthogonal polynomials can be found in [169] by P. Nevai which covers the history until the middle of the eighties.

Theorems 1.1 and 1.2 were proved by E. A. Rakhmanov [189]. Formula (1.2) was conjectured by G. Freud [49, 50]. Independently of Rakhmanov, H. N. Mhaskar and E. B. Saff [157] derived the zero distribution of the orthogonal polynomials. The moments of the asymptotic distribution were given by Nevai and J. S. Dehesa [170], and the exact form of the asymptotic distribution was given by J. L. Ullman [224]. The simple method of Section VII.1 seems to appear for the first time.

Section VII.2

The results of this section are due to D. S. Lubinsky and E. B. Saff [142]. Corollary 2.3 was conjectured by G. Freud (see [169] for its history). This famous Freud problem has attracted much interest in the eighties. Partial results were achieved by several authors, among whom A. Magnus [147, 148], who proved the conjecture for even λ's, must be mentioned. The final solution was given in a series of papers by P. Nevai, A. Knopfmacher, D. S. Lubinsky, H. N. Mhaskar and E. B. Saff [103, 143, 141]. The so-called strong asymptotic formula given in Theorem 2.1, which is the deepest result concerning Freud–type orthogonal polynomials, was proved in the monograph [142] of Lubinsky and Saff for rather general weights that include all the weights $\exp(-|x|^\lambda)$ for any $\lambda > 0$. A short proof for the same formula was given by V. Totik in [220]. The proof here follows that of [220], which has a lot in common with the original argument of Lubinsky and Saff, but the approximation part in it was replaced by the results of Section VI.1.

For related results concerning orthogonal polynomials with respect to so-called Erdős weights $W = \exp(-Q)$, where Q grows faster than any polynomial around infinity, see the works [140] and [133] by D. S. Lubinsky and A. L. Levin and D. S. Lubinsky and T. Z. Mthembu. For other problems connected with exponential weights (such as asymptotics for Christoffel functions, Markoff–Bernstein inequalities, etc.) see the papers [132] and [130] by Levin and Lubinsky.

Section VII.3

The notions "minimal carrier" etc. appearing in this section were introduced and used by J. Ullman [223]–[225], who (later together with M. F. Wyneken and L. Ziegler, see [226] and [227]) systematically studied the basic properties of orthogonal polynomials with respect to general measures. General orthogonal polynomials is the subject of the monograph [212] by H. Stahl and V. Totik. For more on the zeros of general orthogonal polynomials see [212, Chapter 2], where Theorem 3.1 was also taken from.

Chapter VIII. Signed Measures

In this chapter the energy problem is generalized to the signed measure case. Suppose that our set Σ (called a condenser) consists of a finite number of subsets Σ_i on each of which there is a sign and a total charge prescribed for the signed measure. First we show that under these restrictions the energy problem with an external field has a unique solution that is characterized by a self duality property. The weighted equilibrium potential satisfies analogous inequalities and characterization as in the positive measure case.

When there are only two plates, a positive and a negative one, we introduce the analogue of Fekete and Leja points, which now maximize certain rational expressions. It is shown that weighted Fekete points are asymptotically distributed according to the equilibrium distribution, and the same is true of Leja points at least when there is no external field.

Exactly as weighted potentials are connected with weighted polynomial approximation and weighted polynomial inequalities, the present signed measure case is connected to rational approximation and inequalities for rational functions. In particular, the solution of the energy problem permits us to resolve Zolotarjov's problem for pairs of disjoint compact sets (Σ_1, Σ_2): namely, how large can the minimum of a weighted rational function $w^n r_n$ be (in n-th root sense) on Σ_1 if its maximum is at most one on Σ_2? The solution of this problem leads to the determination of the rate of (weighted) approximation of signum-type functions, with nearly optimal rational functions obtained from Fekete or Leja points. Explicit solutions for this problem are given in some concrete cases, such as when both Σ_1 and Σ_2 are the union of finitely many intervals.

In the present signed measure case the extremal point method (based on Fekete or Leja points) yields an algorithm for finding the conformal map of a doubly connected domain onto an annulus (just as in the positive measure case the extremal point method leads to conformal maps of simply connected domains onto disks). The method is based on the observation that the minimal energy is a conformal invariant.

In the final section of the chapter we give an application to discrepancy theorems by deriving the best possible estimate for the discrepancy of the distribution of zeros of polynomials having simple zeros. Here both the direct and the converse parts use the theory developed for the weighted energy problem with signed measures.

VIII.1 The Energy Problem for Signed Measures

The analogy with physical problems suggests considering the following energy problem: suppose there are finitely many closed sets $\Sigma_1, \ldots, \Sigma_N (N \geq 2)$ on $\mathbf{C}$ of positive distance from one another (the conductors), and to each $1 \leq j \leq N$ there is assigned a sign $\varepsilon_j = \pm 1$ (the sign of the charge) and a positive number m_j (the amount of the charge on Σ_j). Furthermore, on $\Sigma := \cup \Sigma_j$, there is given a weight function $w(x) = \exp(-Q(x))$. We want to minimize the weighted energy

$$\iint \log \frac{1}{|z-t|} d\sigma(z) d\sigma(t) + 2 \int Q(t)\, d\sigma(t)$$

for all signed measures of the form $\Sigma \varepsilon_j \sigma_j$, where σ_j is a compactly supported positive measure on Σ_j of total mass m_j. The exact assumptions on Σ_j, Q, etc. are as follows.

In this chapter Σ_j, $j = 1, \ldots, N$, will always denote (not necessarily bounded) closed subsets of $\mathbf{C}$, such that the distance between any two different Σ_j is positive. We set

$$\Sigma := \bigcup_{j=1}^{N} \Sigma_j.$$

As above, to each Σ_j we associate two numbers $\varepsilon_j = 1$ or -1 and $m_j > 0$, and we set

$$\varepsilon(t) := \varepsilon_j \quad \text{if} \quad t \in \Sigma_j, \quad j = 1, \ldots, N.$$

The weight function will be again of the form $w(x) = \exp(-Q(x))$ and for the portions $Q_j = Q\big|_{\Sigma_j}$ of Q supported on Σ_j we assume that each Q_j, $j = 1, \ldots, N$ satisfies the following conditions:

(i) $\varepsilon_j Q_j$ is lower semi-continuous on Σ_j;

(ii) $\varepsilon_j Q_j(x) < \infty$ on a set of positive capacity;

(iii) $\varepsilon_j Q_j(x)/\log|x| \to \infty$ as $|x| \to \infty$, $x \in \Sigma_j$, in case Σ_j is unbounded.

We remark that these assumptions guarantee that the weight function $\exp(-\varepsilon_j Q_j(x))$ is admissible in the sense of the first part of the book, i.e. in the sense of Definition I.1.1. Actually, (i) and (ii) are the same conditions that were imposed on an admissible weight in (I.1.10), but in (I.1.10) condition (iii), which is equivalent to $\varepsilon_j Q_j(x) - \log|x| \to \infty$ as $|x| \to \infty$, $x \in \Sigma_j$, is now replaced by $\varepsilon_j Q_j(x)/\log|x| \to \infty$. However, if Σ_j is compact, then condition (iii) is empty and (i)–(iii) are equivalent to the admissibility of $\exp(-\varepsilon_j Q_j)$ on Σ_j.

Let $\mathcal{M}$ be the set of all signed measures of the form

$$\sigma = \sum_{j=1}^{N} \varepsilon_j \sigma_j,$$

where each σ_j is a positive Borel measure on Σ_j of compact support and of total mass $\|\sigma_j\| = m_j$. When $\sigma \in \mathcal{M}$, then σ_j will always be used in this sense, i.e. σ_j

denotes ε_j times the restriction of σ to Σ_j. With these notations we consider the energy problem

$$V_w := \inf_{\sigma\in\mathcal{M}} I_w(\sigma) := \inf_{\sigma\in\mathcal{M}} \left(\iint \log\frac{1}{|z-t|}\, d\sigma(z)d\sigma(t) + 2\int Q(t)\, d\sigma(t) \right). \tag{1.1}$$

Remark 1.1. The compactness of the support of σ is needed only to guarantee the existence of the double integral. It turns out that in a sense this is not needed, the optimal σ will automatically be of compact support (see the proof of Theorem 1.4 below).

Remark 1.2. That (1.1) and not

$$\inf_{\sigma\in\mathcal{M}} \iint \left(\log\frac{1}{|z-t|} + Q(z) + Q(t) \right) d\sigma(z)d\sigma(t)$$

is the correct formulation of the energy problem (cf. Section I.1) is shown by the fact that in case $\Sigma\varepsilon_j m_j = 0$, the function Q is actually eliminated from the last expression.

Remark 1.3. Above we have taken into account the signs ε_j both in the double and in the single integrals because

$$\begin{aligned} &\iint \log\frac{1}{|z-t|}\, d\sigma(z)d\sigma(t) + 2\int Q(t)\, d\sigma(t) \\ &= \sum_{1\le j\le N} \iint \log\frac{1}{|z-t|}\, d\sigma_j(z)d\sigma_j(t) \\ &\qquad +2 \sum_{1\le i<j\le N} \varepsilon_i\varepsilon_j \iint \log\frac{1}{|z-t|}\, d\sigma_i(z)d\sigma_j(t) \\ &\qquad +2 \sum_{1\le j\le N} \varepsilon_j \int_{\Sigma_j} Q(t)\, d\sigma_j(t), \end{aligned} \tag{1.2}$$

i.e. both in the mutual energy of the components of the charge and towards the external field Q, the charge acts with its sign taken into account. There are, however, external fields (like gravitation), which act on positive and negative charges in the same way. Such fields can also be handled within the above framework by changing the sign of Q on Σ_j appropriately, namely consider $\varepsilon_j Q_j$ instead of Q_j on Σ_j (at least for compact Σ_j's and continuous Q).

Theorem 1.4. *Let Σ and Q be as above. Then*

(a) V_w *is finite.*

(b) *There is a unique signed measure $\mu^* \in \mathcal{M}$ (called the equilibrium measure) minimizing the energy in* (1.1), *i.e. for which* $I_w(\mu^*) = V_w$.

(c) *μ^* has finite logarithmic energy and both Q and U^{μ^*} are bounded on the support of μ^*. Hence, U^{μ^*} is bounded on compact subsets of* $\mathbf{C}$.

(d) *For each j there are constants F_j such that*

$$\varepsilon_j(U^{\mu^*}(z) + Q(z)) \geq F_j$$

for quasi-every $z \in \Sigma_j$ and

$$\varepsilon_j(U^{\mu^*}(z) + Q(z)) \leq F_j$$

for every $z \in$ supp$(\mu^*) \cap \Sigma_j$.

If we introduce the function

$$F(t) := F_j \quad \text{if} \quad t \in \Sigma_j, \quad j = 1, \dots, N, \tag{1.3}$$

then (d) can be expressed as

$$\varepsilon(z)(U^{\mu^*}(z) + Q(z)) \geq F(z) \quad \text{for q.e.} \quad z \in \Sigma$$

and

$$\varepsilon(z)(U^{\mu^*}(z) + Q(z)) \leq F(z) \quad \text{for} \quad z \in \operatorname{supp}(\mu^*).$$

Proof of Theorem 1.4. If we separately solve the weighted energy problem (for positive measures of total mass m_j) with weight $\exp(-\varepsilon_j Q_j)$ on each Σ_j, then from the solutions we get an element of $\mathcal{M}$ for which the energy expression in (1.1) is finite (recall that the Σ_j's are of positive distance apart); hence $V_w < \infty$.

Next we note that for each $c > 0$ there exists a C such that for every $1 \leq i \neq j \leq N$, $z \in \Sigma_i$, $t \in \Sigma_j$, one has

$$\big|\log|z - t|\big| \leq c(\varepsilon_i Q_i(z) + \varepsilon_j Q_j(t)) + C. \tag{1.4}$$

In fact, this immediately follows from the assumption that $\varepsilon_i Q_i(z)$ tends faster to infinity than any constant times $\log|z|$ as $z \to \infty$, $z \in \Sigma_i$, from the lower boundedness of $\varepsilon_i Q_i$ and from the fact that z and t cannot get close to each other. In a similar fashion, if $z, t \in \Sigma_i$, then

$$\log\frac{1}{|z - t|} \geq -c(\varepsilon_i Q_i(z) + \varepsilon_i Q_i(t)) - C. \tag{1.5}$$

If $c > 0$ is some small number and we integrate these inequalities with respect to $d\sigma_i(z)d\sigma_j(t)$ and $d\sigma_i(z)d\sigma_i(t)$, respectively, then it follows from (1.2) that uniformly for $\sigma \in \mathcal{M}$

$$I_w(\sigma) \geq \frac{1}{2}\sum_{1\leq j\leq N}\int_{\Sigma_j} \varepsilon_j Q(t)\, d\sigma_j(t) - C_1. \tag{1.6}$$

Since $\varepsilon_j Q$ is bounded from below on Σ_j, it immediately follows that $V_w > -\infty$. Thus, (a) is proved.

We can see from (1.6), that in (1.1) we can restrict our attention to weights σ satisfying

$$\int Q(t)\,d\sigma(t) = \sum_{1\le j\le N} \int_{\Sigma_j} \varepsilon_j Q(t)\,d\sigma_j(t) \le 2(V_w + C_1 + 1), \tag{1.7}$$

because for other weights the weighted energy is at least $V_w + 1$. Now let $\mathcal{M}_\infty$ be the set of signed Borel measures of the form $\sigma = \Sigma\varepsilon_j\sigma_j$ such that (1.7) is satisfied and each σ_j is a positive Borel measure on Σ_j with total mass m_j. Thus, $\mathcal{M}$ and $\mathcal{M}_\infty$ differ in that the compactness assumption concerning the support of σ is replaced by (1.7). If

$$V_w^{(\infty)} = \inf_{\sigma\in\mathcal{M}_\infty} I_w(\sigma), \tag{1.8}$$

then obviously $V_w^{(\infty)} \le V_w$ (cf. (1.6)). We show that actually equality holds here. Note also that $I_w(\sigma)$ is well defined for all $\sigma \in \mathcal{M}_\infty$ because of (1.4) and (1.5).

First we show that the infimum defining $V_w^{(\infty)}$ is attained. In fact, let $\{\sigma^{(k)}\}$ be a sequence from $\mathcal{M}_\infty$ for which $I_w(\sigma^{(k)})$ tends to $V_w^{(\infty)}$. We assume without loss of generality that $\sigma^{(k)}$ tends to a signed measure σ in the weak* topology on all signed Borel measures on the closed Riemann sphere $\overline{\mathbf{C}}$. Since the function $\varepsilon(z)Q(z)$ is bounded from below on Σ and tends to infinity as $|z|$ does so, (1.7) implies that for each j we have $\sigma_j^{(k)}(\{z \mid |z| \ge R\}) \to 0$ as $R \to \infty$ uniformly in k. To see that σ also belongs to $\mathcal{M}_\infty$, it remains to show that (1.7) is satisfied. But this follows from the lower semi-continuity of the function εQ: there is an increasing sequence $\{g_n\}$ of continuous functions converging to εQ at every point of Σ such that g_1 is bounded from below on Σ; hence by the monotone convergence theorem

$$\begin{aligned}\sum_{1\le j\le N} \int_{\Sigma_j} \varepsilon_j Q(t)\,d\sigma_j(t) &= \lim_{n\to\infty} \sum_{1\le j\le N} \int_{\Sigma_j} g_n(t)\,d\sigma_j(t)\\ &= \lim_{n\to\infty}\lim_{k\to\infty} \sum_{1\le j\le N} \int_{\Sigma_j} g_n(t)\,d\sigma_j^{(k)}(t)\\ &\le \liminf_{k\to\infty} \sum_{1\le j\le N} \int_{\Sigma_j} \varepsilon_j Q(t)\,d\sigma_j^{(k)}(t) \le 2(V_w + C_1 + 1).\end{aligned}$$

Similar computations show that $I_w(\sigma) = V_w^{(\infty)}$:

$$\begin{aligned}&\iint \log\frac{1}{|z-t|}\,d\sigma(z)d\sigma(t) + 2\int Q(t)\,d\sigma(t)\\ &= \lim_{n\to\infty}\left(\sum_{1\le j\le N} \iint \min\left(n, \log\frac{1}{|z-t|}\right) d\sigma_j(z)d\sigma_j(t)\right.\end{aligned}$$

$$+2\sum_{1\le i<j\le N}\varepsilon_i\varepsilon_j\iint\min\left(n,\log\frac{1}{|z-t|}\right)d\sigma_i(z)d\sigma_j(t)$$

$$\left.+\,2\sum_{1\le j\le N}\int_{\Sigma_j}g_n(t)\,d\sigma_j(t)\right)$$

$$=\lim_{n\to\infty}\lim_{k\to\infty}\left(\sum_{1\le j\le N}\iint\min\left(n,\log\frac{1}{|z-t|}\right)d\sigma_j^{(k)}(z)d\sigma_j^{(k)}(t)\right.$$

$$+2\sum_{1\le i<j\le N}\varepsilon_i\varepsilon_j\iint\min\left(n,\log\frac{1}{|z-t|}\right)d\sigma_i^{(k)}(z)d\sigma_j^{(k)}(t)$$

$$\left.+2\sum_{1\le j\le N}\int_{\Sigma_j}g_n(t)\,d\sigma_j^{(k)}(t)\right)$$

$$\le\liminf_{k\to\infty}\left(\sum_{1\le j\le N}\iint\log\frac{1}{|z-t|}d\sigma_j^{(k)}(z)\,d\sigma_j^{(k)}(t)\right.$$

$$+2\sum_{1\le i<j\le N}\varepsilon_i\varepsilon_j\iint\log\frac{1}{|z-t|}\,d\sigma_i^{(k)}(z)d\sigma_j^{(k)}(t)$$

$$\left.+2\sum_{1\le j\le N}\varepsilon_j\int_{\Sigma_j}Q(t)\,d\sigma_j^{(k)}(t)\right)$$

$$=\lim_{k\to\infty}I_w(\sigma^{(k)})=V_w^{(\infty)},$$

where in the next to last step we used the fact that $\log(1/|z-t|)$ is bounded above if z and t belong to different Σ_j's, and hence uniformly for such z and t

$$\min\left(n,\log\frac{1}{|z-t|}\right)=\log\frac{1}{|z-t|}$$

for large n. Thus, our claim concerning the existence of an optimal signed measure σ^* in (1.8) has been verified.

Now we return to the space $\mathcal{M}$. For $\sigma\in\mathcal{M}_\infty$ let

$$\overline{\sigma}_j=\sum_{i\ne j}\varepsilon_i\sigma_i,$$

and set for $z\in\Sigma_j$

$$w_j^{(\sigma)}(z)=\exp(-\varepsilon_j(Q(z)+U^{\overline{\sigma}_j}(z))/m_j).$$

We claim that if σ^* is an optimal element of $\mathcal{M}_\infty$, the existence of which has been verified above, then

$$\sigma_j^* = m_j \mu_{w_j^{(\sigma^*)}}, \tag{1.9}$$

where $\mu_{w_j^{(\sigma^*)}}$ is the solution of the energy problem in the sense of Chapter I on Σ_j corresponding to the weight function $w_j^{(\sigma^*)}$. That is, we claim that if we fix every component σ_i^* of σ^* except for the j-th one and then minimize the energy with respect to this j-th component, then the minimum is attained for σ_j^*. From (1.4) and (1.7) it follows that the potential $U^{\overline{\sigma}_j^*}(z)$ is continuous and finite on Σ_j (note that Σ_j and $\Sigma \setminus \Sigma_j$ are of positive distance apart) and that $w_j^{(\sigma^*)}(z)$ is an admissible weight on Σ_j. But then the fact that each component of σ^* must be optimal when the others are kept fixed is obvious from the optimality of σ^* and this is exactly the identity (1.9) because the equilibrium measures are unique.

With (1.9) the existence question is readily settled. In fact, that every σ_j^*, and hence also σ^* has compact support, follows from (1.9) and Theorem I.1.3. But then $\sigma^* \in \mathcal{M}$, and so $V_w = V_w^{(\infty)}$ and σ^*, an optimal element of $\mathcal{M}_\infty$, is optimal in $\mathcal{M}$, as well. This proves the existence of $\mu^*(= \sigma^*)$.

Next we prove its uniqueness. Suppose that $\sigma \in \mathcal{M}$ is another measure with $I_w(\sigma) = V_w$. Then with the notation

$$(\nu, \lambda) := \iint \log \frac{1}{|z-t|} d\nu(z) d\lambda(t) \tag{1.10}$$

we get

$$V_w = \tfrac{1}{2}(I_w(\sigma^*) + I_w(\sigma)) = I_w\left(\tfrac{1}{2}(\sigma^* + \sigma)\right) + \left(\tfrac{1}{2}(\sigma^* - \sigma), \tfrac{1}{2}(\sigma^* - \sigma)\right). \tag{1.11}$$

Here the signed measure $(1/2)(\sigma^* - \sigma)$ satisfies $(1/2)(\sigma^* - \sigma)(\mathbf{C}) = 0$ and its total variation

$$\tfrac{1}{2}|\sigma^* - \sigma| \le \tfrac{1}{2} \textstyle\sum_{j=1}^n (\sigma_j + \sigma_j^*)$$

has finite logarithmic energy (cf. the above consideration concerning (1.4) and (1.5)); hence Lemma I.1.8 can be applied and we get

$$\left(\tfrac{1}{2}(\sigma^* - \sigma), \tfrac{1}{2}(\sigma^* - \sigma)\right) \ge 0,$$

where equality can only hold for $\sigma = \sigma^*$. Since we also have

$$V_w \le I_w\left(\tfrac{1}{2}(\sigma^* + \sigma)\right),$$

(1.11) can hold only if

$$\left(\tfrac{1}{2}(\sigma^* - \sigma), \tfrac{1}{2}(\sigma^* - \sigma)\right) = 0,$$

i.e. only if $\sigma^* = \sigma$. This proves the uniqueness of the optimal measure in $\mathcal{M}$. With this we have proved (b).

In a similar fashion, (c) follows from the formula (1.9) and Theorem I.1.3(c) (see also Theorem I.4.3 and Remark I.1.4). Finally, (d) also follows from (1.9) with the help of Theorem I.1.3(d) and (e). In fact, if F_j denotes the constant $m_j F_{w_j^{(\mu^*)}}$, then by Theorem I.1.3(d) and (e)

$$U^{\mu_{w_j^{(\mu^*)}}}(z) + \varepsilon_j(Q(z) + U^{\overline{\mu}_j^*}(z))/m_j \geq F_j/m_j \tag{1.12}$$

for q.e. $z \in \Sigma_j$ and

$$U^{\mu_{w_j^{(\mu^*)}}}(z) + \varepsilon_j(Q(z) + U^{\overline{\mu}_j^*}(z))/m_j \leq F_j/m_j \tag{1.13}$$

for every $z \in \operatorname{supp}(\mu_{w_j^{(\mu^*)}})$. By (1.9)

$$U^{\mu_{w_j^{(\mu^*)}}}(z) = \varepsilon_j U^{\varepsilon_j \mu_j^*}(z)/m_j;$$

hence

$$U^{\mu_{w_j^{(\mu^*)}}}(z) + \varepsilon_j U^{\overline{\mu}_j^*}(z)/m_j = \varepsilon_j U^{\mu^*}(z)/m_j,$$

and we obtain the inequalities in (d) from (1.12) and (1.13) by multiplying through by m_j. □

To conclude this section, we mention a formula connecting the minimal energy, Q and the constants F_j that will be frequently used hereafter:

$$V_w - \int Q d\mu^* = \sum_{j=1}^{N} F_j m_j. \tag{1.14}$$

This can be obtained from Theorem 1.4(d) by integrating with respect to $d\mu_j^*$ and adding the resulting equalities together.

VIII.2 Basic Theorems for Equilibrium Potentials and Measures Associated with Signed Measures

In this section we establish some basic facts concerning the solution of the weighted energy problem discussed in the preceding section. We shall keep all the assumptions and notations from the previous section. In particular, μ^* denotes the equilibrium measure shown to uniquely exist in Theorem 1.4. We start with the following characterization of these equilibrium measures.

For $\sigma \in \mathcal{M}$ let

$$\overline{\sigma}_j := \sum_{i \neq j} \varepsilon_i \sigma_i, \quad 1 \leq j \leq N,$$

and set for $z \in \Sigma_j$

$$w_j^{(\sigma)}(z) := \exp(-\varepsilon_j(Q(z) + U^{\overline{\sigma}_j}(z))/m_j).$$

Theorem 2.1. *For all j we have*

$$\mu_j^* = m_j \mu_{w_j^{(\mu^*)}}, \tag{2.1}$$

where $\mu_{w_j^{(\mu^*)}}$ is the solution of the energy problem in the sense of Chapter I *on Σ_j corresponding to $w_j^{(\mu^*)}$.*

Conversely, if for some $\sigma \in \mathcal{M}$ we have

$$\sigma_j = m_j \mu_{w_j^{(\sigma)}}, \qquad j = 1, \ldots, N, \tag{2.2}$$

then $\sigma = \mu^*$.

This theorem says the following: if we fix every component σ_i of $\sigma = \mu^*$ except for the j-th one and then minimize the energy with respect to this j-th component, then the minimum is attained for μ_j^*, and conversely, no other measure has this "self dual optimum" property. In other words, if the measure σ is such that each of its component is optimal when the others are kept fixed, then σ is globally optimal, i.e. $\sigma = \mu^*$.

Proof of Theorem 2.1. That each component of μ^* must be optimal when the others are kept fixed is obvious from the optimality of μ^* and this is exactly the identity (2.1) because the equilibrium measures are unique (see also the proof of Theorem 1.4, where (2.1) has already been established).

Suppose now that (2.2) holds. From Theorem I.1.3 it follows that σ has finite logarithmic energy and there are constants τ_j such that

$$U^{\sigma_j}(z) + \varepsilon_j(U^{\overline{\sigma}_j}(z) + Q(z)) \geq \tau_j$$

quasi-everywhere on Σ_j, and equality holds quasi-everywhere on $\operatorname{supp}(\sigma_j)$. This can also be written in the form

$$\varepsilon_j(U^{\sigma}(z) + Q(z)) \geq \tau_j \quad \text{q.e. on} \quad \Sigma_j, \quad j = 1, \ldots, N, \tag{2.3}$$

with equality quasi-everywhere on each $\operatorname{supp}(\sigma_j)$. Since each μ_j^* and σ_j has finite logarithmic energy, it follows that (2.3) holds μ_j^*-almost everywhere, and equality holds σ_j-almost everywhere. Integrating the inequalities (2.3) with respect to $\mu_1^*, \ldots, \mu_N^*$, respectively, and adding them together we get

$$(\sigma, \mu^*) + \int Q \, d\mu^* \geq \sum_{j=1}^{N} \tau_j m_j,$$

where the inner product is defined as in (1.10), i.e.

$$(\sigma, \mu^*) := \iint \log \frac{1}{|z - t|} \, d\sigma(z) d\mu^*(t) = (\mu^*, \sigma).$$

On the other hand, if we integrate (2.3) with respect to $\sigma_1, \ldots, \sigma_N$ and add the resulting inequalities together we get

$$(\sigma, \sigma) + \int Q\, d\sigma = \sum_{j=1}^{N} \tau_j m_j.$$

Thus,

$$(\sigma, \mu^*) + \int Q\, d\mu^* \geq (\sigma, \sigma) + \int Q\, d\sigma.$$

But here the roles of σ and μ^* can be reversed because the analogue of (2.3) holds with $\sigma = \mu^*$ as well (see Theorem 1.4); hence we have

$$(\sigma, \mu^*) + \int Q\, d\sigma \geq (\mu^*, \mu^*) + \int Q\, d\mu^*.$$

Finally by adding the last two inequalities together we get

$$(\sigma - \mu^*, \sigma - \mu^*) \leq 0,$$

which, in view of the fact that the signed measure $\sigma - \mu^*$ is orthogonal to constants, is possible only for $\sigma = \mu^*$ (see Lemma I.1.8). □

As a by-product of the proof (or apply Theorem I.3.3 and Theorem 2.1) we get the following analogue of Theorem I.3.3.

Theorem 2.2. *If $\sigma \in \mathcal{M}$ is a signed measure with finite energy such that for all j*

$$\text{“}\inf_{x \in \Sigma_j}\text{”} \; \varepsilon_j (U^\sigma(x) + Q(x)) = \varepsilon_j (U^\sigma(z) + Q(z)) \quad \textit{for q.e. } z \in \operatorname{supp}(\sigma_j),$$

then $\sigma = \mu^*$. That is, if there are constants τ_j such that

$$\varepsilon_j (U^\sigma(z) + Q(z)) \geq \tau_j$$

quasi-everywhere on Σ_j and equality holds for quasi-every $z \in \operatorname{supp}(\sigma_j)$, $j = 1, \ldots, N$, then $\sigma = \mu^*$ (and of course $\tau_j = F_j$ for all j).

Here “inf” means infimum neglecting sets of zero capacity. The notation “sup” has the analogous meaning.

Next we verify another important extremality property of the equilibrium measure.

Theorem 2.3. *For any $\sigma \in \mathcal{M}$*

$$\text{“}\inf_{z \in \Sigma}\text{”} \; \varepsilon(z)(U^\sigma(z) + Q(z) - \varepsilon(z)F(z)) \leq 0, \tag{2.4}$$

and for $\sigma \in \mathcal{M}$ with finite logarithmic energy

$$\text{“}\sup_{z \in \operatorname{supp}(\sigma)}\text{”} \; \varepsilon(z)(U^\sigma(z) + Q(z) - \varepsilon(z)F(z)) \geq 0. \tag{2.5}$$

If equality holds both in (2.4) and (2.5), then $\sigma = \mu^*$.

Recall that $F(z)$ was defined in (1.3).

For example, (2.4) says that for some j we must have

$$\text{“inf”}_{z\in\Sigma_j}(U^\sigma(z) + Q(z)) \le F_j$$

if $\varepsilon_j = 1$ or

$$\text{“sup”}_{z\in\Sigma_j}(U^\sigma(z) + Q(z)) \ge -F_j$$

if $\varepsilon_j = -1$. Note, however, that for other j's these need not be satisfied.

The proof is based on the following fundamental lemma which will also be used elsewhere. In its formulation we set for a measure $\mu \in \mathcal{M}$,

$$\mu_+ := \sum_{\varepsilon_j=1} \mu_j, \quad \mu_- := \sum_{\varepsilon_j=-1} \mu_j,$$

$$\Sigma_+ := \bigcup_{\varepsilon_j=1} \Sigma_j, \quad \Sigma_- := \bigcup_{\varepsilon_j=-1} \Sigma_j,$$

i.e. μ_+ and μ_- are the positive and negative parts of μ, and Σ_+ and Σ_- are the corresponding parts of Σ.

Lemma 2.4 (Domination Lemma). *Suppose that $\sigma, \mu \in \mathcal{M}$ and μ has finite logarithmic energy. Then,*

$$\text{“inf”}_{z\in\text{supp}(\mu_+)} (U^\sigma(z) - U^\mu(z)) \le \text{“sup”}_{z\in\text{supp}(\mu_-)} (U^\sigma(z) - U^\mu(z)) \tag{2.6}$$

or, equivalently, for every constant c,

$$\text{“inf”}_{z\in\text{supp}(\mu)} \varepsilon(z)(U^\sigma(z) - U^\mu(z) + c) \le 0. \tag{2.7}$$

Proof. Assume to the contrary that there exist constants δ_1, δ_2 with $\delta_1 > \delta_2$ such that

$$U^\sigma(z) - U^\mu(z) \ge \delta_1 \quad \text{for q.e.} \quad z \in \text{supp}(\mu_+) \tag{2.8}$$

and

$$U^\sigma(z) - U^\mu(z) \le \delta_2 \quad \text{for q.e.} \quad z \in \text{supp}(\mu_-). \tag{2.9}$$

Since μ has finite logarithmic energy, both $E_1 := \text{supp}(\mu_+)$ and $E_2 := \text{supp}(\mu_-)$ have positive capacity. Let $\lambda^* = \lambda_1^* - \lambda_2^*$ be the extremal measure for the unweighted ($Q \equiv 0$) minimum energy problem with $N = 2$, $\Sigma = E_1 \cup E_2$, $\varepsilon_1 = 1$, $\varepsilon_2 = -1$, and $m_1 = m_2 = 1$. Then, by Theorem 1.4, there exist constants A_1, A_2 such that

$$U^{\lambda^*}(z) \ge A_1 \quad \text{for q.e. } z \in E_1,$$

$$U^{\lambda^*}(z) \le -A_2 \quad \text{for q.e. } z \in E_2,$$

with equality quasi-everywhere on $\text{supp}(\lambda_1^*)$ and $\text{supp}(\lambda_2^*)$, respectively. The principle of domination (Theorem II.3.2) then implies that $U^{\lambda^*} = A_1$ q.e. on E_1, $U^{\lambda^*} = -A_2$ q.e. on E_2, and $-A_2 \le U^{\lambda^*}(z) \le A_1$ for all $z \in \mathbf{C}$.

From (2.8) and (2.9) we have

$$\begin{aligned}\int (U^\sigma - U^\mu)\, d\lambda^* &= \int_{E_1} (U^\sigma - U^\mu)\, d\lambda_1^* - \int_{E_2} (U^\sigma - U^\mu)\, d\lambda_2^* \\ &\geq \delta_1 - \delta_2 > 0. \end{aligned} \tag{2.10}$$

On the other hand, since $-A_2 \leq U^{\lambda^*} \leq A_1$ on supp(σ) and the equalities $U^{\lambda^*} = A_1$, $U^{\lambda^*} = -A_2$ hold, respectively, μ_+ and μ_- almost everywhere, we have

$$\begin{aligned}\int (U^\sigma - U^\mu)\, d\lambda^* &= \int U^{\lambda^*} d\sigma - \int U^{\lambda^*} d\mu \\ &\leq A_1 ||\sigma_+|| + A_2 ||\sigma_-|| - A_1 ||\mu_+|| - A_2 ||\mu_-|| \\ &= 0,\end{aligned}$$

which contradicts (2.10). □

Remark 2.5. Notice that in the above proof we only used the fact that $\sigma = \sigma_+ - \sigma_-$ is the difference of positive measures with $||\sigma_+|| \leq ||\mu_+||$ and $||\sigma_-|| \leq ||\mu_-||$. Thus, Lemma 2.4 is valid for any such σ, regardless of the location of its support. As a consequence, inequality (2.4) of Theorem 2.3 holds with this weaker assumption (see the argument below).

We can now prove Theorem 2.3.

Proof of Theorem 2.3. From Theorem 1.4 we have

$$-\varepsilon(z)\, U^{\mu^*}(z) = \varepsilon(z)(Q(z) - \varepsilon(z) F(z))$$

q.e. on supp(μ^*). Hence, if $\sigma \in \mathcal{M}$, we deduce from Lemma 2.4 that

$$\text{“}\inf_{z\in\Sigma}\text{”}\, \varepsilon(z)(U^\sigma(z) + Q(z) - \varepsilon(z)F(z)) \leq \text{“}\inf_{z\in \text{supp}(\mu^*)}\text{”}\, \varepsilon(z)(U^\sigma(z) - U^{\mu^*}(z)) \leq 0,$$

which proves (2.4).

Similarly , if $\sigma \in \mathcal{M}$ has finite energy, then since

$$\varepsilon(z)(Q(z) - \varepsilon(z)F(z)) \geq -\varepsilon(z) U^{\mu^*}(z) \quad \text{for q.e.} \quad z \in \Sigma,$$

we again have from the domination lemma that

$$\begin{aligned}&\text{“}\sup_{z\in\text{supp}(\sigma)}\text{”}\, \varepsilon(z)(U^\sigma(z) + Q(z) - \varepsilon(z)F(z)) \\ &\qquad \geq \text{“}\sup_{z\in\text{supp}(\sigma)}\text{”}\, \varepsilon(z)(U^\sigma(z) - U^{\mu^*}(z)) \\ &\qquad = -\text{“}\inf_{z\in\text{supp}(\sigma)}\text{”}\, \varepsilon(z)(U^{\mu^*}(z) - U^\sigma(z)) \geq 0,\end{aligned}$$

which proves (2.5).

Finally, if equality holds in both (2.4) and (2.5), then $\sigma = \mu^*$ follows from Theorem 2.2. □

We now formalize the result that was used in the proof of the domination lemma and which we will apply several more times throughout this chapter.

Theorem 2.6. *Suppose that Σ is compact, $w \equiv 1$, $N = 2$, $\varepsilon_1 = 1$, $\varepsilon_2 = -1$, and $m_1 = m_2 = 1$. Then μ_1^* is the balayage of μ_2^* onto Σ_1 and μ_2^* is the balayage of μ_1^* onto Σ_2. Conversely, if $\sigma \in \mathcal{M}$ is such that each of σ_j is the balayage of the other one onto Σ_j, $j = 1, 2$, then $\sigma = \mu^*$.*

Furthermore, in this case we have

$$U^{\mu^*}(z) = F_1 \qquad \textit{for q.e. } z \in \Sigma_1, \tag{2.11}$$

$$U^{\mu^*}(z) = -F_2 \qquad \textit{for q.e. } z \in \Sigma_2, \tag{2.12}$$

and for every $z \in \mathbf{C}$

$$-F_2 \le U^{\mu^*}(z) \le F_1. \tag{2.13}$$

Recall that the sets Σ_1 and Σ_2 are assumed to be of positive capacity; therefore, the balayages in question are well defined, and they are unique.

In this special case U^{μ^*} is often referred to as the *condenser potential* and $1/I_w(\mu^*)$ as the *condenser capacity* $C(\Sigma_1, \Sigma_2)$ (of the condenser (Σ_1, Σ_2) with plates Σ_1 and Σ_2).

Proof of Theorem 2.6. The first part of the theorem is an immediate consequence of Theorem 1.4 and Theorems II.4.4 and II.4.7. On the other hand, if σ is assumed to have the stated properties, then σ has finite logarithmic energy (see Section II.4), and U^σ is constant quasi-everywhere on each of Σ_1 and Σ_2 (see Theorem II.4.7). Thus, Theorem 2.2 can be applied to conclude $\sigma = \mu^*$.

The last assertion of the theorem was previously addressed in the proof of Lemma 2.4. □

The next corollary relates the minimum logarithmic energy problem for signed measures on two compacta to the minimum Green energy problem of Section II.5 for positive measures.

Corollary 2.7. *If Σ_1, Σ_2 are disjoint compact sets having positive capacity and Σ_2 has connected complement G, then*

$$\mu_{\Sigma_1}^G = \mu_1^*,$$

where $\mu_{\Sigma_1}^G$ is the Green equilibrium measure for the set Σ_1 relative to G and $\mu^* = \mu_1^* - \mu_2^*$ is the equilibrium measure in Theorem 2.6.

Proof. From Theorem 2.6, we have that μ_2^* is the balayage of μ_1^* onto Σ_2 and hence onto ∂G. Thus, by the representation theorem for Green potentials (see Theorem II.5.1(iii)), and Theorem 2.6 we deduce that $U_G^{\mu_1^*} = U^{\mu^*} +$ const. is constant for quasi-every $z \in \Sigma_1$. Hence $\mu_{\Sigma_1}^G = \mu_1^*$ follows from Theorem II.5.12. □

VIII.3 Rational Fekete Points and a Weighted Variant of a Problem of Zolotarjov

In this section we will consider a special case of the general setup discussed previously which has applications to rational approximation.

Let $N = 2$, $\varepsilon_1 = 1$, $\varepsilon_2 = -1$, $m_1 = m_2 = 1$, and assume that Σ_1 and Σ_2 are disjoint compact sets of positive capacity. Thus, we consider signed measures σ of the form $\sigma = \sigma_1 - \sigma_2$, where σ_1 and σ_2 are probability measures on Σ_1 and Σ_2, respectively, and $\mathcal{M}$ is the collection of all such signed measures. F_1 and F_2 are then the constants in

$$U^{\mu^*}(z) \geq -Q(z) + F_1 \quad \text{for q.e.} \quad z \in \Sigma_1$$

and

$$U^{\mu^*}(z) \leq -Q(z) - F_2 \quad \text{for q.e.} \quad z \in \Sigma_2$$

with equality quasi-everywhere on $\operatorname{supp}(\mu_1^*)$ and $\operatorname{supp}(\mu_2^*)$, respectively, where $\mu^* = \mu_1^* - \mu_2^*$ is the optimal signed measure in the weighted energy problem

$$\min_{\sigma \in \mathcal{M}} \left\{ \int\int \log \frac{1}{|z-t|} d\sigma(z) d\sigma(t) + 2 \int Q(t)\, d\sigma(t) \right\}.$$

Theorem 3.1. *We have, for any signed measure σ of compact support and of the form $\sigma = \sigma_1 - \sigma_2$ with $\|\sigma_1\| = \|\sigma_2\| = 1$, the inequality*

$$\text{“}\inf_{z \in \Sigma_1}\text{”}(U^\sigma(z) + Q(z)) - \text{“}\sup_{z \in \Sigma_2}\text{”}(U^\sigma(z) + Q(z)) \leq F_1 + F_2 \tag{3.1}$$

with equality for $\sigma = \mu^*$. In particular, if

$$r(z) = \frac{z^n + \cdots}{z^n + \cdots}$$

is any rational function of the indicated form, then

$$\left\{ \left(\sup_{z \in \Sigma_1} |r(z)| w^n(z) \right) \Big/ \left(\inf_{z \in \Sigma_2} |r(z)| w^n(z) \right) \right\}^{1/n} \geq e^{-(F_1 + F_2)}. \tag{3.2}$$

We shall shortly see that inequality (3.2) is asymptotically sharp.

The determination of the expression on the left-hand side of (3.2) in the special case $w \equiv 1$ is called the third problem of Zolotarjov (see [62]).

Corollary 3.2. *Inequality* (3.1) *also holds for any* σ *of the form* $\sigma = \sigma_1 - \sigma_2$, $\|\sigma_1\| \leq 1$, $\|\sigma_2\| \leq 1$. *In a similar fashion,* (3.2) *holds for any rational function with numerator and denominator degrees at most* n.

Remark 3.3. Note that in the first statement we do not restrict σ to be in $\mathcal{M}$ (i.e. we do not require $\text{supp}(\sigma_1) \subset \Sigma_1$, $\text{supp}(\sigma_2) \subset \Sigma_2$). However, if $\sigma \in \mathcal{M}$, then the "inf" and "sup" in (3.1) can be replaced by inf and sup, respectively.

It is not true that μ^* is the only measure for which equality can hold in (3.1). In fact, let $\Sigma_1 = \{z \mid |z| \leq 1\}$, $\Sigma_2 = \{z \mid |z| = 2\}$, $Q \equiv 0$. By the unicity of the equilibrium measures, μ^* is circular symmetric; hence μ_2^* is the normalized Lebesgue measure on Σ_2. But then the potential field of μ_2^* is constant (equal to $-\log 2$) on Σ_1; hence (see Theorem 2.1) μ_1^* will be the normalized Lebesgue measure on $\partial\Sigma_1$. But if $\sigma = \sigma_{1/2} - \mu_2^*$ where $\sigma_{1/2}$ is the normalized Lebesgue measure on $\{z \mid |z| = 1/2\}$ then we have

$$\text{“inf”}_{z\in\Sigma_1} U^\sigma(z) - \text{“sup”}_{z\in\Sigma_2} U^\sigma(z) = \log 2 - (-\log 2 + \log 2) = \log 2,$$

which is of course the same as $F_1 + F_2$.

Proof of Theorem 3.1. From inequality (2.6) of the domination lemma and Remark 2.5, we have

$$\begin{aligned}
&\text{“inf”}_{z\in\Sigma_1} (U^\sigma(z) + Q(z) - F_1) - \text{“sup”}_{z\in\Sigma_2} (U^\sigma(z) + Q(z) + F_2) \\
&\leq \text{“inf”}_{z\in\text{supp}(\mu_1^*)} (U^\sigma(z) + Q(z) - F_1) - \text{“sup”}_{z\in\text{supp}(\mu_2^*)} (U^\sigma(z) + Q(z) + F_2) \\
&= \text{“inf”}_{z\in\text{supp}(\mu_1^*)} (U^\sigma(z) - U^{\mu^*}(z)) - \text{“sup”}_{z\in\text{supp}(\mu_2^*)} (U^\sigma(z) - U^{\mu^*}(z)) \leq 0,
\end{aligned}$$

from which (3.1) follows.

The last statement concerning rational functions immediately follows from the first part of the theorem because the modulus of every such rational function can be written as the exponential of the negative of a potential corresponding to a discrete measure associated with the poles and zeros of the rational function in question. □

Proof of Corollary 3.2. If we add the missing masses to σ_1 and σ_2 (to get probability measures) in the form $(1 - \|\sigma_1\|)\delta_{x_1}$ and $(1 - \|\sigma_2\|)\delta_{x_2}$ and let x_1 and x_2 tend to infinity, then we get this stronger form of the first result in Theorem 3.1 from the original one. The proof of the second statement is analogous. □

Now we show that the lower bound in (3.2) is achieved in the limit. In analogy with the polynomial case let us call point systems $\mathcal{F}_j^{(n)} = \{z_k^{(j,n)}\}_{k=1}^n \subset \Sigma_j$, $j = 1, 2$, maximizing the expression

$$\mathcal{F}\left(\{\zeta_k^{(1,n)}\}_{k=1}^n, \{\zeta_k^{(2,n)}\}_{k=1}^n\right) :=$$

$$\prod_{1\le i,\,k\le n,\,i\ne k} \frac{\left|\zeta_i^{(1,n)} - \zeta_k^{(1,n)}\right|\left|\zeta_i^{(2,n)} - \zeta_k^{(2,n)}\right| w\left(\zeta_i^{(1,n)}\right) w\left(\zeta_k^{(1,n)}\right)}{\left|\zeta_i^{(1,n)} - \zeta_k^{(2,n)}\right|\left|\zeta_i^{(2,n)} - \zeta_k^{(1,n)}\right| w\left(\zeta_i^{(2,n)}\right) w\left(\zeta_k^{(2,n)}\right)}, \tag{3.3}$$

$\{\zeta_k^{(j,n)}\}_{k=1}^n \subset \Sigma_j$, $j = 1, 2$, *rational Fekete sets* and the points in them *rational Fekete points*. If the maximum in question is $\delta_n^{n(n-1)}$, then we have $\delta_{n+1} \le \delta_n$ and

$$\delta_n \ge \exp(-V_w) = \exp(-I_w(\mu^*)) =: c(w, \Sigma_1, \Sigma_2), \tag{3.4}$$

$n = 1, 2, \ldots$, where $V_w = I_w(\mu^*)$ is the minimal weighted energy and $c(w, \Sigma_1, \Sigma_2)$ might be called the corresponding *weighted "signed" capacity*. In fact, $\delta_{n+1} \le \delta_n$ follows from

$$\delta_n^{n(n-1)} \ge \mathcal{F}\left(\{z_k^{(1,n+1)}\}_{k=1,k\ne i}^{n+1}, \{z_k^{(2,n+1)}\}_{k=1,k\ne i}^{n+1}\right), \quad i = 1, \ldots, n+1,$$

by multiplying these inequalities together. On the other hand,

$$\delta_n \ge \exp(-I_w(\mu^*))$$

is a consequence of

$$n(n-1)\log\delta_n \ge \log\left(\mathcal{F}\left(\{y_k\}_{k=1}^n, \{x_k\}_{k=1}^n\right)\right),$$

$\{y_k\}_{k=1}^n \subset \Sigma_1$, $\{x_k\}_{k=1}^n \subset \Sigma_2$, if we integrate this inequality with respect to

$$d\mu_1^*(y_1)\cdots d\mu_1^*(y_n)d\mu_2^*(x_1)\cdots d\mu_2^*(x_n).$$

Next, we show

$$\lim_{n\to\infty} \delta_n = c(w, \Sigma_1, \Sigma_2), \tag{3.5}$$

and that the asymptotic distribution of the Fekete points in the Fekete sets $\mathcal{F}_j^{(n)}$ is μ_j^*, $j = 1, 2$. In fact, let $\nu_j^{(n)}$ be the discrete measure with mass $1/n$ at each point of $\mathcal{F}_j^{(n)}$ and let σ_j be a weak* limit of some sequence $\{\nu_j^{(n)}\}_{n\in\mathcal{N}}$ simultaneously for $j = 1, 2$. Set $\sigma = \sigma_1 - \sigma_2$, $\nu^{(n)} = \nu_1^{(n)} - \nu_2^{(n)}$. It easily follows from the monotone convergence theorem and the disjointness of Σ_1 and Σ_2 that

$$\begin{aligned} I_w(\sigma) &= \lim_{M\to\infty} \iint \min\left(M, \log\frac{1}{|z-t|}\right) d\sigma(z)d\sigma(t) + 2\int Q(t)\, d\sigma(t) \\ &\le \lim_{M\to\infty} \liminf_{\substack{n\to\infty \\ n\in\mathcal{N}}} \left(\iint \min\left(M, \log\frac{1}{|z-t|}\right) d\nu^{(n)}(z)d\nu^{(n)}(t)\right. \\ &\qquad \left. + 2\int Q(t)\, d\nu^{(n)}(t)\right) \end{aligned}$$

$$= \lim_{M\to\infty} \liminf_{\substack{n\to\infty \\ n\in\mathcal{N}}} \left(\iint_{z\neq t} \min\left(M, \log\frac{1}{|z-t|}\right) d\nu^{(n)}(z) d\nu^{(n)}(t) \right.$$

$$\left. + 2\int Q(t)\, d\nu^{(n)}(t) \right)$$

$$\leq \lim_{\substack{n\to\infty \\ n\in\mathcal{N}}} \frac{-1}{n(n-1)} \log\left(\mathcal{F}\left(\{z_k^{(1,n)}\}_{k=1}^n, \{z_k^{(2,n)}\}_{k=1}^n\right)\right)$$

$$= \lim_{\substack{n\to\infty \\ n\in\mathcal{N}}} \log(1/\delta_n) \leq I_w(\mu^*), \tag{3.6}$$

where, in the last step, we used (3.4), and in the first inequality we used the fact that by the lower semi-continuity of $\varepsilon(t)Q(t)$, we have

$$\int Q(t)\, d\sigma(t) \leq \liminf_{n\to\infty} \int Q(t)\, d\nu^{(n)}(t)$$

(see Lemma III.1.5). Hence, the unicity of the measure μ^* implies that $\sigma = \mu^*$, and since this is true for any weak* limit point of the sequence $\{\nu^{(n)}\}$, our claim concerning the distribution of the Fekete points has been verified. The preceding inequalities then become equalities and (3.5) also follows.

Another consequence of the fact that all inequalities become equalities is that

$$\int Q(t)\, d\mu^*(t) = \lim_{n\to\infty} \int Q(t)\, d\nu^{(n)}(t). \tag{3.7}$$

Now consider the rational functions

$$r_n(z) = \prod_{i=1}^n \frac{z - z_i^{(1,n)}}{z - z_i^{(2,n)}}. \tag{3.8}$$

For $y \in \Sigma_1$, $x \in \Sigma_2$, we have from the definition of the quantities δ_n

$$(|r_n(y)|w^n(y)/|r_n(x)|w^n(x))^2 =$$

$$\mathcal{F}\left(\{z_k^{(1,n)}\}_{k=1}^n \cup \{y\}, \{z_k^{(2,n)}\}_{k=1}^n \cup \{x\}\right) \delta_n^{-n(n-1)} \left(\prod_{i=1}^n \frac{w\left(z_i^{(1,n)}\right)}{w\left(z_i^{(2,n)}\right)}\right)^{-2}$$

$$\leq \delta_{n+1}^{n(n+1)} \delta_n^{-n(n-1)} \exp\left(-2\sum_{i=1}^n \left(-Q\left(z_i^{(1,n)}\right) + Q\left(z_i^{(2,n)}\right)\right)\right)$$

$$\leq \delta_n^{2n} \exp\left(-2\sum_{i=1}^n \left(-Q\left(z_i^{(1,n)}\right) + Q\left(z_i^{(2,n)}\right)\right)\right)$$

where, in the very last step, we used the monotonicity of the sequence $\{\delta_n\}$. Taking $2n$-th roots and applying (3.5)–(3.7) we arrive at

$$\limsup_{n\to\infty}\left\{\left(\sup_{z\in\Sigma_1}|r_n(z)|w^n(z)\right)\Big/\left(\inf_{z\in\Sigma_2}|r_n(z)|w^n(z)\right)\right\}^{1/n}$$
$$\le \exp\left(-I_w(\mu^*)+\int Q\,d\mu^*\right)=e^{-(F_1+F_2)}$$

(see (1.14)). If we compare this with the last statement of Theorem 3.1 and Corollary 3.2 we can see that the *weighted Zolotarjov constants*

$$\tau_n := \inf_{\deg(r)\le n}\left\{\left(\sup_{z\in\Sigma_1}|r(z)|w^n(z)\right)\Big/\left(\inf_{z\in\Sigma_2}|r(z)|w^n(z)\right)\right\}^{1/n},$$

where the infimum is taken for all rational functions with numerator and denominator degree at most n, satisfy the limit relationship

$$\lim_{n\to\infty}\tau_n = e^{-(F_1+F_2)} = c(w,\Sigma_1,\Sigma_2)\exp\left(\int Q\,d\mu^*\right).$$

This minimum problem is the analogue of the Chebyshev problem for rational functions and is called the *weighted Zolotarjov problem.*

In particular, when $w\equiv 1$, we get that these Zolotarjov constants converge to the "signed" capacity $c(1,\Sigma_1,\Sigma_2)$ of Σ_1 and Σ_2 (also called the *modulus* of the condenser) defined by

$$\log\frac{1}{c(1,\Sigma_1,\Sigma_2)} := \inf\left\{\int\!\!\int\log\frac{1}{|z-t|}\,d\sigma(z)d\sigma(t)\,\middle|\,\sigma=\sigma_1-\sigma_2,\right.$$
$$\left.\operatorname{supp}(\sigma_j)\subset\Sigma_j,\ \ \|\sigma_j\|=1\right\}. \tag{3.9}$$

In summary, we have proved the following.

Theorem 3.4. *With the above notation, the rational Fekete numbers δ_n, the weighted "signed" capacity $c(w,\Sigma_1,\Sigma_2)$, and the weighted Zolotarjov numbers τ_n are related by*

$$\lim_{n\to\infty}\delta_n = c(w,\Sigma_1,\Sigma_2)$$

$$=\left(\lim_{n\to\infty}\tau_n\right)\exp\left(-\textstyle\int Q\,d\mu^*\right)=e^{-(F_1+F_2)}\exp\left(-\textstyle\int Q\,d\mu^*\right).$$

The Fekete points $\mathcal{F}_j^{(n)}$ have asymptotic distribution μ_j^*, $j=1,2$, and the Fekete rational functions r_n defined in (3.8) *are asymptotically optimal in the weighted Zolotarjov (or rational Chebyshev) problem.*

In particular, if $w\equiv 1$, then the three quantities: rational Fekete constant (= limit of the δ_n's), rational Chebyshev constant (= limit of the τ_n's), and the "signed" capacity associated with the pair (Σ_1,Σ_2), are the same.

It also follows that the asymptotic behavior of the weighted Zolotarjov numbers does not change if, instead of free zeros and poles, we require that the zeros of r_n lie in Σ_1 while its poles lie in Σ_2.

The last statement in the theorem is the analogue for rational functions of the well known result of Fekete and Szegő (see Sections III.1 and III.3). We also mention that these results actually contain those of Fekete and Szegő, because, by moving Σ_2 to infinity as in the proof of Corollary 3.2, we can derive all the classical theorems concerning polynomials from results on rational functions.

It is quite a difficult problem to numerically determine the rational Fekete sets because one has to maximize the expression in question for all $2n$-tuples. Therefore, in the spirit of Section V.1, we investigate whether there is a simpler procedure to generate asymptotically optimal rational functions in the Zolotarjov problem.

Fix two closed subsets E_1, E_2 of Σ_1, Σ_2, respectively. Starting with any two points $a_0 \in E_1, b_0 \in E_2$, we successively define the points $a_n \in E_1$ and $b_n \in E_2$ as points where the expression

$$\frac{w(a)^n \left|\prod_{j=0}^{n-1} \frac{a-a_j}{a-b_j}\right|}{w(b)^n \left|\prod_{j=0}^{n-1} \frac{b-a_j}{b-b_j}\right|} \tag{3.10}$$

takes its maximum A_n^n for $a \in E_1$ and $b \in E_2$. These points are the analogues of the Leja points for the weighted rational case relative to the two sets E_1, E_2. Let $\sigma_n^{(1)}, \sigma_n^{(2)}$ be respectively the normalized counting measures for the sets $\{a_k\}_{k=0}^{n-1}$, $\{b_k\}_{k=0}^{n-1}$, and set $\sigma_n := \sigma_n^{(1)} - \sigma_n^{(2)}$. We shall show that for a suitable choice of the sets E_1, E_2 we have $\sigma_n \xrightarrow{*} \mu^*$ as $n \to \infty$.

Theorem 3.5. *If* $E_1 = \mathrm{supp}(\mu_1^*)$ *and* $E_2 = \mathrm{supp}(\mu_2^*)$, *then the points* $\{a_k\}$ *and* $\{b_k\}$ *defined above have normalized asymptotic distribution* μ_1^* *and* μ_2^*, *respectively. Furthermore, the corresponding rational functions*

$$r_n(z) := \prod_{j=0}^{n-1} \frac{z-a_j}{z-b_j}$$

satisfy

$$\begin{aligned}\lim_{n\to\infty}\left(\frac{\max_{z\in E_1}\{|r_n(z)|w^n(z)\}}{\min_{z\in E_2}\{|r_n(z)|w^n(z)\}}\right)^{1/n} &= e^{-(F_1+F_2)} \\ &= c(w,\Sigma_1,\Sigma_2)\exp\left(\int Q\,d\mu^*\right).\end{aligned} \tag{3.11}$$

Moreover, in the unweighted case ($w \equiv 1$), *the above assertions also hold for* $E_1 = \Sigma_1$ *and* $E_2 = \Sigma_2$.

Remark 3.6. We shall prove Theorem 3.5 in the more general setting when the closed sets E_1, E_2 satisfy

$$\operatorname{supp}(\mu_1^*) \subset E_1 \subset \{z \in \Sigma_1 | U^{\mu^*}(z) + Q(z) \le F_1\}, \tag{3.12}$$

$$\operatorname{supp}(\mu_2^*) \subset E_2 \subset \{z \in \Sigma_2 | U^{\mu^*}(z) + Q(z) \ge -F_2\}. \tag{3.13}$$

Notice that for $Q \equiv 0$, we can then take $E_1 = \Sigma_1$ and $E_2 = \Sigma_2$ because of Theorem 2.6. Furthermore, for certain Q it may well be possible to explicitly determine $\operatorname{supp}(\mu_i^*)$, $i = 1, 2$, especially when Σ_1, Σ_2 are subintervals of $\mathbf{R}$.

Proof of Theorem 3.5. Assume that the two closed sets E_1, E_2 satisfy (3.12) and (3.13). Since U^{μ^*} is bounded near infinity and $(-1)^{j-1}Q(z) \to \infty$ as $|z| \to \infty$, $z \in \Sigma_j$, the sets on the right-hand sides of (3.12) and (3.13) are bounded; hence E_1 and E_2 are compact.

From the definition of A_k^k (the maximum value of (3.10) over E_1, E_2 when $n = k$), we have

$$\begin{aligned}
\frac{1}{k}\log\frac{1}{A_k^k} &= U^{\sigma_k}(a_k) + Q(a_k) - U^{\sigma_k}(b_k) - Q(b_k) \\
&= \inf_{a\in E_1}(U^{\sigma_k} + Q)(a) - \sup_{b\in E_2}(U^{\sigma_k} + Q)(b) \\
&\le \int (U^{\sigma_k} + Q)\, d\mu^*,
\end{aligned} \tag{3.14}$$

where, in the last inequality, we use the facts that $\operatorname{supp}(\mu_j^*) \subset E_j$, $j = 1, 2$, and $||\mu_1^*|| = ||\mu_2^*|| = 1$. Furthermore, since $(-1)^{j-1}(U^{\mu^*} + Q) \le F_j$ on E_j, $j = 1, 2$, it follows that

$$\begin{aligned}
\int (U^{\sigma_k} + Q)\, d\mu^* &= \int Q\, d\mu^* + \int U^{\mu^*} d\sigma_k^{(1)} - \int U^{\mu^*} d\sigma_k^{(2)} \\
&\le \int Q\, d\mu^* + \int (F_1 - Q)\, d\sigma_k^{(1)} + \int (F_2 + Q)\, d\sigma_k^{(2)} \\
&= \int Q\, d\mu^* + F_1 + F_2 - \int Q\, d\sigma_k^{(1)} + \int Q\, d\sigma_k^{(2)} \\
&= V_w - \int Q\, d\sigma_k^{(1)} + \int Q\, d\sigma_k^{(2)}
\end{aligned}$$

(recall formula (1.14)). Combining this with (3.14) and multiplying by k we get

$$kV_w \ge \log\frac{1}{A_k^k} + k\int Q\, d\sigma_k^{(1)} - k\int Q\, d\sigma_k^{(2)} = \log\frac{1}{A_k^k} + \sum_{j=0}^{k-1} Q(a_j) - \sum_{j=0}^{k-1} Q(b_j). \tag{3.15}$$

When we add together these inequalities for $k = 1, \ldots, n-1$ we obtain

$$\begin{aligned}
\frac{n(n-1)}{2} V_w &\geq \log \frac{1}{A_1 A_2^2 \cdots A_{n-1}^{n-1}} + \sum_{l=0}^{n-2} (n-1-l)\,[Q(a_l) - Q(b_l)] \\
&= \sum_{j<k} \log \frac{1}{|a_k - a_j|} + \sum_{j<k} \log \frac{1}{|b_k - b_j|} \\
&\quad - \sum_{j<k} \log \frac{1}{|b_k - a_j|} - \sum_{j<k} \log \frac{1}{|a_k - b_j|} \\
&\quad + (n-1) \sum_{l=0}^{n-2} [Q(a_l) - Q(b_l)] \\
&= -\frac{1}{2} \log \mathcal{F}(\{a_i\}_{i=0}^{n-1}, \{b_i\}_{i=0}^{n-1}), \qquad (3.16)
\end{aligned}$$

where $\mathcal{F}$ is the Fekete expression defined in (3.3). Thus

$$-\frac{1}{n(n-1)} \log \mathcal{F}(\{a_i\}_{i=0}^{n-1}, \{b_i\}_{i=0}^{n-1}) \leq V_w = I(\mu^*),$$

and so it follows by the same reasoning used in (3.6) that every weak* limit measure σ of the sequence σ_n equals μ^*. Hence $\sigma_n \xrightarrow{*} \mu^*$ as $n \to \infty$.

To establish (3.11) we need to show that

$$\lim_{k \to \infty} \log \frac{1}{A_k} = F_1 + F_2. \qquad (3.17)$$

For this purpose, we first observe that from the preceding argument it follows that

$$\frac{1}{n(n-1)} \log \mathcal{F}_n := \frac{1}{n(n-1)} \log \mathcal{F}(\{a_i\}_{i=0}^{n-1}, \{b_i\}_{i=0}^{n-1}) \to -V_w \qquad (3.18)$$

as $n \to \infty$, and that

$$\lim_{k \to \infty} \int Q\, d\sigma_k = \int Q\, d\mu^* \qquad (3.19)$$

(compare (3.7)). Inequality (3.15) together with (3.19) yield the upper bound

$$\limsup_{k \to \infty} \log \frac{1}{A_k} \leq V_w - \int Q\, d\mu^* = F_1 + F_2. \qquad (3.20)$$

Next, from (3.16), we obtain

$$\sum_{k=1}^{n-1} k \log \frac{1}{A_k} \geq -\frac{1}{2} \log \mathcal{F}_n - \sum_{l=0}^{n-2} (n-1-l)[Q(a_l) - Q(b_l)]. \qquad (3.21)$$

Since the first Cesàro means

$$\int Q\,d\sigma_n = \frac{1}{n}\sum_{l=0}^{n-1}[Q(a_l) - Q(b_l)]$$

converge to $\int Q\,d\mu^*$ as $n \to \infty$ by (3.19), so do the second Cesàro means

$$\frac{2}{n(n-1)}\sum_{l=0}^{n-2}(n-1-l)[Q(a_l) - Q(b_l)].$$

Hence from (3.21) and (3.18) it follows that

$$\liminf_{n\to\infty} \frac{2}{n(n-1)}\sum_{k=1}^{n-1} k \log \frac{1}{A_k} \geq V_w - \int Q\,d\mu^* = F_1 + F_2.$$

Thus, with (3.20) we get

$$\lim_{n\to\infty} \frac{2}{n(n-1)}\sum_{k=1}^{n} k \log \frac{1}{A_k} = F_1 + F_2. \tag{3.22}$$

Finally we observe from the definition of A_k^k that

$$A_k^k \leq L A_{k-1}^{k-1} \quad \text{with} \quad L := M_1/M_2,$$

where

$$M_1 := (\max_{E_1} w)\frac{\operatorname{diam}(E_1)}{\operatorname{dist}(E_1, E_2)}, \qquad M_2 := (\min_{E_2} w)\frac{\operatorname{dist}(E_1, E_2)}{\operatorname{diam}(E_2)}.$$

From this, (3.20) and (3.22) we can conclude that (3.17) holds by the same reasoning that was used in the proof of Theorem V.1.2. □

Note that the n-point optimization problem of determining the rational Fekete points is replaced here by the much easier task of successively computing the maxima and minima of r_n on E_1 and E_2, respectively. That this is a numerically stable procedure is shown by

Theorem 3.7. *Let the closed sets E_1, E_2 satisfy* (3.12) *and* (3.13) *and assume w is continuous on $E_1 \cup E_2$. For each n, let $S_1^{(n)}$ and $S_2^{(n)}$ be finite subsets of E_1 and E_2 in such a way that the distance of any point of E_j from the corresponding set $S_j^{(n)}$ is at most ε_n. Starting from two points $\hat{a}_0 \in E_1$ and $\hat{b}_0 \in E_2$ define successively $\hat{a}_n \in S_1^{(n)}$ and $\hat{b}_n \in S_2^{(n)}$ as points where the function*

$$w(a)^n \left|\prod_{j=0}^{n-1}\frac{a-\hat{a}_j}{a-\hat{b}_j}\right| \Bigg/ w(b)^n \left|\prod_{j=0}^{n-1}\frac{b-\hat{a}_j}{b-\hat{b}_j}\right|$$

attains its maximum for $a \in S_1^{(n)}$ and $b \in S_2^{(n)}$. Assume further that

$$\lim_{n\to\infty} \varepsilon_n^{1/n} = 0. \tag{3.23}$$

Then the asymptotic distribution of the points $\{\hat{a}_n\}$ is μ_1^*, that of $\{\hat{b}_n\}$ is μ_2^*, and the rational functions

$$\hat{r}_n(z) := \prod_{j=0}^{n-1} \frac{z - \hat{a}_j}{z - \hat{b}_j}$$

are asymptotically optimal in the Zolotarjov problem:

$$\lim_{n\to\infty} \left(\max_{z\in E_1}\{|\hat{r}_n(z)|w^n(z)\} \Big/ \min_{z\in E_2}\{|\hat{r}_n(z)/w^n(z)|\} \right)^{1/n} = e^{-(F_1+F_2)}$$

$$= c(w, \Sigma_1, \Sigma_2) \exp(\int Q \, d\mu^*).$$

In regular cases the condition (3.23) can be relaxed to $\varepsilon_n = O(n^{-\alpha})$ for some $\alpha > 0$. See the discussion at the end of Section V.1.

Proof of Theorem 3.7. Since all the potentials U^σ, where σ is a probability measure on E_2, are uniformly equicontinuous on E_1, it easily follows (or verify it directly) from the proof of Theorem V.1.4, see especially (V.1.11), that

$$\limsup_{n\to\infty} \left(\| w^n \rho_n \|_{E_1} / \| w^n \rho_n \|_{S_1^{(n)}} \right)^{1/n} = 1$$

for every sequence $\{\rho_n\}$ of rational functions of corresponding degree at most n such that ρ_n has all its zeros in E_1 and all its poles in E_2. Reversing the role of E_1 and E_2 and considering $1/\rho_n$ instead of ρ_n we get that the limsup of the ratio of

$$\left(\max_{z\in E_1} |\rho_n(z)|w^n(z) / \min_{z\in E_2} |\rho_n(z)|w^n(z) \right)^{1/n}$$

and

$$\left(\max_{z\in S_1^{(n)}} |\rho_n(z)|w^n(z) / \min_{z\in S_2^{(n)}} |\rho_n(z)|w^n(z) \right)^{1/n}$$

is 1 uniformly for all such sequences $\{\rho_n\}$. Now with this relation at hand we can repeat the proof of Theorem 3.5 word for word in the present situation. □

VIII.4 Examples

In this section we illustrate the general theory discussed in the preceding three sections by three examples. The first one will involve a weight function. We shall need these examples in subsequent sections to compute the rate of best rational approximation of signum like functions and to determine conformal mappings of ring domains.

Example 4.1. Suppose $0 < r < \rho < R$, and C_r is the circle around the origin of radius r. We resolve the following problem: what is the necessary and sufficient condition for the positive values w_r, w_ρ, and w_R that ensures that there are rational functions r_n of numerator and denominator degree at most n that are asymptotically as small as w_r^n and w_R^n on C_r and C_R, respectively, while asymptotically as large as w_ρ^n on C_ρ? In other words, if w is the weight function on $\Sigma = C_r \cup C_\rho \cup C_R$ that is $1/w_r$ on C_r, $1/w_\rho$ on C_ρ, and $1/w_R$ on C_R, then we want rational functions r_n such that

$$\limsup_{n\to\infty} \|r_n w^n\|_{C_r \cup C_R}^{1/n} \le 1,$$

while

$$\liminf_{n\to\infty} \|r_n w^n\|_{C_\rho}^{1/n} \ge 1.$$

Set $N = 2$, $\Sigma_1 = C_r \cup C_R$, $\Sigma_2 = C_\rho$, $\varepsilon_1 = 1$, $\varepsilon_2 = -1$, and $m_1 = m_2 = 1$. We know from Theorem 3.1 that then we must have

$$1 \ge \limsup_{n\to\infty} \left(\|r_n w^n\|_{\Sigma_1}^{1/n} \Big/ \inf_{z\in\Sigma_2} |r_n(z) w^n(z)|^{1/n} \right) \ge e^{-(F_1+F_2)},$$

i.e. $F_1 + F_2 \ge 0$. Furthermore, by normalizing the Fekete rational functions r_n introduced in Section VIII.3 so that they have weighted norm $\|r_n w^n\|_{\Sigma_1} = 1$, we can deduce from Theorem 3.4 that $F_1 + F_2 \ge 0$ is not only necessary but also sufficient for the existence of the rational functions r_n. Thus, we need necessary and sufficient conditions on w that ensure $F_1 + F_2 \ge 0$.

By the unicity of the extremal measure μ^* it follows from the circular symmetry of the problem that μ^* is also circular symmetric; hence on each circle it coincides with a constant times the normalized arc measure on this circle (normalized to have total mass 1). If for C_r this constant is $0 \le \alpha \le 1$, then for C_R it is $1 - \alpha$, and of course for C_ρ it is -1. Then we can easily compute the potential U^{μ^*} on the three circles (see formula (0.5.5)), and Theorem 1.4 says that we must have with some constants F_1 and F_2

$$\alpha \log 1/r + (1-\alpha) \log 1/R - \log 1/\rho + \log w_r \ge F_1, \tag{4.1}$$

$$\alpha \log 1/\rho + (1-\alpha) \log 1/R - \log 1/\rho + \log w_\rho = -F_2, \tag{4.2}$$

$$\alpha \log 1/R + (1-\alpha) \log 1/R - \log 1/R + \log w_R \ge F_1, \tag{4.3}$$

and in the first inequality, that corresponds to C_r, the "=" sign must hold unless $\alpha = 0$, and similarly, in the third one, "=" holds unless $\alpha = 1$. Conversely, if α has all these properties, then the corresponding measure will be the equilibrium measure μ^* (see Theorem 2.2).

Now $\alpha = 0$ is a solution if and only if $\rho w_r / R w_R \ge 1$, because then equality must hold in (4.3) (see also (4.1) and (4.2)), and in this case $F_1 + F_2 \ge 0$ if and only if $R w_R / \rho w_\rho \ge 1$, which is obtained by subtracting (4.2) from (4.3). The value $\alpha = 1$ is a solution if and only if $r w_R / \rho w_r \ge 1$, and then $F_1 + F_2 \ge 0$ if and only if $\rho w_r / r w_\rho \ge 1$. Finally, if both $\rho w_r / R w_R < 1$ and $r w_R / \rho w_r < 1$, then we must have equality both in (4.1) and (4.3) and the solution is $\alpha =$

$(\log(Rw_R/\rho w_r))/(\log R/r)$ (which lies between 0 and 1) and $F_1 + F_2 \geq 0$ if and only if

$$\log(Rw_R/\rho w_r) \cdot \log(\rho/r) \geq \log(w_\rho/w_r) \cdot \log(R/r).$$

These formulae solve the problem, and we can also easily answer the question: How large can w_ρ be if w_r and w_R are fixed? In fact, from the discussion above we can easily see that if $\rho w_r/Rw_R \geq 1$, then the largest value of w_ρ is Rw_R/ρ; if $rw_R/\rho w_r \geq 1$, then it is $\rho w_r/r$; and otherwise it is

$$w_r \exp\left(\frac{\log(Rw_R/\rho w_r) \cdot \log(\rho/r)}{\log(R/r)}\right).$$

□

In the next example the symmetry properties of the set Σ are relaxed.

Example 4.2. Let $w \equiv 1$, $N = 2$, $\varepsilon_1 = 1$, $\varepsilon_2 = -1$, $m_1 = m_2 = 1$, and suppose that Σ_1 and Σ_2 are two disjoint circles (or disks) with center at o_j and of radius r_j. If Σ_1 and Σ_2 are concentric circles then it immediately follows from Theorem 2.2 that μ^* is simply the difference of the normalized arc measures on Σ_1 and Σ_2; therefore, $F_1 + F_2 = |\log(r_2/r_1)|$. Thus, in what follows we can assume that $o_1 \neq o_2$ and that these centers lie on the real line. In this case there is a unique pencil of circles, with degenerate circles at some points z_1 and z_2, containing Σ_1 and Σ_2 (see Figure 4.1). The points z_1 and z_2 are characterized by the property that if we apply inversion onto either of Σ_j, then the image of z_1 is z_2. In other words, z_1 is one of the solutions of the equation

$$z_1 = r_2^2/(r_1^2/(z_1 - o_1) + o_1 - o_2) + o_2$$

and

$$z_2 = r_1^2/(z_1 - o_1) + o_1,$$

is the other one. Since each z_j must be on the line connecting o_1 and o_2, i.e. on the real axis, it is a matter of solving a second order algebraic equation to determine the points z_1, z_2. Let z_1 be in that component of $\mathbf{C} \setminus \Sigma_1$ which does not contain Σ_2 (with this we have given a definite ordering to the points $\{z_1, z_2\}$ which have played a symmetric role until now).

By the definition of z_j, each of the circles Σ_1 and Σ_2 are Apollonius circles for the pair $\{z_1, z_2\}$, i.e. the ratio $|z - z_1|/|z - z_2|$ is constant on each Σ_j. Thus, if $\sigma = \delta_{z_1} - \delta_{z_2}$, then the potential $U^\sigma(z) = -\log|z - z_1| + \log|z - z_2|$ is constant on each Σ_j. This implies that if μ_j is the balayage of δ_{z_j} onto Σ_j and $\mu = \mu_1 - \mu_2$, then

$$U^\mu = U^\sigma + c = -\log|z - z_1| + \log|z - z_2| + c$$

is also constant on each Σ_j; therefore (see Theorem 2.2), $\mu^* = \mu$. If

$$\Sigma_j = \{o_j + r_j e^{it} \,|\, t \in [0, 2\pi]\}$$

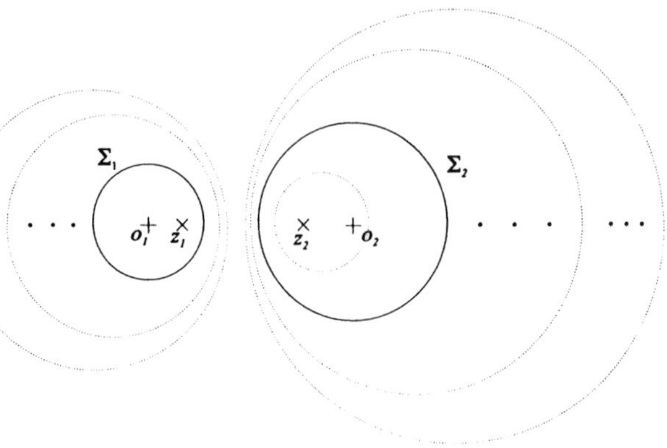

Fig. 4.1

is the standard parametric representation of Σ_j, and, say, z_j is inside Σ_j, then the above balayage measure is given by

$$\frac{1}{2\pi}\frac{r_j^2 - |z_j - o_j|^2}{r_j^2 - 2r_j|z_j - o_j|\cos t + |z_j - o_j|^2}dt$$

(see Example II.4.3) and the minimal energy $F_1 + F_2$ turns out to be $\log(r_1 r_2/|o_1 - z_1||o_2 - z_2|)$. □

Finally, we present a much less trivial example for the actual determination of the equilibrium measure and related quantities.

Example 4.3. Again, let $N = 2$, $\varepsilon_1 = 1$, $\varepsilon_2 = -1$, $m_1 = m_2 = 1$. Let C_r be the circle of radius r with center at the origin, and suppose that

$$\Sigma_1 = \bigcup_{j=1}^{k} C_{r_j}, \quad \Sigma_2 = \bigcup_{j=1}^{k} C_{\rho_j},$$

where

$$0 < r_1 < \rho_1 < r_2 < \rho_2 < \cdots < r_k < \rho_k.$$

By the circular symmetry of the problem and by the unicity of the equilibrium measure μ^* we can conclude that μ^* must be circular symmetric, i.e. on each circle C_r of Σ the signed measure μ^* must be a constant times the normalized arc measure m_r normalized in such a way as to have total mass 1. Therefore, there are nonnegative numbers α_j, β_j such that

$$\mu^* = \sum_{j=1}^{k} \alpha_j m_{r_j} - \sum_{j=1}^{k} \beta_j m_{\rho_j}.$$

Furthermore, we must have $\Sigma\alpha_j = 1$, $\Sigma\beta_j = 1$. Thus, for the potential of μ^* we have for $r_j \le |z| \le \rho_j$ from formula (0.5.5):

$$\begin{aligned} U^{\mu^*}(z) &= \left(\sum_{i=1}^{j}\alpha_i - \sum_{i=1}^{j-1}\beta_i\right)\log 1/|z| \\ &\quad + \left(\sum_{i=j+1}^{k}\alpha_i \log 1/r_i - \sum_{i=j}^{k}\beta_i \log 1/\rho_i\right), \end{aligned}$$

and for $\rho_j \le |z| \le r_{j+1}$:

$$\begin{aligned} U^{\mu^*}(z) &= \left(\sum_{i=1}^{j}\alpha_i - \sum_{i=1}^{j}\beta_i\right)\log 1/|z| \\ &\quad + \left(\sum_{i=j+1}^{k}\alpha_i \log 1/r_i - \sum_{i=j+1}^{k}\beta_i \log 1/\rho_i\right). \end{aligned}$$

We claim that the support of μ^* is the whole set Σ. In fact, suppose that for some j the measure μ^* does not have mass on, say, C_{ρ_j}, i.e. $\beta_j = 0$. Since $F_1 + F_2$ equals the minimal energy, and the latter is positive (see Lemma I.1.8), we have $F_1 + F_2 > 0$. But we know that $U^{\mu^*}(z) \ge F_1$ for all $z \in \Sigma_1 = \cup C_{r_j}$, while $U^{\mu^*}(z) \le -F_2$ for all $z \in \Sigma_2 = \cup C_{\rho_j}$. It follows from the above expression of the equilibrium potential that in case of $\beta_j = 0, 0 \le j \le k-1$, the value $U^{\mu^*}(\rho_j)$ is in between the values $U^{\mu^*}(r_j)$ and $U^{\mu^*}(r_{j+1})$; hence $U^{\mu^*}(\rho_j) \ge F_1$ which is impossible because $U^{\mu^*}(\rho_j) \le -F_2 < F_1$. When $j = k$, this argument needs only slight modification: in this case we would have $U^{\mu^*}(r_k) = U^{\mu^*}(\rho_k) = 0$, which is again a contradiction. The argument that each α_j must be positive is similar.

From the fact that the support of μ^* is Σ it follows with Theorem 1.4 that we must have $U^{\mu^*}(z) = F_1$ for all $z \in \Sigma_1 = \cup C_{r_j}$ and $U^{\mu^*}(z) = -F_2$ for all $z \in \Sigma_2 = \cup C_{\rho_j}$. From the equality $U^{\mu^*}(r_{j+1}) = U^{\mu^*}(r_j)$ we get the equation

$$\left(\sum_{i=1}^{j}(\alpha_i - \beta_i)\right)\log r_{j+1}/r_j = -\beta_j \log \rho_j/r_j, \quad j = 1, \ldots, k-1,$$

while from $U^{\mu^*}(\rho_{j+1}) = U^{\mu^*}(\rho_j)$

$$\left(\sum_{i=1}^{j}(\alpha_i - \beta_i)\right)\log \rho_{j+1}/\rho_j = -\alpha_{j+1}\log \rho_{j+1}/r_{j+1}, \quad j = 1, \ldots, k-1, \tag{4.4}$$

follows. Thus,

$$\beta_j = \alpha_{j+1}\frac{(\log \rho_{j+1}/r_{j+1})\log r_{j+1}/r_j}{(\log \rho_j/r_j)\log \rho_{j+1}/\rho_j} =: \alpha_{j+1}\kappa_j, \quad j = 1, \ldots, k-1. \tag{4.5}$$

From (4.4) for j and $j-1$ we get

$$-\alpha_{j+1}\frac{\log \rho_{j+1}/r_{j+1}}{\log \rho_{j+1}/\rho_j}+\alpha_j\frac{\log \rho_j/r_j}{\log \rho_j/\rho_{j-1}}=\alpha_j-\alpha_{j+1}\kappa_j,$$

which yields

$$\alpha_{j+1}=\alpha_j\frac{\log \rho_{j+1}/\rho_j}{\log \rho_j/\rho_{j-1}}\,\frac{\log \rho_j/r_j}{\log \rho_{j+1}/r_{j+1}}\,\frac{\log r_j/\rho_{j-1}}{\log r_{j+1}/\rho_j},\qquad j=2,\ldots,k-1.$$

In a similar fashion we get

$$\alpha_2=\alpha_1\frac{\log \rho_2/\rho_1}{1}\,\frac{\log \rho_1/r_1}{\log \rho_2/r_2}\,\frac{1}{\log r_2/\rho_1}.$$

Repeated application of these formulae show that

$$\alpha_j=\alpha_1\left(\log\frac{\rho_1}{r_1}\right)\frac{\log \rho_j/\rho_{j-1}}{\log \rho_j/r_j}\,\frac{1}{\log r_j/\rho_{j-1}},\qquad j=2,\ldots,k.$$

Thus, we obtain from $\Sigma\alpha_j=1$ that

$$\alpha_1=\left(1+\log\frac{\rho_1}{r_1}\sum_{i=2}^{k}\frac{\log \rho_i/\rho_{i-1}}{\log \rho_i/r_i}\,\frac{1}{\log r_i/\rho_{i-1}}\right)^{-1}$$

and

$$\alpha_j=\log\frac{\rho_1}{r_1}\,\frac{\log \rho_j/\rho_{j-1}}{\log \rho_j/r_j}\,\frac{1}{\log r_j/\rho_{j-1}}\times$$

$$\times\left(1+\log\frac{\rho_1}{r_1}\sum_{i=2}^{k}\frac{\log \rho_i/\rho_{i-1}}{\log \rho_i/r_i}\,\frac{1}{\log r_i/\rho_{i-1}}\right)^{-1}$$

if $2\le j\le k$. From (4.5) we then get

$$\beta_j=\log\frac{\rho_1}{r_1}\,\frac{\log r_{j+1}/r_j}{\log r_{j+1}/\rho_j}\,\frac{1}{\log \rho_j/r_j}\times$$

$$\times\left(1+\log\frac{\rho_1}{r_1}\sum_{i=2}^{k}\frac{\log \rho_i/\rho_{i-1}}{\log \rho_i/r_i}\,\frac{1}{\log r_i/\rho_{i-1}}\right)^{-1}$$

if $1\le j\le k-1$, while for β_k we obtain from

$$\beta_k=1-\sum_{j=1}^{k-1}\beta_j$$

and

$$\sum_{i=2}^{k}\left(\frac{\log \rho_i/\rho_{i-1}}{\log \ \rho_i/r_i} - \frac{\log r_i/r_{i-1}}{\log \rho_{i-1}/r_{i-1}}\right)\frac{1}{\log r_i/\rho_{i-1}}$$

$$= \sum_{i=2}^{k}\left(\frac{\log r_i/\rho_{i-1}}{\log \rho_i/r_i} - \frac{\log r_i/\rho_{i-1}}{\log \rho_{i-1}/r_{i-1}}\right)\frac{1}{\log r_i/\rho_{i-1}}$$

$$= \frac{1}{\log r_k/\rho_k} - \frac{1}{\log r_1/\rho_1}$$

that

$$\beta_k = \frac{\log \rho_1/r_1}{\log \rho_k/r_k}\left(1 + \log\frac{\rho_1}{r_1}\sum_{i=2}^{k}\frac{\log \rho_i/\rho_{i-1}}{\log \rho_i/r_i}\frac{1}{\log r_i/\rho_{i-1}}\right)^{-1}.$$

With this, we have completely determined the measure μ^*. As for the minimal energy, it is the same as

$$F_1 + F_2 = U^{\mu^*}(r_k) - U^{\mu^*}(\rho_k) = (1 - (1 - \beta_k)) \log \rho_k/r_k$$

$$= \left(\log\frac{\rho_1}{r_1}\right)\left(1 + \log\frac{\rho_1}{r_1}\sum_{i=2}^{k}\frac{\log \rho_i/\rho_{i-1}}{\log \rho_i/r_i}\frac{1}{\log r_i/\rho_{i-1}}\right)^{-1}.$$

□

VIII.5 Rational Approximation of Signum Type Functions

In this section we associate another quantity with those introduced in the preceding sections, namely the order of best rational approximation of signum-like functions. Partitions of unities consisting of rational functions naturally appear in many questions and their basic building blocks are rational functions that are close to 1 on some set and close to 0 on another one. For simplicity, and because of its importance, we shall first consider the unweighted case $w \equiv 1$. The case of weighted approximation, where the weight depends exponentially on the degree, will be dealt with at the end of the section.

Thus we consider the best rational approximation of a given degree of the function

$$\chi(z) = \begin{cases} 1 & \text{if} \quad z \in \Sigma_1 \\ 0 & \text{if} \quad z \in \Sigma_2, \end{cases}$$

where, as before, Σ_1 and Σ_2 are disjoint compact sets of positive capacity. Let the error in this best approximation be

$$\varepsilon_n = \inf_r \{\|\chi - r\|_{L^\infty(\Sigma_1 \cup \Sigma_2)}\},$$

where the infimum is taken for all rational functions r with numerator and denominator degrees at most n. If $c(1, \Sigma_1, \Sigma_2)$ is the "signed" capacity of the pair (Σ_1, Σ_2) introduced in (3.9), then we have the following result.

Theorem 5.1. *With the notations above,*

$$\lim_{n\to\infty} \varepsilon_n^{1/n} = c(1, \Sigma_1, \Sigma_2)^{1/2}.$$

Since for nonzero signed measures that are orthogonal to constant functions the logarithmic energy is positive (see Lemma I.1.8), it follows from the definition of $c(1, \Sigma_1, \Sigma_2)$ that it is always less than 1 (note that the minimal energy is attained), i.e. every signum-type function can be approximated geometrically fast by rational functions of degree $n = 1, 2, \ldots$.

For weighted approximation, see Theorem 5.4 below.

Proof of Theorem 5.1. Let $\eta > 0$ be such that

$$c_0 := \log(1/c(1, \Sigma_1, \Sigma_2)) - \eta > 0.$$

In Section VIII.3 we verified that for each n, there are rational functions r_n of degree n, such that

$$|r_n(z)| \le e^{-c_0 n} \quad \text{for} \quad z \in \Sigma_1$$

and

$$|r_n(z)| \ge 1 \quad \text{for} \quad z \in \Sigma_2$$

provided n is sufficiently large. But then for

$$r_n^*(z) := \frac{1}{1 + r_n(z)e^{c_0 n/2}}$$

and sufficiently large n we obtain

$$|r_n^*(z) - 1| \le 2e^{-c_0 n/2} \quad \text{for} \quad z \in \Sigma_1$$

and

$$|r_n^*(z)| \le 2e^{-c_0 n/2} \quad \text{for} \quad z \in \Sigma_2.$$

Letting n tend to infinity and η tend to zero, we get

$$\limsup_{n\to\infty} \varepsilon_n^{1/n} \le c(1, \Sigma_1, \Sigma_2)^{1/2}.$$

To prove the converse inequality, suppose that for some constant $d > 0$ there are rational functions r_n^* of degree at most n such that for large n

$$|r_n^*(z) - 1| \le e^{-dn} \quad \text{for} \quad z \in \Sigma_1$$

and

$$|r_n^*(z)| \le e^{-dn} \quad \text{for} \quad z \in \Sigma_2.$$

Then for

$$r_n(z) = \frac{1}{r_n^*(z)} - 1$$

we have

$$\left(\sup_{z\in\Sigma_1}|r_n(z)|\right)\Big/\left(\inf_{z\in\Sigma_2}|r_n(z)|\right)\leq 4e^{-2dn},$$

and so from (3.2) we obtain $d \leq (F_2+F_1)/2$. Since in the unweighted case ($w\equiv 1$) the "signed" capacity $c(1,\Sigma_1,\Sigma_2)$ is just $\exp(-(F_1+F_2))$, the proof is done. □

Now we apply the above theorem to calculate the order of best rational approximation to χ when Σ_1 and Σ_2 consist of finitely many intervals on the real line. Thus, let $\Sigma=\Sigma_1\cup\Sigma_2$ be the union of the intervals $[a_j,b_j]$, $j=1,\dots,m$, where $m\geq 2$. We assume $b_j<a_{j+1}$ for $j=1,\dots,m-1$. Let $\mu^*=\mu_1^*-\mu_2^*$, $\|\mu_i^*\|=1$, $\operatorname{supp}(\mu_i^*)\subset\Sigma_i$, be the equilibrium measure from the energy problem (with weight $w\equiv 1$). We know that each of μ_1^* and μ_2^* is the balayage of the other one onto Σ_1 and Σ_2, respectively (see Theorem 2.6); hence the support of μ_1^* is Σ_1 and that of μ_2^* is Σ_2. Then U^{μ^*} equals some constant F_1 on Σ_1 and some $-F_2$ on Σ_2 (use that by Theorem 2.1 and Theorem I.4.8 the potential U^{μ^*} is continuous), and by Theorem 5.1

$$\lim_{n\to\infty}\varepsilon_n^{1/n}=c(1,\Sigma_1,\Sigma_2)^{1/2}=e^{-(F_1+F_2)/2}.$$

Thus, all we have to do is to find F_1+F_2, which is nothing else than the difference of the values of U^{μ^*} on Σ_1 and on Σ_2. Furthermore (see Theorem 2.2), μ^* is the only measure in $\mathcal{M}$ with the property that its potential is constant on each of Σ_1 and Σ_2. Using these facts, we now determine the signed measure μ^*.

First we show that if ω_K denotes the equilibrium measure of a compact set K, then there are positive constants c, C such that

$$c\omega_{\Sigma_j}\leq\mu_j^*\leq C\omega_{\Sigma_j},\quad j=1,2. \tag{5.1}$$

Let h be an arbitrary nonnegative continuous function on Σ_1. Since Σ_1 is regular with respect to the solution of Dirichlet problem in $\overline{\mathbb{C}}\setminus\Sigma_1$, h can be extended to a nonnegative harmonic function to $\overline{\mathbb{C}}\setminus\Sigma_1$, which we continue to denote by h, so that h is continuous on the whole Riemann sphere. Using that μ_1^* is the balayage of μ_2^* onto Σ_1 we have (see Theorem II.4.7(c))

$$\int h\,d\mu_2^*=\int h\,d\mu_1^*.$$

In a similar fashion,

$$h(\infty)=\int h\,d\omega_{\Sigma_1}$$

because the balayage of the Dirac measure δ_∞ onto Σ_1 is the equilibrium measure of Σ_1 (see (II.4.10)). Now Harnack's inequality for nonnegative harmonic functions implies that there are positive constants c, C independent of h such that

$$ch(\infty)\leq h(t)\leq Ch(\infty)$$

for $t \in \operatorname{supp}(\mu_2^*) = \Sigma_2$. On integrating this inequality with respect to μ_2^* and taking into account the preceding relations, we arrive at

$$\int h\, d(\mu_1^* - c\omega_{\Sigma_1}) \geq 0, \quad \int h\, d(C\omega_{\Sigma_1} - \mu_1^*) \geq 0.$$

The signed measures with respect to which the integrals are taken are supported on Σ_1, and since these inequalities hold for all nonnegative continuous function h on Σ_1, we can conclude that the signed measures $\mu_1^* - c\omega_{\Sigma_1}$ and $C\omega_{\Sigma_1} - \mu_1^*$ are actually nonnegative measures and this is the inequality (5.1) for $j = 1$. When $j = 2$, the proof is similar.

Next we need a representation for the equilibrium measures ω_{Σ_j}. Namely it is known that they are absolutely continuous with respect to Lebesgue measure on Σ_j and if

$$\Sigma_j = \bigcup_{k=1}^{l_j} \left[a_k^{(j)}, b_k^{(j)}\right],$$

then there are numbers $y_k^{(j)} \in \left(b_k^{(j)}, a_{k+1}^{(j)}\right)$, $k = 1, \ldots, l_j - 1$, such that

$$d\omega_{\Sigma_j}(t) = \frac{S_j(t)}{\pi\sqrt{|R_j(t)|}} dt, \ t \in \Sigma_j,$$

where

$$R_j(t) = \prod_{k=1}^{l_j} \left(t - a_k^{(j)}\right)\left(t - b_k^{(j)}\right)$$

and

$$S_j(t) = \prod_{k=1}^{l_j-1} \left|t - y_k^{(j)}\right|$$

(see e.g. [212, Lemma 4.4]).

From this representation of the equilibrium measure and from (5.1), it easily follows that the function

$$H(z) = \left(\int \frac{d\mu^*(t)}{z-t}\right)^2$$

has a simple pole at each a_j, b_j. We claim that elsewhere H is analytic. This is obvious in $\overline{\mathbf{C}} \setminus \Sigma$, and the analyticity on each of (a_j, b_j) can be proved as follows. If we cut $\mathbf{C}$ along Σ, then

$$\int \frac{d\mu^*(t)}{z-t} \tag{5.2}$$

is purely imaginary on the cut because the real part of

$$\int \log(z-t)\, d\mu^*(t) \tag{5.3}$$

is the negative of the potential $U^{\mu^*}(z)$, and so it is constant on each interval of Σ; hence the real part of the derivative of (5.3) vanishes on Σ. Furthermore, the function in (5.2) takes conjugate values for conjugate arguments; therefore, this function takes opposite values on the upper and lower parts of the cut. Squaring these opposite values as in H we get that H is real on the cut and takes conjugate values for conjugate arguments on the upper and lower parts of the cut; hence the analyticity of H on $\cup(a_j, b_j)$ follows from the continuation principle for analytic functions. Of course, to do all these deductions we need that H, which on $\bigcup_{j=1}^m (a_j, b_j)$ must be understood in principal value sense, is continuous on the cut. Seeing however that e.g. on Σ_1 the measure μ_1^* is given as the balayage of μ_2^* onto Σ_1, the density function of μ^* is a C^∞ function on $\cup(a_j, b_j)$ (this follows from Corollary II.4.11 if we apply it twice), from which the claimed continuity easily follows.

In summary, the function H is a rational function. Obviously, H has a zero at infinity with multiplicity 4 (recall that μ^* is orthogonal to constants) and each of its zeros is of even multiplicity; hence H is of the form

$$H(z) = (P_{m-2}(z))^2/R(z),$$

where

$$R(z) = \prod_{k=1}^{m}(z - a_k)(z - b_k),$$

and

$$P_{m-2}(z) = c_{m-2}z^{m-2} + \cdots + c_0$$

is a polynomial of degree at most $m - 2$. Thus, by multiplying P_{m-2} by -1 if necessary we can conclude that

$$\int \frac{d\mu^*(t)}{z - t} = \frac{P_{m-2}(z)}{\sqrt{R(z)}}, \qquad z \in \mathbf{C} \setminus \Sigma.$$

Here and in what follows we take that branch of the square root that is positive on the positive part of the real line. From Cauchy's formula applied to $\overline{\mathbf{C}} \setminus \Sigma$ we can see that

$$\frac{P_{m-2}(z)}{\sqrt{R(z)}} = \frac{1}{2\pi i} \oint_{\Sigma} \frac{P_{m-2}(\xi)}{\sqrt{R(\xi)}} \frac{1}{\xi - z} d\xi = \int_{\Sigma} \frac{P_{m-2}(t)}{\pi i \sqrt{R(t)}} \frac{1}{t - z} dt,$$

where the first integral is taken on the cut in the clockwise direction and the second integral is an ordinary Lebesgue integral and the values of $\sqrt{R(t)}$ in it are taken on the upper part of the cut. Since Cauchy transforms determine the measures if their support has zero two-dimensional Lebesgue measure (see [14], in fact this follows from Theorem II.2.1 by integration and then taking real part), it follows from the preceding two formulae that

$$d\mu^*(t) = \frac{P_{m-2}(t)}{-\pi i \sqrt{R(t)}} dt.$$

Since $i\sqrt{R(t)}$ is real on the upper part of the cut, we can also conclude that P_{m-2} has real coefficients.

Now let x and y belong to the same interval $[a_j, b_j]$. Then the function

$$\frac{P_{m-2}(z)}{-\pi i\sqrt{R(z)}}\log\frac{x-z}{y-z}$$

is analytic on $\overline{\mathbf{C}}\setminus\Sigma$ and has at least a double zero at infinity; hence

$$\oint_{\Sigma}\frac{P_{m-2}(\xi)}{-\pi i\sqrt{R(\xi)}}\log\frac{x-\xi}{y-\xi}d\xi=0. \tag{5.4}$$

Taking real parts, we see that whatever the real polynomial P_{m-2} of degree at most $m-2$ is, the potential of the (signed) measure

$$d\sigma(t)=\frac{P_{m-2}(t)}{-\pi i\sqrt{R(t)}}dt$$

is constant on each interval $[a_j, b_j]$; in particular,

$$U^{\sigma}(a_j)=U^{\sigma}(b_j). \tag{5.5}$$

Next we compute $U^{\sigma}(b_j)-U^{\sigma}(a_{j+1})$, $1\le j\le m-1$. If $L=\Sigma\cup[b_j, a_{j+1}]$, then (5.4) with $x=b_j$ and $y=a_{j+1}$ holds again if the integration on Σ is replaced by integration around L, and for the same reason. Taking again real parts we can see from the facts that $\sqrt{R(t)}$ is real on (b_j, a_{j+1}) and

$$\log\frac{a_{j+1}-t}{b_j-t}=\log\left|\frac{a_{j+1}-t}{b_j-t}\right|+i\pi$$

there, that

$$\operatorname{Re}\left(\log\frac{a_{j+1}-t}{b_j-t}\,\frac{P_{m-2}(t)}{-\pi i\sqrt{R(t)}}\right)=-\frac{P_{m-2}(t)}{\sqrt{R(t)}}$$

on the upper part of the cut along L on (b_j, a_{j+1}); therefore,

$$\int_{\Sigma}\log\left|\frac{a_{j+1}-t}{b_j-t}\right|\frac{P_{m-2}(t)}{-\pi i\sqrt{R(t)}}dt=\int_{b_j}^{a_{j+1}}\frac{P_{m-2}(t)}{\sqrt{R(t)}}dt. \tag{5.6}$$

It follows from (5.5) and (5.6) that for any $l\ge j$

$$\int_{\Sigma}\log\left|\frac{a_{l+1}-t}{b_j-t}\right|\frac{P_{m-2}(t)}{-\pi i\sqrt{R(t)}}dt \tag{5.7}$$

$$=\left(\int_{b_j}^{a_{j+1}}+\int_{b_{j+1}}^{a_{j+2}}+\cdots+\int_{b_l}^{a_{l+1}}\right)\frac{P_{m-2}(t)}{\sqrt{R(t)}}dt.$$

From these formulae we can easily derive necessary and sufficient conditions for the potential U^{σ} to be constant on Σ_1 and constant on Σ_2. In fact, let j_1 and

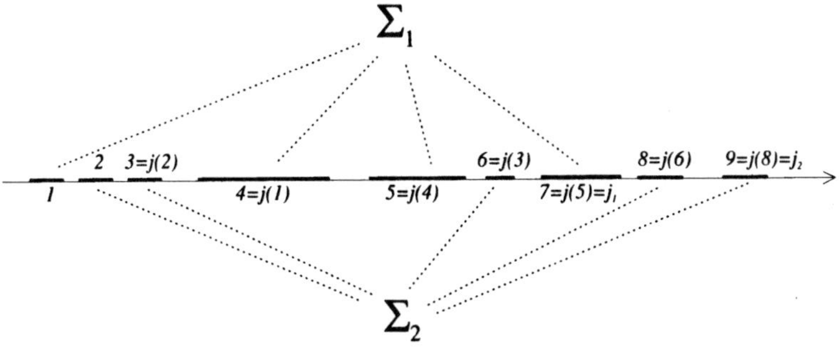

Fig. 5.1

j_2 be the indices of the last (right-most) intervals of Σ_1 and Σ_2, respectively, and set

$$\mathcal{I} = \{j \mid [a_j, b_j] \subset \Sigma_1 \text{ and } [a_{j+1}, b_{j+1}] \subset \Sigma_2 \text{ or } [a_j, b_j] \subset \Sigma_2 \text{ and } [a_{j+1}, b_{j+1}] \subset \Sigma_1,\ j \neq j_1, j_2\}$$

and

$$\mathcal{J} = \{j \mid [a_j, b_j] \subset \Sigma_1 \text{ and } [a_{j+1}, b_{j+1}] \subset \Sigma_1 \text{ or } [a_j, b_j] \subset \Sigma_2 \text{ and } [a_{j+1}, b_{j+1}] \subset \Sigma_2\}.$$

Then $\mathcal{I} \cup \mathcal{J}$ has $m-2$ elements because the indices of the last intervals of Σ_1 and Σ_2 do not appear in $\mathcal{I} \cup \mathcal{J}$ (see Figure 5.1). If $j \in \mathcal{J}$ and U^σ is constant on Σ_1 and on Σ_2, then U^σ must take the same value on $[a_j, b_j]$ and $[a_{j+1}, b_{j+1}]$; hence by (5.6)

$$\int_{b_j}^{a_{j+1}} \frac{P_{m-2}(t)}{\sqrt{R(t)}} dt = 0, \quad j \in \mathcal{J} \tag{5.8}$$

(note that the left-hand side in (5.6) is $U^\sigma(b_j) - U^\sigma(a_{j+1})$). Let now $j \in \mathcal{I}$, and let $l(j) \geq j$ be the smallest index such that $[a_j, b_j]$ and $[a_{l(j)+1}, b_{l(j)+1}]$ belong to the same set Σ_1 or Σ_2 (there is such an $l(j)$ because we omitted the right-most intervals from Σ_1 and Σ_2). The condition $j \in \mathcal{I}$ means that $l(j) > j$. If U^σ is constant on Σ_1 and on Σ_2, then U^σ must take the same value on $[a_j, b_j]$ and $[a_{l(j)+1}, b_{l(j)+1}]$; hence by (5.7)

$$\left(\int_{b_j}^{a_{j+1}} + \int_{b_{j+1}}^{a_{j+2}} + \cdots + \int_{b_{l(j)}}^{a_{l(j)+1}} \right) \frac{P_{m-2}(t)}{\sqrt{R(t)}} dt = 0.$$

But the indices $j+1, j+2, \ldots, l(j)-1$ then belong to $\mathcal{J}$; hence in view of (5.8) we see that this is the same as

$$\left(\int_{b_j}^{a_{j+1}} + \int_{b_{l(j)}}^{a_{l(j)+1}} \right) \frac{P_{m-2}(t)}{\sqrt{R(t)}} dt = 0. \tag{5.9}$$

Equalities (5.8) and (5.9) give $m-2$ equations for the $m-1$ coefficients of P_{m-2}. The $(m-1)$-st condition is

$$\int_{\Sigma_1} \frac{P_{m-2}(t)}{-\pi i\sqrt{R(t)}}dt = 1 \tag{5.10}$$

because we only consider signed measures that have total mass 1 on Σ_1. From Cauchy's formula it then follows from (5.10) that

$$\int_{\Sigma_2} \frac{P_{m-2}(t)}{-\pi i\sqrt{R(t)}}dt = -1$$

as is required of the measures.

From the discussion it is clear that if the coefficients of P_{m-2} are chosen to satisfy (5.8)–(5.10), then

$$d\sigma(t) = \frac{P_{m-2}(t)}{-\pi i\sqrt{R(t)}}dt$$

is a signed measure on Σ such that $\sigma(\Sigma_1) = 1$, $\sigma(\Sigma_2) = -1$, and U^σ is constant on each of Σ_1 and Σ_2. We claim that then σ must be μ^*, i.e. the equilibrium measure for the signed energy problem with weight $w \equiv 1$. In fact, since $\mu^* = \mu_1^* - \mu_2^*$ also has these properties, it follows that there are constants α and β such that the potential of the signed measure $\sigma - \alpha\mu^*$ is identically equal to β on Σ (recall that $F_1 + F_2 > 0$, so U^{μ^*} does not take the same value on Σ_1 and Σ_2). Thus, if $\sigma = \sigma_1 - \sigma_2$, where $(-1)^{j+1}\sigma_j$ denotes the restriction of σ to Σ_j, and if ν_+ denotes the positive part of a measure ν, then we have for all $z \in \Sigma$

$$U^{\sigma_{1+}+\sigma_{2-}+\alpha\mu_2^*}(z) = U^{\sigma_{2+}+\sigma_{1-}+\alpha\mu_1^*}(z) + \beta. \tag{5.11}$$

Here, for the positive measures $\sigma_{1+} + \sigma_{2-} + \alpha\mu_2^*$ and $\sigma_{2+} + \sigma_{1-} + \alpha\mu_1^*$, we have

$$\|\sigma_{1+} + \sigma_{2-} + \alpha\mu_2^*\| = \|\sigma_{2+} + \sigma_{1-} + \alpha\mu_1^*\|$$

because $\|\sigma_{1+}\| - \|\sigma_{1-}\| = \sigma(\Sigma_1) = 1$, $\|\sigma_{2+}\| - \|\sigma_{2-}\| = \sigma(\Sigma_2) = 1$ and $\|\mu_1^*\| = \|\mu_2^*\|$. Furthermore, they have finite logarithmic energy; hence it follows from the principle of domination (Theorem II.3.2) that (5.11) is true for all z. Then Theorem II.2.1 yields $\sigma_{1+} + \sigma_{2-} + \alpha\mu_2^* = \sigma_{2+} + \sigma_{1-} + \alpha\mu_1^*$, i.e. $\sigma = \alpha\mu^*$, and since $\sigma(\Sigma_1) = 1 = \mu^*(\Sigma_1)$, we get $\sigma = \mu^*$ as was claimed above.

Finally we compute $F_1 + F_2$. Since this is the difference of the potential values taken on Σ_1 and on Σ_2, formula (5.6) yields

$$F_1 + F_2 = \left| \int_{b_j}^{a_{j+1}} \frac{P_{m-2}(t)}{\sqrt{R(t)}}dt \right|,$$

where j is an index such that b_j and a_{j+1} belong to different sets Σ_1 and Σ_2.

In summary, using Theorem 5.1, we have proved

Theorem 5.2. *Let Σ_1 and Σ_2 consist of intervals on the real line, $\Sigma = \Sigma_1 \cup \Sigma_2 = \bigcup_{j=1}^m [a_j, b_j]$, and let ε_n be the error in best approximation of the function that is 1 on Σ_1 and 0 on Σ_2, by rational functions of numerator and denominator degrees at most n. Then*

$$\lim_{n\to\infty} \varepsilon_n^{1/n} = \exp\left(-\frac{1}{2}\left|\int_{b_j}^{a_{j+1}} \frac{P_{m-2}(t)}{\sqrt{R(t)}}dt\right|\right),$$

where

$$R(z) = \prod_{k=1}^{m} (z-a_k)(z-b_k),$$

j is an index such that b_j and a_{j+1} belong to different sets Σ_1 and Σ_2, and where the coefficients of the polynomial

$$P_{m-2}(t) = c_{m-2}t^{m-2} + \cdots + c_0$$

are the solutions of the linear system of equations

$$\left(\int_{b_j}^{a_{j+1}} + \int_{b_{l(j)}}^{a_{l(j)+1}}\right) \frac{P_{m-2}(t)}{\sqrt{R(t)}}dt = 0, \quad j \neq j_1, j_2,$$

$$\int_{\Sigma_1} \frac{P_{m-2}(t)}{-\pi i\sqrt{R(t)}}dt = 1.$$

In this system for $1 \le j \le m$ the number $l(j) \ge j$ denotes the smallest index for which the intervals $[a_j, b_j]$ and $[a_{l(j)+1}, b_{l(j)+1}]$ belong to the same set Σ_1 or Σ_2, and j_1 and j_2 denote those two j's for which such an $l(j)$ does not exist. This system of equations is a real system and has a unique solution.

Corollary 5.3. *The errors in best rational approximation of the function*

$$\chi(t) = \begin{cases} 1 & on \quad [a,b] \\ 0 & on \quad [c,d], \end{cases}$$

where $b < c$, satisfy

$$\lim_{n\to\infty} \varepsilon_n^{1/n} = \exp\left(-\frac{\pi}{2}\int_b^c ((t-a)(t-b)(c-t)(d-t))^{-1/2}dt \Big/ \int_a^b ((t-a)(b-t)(c-t)(d-t))^{-1/2}dt\right).$$

Using the examples from the preceding section, we can give other applications of Theorem 5.1. Consider, for example, the best approximation error ε_n of the function

$$\chi(z) = \begin{cases} 1 & \text{if} \quad |z| = r_j, \quad j = 1, \ldots, k \\ 0 & \text{if} \quad |z| = \rho_j, \quad j = 1, \ldots, k, \end{cases}$$

by rational functions of degree at most n, where $0 < r_1 < \rho_1 < r_2 < \cdots < r_k < \rho_k$. From Theorem 5.1 and Example 4.3, we immediately get that

$$\lim_{n\to\infty} \varepsilon_n^{1/n} = \exp\left(-\frac{1}{2}(F_1 + F_2)\right) = \left(\frac{r_1}{\rho_1}\right)^{\tau},$$

where

$$\tau = \left(2 + 2\log\frac{\rho_1}{r_1}\sum_{i=2}^{k}\frac{\log\rho_i/\rho_{i-1}}{\log\rho_i/r_i}\,\frac{1}{\log r_i/\rho_{i-1}}\right)^{-1}.$$

In a similar fashion, if

$$\chi(z) = \begin{cases} 1 & \text{if } |z - o_1| \le r_1 \\ 0 & \text{if } |z - o_2| \le r_2, \end{cases}$$

where the disks $\{z \mid |z - o_j| \le r_j\}$, $j = 1, 2$, are disjoint, and ε_n is the corresponding best rational approximation error, then it follows from Example 4.1 that

$$\lim_{n\to\infty} \varepsilon_n^{1/n} = \sqrt{\frac{|z_1 - o_1||z_2 - o_2|}{r_1 r_2}},$$

where the points z_1 and z_2 are the points introduced in Example 4.1.

Next we consider weighted approximation of signum like functions. Let $\Sigma = \Sigma_1 \cup \Sigma_2$ be as above, where Σ_1 and Σ_2 are compact sets of positive capacity, and suppose that on Σ there is given a positive continuous weight

$$W(x) = e^{-Q(x)}. \tag{5.12}$$

We set $Q_j = Q\big|_{\Sigma_j}$, $j = 1, 2$, and define the weight w as

$$w(x) = \begin{cases} e^{-Q_1(x)} & \text{if } x \in \Sigma_1 \\ e^{Q_2(x)} & \text{if } x \in \Sigma_2 \end{cases} \tag{5.13}$$

(note the sign change!). If we also set $\varepsilon_1 = 1$, $\varepsilon_2 = -1$, then this w satisfies the conditions in Section VIII.1 and if F_j , $j = 1, 2$, are the constants from Theorem 1.4(d), then we set

$$F_w^c := F_1 + F_2.$$

We want to determine the rate of best rational approximation to the function

$$\chi(x) = \begin{cases} 1 & \text{if } x \in \Sigma_1 \\ 0 & \text{if } x \in \Sigma_2, \end{cases}$$

with weight W^n, i.e. the asymptotic behavior of

$$e_n^W := \inf_{r_n} \|W^n(\chi - r_n)\|_\Sigma,$$

where the infimum is taken for all rational functions with numerator and denominator degrees at most n. The extension of Theorem 5.1 to the present weighted case is given by

Theorem 5.4. *With the above notations, suppose that*

$$\max_{x\in\Sigma} Q(x) - \min_{x\in\Sigma} Q(x) < \frac{1}{2}\log\frac{1}{c(1,\Sigma_1,\Sigma_2)}. \tag{5.14}$$

Then

$$\lim_{n\to\infty} (e_n^W)^{1/n} = e^{-F_w^c/2}. \tag{5.15}$$

Note that F_w^c is defined via w, i.e. via the weight (5.13), while in e_n^W the weight function is the one in (5.12).

If $W \equiv 1$, then $w \equiv 1$ and $e^{-F_w^c} = c(1, \Sigma_1, \Sigma_2)$; so in this special case Theorem 5.4 reduces to Theorem 5.1.

We remark that the conclusion can be false if the assumption (5.14) is dropped. Indeed, let

$$Q(x) = \begin{cases} M & \text{if } x \in \Sigma_1 \\ -M & \text{if } x \in \Sigma_2, \end{cases}$$

where

$$M > \tfrac{1}{2}\log(1/c(1,\Sigma_1,\Sigma_2)).$$

Then w is constant, and so

$$c(w,\Sigma_1,\Sigma_2) = c(1,\Sigma_1,\Sigma_2) = e^{-F_w^c},$$

and (5.15) takes the form

$$\lim_{n\to\infty} (e_n^W)^{1/n} = c(1,\Sigma_1,\Sigma_2)^{1/2}. \tag{5.16}$$

By considering the rational function that is identically zero, we see that for all n

$$e_n^W \le e^{-Mn},$$

and so (5.16) is false by the choice of M.

Proof of Theorem 5.4. The proof will be partly a repetition of the proof of Theorem 5.1, but first we have to make some preparations.

Let

$$W^* = W \cdot M,$$

where the positive constant M will be chosen in a moment. If the quantities defined above for W are marked by a "$*$" for W^*, then it is obvious that

$$e_n^{W^*} = M^n e_n^W, \quad Q^* = Q - \log M,$$

$$w^*(x) = \begin{cases} Mw(x) & \text{if } x \in \Sigma_1 \\ w(x)/M & \text{if } x \in \Sigma_2, \end{cases}$$

and so

$$\exp(-F_{w^*}^c) = M^2 \exp(-F_w^c).$$

Hence, the limit relations (5.15) for W and W^* are equivalent. Now choosing M appropriately we can assume without loss of generality (see (5.14)) that

$$\max_{x\in\Sigma} |Q(x)| < \frac{1}{4}\log\frac{1}{c(1,\Sigma_1,\Sigma_2)}. \tag{5.17}$$

Using (3.2) and Theorem 3.4 we can see that

$$\begin{aligned} e^{-F_w} &\leq \exp\left(\max_{z\in\Sigma_1}(-Q_1(x)) - \min_{x\in\Sigma_2}(Q_2(x))\right) c(1,\Sigma_1,\Sigma_2) \\ &\leq c(1,\Sigma_1,\Sigma_2)^{1/2}, \end{aligned} \tag{5.18}$$

where we have also used (5.17). For every $\varepsilon > 0$, there are rational functions r_n of degree at most n (see Theorem 3.4) such that for large n

$$|r_n(z)| \leq e^{nQ_1(z)} e^{-n(F_w^c-\varepsilon)/2}, \quad z \in \Sigma_1, \tag{5.19}$$

and

$$|r_n(z)| \geq e^{-nQ_2(z)} e^{n(F_w^c-\varepsilon)/2}, \quad z \in \Sigma_2. \tag{5.20}$$

Now it follows from (5.17) and (5.18) that for small $\varepsilon > 0$ the right-hand side of (5.19) is exponentially small while the right-hand side of (5.20) is exponentially large. Hence, for

$$r_n^*(z) = \frac{1}{1 + r_n(z)}$$

and sufficiently large n, we get exactly as in the proof of Theorem 5.1 the estimate

$$|\chi - r_n^*| \leq 2e^{nQ - n(F_w^c-\varepsilon)/2},$$

which proves

$$\limsup_{n\to\infty} \left(e_n^W\right)^{1/n} \leq e^{-F_w^c/2}.$$

Now if for some $d \geq F_w^c/2$ we have

$$\liminf_{n\to\infty} \left(e_n^W\right)^{1/n} < e^{-d},$$

then for infinitely many n there would be rational functions r_n^* of degree at most n satisfying

$$|\chi - r_n^*| < e^{nQ - nd},$$

and by the discussion above, the right-hand side of this inequality is exponentially small. Hence for the rational functions

$$r_n(z) = \frac{1}{r_n^*(z)} - 1,$$

we get

$$|r_n(z)| \leq 2e^{nQ_1(z) - nd}, \quad z \in \Sigma_1,$$

and

$$|r_n(z)| \geq \frac{1}{2}e^{-nQ_2(z)+nd}, \quad z \in \Sigma_2,$$

which shows that

$$\left(\max_{z\in\Sigma_1} w^n(z)|r_n(z)|\right) \Big/ \left(\min_{z\in\Sigma_2} w^n(z)|r_n(z)|\right) < 4e^{-2nd}$$

and so (see (3.2)) $d \leq F_w^c/2$. This proves the theorem. □

VIII.6 Conformal Mapping of Ring Domains

Let G be a doubly connected domain in $\mathbf{C}$ such that each component of the complement of G contains at least two points. It is well known that G can be conformally mapped onto a ring

$$D(r_1, r_2) = \{z \,|\, r_1 \leq |z| \leq r_2\}, \quad 0 < r_1 < r_2.$$

The ratio r_2/r_1 of the radii of the boundary circles of such rings is a conformal invariant and it is called the *modulus* of the ring domain G. In this section we show that the potential belonging to the equilibrium distribution of the energy problem for signed measures on the complementary components of G is closely connected to these conformal mappings.

Let $\mathbf{C} \setminus G$ have connected components Σ_1 and Σ_2. Consider the unweighted ($w \equiv 1$) energy problem on $\partial\Sigma = \partial\Sigma_1 \cup \partial\Sigma_2$ in the case $\varepsilon_1 = 1$, $\varepsilon_2 = -1$, $m_1 = m_2 = 1$, and let μ^* be the corresponding extremal measure. Recall from (3.9) that then the "signed" capacity of the pair $\partial\Sigma_1, \partial\Sigma_2$ is

$$c(1, \partial\Sigma_1, \partial\Sigma_2) = \exp(-(F_1 + F_2)),$$

where $F_1 + F_2$ is the minimal energy. To get a well defined problem we have to assume that $\partial\Sigma$ is compact (recall that $w = \exp(-\varepsilon Q)$ must be admissible on the set where the energy problem is considered), i.e. that either G contains the point infinity, or is bounded. This can always be achieved by a Möbius transformation.

Theorem 6.1. *With the above notation and assumptions,*

$$\Phi(z) := \exp\left(\int \log(z - t)\, d\mu^*(t)\right)$$

is a single-valued analytic function on G and it maps G conformally onto a ring $D(r_1, r_2)$ with boundary circles C_{r_1} and C_{r_2}. The reciprocal of the "signed" capacity $1/c(1, \partial\Sigma_1, \partial\Sigma_2) = \exp(F_1 + F_2)$ is equal to r_2/r_1, i.e. to the modulus of the ring domain G. Furthermore, if $\partial\Sigma_1$ and $\partial\Sigma_2$ are piecewise C^1 Jordan curves, then μ_1^* and μ_2^* are the inverse images of the normalized arc measures on C_{r_1} and C_{r_2}, respectively, under the mapping Φ.

In the latter statement we used that Φ can be continuously extended to the boundary of G.

Proof of Theorem 6.1. First we assume that $\partial\Sigma_1$ and $\partial\Sigma_2$ are three times continuously differentiable simple Jordan curves and φ is a conformal mapping of G onto a $\hat{G}$ with boundary components $\hat{\Sigma}_1$ and $\hat{\Sigma}_2$, where we assume that $\partial\hat{\Sigma}$ is compact and is likewise three times continuously differentiable. In this case φ can be extended to a continuously differentiable function on $\overline{G}$. Our first aim is then to prove that the minimal energy with respect to the pair $\partial\Sigma_1, \partial\Sigma_2$ is the same as with respect to $\partial\hat{\Sigma}_1, \partial\hat{\Sigma}_2$.

Let ν be the image measure of μ^* under the mapping φ, i.e. for any Borel set E we set $\nu(E) = \mu^*(\varphi^{-1}(E))$. Then the logarithmic energy of ν can be written as

$$\iint \log\frac{1}{|\zeta-\tau|}d\nu(\zeta)d\nu(\tau) = \iint \log\frac{1}{|\varphi(z)-\varphi(t)|}d\mu^*(z)d\mu^*(t)$$
$$= \iint \log\frac{|z-t|}{|\varphi(z)-\varphi(t)|}d\mu^*(z)d\mu^*(t) + \iint \log\frac{1}{|z-t|}d\mu^*(z)d\mu^*(t).$$

The last expression is the minimal energy with respect to $\partial\Sigma_1, \partial\Sigma_2$, and if we can show that the first term on the right is zero, then we can conclude that under conformal mapping the minimal energy does not increase. But then applying this to the inverse mapping, we get that it is actually conformal invariant, and this is what we wanted to establish in this first part of the proof.

Thus, it is left to show that

$$\iint \log\frac{|\varphi(z)-\varphi(t)|}{|z-t|}d\mu^*(z)d\mu^*(t) = 0. \tag{6.1}$$

Let $\gamma \subset G$ be any simple, piecewise continuously differentiable oriented closed Jordan curve. Define the index of a point z with respect to γ as

$$\mathrm{ind}_\gamma(z) := \frac{1}{2\pi i}\oint_\gamma \frac{1}{\xi - z}d\xi,$$

where the integration is taken in the direction of γ. We claim that

$$\tau_{\varphi,\gamma}(z) := \mathrm{ind}_{\varphi(\gamma)}(\varphi(z)) - \mathrm{ind}_\gamma(z),$$

where the orientation of $\varphi(\gamma)$ is the one induced by φ and γ, is constant on $\overline{G}\setminus\gamma$ (note that the index is not a conformal invariant, so $\tau_{\varphi,\gamma}$ need not be zero). In fact, it is clear that $\tau_{\varphi,\gamma}$ is constant on each (of the two) components of $\overline{G}\setminus\gamma$; furthermore $\mathrm{ind}_\gamma(z)$ (or $\mathrm{ind}_{\varphi(\gamma)}\varphi(z)$) is zero on the unbounded component of $\mathbf{C}\setminus\gamma$ ($\mathbf{C}\setminus\varphi(\gamma)$) and it is ± 1 on the bounded component of $\mathbf{C}\setminus\gamma$ ($\mathbf{C}\setminus\varphi(\gamma)$) depending on whether that bounded component is on the left- or right-hand side of γ ($\varphi(\gamma)$) viewed from the direction of γ ($\varphi(\gamma)$). Thus, we only have to show that $\tau_{\varphi,\gamma}$ does not change if we cross γ along a small segment $\overrightarrow{ab}$. We can assume that γ intersects $\overrightarrow{ab}$ in exactly one point $P = \gamma(t_0)$ and that γ and $\overrightarrow{ab}$ are perpendicular at P. If $\arg(\gamma'(t_0)/(b-a)) = \pi/2$, i.e. if at P the direction of γ is obtained from

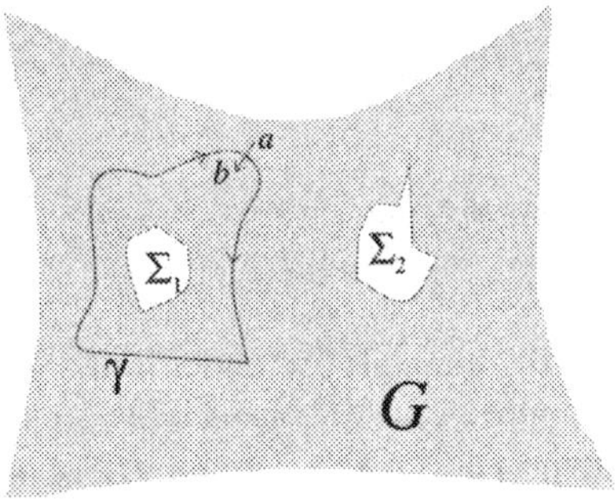

Fig. 6.1

the direction of $\overrightarrow{ab}$ by 90° rotation in the counterclockwise direction, then $\operatorname{ind}_\gamma(z)$ decreases by 1 if we cross γ along $\overrightarrow{ab}$ (see Figure 6.1).

However, because of the conformality of φ, in this case the direction of $\varphi(\gamma)$ at $\varphi(P)$ is obtained from that of $\varphi(\overrightarrow{ab})$ again by 90° rotation in the counterclockwise direction, so $\operatorname{ind}_{\varphi(\gamma)}(z)$ also decreases by 1 as we move along $\varphi(\overrightarrow{ab})$. Thus, $\tau_{\varphi,\gamma}$ does not change. The other case when $\arg(\gamma'(t_0)/(b-a)) = -\pi/2$ is similar.

The fact that $\tau_{\varphi,\gamma}$ is constant implies that if $a \in \partial\Sigma_1$ and $b \in \partial\Sigma_2$, then the total change of the argument of

$$\frac{\varphi(z)-\varphi(a)}{z-a} \Big/ \frac{\varphi(z)-\varphi(b)}{z-b}$$

is zero along every closed broken line in G; hence

$$\log\left(\frac{\varphi(z)-\varphi(a)}{z-a} \Big/ \frac{\varphi(z)-\varphi(b)}{z-b}\right)$$

is a single-valued analytic function on G and it is uniformly equicontinuous on $\overline{G}$ in $a \in \Sigma_1$ and $b \in \Sigma_2$ (recall that φ is continuously differentiable up to the boundary). By integration with respect to $d\mu_1^*(a)$ and $d\mu_2^*(b)$ we can see that

$$\begin{aligned} H(z) &:= \int \log\left(\frac{\varphi(z)-\varphi(t)}{z-t}\right) d\mu^*(t) \qquad (6.2)\\ &= \int_{\Sigma_1}\int_{\Sigma_2}\left(\log\frac{\varphi(z)-\varphi(a)}{z-a} - \log\frac{\varphi(z)-\varphi(b)}{z-b}\right) d\mu_1^*(a)d\mu_2^*(b) \end{aligned}$$

is a single-valued analytic function on G and it is continuous on $\overline{G}$ with real part

$$h(z) = \int \log\left|\frac{\varphi(z)-\varphi(t)}{z-t}\right| d\mu^*(t). \qquad (6.3)$$

From Cauchy's formula we get

$$H(z) = \frac{1}{2\pi i}\oint_{\partial\Sigma_1} \frac{H(\xi)}{\xi - z}d\xi + \frac{1}{2\pi i}\oint_{\partial\Sigma_2} \frac{H(\xi)}{\xi - z}d\xi$$

if G is bounded and

$$H(z) = \frac{1}{2\pi i}\oint_{\partial\Sigma_1} \frac{H(\xi)}{\xi - z}d\xi + \frac{1}{2\pi i}\oint_{\partial\Sigma_2} \frac{H(\xi)}{\xi - z}d\xi + H(\infty)$$

if G is unbounded, where the integrals are taken in the direction so that G stays on the left-hand side. In either case we get a representation $H(z) = H_1(z) + H_2(z)$, where H_1 is holomorphic in $\overline{\mathbf{C}} \setminus \Sigma_1$ and H_2 is holomorphic in $\overline{\mathbf{C}} \setminus \Sigma_2$. Furthermore, these H_j's, which are given by the integrals on the right sides of the preceding two formulae, are continuous on $\overline{\mathbf{C} \setminus \Sigma_j}$. Taking real parts, we arrive at the representation

$$h(z) = h_1(z) + h_2(z), \tag{6.4}$$

where h_1 and h_2 are continuous on $\overline{\mathbf{C} \setminus \Sigma_1}$ and $\overline{\mathbf{C} \setminus \Sigma_2}$ and harmonic in $\mathbf{C} \setminus \Sigma_1$ and $\mathbf{C} \setminus \Sigma_2$, respectively.

We shall now make use of the fact that each of μ_1^* and μ_2^* is the balayage of the other one onto $\partial\Sigma_1$ and $\partial\Sigma_2$, respectively (see Theorem 2.6); hence (see Theorem II.4.7(c))

$$\int_{\partial\Sigma_2} h_1\,d\mu_2^* = \int_{\partial\Sigma_1} h_1\,d\mu_1^*, \quad \int_{\partial\Sigma_1} h_2\,d\mu_1^* = \int_{\partial\Sigma_2} h_2\,d\mu_2^*.$$

Thus,

$$\int_{\partial\Sigma} h\,d\mu^* = 0$$

(cf. (6.3) and (6.4)), and this is exactly (6.1), so our claim concerning the conformal invariance of the minimal energy, and hence that of the "signed" capacity of the boundary components, has been established. The proof above also shows that ν must be the equilibrium distribution for the pair $\hat{\Sigma}_1, \hat{\Sigma}_2$, by which the last statement in the theorem is also established (at least for the case when the boundary is three times continuously differentiable).

Now let φ map G onto a $D(r_1, r_2)$ conformally. By Example 4.2, for the pair C_{r_1}, C_{r_2} the difference of the two normalized arc measures on C_{r_1} and C_{r_2} serves as the extremal measure, and the minimal energy is $\log(r_2/r_1)$. The minimal energy for the original problem (i.e. for the pair $\partial\Sigma_1, \partial\Sigma_2$) is (see (1.14)) $F_1 + F_2$, where F_1 and $-F_2$ are the values of the potential U^{μ^*} on $\partial\Sigma_1$ and $\partial\Sigma_2$, respectively (see also Theorem 2.6, which implies that $\partial\Sigma$ is the support of μ^*, and also Theorems 2.1 and I.4.8 which give that U^{μ^*} must be constants on $\partial\Sigma_1$ and $\partial\Sigma_2$). Thus $F_1 + F_2 = \log(r_2/r_1)$.

With

$$\psi(z) = \int \log(z - t)\,d\mu^*(t)$$

it is clear that $\exp(\psi(z))$ is single-valued and analytic on G; furthermore it follows from the just established relation $\log(r_2/r_1) = F_1 + F_2$ that one of the functions

$|\varphi(z)|\exp(\pm\mathrm{Re}\psi(z))$ is constant on the boundary of G for one of the $\pm$ signs. Thus, $\varphi(z)/\exp(\pm\psi(z))$ must also be constant, and this proves the theorem under the assumption that the boundary curves of G are three times differentiable.

We can get eliminate this smoothness condition in a standard way. Let $\{G^{(k)}\}$ be an increasing sequence of doubly connected domains for which the boundaries of the connected components $\Sigma_1^{(k)}$ and $\Sigma_2^{(k)}$ of the corresponding complements are three times continuously differentiable simple closed Jordan curves and the Hausdorff distance between the sets $\partial\Sigma_j^{(k)}$ and $\partial\Sigma_j$ is smaller than $1/k$. It is easy to see that then the minimal (signed) energy for the pair $\partial\Sigma_1^{(k)}, \partial\Sigma_2^{(k)}$ is not bigger than the minimal energy corresponding to the pair $\partial\Sigma_1, \partial\Sigma_2$. In fact, by applying a Möbius transformation we can assume that G contains the point infinity. But then it follows from Theorem 2.6 that the solution of the signed energy problem for the pairs $\Sigma_1^{(k)}, \Sigma_2^{(k)}$ (or Σ_1, Σ_2) is the same as that of for the pairs $\partial\Sigma_1^{(k)}, \partial\Sigma_2^{(k)}$. $(\partial\Sigma_1, \partial\Sigma_2)$, and since $\Sigma_j \subset \Sigma_j^{(k)}$, the monotonicity of the minimal energy follows from its definition. Thus, if $\mu^{(k)*}$ denotes the corresponding extremal measures and μ is a weak* limit point of them, then it follows from the principle of descent (see also the reasoning given in the proof of Theorem 1.4(b)) that μ must be the extremal measure for the pair $\partial\Sigma_1, \partial\Sigma_2$, i.e. the whole sequence must converge to μ^* in the weak* topology. If we normalize the conformal mappings of $G^{(k)}$ onto a ring $D(1, r^{(k)})$ suitably (prescribe for example, that the image of a given boundary point be 1, cf. [151, Theorem III.1.1]), then they will converge to a conformal mapping of G as $k \to \infty$, and the statements of the theorem in the general case then follow from the smooth boundary case verified above. □

If we combine the results proved in Theorems 3.5 and 3.7 with those of Theorem 6.1, then we get the following numerical procedure for finding a conformal mapping of a ring domain onto a ring. This procedure is numerically stable and can be used regardless of smoothness assumptions on the boundary of the sets involved, but for the same reason, we cannot expect it to converge very rapidly even if the boundary is smooth.

Theorem 6.2. *Let G be as above with connected and bounded boundary components $\partial\Sigma_1, \partial\Sigma_2$. Starting from any points $a_0 \in \partial\Sigma_1$ and $b_0 \in \partial\Sigma_2$ we successively define the points $a_n \in \partial\Sigma_1$ and $b_n \in \partial\Sigma_2$ as points where the expression*

$$\left|\prod_{j=0}^{n-1}\frac{a-a_j}{a-b_j}\right| \Bigg/ \left|\prod_{j=0}^{n-1}\frac{b-a_j}{b-b_j}\right|$$

takes its maximum A_n^n for $a \in \partial\Sigma_1$ and $b \in \partial\Sigma_2$. Then with that branch of the n-th root that is positive for positive values, the expression

$$\left(\prod_{j=0}^{n-1}\frac{z-a_j}{z-b_j}\right)^{1/n}$$

converges to a conformal mapping of G onto a ring $D(r_1, r_2)$ as $n \to \infty$, and the modulus r_2/r_1 of the ring domain G can be determined from

$$\lim_{n\to\infty} A_n = r_1/r_2.$$

Moreover, suppose that for each n the (discrete) sets $S_1^{(n)}$ and $S_2^{(n)}$ are subsets of $\partial\Sigma_1$ and $\partial\Sigma_2$ in such a way that the distance of any point of $\partial\Sigma_j$ from the corresponding set $S_j^{(n)}$ is at most ε_n. Starting from two points $\hat{a}_0 \in \Sigma_1$ and $\hat{b}_0 \in \Sigma_2$ define successively $\hat{a}_n \in S_1^{(n)}$ and $\hat{b}_n \in S_2^{(n)}$ where the function

$$\left|\prod_{j=0}^{n-1} \frac{a-\hat{a}_j}{a-\hat{b}_j}\right| \Bigg/ \left|\prod_{j=0}^{n-1} \frac{b-\hat{a}_j}{b-\hat{b}_j}\right|$$

attains its maximum $\hat{A}_n^n$ for $a \in S_1^{(n)}$ and $b \in S_2^{(n)}$. Assume further, that for ε_n we have

$$\lim_{n\to\infty} \varepsilon_n^{1/n} = 0.$$

Then the function

$$\left(\prod_{j=0}^{n-1} \frac{z-\hat{a}_j}{z-\hat{b}_j}\right)^{1/n}$$

converges to a conformal mapping of G onto a ring $D(r_1, r_2)$ as $n \to \infty$, and the modulus r_2/r_1 of the ring domain G can be determined from

$$\lim_{n\to\infty} \hat{A}_n = r_1/r_2.$$

In many cases one can relax the condition concerning the sequence $\{\varepsilon_n\}$; see the discussion at the end of Sections V.1 and VIII.3.

VIII.7 A Discrepancy Theorem for Simple Zeros of Polynomials

We have seen in Section III.4, Theorem III.4.2, that if $\{P_n\}$ is a sequence of monic polynomials of degree n with the property

$$\lim_{n\to\infty} \|P_n\|_{[-1,1]}^{1/n} = \frac{1}{2},$$

then the zeros of the P_n's are distributed according to the arcsine distribution

$$d\omega(x) := \frac{1}{\pi\sqrt{1-x^2}}\, dx, \qquad x \in [-1, 1].$$

In particular, if all the zeros of the P_n's are real, then

$$\lim_{n\to\infty} |(\nu_{P_n} - \omega)([a, b])| = 0 \tag{7.1}$$

uniformly for $[a, b] \subseteq [-1, 1]$, where ν_{P_n} denotes the normalized zero counting measure of P_n. The supremum of the expression on the left-hand side for all intervals $[a, b] \subseteq [-1, 1]$ is called the *discrepancy* of the zeros of P_n.

Hereafter we shall assume that all the zeros $x_{i,n}$ of the P_n's are real and lie in $[-1, 1]$.

In [40] P. Erdős and P. Turán gave a quantitative version of the convergence in (7.1) in the form

$$|(\nu_{P_n} - \omega)([a, b])| \leq \frac{8}{\log 3}\sqrt{\frac{\log A_n}{n}}$$

for any interval $[a, b] \subseteq [-1, 1]$, which holds whenever the sup norm of P_n satisfies

$$\|P_n\|_{[-1,1]} \leq A_n \frac{1}{2^n}. \tag{7.2}$$

This result is sharp except perhaps for the constant $8/\log 3$.

This basic estimate has been applied to obtain various discrepancy theorems. However, in some applications we have more *a priori* information on the polynomial, namely a lower bound for the absolute value of the derivative at the zeros (as a typical situation consider the distribution of Fekete points, at which the derivative of the corresponding Fekete polynomials is bounded from below by $\text{cap}([-1, 1])^{n-1}$ because of the extremal property of these points). In such situations we can considerably improve the aforementioned estimate.

Thus, let us assume that

$$|P_n'(x_{i,n})| \geq \frac{1}{B_n}\frac{1}{2^n}, \qquad 1 \leq i \leq n. \tag{7.3}$$

Then the following holds.

Theorem 7.1. *Let P_n be monic polynomials with zeros in $[-1, 1]$ satisfying the conditions (7.2) and (7.3). Then there exists a constant C (independent of n) such that*

$$|(\nu_{P_n} - \omega)([a, b])| \leq C\frac{\log C_n}{n}\log\frac{n}{\log C_n} \tag{7.4}$$

for any interval $[a, b] \subset [-1, 1]$, where

$$C_n := \max(A_n, B_n, n).$$

This result is best possible in the following sense.

Theorem 7.2. *Let $\{C_n\}$ be an arbitrary sequence with the property that $n \leq C_n \leq e^{n/2}$. Then there are monic polynomials P_n of corresponding degrees $n = 1, 2, \ldots$ such that*

$$\|P_n\|_{[-1,1]} \leq C_n\frac{1}{2^n}; \tag{7.5}$$

for every zero $x_{i,n}$ of P_n

$$|P_n'(x_{i,n})| \geq \frac{1}{C_n}\frac{1}{2^n}, \quad 1 \leq i \leq n; \tag{7.6}$$

and such that for some intervals $[a_n, b_n]$ *of* $[-1, 1]$ *the estimate*

$$|(\nu_{P_n} - \omega)([a_n, b_n])| \geq c\frac{\log C_n}{n}\log\frac{n}{\log C_n} \tag{7.7}$$

holds with some positive c *independent of* n.

Proof of Theorem 7.1. Let $g(z) = \log|z + \sqrt{z^2-1}|$ be the Green function of $\overline{\mathbf{C}} \setminus [-1, 1]$ with pole at infinity (where we take that branch of $\sqrt{z}$ that is positive for positive z).

Inequality (7.2) and the principle of domination (see also the representation of the Green function in Section II.4 via the equilibrium potential) yield

$$\frac{1}{n}\log|P_n(z)| - g(z) - \log\frac{1}{2} \leq \frac{\log A_n}{n} \quad \text{for all} \quad z \in \mathbf{C}. \tag{7.8}$$

We also need a matching lower estimate on the left-hand side. Lagrange's interpolation formula shows that

$$1 = \sum_{i=1}^{n} \frac{P_n(z)}{P_n'(x_{i,n})(z - x_{i,n})}.$$

For $z \notin [-1, 1]$, let $d(z)$ denote the distance from the point z to the interval $[-1, 1]$. Then the preceding equality and (7.3) yield

$$1 \leq n\frac{|P_n(z)|}{d(z)}B_n 2^n,$$

i.e.

$$|P_n(z)| \geq \frac{1}{n}\frac{d(z)}{B_n}\frac{1}{2^n}.$$

Let $\Gamma_\kappa = \{z \in \mathbf{C} \mid g(z) = \log\kappa\}$, $\kappa > 1$, be a level curve of the Green function $g(z)$. Then Γ_κ is an ellipse with foci at ± 1 and major axis $\kappa + \frac{1}{\kappa}$. Hence,

$$\inf_{z \in \Gamma_\kappa} d(z) = \frac{1}{2}\left(\kappa + \frac{1}{\kappa}\right) - 1.$$

Choosing

$$\kappa = \kappa_n := 1 + n^{-12}$$

in the last inequality leads to

$$\frac{1}{n}\log|P_n(z)| - g(z) - \log\frac{1}{2} \geq -d\frac{\log C_n}{n} \tag{7.9}$$

for $z \in \Gamma_{\kappa_n}$, where $d > 0$ is an absolute constant independent of n. The minimum principle for harmonic functions shows that (7.9) is actually satisfied for all z with $g(z) \geq \log\kappa_n$.

Inequalities (7.8) and (7.9) together show that

$$\left|\frac{1}{n}\log|P_n(z)| - g(z) - \log\frac{1}{2}\right| \le D\frac{\log C_n}{n} \tag{7.10}$$

for all z where $g(z) \ge \log \kappa_n$.

Since $-\frac{1}{n}\log|P_n(z)|$ is the logarithmic potential $U^{\nu_{P_n}}$ of the measure ν_{P_n}, and $-g(z) - \log\frac{1}{2}$ is the logarithmic potential $U^{\omega}(z)$ of the arcsine distribution ω, (7.10) can be written as

$$|U^{\nu_{P_n}}(z) - U^{\omega}(z)| \le D\frac{\log C_n}{n}$$

for all z with $g(z) \ge \log \kappa_n$.

Now Theorem 7.1 follows from the last estimate and from the next theorem if we set $\sigma = \nu_{P_n} - \omega$, $\varepsilon = (\log C_n)/n$, and $\Delta = 1/2$ in it. □

Theorem 7.3. *Let $\sigma = \sigma_+ - \sigma_-$ be a signed measure such that $\sigma_\pm$ are probability measures on $[-1, 1]$ with the property that for some $0 < \Delta \le 1$ the estimate*

$$\sigma_-(E) \le C_0 m(E)^{\Delta} \tag{7.11}$$

holds for every interval E, where m denotes the linear Lebesgue measure. Suppose that

$$|U^{\sigma}(z)| \le C_1\varepsilon$$

for every z with

$$\mathrm{dist}(z, [-1, 1]) \ge \varepsilon^L,$$

where $L := 5/\Delta + 2$. Then

$$|\sigma([a, b])| \le C_2\varepsilon\log\frac{1}{\varepsilon}$$

holds for every interval $[a, b]$, where the constant C_2 depends exclusively on C_0, C_1, and Δ.

We continue with the

Proof of Theorem 7.3. The idea of the proof can be explained as follows. Let $[a, b] \subseteq [-1, 1]$ be arbitrary. Denote by $\chi_{[a,b]}$ the characteristic function of the interval $[a, b]$:

$$\chi_{[a,b]} := \begin{cases} 1 & \text{if } x \in [a, b] \\ 0 & \text{if } x \notin [a, b]. \end{cases}$$

Suppose we had a signed measure μ of compact support lying at a distance $\ge \varepsilon^L$ from $[-1, 1]$ such that $\|\mu\| \le 2$ and for some constant c,

$$U^{\mu}(x) = c + \tau_\varepsilon\chi_{[a,b]}, \quad \tau_\varepsilon := 1/(\log 1/\varepsilon),$$

for all $x \in [-1, 1]$. Then, using Fubini's theorem, we could write

$$2C_1\varepsilon \geq \left|\int U^\sigma d\mu\right| = \left|\int U^\mu d\sigma\right| = \left|\int_a^b \tau_\varepsilon d\sigma\right| = \tau_\varepsilon|\sigma([a,b])|,$$

from which

$$|\sigma([a,b])| \leq 2C_1\varepsilon \log\frac{1}{\varepsilon}$$

follows immediately, and this is what we need to prove.

Unfortunately, a signed measure μ with the above properties does not exist. We can, however, find a measure having properties close to the ones above; hence this measure can serve as a substitute. The rest of the proof is devoted to the construction of that measure and to showing that the weaker properties it possesses are nonetheless sufficient for our purposes.

More precisely, we will construct a μ with the following properties.

Lemma 7.4. *Let L and Δ be the numbers from Theorem 7.3, and let $[a,b] \subseteq [-1,1]$, and $0 < \varepsilon < 1/2$ be arbitrary with $b - a \geq 2\varepsilon^{1/\Delta}$. Then there is a signed measure $\mu = \mu_{\varepsilon,a,b}$ and two numbers $c = c_{\varepsilon,a,b}$ and $\tau = \tau_{\varepsilon,a,b}$ with the following properties:*

1) $\operatorname{supp}(\mu)$ *is compact and is of distance $\geq \varepsilon^L$ from $[-1,1]$;*
2) $\|\mu\| \leq 2$;
3) $c \leq U^\mu(x) \leq c + \tau$ *for every $x \in [-1,1]$;*
4) *for $x \in [a,b]$ and $x \in [-1,1] \setminus (a - \varepsilon^{2/\Delta}, b + \varepsilon^{2/\Delta})$, there holds*
$$\left|U^\mu(x) - \tau\chi_{[a,b]}(x) - c\right| \leq C_3\varepsilon;$$
5) $1/C_3 \leq \tau\log(1/\varepsilon) \leq C_3$.

Furthermore, here C_3 is an absolute constant.

The proof of Lemma 7.4 will follow from the results of the present chapter, but before we set out to prove it we show how Theorem 7.3 can be obtained from it. So let us return to the proof of Theorem 7.3.

First we simplify the problem; namely, it is enough to prove the inequality

$$\sigma([a,b]) \leq C_2\varepsilon\log\frac{1}{\varepsilon} \tag{7.12}$$

with some constant C_2 for all $[a,b] \subseteq [-1,1]$. In fact, then by applying (7.12) to the intervals $[-1,a]$ and $[b,1]$ instead of $[a,b]$ and using that $\sigma([-1,1]) = 0$, we obtain the counterpart

$$\sigma([a,b]) \geq -2C_2\varepsilon\log\frac{1}{\varepsilon}$$

of (7.12), and with (7.12) this proves the claim.

Next we observe that we may assume without loss of generality that $b - a \geq 2\varepsilon^{1/\Delta}$. In fact, suppose (7.12) has been verified in this case. Then if a and b are closer than $2\varepsilon^{1/\Delta}$, we can enlarge $[a,b]$ to have length $2\varepsilon^{1/\Delta}$. If the enlarged interval is $[a',b']$, then we can apply (7.12) to $[a',b']$ instead of $[a,b]$ to get

$$\sigma_+([a,b]) \le \sigma_+([a',b']) \le \sigma_-([a',b']) + C_2\varepsilon \log\frac{1}{\varepsilon} \le 2C_0\varepsilon + C_2\varepsilon\log\frac{1}{\varepsilon},$$

where, in the last step we applied (7.11). This proves (7.12) for all $[a,b]$ (with a possibly bigger constant).

Now we can apply Lemma 7.4. With the signed measure μ obtained there and with $\delta := \varepsilon^{2/\Delta}$ we get exactly as in the sketch above

$$\begin{aligned} 2C_1\varepsilon &\ge \int U^\sigma d\mu = \int U^\mu d\sigma \\ &= \int (U^\mu - c)\,d\sigma = \int_{[a,b]} + \int_{[-1,1]\setminus[a-\delta,b+\delta]} + \int_{(a-\delta,a)\cup(b,b+\delta)}, \end{aligned}$$

where the domain of the last integral has to be appropriately adjusted if $a-\delta < -1$ or $b+\delta > 1$. Using properties of U^μ we can continue this inequality as

$$2C_1\varepsilon \ge \int_{[a,b]} \tau d\sigma - \int_{(a-\delta,a)\cup(b,b+\delta)} \tau d\sigma_- - 2C_3\varepsilon \ge \tau\sigma[a,b] - 2\tau C_0\delta^\Delta - 2C_3\varepsilon,$$

where we have used again (7.11). Since $\delta^\Delta = \varepsilon^2$, and by property (5) of the measure μ

$$\tau \sim \frac{1}{\log 1/\varepsilon},$$

where $\sim$ means that the ratio of the two sides is bounded away from 0 and ∞ by two absolute constants, we immediately arrive at (7.12) from this estimate. □

The preceding proof was based on the validity of Lemma 7.4, which we now prove.

Proof of Lemma 7.4. We shall appeal to Theorem 5.2; more precisely the following corollary of that result.

Corollary 7.5. *In Theorem 5.2 let $\Sigma_1 = [-\alpha,\alpha]$ and $\Sigma_2 = [-2-\alpha, -\alpha-\eta] \cup [\alpha+\eta, 2+\alpha]$ with some $0 < \eta \le \alpha^2 \le 1/4$. Then*

$$F_1 + F_2 \sim 1/\log\frac{1}{\eta}. \tag{7.13}$$

Furthermore, the signed measure μ^* is absolutely continuous with respect to Lebesgue measure, and if we set

$$d\mu^*(t) = v(t)dt,$$

then for $j = 1,2,3$ and $t \in [a_j, b_j]$

$$|v(t)| \le \frac{1}{\eta}v_j(t) := \frac{1}{\eta}\frac{1}{\pi}\frac{1}{\sqrt{(t-a_j)(b_j-t)}}. \tag{7.14}$$

Recall that here $[a_j, b_j]$ denote the intervals of $\Sigma = \Sigma_1 \cup \Sigma_2$.
First let us show how to get this from Theorem 5.2.

Proof of Corollary 7.5. According to Theorem 5.2 we have to solve the system of equations

$$\left(\int_{-\alpha-\eta}^{-\alpha} + \int_{\alpha}^{\alpha+\eta}\right) \frac{c_1 t + c_0}{\sqrt{(t^2-(2+\alpha)^2)(t^2-(\alpha+\eta)^2)(t^2-\alpha^2)}}\, dt = 0 \qquad (7.15)$$

$$\int_{-\alpha}^{\alpha} \frac{1}{\pi i} \frac{c_1 t + c_0}{\sqrt{(t^2-(2+\alpha)^2)(t^2-(\alpha+\eta)^2)(t^2-\alpha^2)}}\, dt = 1 \qquad (7.16)$$

(we have incorporated the " $-$ " sign from $-i\pi$ in (7.16) into c_1 and c_0 in order to get positive c_0 below). Since the denominator in (7.15) takes opposite signs on $[-\alpha-\eta, -\alpha]$ and $[\alpha, \alpha+\eta]$, we get that c_1 must be zero. Then c_0 is obtained from the second equation:

$$c_0 = 1 \Big/ \int_{-\alpha}^{\alpha} \frac{1}{\pi} \frac{1}{\sqrt{((2+\alpha)^2-t^2)((\alpha+\eta)^2-t^2)(\alpha^2-t^2)}}\, dt.$$

This easily yields

$$c_0 \sim \alpha/\log\frac{\alpha}{\eta} \sim \alpha/\log\frac{1}{\eta}.$$

But

$$\begin{aligned} F_1 + F_2 &= \left| \int_{\alpha}^{\alpha+\eta} \frac{c_1 t + c_0}{\sqrt{(t^2-(2+\alpha)^2)(t^2-(\alpha+\eta)^2)(t^2-\alpha^2)}}\, dt \right| \\ &\sim \frac{\alpha}{\log 1/\eta} \int_{\alpha}^{\alpha+\eta} \frac{1}{\sqrt{((2+\alpha)^2-t^2)((\alpha+\eta)^2-t^2)(t^2-\alpha^2)}}\, dt, \end{aligned}$$

and if we use that

$$\int_{\alpha}^{\alpha+\eta} \frac{1}{\pi} \frac{1}{\sqrt{(t-\alpha)(\alpha+\eta-t)}}\, dt = 1,$$

we get (7.13).

Since

$$v(t) = \frac{1}{\pi i} \frac{c_0}{\sqrt{(t^2-(2+\alpha)^2)(t^2-(\alpha+\eta)^2)(t^2-\alpha^2)}}$$

if $t \in \Sigma = [-2-\alpha, -\alpha-\eta] \cup [\alpha+\eta, \alpha+2] \cup [-\alpha, \alpha]$, while

$$v_j(t) = \frac{1}{\pi} \frac{1}{\sqrt{(t-a_j)(b_j-t)}},$$

(7.14) also easily follows. □

Lemma 7.6. *With the assumptions and notations of Corollary 7.5 we have for the potential of the extremal measure μ^* the estimate*

$$\left|U^{\mu^*}(x) - U^{\mu^*}(x \pm i\xi)\right| \leq \frac{6}{\eta}\left(\frac{\xi}{\alpha}\right)^{1/2} \tag{7.17}$$

for every $x \in \mathbf{R}$ and $0 < \xi \leq \alpha/2$.

Proof. Using the second part of Corollary 7.5 we can write

$$\begin{aligned} \left|U^{\mu^*}(x) - U^{\mu^*}(x \pm i\xi)\right| &= \left|\int \log\left|\frac{x-t\pm i\xi}{x-t}\right| d\mu^*(t)\right| \\ &\leq \int \log\left|\frac{x-t\pm i\xi}{x-t}\right| d|\mu^*|(t) \\ &\leq \frac{1}{\eta}\sum_{j=1}^{3}\int_{a_j}^{b_j} \log\left|\frac{x-t\pm i\xi}{x-t}\right| v_j(t)dt \\ &= \frac{1}{\eta}\sum_{j=1}^{3} |U^{v_j}(x) - U^{v_j}(x \pm i\xi)|, \end{aligned}$$

where we have used the self explanatory notation for the potential of a measure given by its density function. But with

$$v(t) = \frac{1}{\pi}\frac{1}{\sqrt{1-t^2}},$$

we have

$$U^{v_j}(z) = U^{v}(y) + \log\frac{2}{b_j - a_j},$$

where z and y are connected by the formula

$$y = \left(z - \frac{a_j + b_j}{2}\right)\frac{2}{b_j - a_j}.$$

Hence the last sum is at most as large as

$$\frac{3}{\eta}\max_{y\in\mathbf{R}}\left|U^{v}(y) - U^{v}\left(y \pm 2i\frac{\xi}{b_j - a_j}\right)\right|. \tag{7.18}$$

Here

$$\left|\frac{2\xi}{b_j - a_j}\right| \leq \frac{2\xi}{2\alpha} \leq \frac{1}{2} \tag{7.19}$$

and

$$U^{v}(z) = -\log\left|z + \sqrt{z^2 - 1}\right| + \log 2.$$

One can easily prove that for fixed real ζ, $|\zeta| \le 1/2$ the function

$$|U^v(y) - U^v(y \pm i\zeta)|$$

attains its maximum at $y = \pm 1$ and this maximum is at most $2\sqrt{|\zeta|}$.

Substituting this into (7.18) we arrive at (7.17) (cf. also (7.19)). □

After these preparations we return to the proof of Lemma 7.4. Suppose first that $[a, b]$ is symmetric with respect to the origin, say $[a, b] = [-\alpha, \alpha]$. Then we set $\eta = \varepsilon^{2/\Delta}$ and choose μ to be equal to the translation of the measure μ^* from the previous two lemmas by $i\varepsilon^L$. With $c = -F_2$ and $\tau = F_1 + F_2 = F_1 - (-F_2)$ the first two properties in Lemma 7.4 follow from the construction, the third one follows from the fact that for every z the potential U^{μ^*} lies between F_1 and $-F_2$ (see Theorem 2.6). Property 4 is a consequence of the properties of U^{μ^*} (see Theorem 2.6, which yields that U^{μ^*} equals F_1 and $-F_2$ on Σ_1 and Σ_2, respectively) and Lemma 7.6 if we also use that by the choice of the parameters we have $\alpha \ge \varepsilon^{1/\Delta}$, and so

$$\frac{1}{\eta}\left(\frac{\varepsilon^L}{\alpha}\right)^{1/2} \le \varepsilon.$$

Note that property 4 actually holds in a wider range, namely for all

$$x \in \Sigma = [-2-\alpha, -\alpha-\eta] \cup [\alpha+\eta, \alpha+2] \cup [-\alpha, \alpha]. \tag{7.20}$$

Finally, the last property was proved in (7.13). These facts prove Lemma 7.4 in the symmetric case.

If $[a, b] \subseteq [-1, 1]$ is arbitrary, then let $[a', b'] = [-(b-a)/2, (b-a)/2]$, and let the just constructed signed measure for $[a', b']$ be μ'. Now we choose μ as the translation of the measure μ' by $(a+b)/2$. Since we have verified property 4 in the larger range (7.20), the translation of which (by $(a+b)/2$) certainly covers the interval $[-1, 1]$, the signed measure μ satisfies all the requirements. □

Finally, we prove Theorem 7.2.

Proof of Theorem 7.2. The proof is different in the ranges $C_n \ge n^4$ and $C_n < n^4$. We shall separate these two cases below. Of course, the sequence $\{C_n\}$ need not satisfy either $C_n \ge n^4$ or $C_n < n^4$ for all n, in which case one has to separate the terms with these two properties, respectively, and apply the two methods below to the appropriate terms.

Proof of Theorem 7.2 in the case when $C_n \ge n^4$. First we need a lemma.

Lemma 7.7. *For any $x \in \mathbf{R}$ and $0 < \eta < \theta$*

$$\left|\int_\eta^\theta \log\left|\frac{x+t}{x-t}\right| \frac{1}{t}\, dt\right| \le 10. \tag{7.21}$$

Proof. By the homogeneity of the integral we can assume that $\eta = 1$ and $x \geq 0$. Furthermore, the ratio

$$\left| \frac{x+t}{x-t} \right| \tag{7.22}$$

is increasing as x increases on $(0, 1)$ for every fixed $t \geq 1$; hence we may also assume $x \geq 1$. Now we divide the domain of integration in (7.21) into three parts: $(1, x/2)$, $(x/2, 2x)$, and $(2x, \theta)$ with the obvious modifications if $x \leq 2$ or $x \geq \theta/2$. On the first part we use that (7.22) is at most

$$1 + \frac{2t}{(x-t)} \leq 1 + \frac{4t}{x},$$

and so the integrand is at most $4/x$, from which the contribution of this part to the left side of (7.21) is at most 2. In a similar manner, on $(2x, \theta)$ we have for (7.22) the upper estimate

$$1 + \frac{2x}{(t-x)} \leq 1 + \frac{4x}{t},$$

so the contribution of the third integral is also at most 2.

Finally,

$$\left| \int_{x/2}^{2x} \log \left| \frac{x+t}{x-t} \right| \frac{1}{t}\, dt \right| \leq \frac{2}{x} \int_{x/2}^{2x} \log \left| \frac{x+t}{x-t} \right| \, dt = \frac{2}{x}\frac{x}{2}(2.5 \log 2 + 3 \log 3) < 6.$$

□

With this technical lemma in hand we can now prove Theorem 7.2 in the case when $C_n \geq n^4$.

Consider for $0 < \varepsilon \leq e^{-13}$ the function

$$v_\varepsilon(t) := \begin{cases} (t - (1-\varepsilon))^{-1} & \text{if } \varepsilon^{3/2} \leq |t - (1-\varepsilon)| \leq \varepsilon \\ 0 & \text{otherwise,} \end{cases}$$

and the signed measure ν_ε that it defines:

$$d\nu_\varepsilon(t) = c_\varepsilon v_\varepsilon(t)\, dt,$$

where the normalizing constant c_ε is chosen so that the total variation of ν_ε is 2, i.e.

$$c_\varepsilon = \frac{2}{\log 1/\varepsilon}.$$

For $\varepsilon^{3/2} \leq |t - (1-\varepsilon)| \leq \varepsilon$ we have

$$|\varepsilon c_\varepsilon v_\varepsilon(t)| \leq \frac{2\varepsilon}{\log 1/\varepsilon} \frac{1}{\varepsilon^{3/2}} \leq \frac{13}{\log 1/\varepsilon} \frac{1}{\pi\sqrt{1-t^2}} \leq \frac{1}{\pi\sqrt{1-t^2}};$$

hence the signed measure

$$\mu := \omega + \varepsilon \nu_\varepsilon$$

is a positive measure of total mass 1 with density less than or equal to

$$\frac{2}{\pi\sqrt{1-t^2}} \tag{7.23}$$

on $[-1, 1]$ (recall that ω denotes the arcsine measure). Furthermore, we can immediately get from Lemma 7.7 and the equality $U^\omega(x) = \log 2$ for $x \in [-1, 1]$ that for such x

$$\begin{aligned} |U^\mu(x) - U^\omega(x)| &= |\varepsilon U^{\nu_\varepsilon}(x)| \\ &= \varepsilon c_\varepsilon \left| \int_{\varepsilon^{3/2}}^{\varepsilon} \log\left|\frac{x+t}{x-t}\right| \frac{1}{t}\, dt \right| \le \frac{20\varepsilon}{\log 1/\varepsilon}. \end{aligned} \tag{7.24}$$

Now we shall utilize the discretization technique of Section VI.4. For an odd integer n let

$$-1 = y_{0,n} < y_{1,n} < \ldots < y_{n,n} = 1$$

be that partition of $[-1, 1]$ for which $\mu([y_{j,n}, y_{j+1,n}]) = 1/n$ for all $0 \le j \le n-1$. Consider the polynomials

$$P_n(x) = \prod_{j=1}^{n-1} (x - y_{j,n})$$

of degree $(n-1)$. By Theorem VI.4.1, if for some constants α and β the inequality

$$\int_{|x-t| \le n^{-\alpha}} |\log|x-t||\, d\mu(t) \le \beta \frac{\log\, n}{n} \tag{7.25}$$

holds, then

$$|P_n(x)| \le n^{\alpha+\beta} \exp(-nU^\mu(x)), \qquad x \in \mathbf{R},$$

and

$$|P_n(x)| \ge \frac{1}{4} \exp(-nU^\mu(x))|x - y_{n_x,n}|,$$

where $y_{n_x,n}$ denotes the closest zero of P_n to x.

If the potential U^μ is continuous on $[-1, 1]$, then the latter inequality immediately implies

$$|P_n'(y_{j,n})| \ge \frac{1}{4} \exp(-nU^\mu(y_{j,n}))$$

for every zero $y_{j,n}$ of P_n (actually this is true without the continuity assumption).

In our case the potential U^μ is obviously continuous; furthermore (7.25) holds with $\alpha = 5/2$ and $\beta = 1/2$ for large n (cf. the estimate (7.23) for the density of μ). On applying (7.24) we can thus write

$$|P_n(x)| \le \exp\left(\frac{20\varepsilon}{\log 1/\varepsilon} n + 3\log n\right) \frac{1}{2^n},$$

and for each $j = 1, \ldots, n$

$$|P_n'(y_{j,n})| \geq \exp\left(-\frac{20\varepsilon}{\log 1/\varepsilon}n - 2\right)\frac{1}{2^n}.$$

Now if $C_n \geq n^4$ is given, then we define $\varepsilon = \varepsilon_n$ by the equality

$$\log C_n = \frac{20\varepsilon}{\log 1/\varepsilon}n + 3\log n. \tag{7.26}$$

Since $\log C_n - 3\log n \geq \frac{1}{4}\log C_n$, we can deduce that

$$\varepsilon_n \sim \frac{\log C_n}{n}\log\frac{n}{\log C_n}. \tag{7.27}$$

Now if we assume that this ε satisfies $\varepsilon \leq e^{-13}$, then we can apply all of the estimates to deduce

$$\|P_n\|_{[-1,1]} \leq \frac{C_n}{2^n}$$

and

$$|P_n'(y_{j,n})| \geq \frac{1}{C_n}\frac{1}{2^n}, \qquad j = 1, \ldots, n.$$

But the polynomial P_n has $[n\mu([1-2\varepsilon, 1-\varepsilon])]$ zeros or at most one more on the interval $[1-2\varepsilon, 1-\varepsilon]$; hence for the discrepancy of its zeros we have

$$|(\nu_{P_n} - \omega)([1-2\varepsilon, 1-\varepsilon])| \geq |\varepsilon\nu_\varepsilon([1-2\varepsilon, 1-\varepsilon])| - \frac{1}{n} \tag{7.28}$$

$$= \varepsilon\|\nu_\varepsilon\|/2 - \frac{1}{n} = \varepsilon - \frac{1}{n} \geq c\frac{\log C_n}{n}\log\frac{n}{\log C_n},$$

with some absolute constant $c > 0$, where at the last step we used (7.27).

These inequalities prove Theorem 7.2 in the case when $C_n \geq n^4$ and the $\varepsilon = \varepsilon_n$ from (7.26) satisfies $\varepsilon \leq e^{-13}$. If the latter condition is not satisfied, then all we have to do to carry out the above argument is to choose $\varepsilon = e^{-13}$, for which the last inequality in (7.28) is still valid with some positive c. □

Proof of Theorem 7.2 in the case $C_n < n^4$. Let

$$T_n(x) = \frac{1}{2^{n-1}}\cos(n\arccos x)$$

be the monic Chebyshev polynomials. T_n has the zeros $\cos((2k-1)\pi/2n)$, $k = 1, \ldots, n$, which are the projections onto $[-1, 1]$ of the equidistant points $\exp((2k-1)\pi i/2n)$, $k = 1, 2, \ldots, 2n$, lying on the unit circumference. This easily implies that the discrepancy of T_n is at most $1/n$. We shall construct the P_n in the theorem by moving some zeros of T_n.

Given n set

$$\varepsilon = \frac{\log^2 n}{n},$$

and $a = 1-2\varepsilon$. The point a will be the center of the movements of zeros: we shall, roughly speaking, reflect some zeros of T_n, distributed according to a logarithmic scale, onto a.

To this end we choose a large constant C that will be specified later (we shall see that actually any $C > 80$ will do the job), and with it we define some numbers $\xi_0, \ldots, \xi_J$ as follows: we set $\xi_0 = \varepsilon^{4/3}$, and for other j's we define ξ_{j+1} in terms of ξ_j via the formula

$$\int_{\xi_j}^{\xi_{j+1}} \frac{1}{t}\,dt = \frac{C}{\log n},$$

and let J be the largest number for which $\xi_{J+1} \le \varepsilon$. Then

$$J \sim \frac{\log^2 n}{C}.$$

Now let x_j be the nearest zero of T_n to $a - \xi_j$ and y_j the nearest zero of T_n' (note the prime!) to $a + \xi_j$, and form the rational function

$$r_J(t) = \prod_{j=0}^{J} \frac{t - y_j}{t - x_j}.$$

We transform the zeros of T_n with the help of r_J, namely we set

$$P_n(t) := T_n(t) r_J(t).$$

We claim that for large enough C and large n the following estimates hold: for every $t \in [-1, 1]$

$$|P_n(t)| \le \frac{n^{3/4}}{2^n}, \tag{7.29}$$

and for every zero θ of P_n

$$|P_n'(\theta)| \ge \frac{1}{n^{3/4}} \frac{1}{2^n}. \tag{7.30}$$

From these estimates Theorem 7.2 immediately follows in the case $n \le C_n \le n^4$. In fact, by the construction we have removed $J + 1$ zeros of T_n from the interval $[1 - 3\varepsilon, 1 - 2\varepsilon]$; hence the discrepancy of P_n is at least as large as

$$\frac{J}{n} \ge c\frac{\log^2 n}{n},$$

which is greater than or equal to

$$c\frac{\log C_n}{n} \log \frac{n}{\log C_n}$$

in the present case.

Thus, it remains to prove (7.29) and (7.30). We start with (7.29). In the proof below, D will denote absolute constants that may vary from line to line, but C is one and the same throughout the proof.

Proof of estimate (7.29). First of all, the definition of the ξ_j's gives

$$\xi_{j+1} - \xi_j = \xi_j \left(e^{C/\log n} - 1\right) = \xi_j \frac{C}{\log n} + O\left(\frac{\xi_j}{\log^2 n}\right). \tag{7.31}$$

This is much larger than the largest distance between consecutive zeros of T_n and T_n' on $[1 - 4\varepsilon, 1 - (\varepsilon/2)]$ which is equivalent to

$$\frac{\sqrt{\varepsilon}}{n} \le D \frac{\log n}{n^{3/2}}.$$

Thus, we immediately get the estimates

$$|y_j - (a + \xi_j)| \le D \frac{\log n}{n^{3/2}}$$

and

$$|x_j - (a - \xi_j)| \le D \frac{\log n}{n^{3/2}}.$$

Since every ratio

$$\left| \frac{t - y_j}{t - x_j} \right|, \qquad j = 0, 1, \ldots, J, \tag{7.32}$$

is increasing on the interval $[-1, x_J]$, and the polynomial T_n attains its maximum on $[x_{J+1}, x_J]$, we can restrict our attention to $t \in [x_{J+1}, 1]$. It is also immediate that for $t \in [a + (\varepsilon^{4/3}/2), 1]$ the rational function $r_J(t)$ is at most 1 in absolute value, so this leaves us to consider the case $t \in [x_{J+1}, a + (\varepsilon^{4/3}/2)]$. We shall prove (7.29) for $t \in [x_J, x_0]$ because the consideration is the same (actually somewhat simpler) for $t \in [x_{J+1}, x_J]$ or $t \in [x_0, a + (\varepsilon^{4/3}/2)]$.

Thus, let $x_{j_0+1} \le t \le x_{j_0}$ for some $j_0 = 0, \ldots, J - 1$. We separate the j_0-th and $(j_0 + 1)$-st terms in r_J, and first estimate the product of the terms with index smaller than j_0 and then with index greater than j_0, respectively.

We write

$$\prod_{j=0}^{j_0 - 1} \frac{t - y_j}{t - x_j} = \prod_{j=0}^{j_0 - 1} \frac{1 - \frac{y_j - (a + \xi_j)}{t - (a + \xi_j)}}{1 - \frac{x_j - (a - \xi_j)}{t - (a - \xi_j)}} \frac{t - (a + \xi_j)}{t - (a - \xi_j)} =: \Pi_1 \Pi_2.$$

Here we have for the denominators

$$|t - (a \pm \xi_j)| \ge \frac{1}{2} |\xi_{j_0} - \xi_{j_0 - 1}| \ge \xi_{j_0 - 1} \frac{C}{3 \log n} \ge \frac{C}{3} \frac{\log^{5/3} n}{n^{4/3}},$$

and so

$$\left|\frac{y_j-(a+\xi_j)}{t-(a+\xi_j)}\right| \le \frac{Dn^{-3/2}\log n}{(C/3)n^{-4/3}\log^{5/3} n} \le \frac{D}{C}n^{-1/6}$$

and

$$\left|\frac{x_j-(a-\xi_j)}{t-(a-\xi_j)}\right| \le \frac{Dn^{-3/2}\log n}{(C/3)n^{-4/3}\log^{5/3} n} \le \frac{D}{C}n^{-1/6}.$$

These inequalities yield

$$|\Pi_1| \le \left(\frac{1+\frac{D}{C}n^{-1/6}}{1-\frac{D}{C}n^{-1/6}}\right)^J \le \exp\left(\frac{D}{C^2}\frac{\log^2 n}{n^{1/6}}\right). \tag{7.33}$$

In the estimate of Π_2 we shall make use of Lemma 7.7. Using the monotonicity of the ratios (7.32) we can write with $\tau := t-a<0$

$$\left|\log\prod_{j=0}^{j_0-1}\left|\frac{\tau-\xi_j}{\tau+\xi_j}\right|\right| = \frac{\log n}{C}\sum_{j=0}^{j_0-1}\log\left|\frac{\tau-\xi_j}{\tau+\xi_j}\right|\int_{\xi_j}^{\xi_{j+1}}\frac{1}{u}\,du \tag{7.34}$$

$$\le \frac{\log n}{C}\int_{\xi_0}^{\xi_{j_0}}\log\left|\frac{\tau-u}{\tau+u}\right|\frac{1}{u}\,du \le \frac{10\log n}{C},$$

where, at the last step, we used Lemma 7.7.

From (7.33) and (7.34) we finally arrive at

$$\prod_{j=0}^{j_0-1}\left|\frac{t-y_j}{t-x_j}\right| = |\Pi_1||\Pi_2| \le \exp\left(\frac{D}{C^2}\frac{\log^2 n}{n^{1/6}}+\frac{10\log n}{C}\right). \tag{7.35}$$

Since

$$\left|\log\prod_{j=j_0+2}^{J}\left|\frac{\tau-\xi_j}{\tau+\xi_j}\right|\right| = \frac{\log n}{C}\sum_{j=j_0+2}^{J}\log\left|\frac{\tau-\xi_j}{\tau+\xi_j}\right|\int_{\xi_j-1}^{\xi_j}\frac{1}{u}\,du$$

$$\le \frac{\log n}{C}\int_{\xi_{j_0+1}}^{\xi_J}\log\left|\frac{\tau-u}{\tau+u}\right|\frac{1}{u}\,du \le \frac{10\log n}{C},$$

we similarly get

$$\prod_{j=j_0+2}^{J}\left|\frac{t-y_j}{t-x_j}\right| \le \exp\left(\frac{D}{C^2}\frac{\log^2 n}{n^{1/6}}+\frac{10\log n}{C}\right). \tag{7.36}$$

As for the remaining two factors

$$\frac{t-y_j}{t-x_j}$$

in r_J with $j=j_0$ and $j=j_0+1$, we note that only one of them can be really large. In fact, t lies either closer to x_{j_0} or closer to x_{j_0+1}. Consider the first case, the other one is similar. Then for the second term we get from (7.31)

$$\left|\frac{t-y_j}{t-x_j}\right| \le \frac{3\xi_{j_0}}{\xi_{j_0} C/(3\log n)} \le \frac{9}{C}\log n. \tag{7.37}$$

Finally, for the other term with $j = j_0$ we get by the mean value theorem

$$\left|T_n(t)\frac{t-y_j}{t-x_j}\right| = \left|\frac{T_n(t)-T_n(x_{j_0})}{t-x_{j_0}}\right| |t-y_{j_0}| = |T_n'(\theta)||t-y_{j_0}|$$

with some $\theta \in [1-3\varepsilon, 1-\varepsilon]$. We can explicitly calculate the derivative of T_n, and with the inequality $|t-y_{j_0}| \le 2\varepsilon$ we finally arrive at

$$\left|T_n(t)\frac{t-y_j}{t-x_j}\right| \le \frac{n}{2^{n-1}}\frac{1}{\sqrt{1-(1-\varepsilon)^2}}2\varepsilon \le \frac{4\sqrt{n}\log n}{2^n}. \tag{7.38}$$

From (7.35)–(7.38) it follows that

$$\begin{aligned} |P_n(t)| &= |T_n(t)r_n(t)| \\ &\le \frac{4\sqrt{n}\log n}{2^n}\frac{9\log n}{C}\exp\left(\frac{D}{C^2}\frac{\log^2 n}{n^{1/6}} + \frac{20\log n}{C}\right) \le \frac{n^{3/4}}{2^n} \end{aligned}$$

if we choose C larger than, say 80, and n is sufficiently large. This proves (7.29). □

Proof of inequality (7.30). Let θ be a zero of P_n. Then θ is either a y_j, or a zero of T_n different from every x_j. Let us consider first the case when $\theta = y_{j_0}$ for some $j_0 \in \{0, \dots, J\}$. Then

$$|P_n'(\theta)| = |T_n'(\theta)r_J(\theta) + T_n(\theta)r_J'(\theta)| = \frac{1}{2^{n-1}}|r_J'(y_{j_0})|$$

because θ is a zero of T_n' by the choice of the numbers y_j, and at every zero of T_n' the value of T_n equals 2^{-n+1}. The derivative of r_J at y_{j_0} equals

$$\prod_{j\ne j_0}\frac{y_{j_0}-y_j}{y_{j_0}-x_j}\cdot\frac{1}{y_{j_0}-x_{j_0}}.$$

In the proof of (7.29) we have verified that

$$\left|\prod_{j\ne j_0}\frac{y_{j_0}-x_j}{y_{j_0}-y_j}\right| \le n^{3/4};$$

more precisely we have proved in (7.35)–(7.38) a similar inequality in which the role of the x_j's and y_j's were switched. Thus, taking reciprocals, we finally get

$$|P_n'(\theta)| \ge \frac{1}{n^{3/4}}\frac{1}{2^n},$$

which is exactly (7.30).

If θ is one of the zeros of T_n, then

$$P_n'(\theta) = T_n'(\theta)r_J(\theta) + T_n(\theta)r_J'(\theta) = T_n'(\theta)r_J(\theta).$$

Here

$$|T_n'(\theta)| \geq \frac{n}{2^{n-1}},$$

while exactly as above

$$|r_J(\theta)| \geq \frac{1}{n^{3/4}},$$

by which (7.30) has been verified. □

VIII.8 Notes and Historical References

Section VIII.1

The energy problem can be generalized to the case when the sign of the attraction between the charges placed on the conductors Σ_j and Σ_k is prescribed for each pair separately. Furthermore, we can also scale this attraction by a number depending on the pair Σ_j, Σ_k. This approach was given by A. A. Gonchar and E. A. Rakhmanov [64] as follows: Let $\Sigma = \cup_{j=1}^n \Sigma_j$, where the Σ_j's are pairwise disjoint compact sets of positive capacity and with empty interior; $m = (m_1, \ldots, m_n)$ a vector with positive coordinates (the mass vector); $A = (a_{ij})$ a symmetric $n \times n$ matrix with positive diagonal elements (the matrix of interaction of the charges) and $Q : \Sigma \to (-\infty, \infty]$ a continuous function on Σ which is not identically infinite on any of the Σ_j's (the external field). By $M = M(\Sigma)$ we denote the set of positive measures supported on Σ such that if we set $\mu_j = \mu|_{\Sigma_j}$, then μ_j has total mass $\|\mu_j\| = m_j$, $j = 1, \ldots, n$. Let the energy of μ be defined by

$$E(\mu) = \sum_{j,k=1}^{n} a_{jk}(\mu_j, \mu_k) + 2\sum_{k=1}^{n} \int Q d\mu_k$$

where, as usual

$$(\mu_j, \mu_k) = \int \log \frac{1}{|z-t|} d\mu_j(t) d\mu_k(z).$$

It is convenient to think of μ and Q as vectors $\mu = (\mu_1, \ldots, \mu_n)$ and $Q = (Q_1, \ldots, Q_j)$ with j-th component related to Σ_j. Then the weighted potential

$$W^{\mu}(z) = \int \log \frac{1}{|z-t|} dA\mu(t) + Q(z)$$

is again a vector with components

$$W_j^{\mu}(z) = \sum_{k=1}^{n} a_{jk} \int \log \frac{1}{|z-t|} d\mu_k(t) + Q_j(z),$$

and the energy can be expressed as

$$E(\mu) = \int (W^\mu + Q)(z)\, d\mu(z).$$

For the case when each $\mathbf{C} \setminus \Sigma_j$ is regular with respect to the Dirichlet problem in $\mathbf{C} \setminus \Sigma_j$, the next result was stated without proof in [64]: The following three statements are pairwise equivalent for a measure $\gamma \in M$:

I. $$W_j^\gamma(z) = \min_{x \in \Sigma_j} W_j^\gamma(x), \quad z \in \operatorname{supp}(\gamma_j),\ j = 1, \ldots, n,$$

II. $$E(\gamma) = \inf_{\mu \in M} E(\mu),$$

III. $$\min_{x \in \Sigma_j} W_j^\gamma(x) = \max_{\mu \in M_j(\gamma)} \min_{x \in \Sigma_j} W_j^\mu(x), \quad j = 1, \ldots, n,$$

where $M_j(\gamma)$ is the set of all measures $\mu \in M$ for which $\mu_k = \gamma_k$ for $k \neq j$. Furthermore, there is a unique γ with these properties, which is called the equilibrium measure for these vector-valued weighted potentials.

Gonchar and Rakhmanov also remark that some of the above conditions can be relaxed; for example, Q need not be continuous.

The discussion given in Theorem 1.4 (see also Theorem 2.1) matches the case when the matrix elements are given by $a_{ij} = \varepsilon_i \varepsilon_j$, $i, j = 1, \ldots, n$. However, the method of the proof can be easily modified to get the more general vector form.

Section VIII.2

As an application of Corollary 2.7, we describe the equilibrium measure $\mu^* = \mu_1^* - \mu_2^*$ in Theorem 2.6 for the case when $\Sigma_1 = [-a, a]$, $0 < a < 1$, and $\Sigma_2 = \{z \mid |z| \geq 1\}$. Indeed, from Example II.5.14 and Corollary 2.7, we have

$$d\mu_1^*(x) = \frac{1}{2K} \frac{dx}{\sqrt{(a^2 - x^2)(1 - a^2 x^2)}}, \quad x \in [-a, a],$$

where

$$K := \int_0^1 \frac{dt}{\sqrt{(1 - t^2)(1 - a^4 t^2)}}.$$

Moreover, μ_2^* is the balayage of μ_1^* onto the unit circle $|z| = 1$, which is given by

$$d\mu_2^*(t) = \frac{1}{2\pi} \int_{-a}^{a} P(t, x)\, d\mu_1^*(x) d\theta, \quad t = e^{i\theta},$$

where $P(\cdot, \cdot)$ is the Poisson kernel (cf. Example II.4.3).

Section VIII.3

The terminology in this section slightly differs from the usual one in that $\log(1/c(1, \Sigma_1, \Sigma_2))$ is usually called the modulus of the condenser (Σ_1, Σ_2) and its reciprocal is called the capacity of the condenser (Σ_1, Σ_2) (the literature is not unanimous in this respect, for sometimes a normalizing factor is used in forming the modulus). By a condenser we mean a pair (Σ_1, Σ_2) of disjoint closed sets, although very often it is assumed that both $\mathbf{C}\setminus\Sigma_1$ and $\mathbf{C}\setminus\Sigma_2$ are connected when one speaks of a condenser. The terminology in Section VIII.3 follows the one used earlier in this book.

In the unweighted case ($w \equiv 1$) there are different approaches to the equilibrium problem with a positive and negative unit charge on Σ_1 and Σ_2, respectively. Consider e.g. the case when both $\mathbf{C}\setminus\Sigma_1$ and $\mathbf{C}\setminus\Sigma_2$ are connected (and of course $\Sigma_1 \cap \Sigma_2 = \emptyset$). By an application of a Möbius transformation we can also assume that $\Omega = \mathbf{C}\setminus\Sigma$ contains the point ∞, i.e. that Σ is compact. Let u be the solution of the Dirichlet's problem in Ω with boundary function 1 on Σ_1 and 0 on Σ_2, and let us choose a smooth closed curve γ in Ω that separates Σ_1 and Σ_2. If $\mathbf{n}$ denotes the normal to γ pointing in the direction of Σ_1, then the integral

$$\frac{1}{2\pi}\int_\gamma \frac{\partial u}{\partial \mathbf{n}} ds$$

turns out to be independent of γ ([62]) and equals $1/(F_1+F_2)$, where the F_j's are the constants from Sections VIII.1–VIII.3 (cf. Theorem 1.4(d)), and $1/(F_1+F_2)$ is nothing else than the reciprocal of the minimal energy (see (1.14)).

Another expression for $1/(F_1+F_2)$ is

$$\inf_V \frac{1}{2\pi}\iint_\Omega \left[\left(\frac{\partial V}{\partial x}(x,y)\right)^2 + \left(\frac{\partial V}{\partial y}(x,y)\right)^2\right] dxdy,$$

where the infinium is taken for all continuously differentiable functions in Ω that have boundary values 1 at Σ_1 and 0 at Σ_2 (cf. [5]).

Finally, we mention the following representation (see Corollary 2.7) for $1/(F_1+F_2)$: Let $g_{\mathbf{C}\setminus\Sigma_1}(x,y)$ be the Green function of the domain $\mathbf{C}\setminus\Sigma_1$ with pole at $y \in \mathbf{C}\setminus\Sigma_1$. Then

$$\frac{1}{F_1+F_2} = \inf_\sigma \iint_{\Sigma_1\Sigma_2} g_{\mathbf{C}\setminus\Sigma_1}(x,y)\, d\sigma(x) d\sigma(y),$$

where the infimum is taken for all probability measures σ with support in Σ_2.

Apparently it was T. Bagby who first noticed the connection of these quantities with the minimal energy for signed measures (see [5]). The analogues of the above relations in the weighted case are not known. Here we only want to point out that it is not possible to describe (F_1+F_2) or the minimal energy (in the general weighted case) via quantities that depend only on Ω and the boundary values of Q on $\partial\Omega$, for the extremal measure may not be supported on $\partial\Omega$.

Bagby [5] used Fekete points to discretize the signed energy problem, as well as Leja points ([6]) for finding good points of interpolation and good poles for rational interpolants to analytic functions. This is why the Fekete/Leja points are often called Bagby points in the rational case. In particular, under some conditions on Σ_1 and Σ_2, he showed that if $\{z_k^{(j,n)}\}_{k=1}^n \subseteq \Sigma_j$, $j = 1, 2$, are points for which the expression

$$\mathcal{F}(\{z_k^{(1,n)}\}_{k=1}^n, \{z_k^{(2,n)}\}_{k=1}^n)^{1/n}$$

(see the formula before (3.4) with $w \equiv 1$) tends to $c(1, \Sigma_1, \Sigma_2)$ (cf. (3.5)), then the following is true: if f is analytic on Σ_1 and r_n is the rational function of degree at most n that has poles at the points $z_k^{(2,n)}$, $k = 1, \ldots, n$, and interpolates f at the points $z_k^{(1,n)}$, $k = 1, \ldots, n$, then r_n uniformly tends to f on Σ_1. Since these relations are symmetric with respect to Σ_1 and Σ_2 if $w \equiv 1$, the roles of Σ_1 and Σ_2 can be reversed here. The Fekete or Leja points always satisfy the above minimality condition; hence one can apply this interpolation result to them.

The original Zolotarjov problem mentioned in Section VIII.3 is the following. Let $E = [-1, 1]$ and $F = \{x \in \mathbf{R} \mid |x| \geq 1/k\}$ where $0 < k < 1$, and set

$$\sigma_n(E, F) = \sup_{r_n} \frac{\min\{|r_n(x)| \mid x \in F\}}{\max\{|r_n(x)| \mid x \in E\}},$$

where the minimum is taken for all rational functions of degree less than or equal to n (both the numerator and denomenator degree). The problem is to find the quantities $\sigma_n(E, F)$. E. I. Zolotarjov [239] gave the exact value of $\sigma_n(E, F)$ for all n, namely

$$\sigma_n = k^{-n}\left(\operatorname{sn}\frac{K}{n}\operatorname{sn}\frac{3K}{n}\cdots\operatorname{sn}\frac{pK}{n}\right)^{-4},$$

where p is the largest odd integer less than n,

$$\operatorname{sn} u = \sin\varphi, \qquad u = \int_0^{\varphi} \frac{dt}{\sqrt{1 - k\sin^2 t}},$$

and

$$K = \int_0^{\pi/2} \frac{dt}{\sqrt{1 - k\sin^2 t}}$$

is the complete elliptic integral of first type of modulus k.

For general sets E and F it was A. A. Gonchar [62] who determined the asymptotic behavior of $\sigma_n(E, F)^{1/n}$ as $n \to \infty$, and this coincides with the $w \equiv 1$ special case of (3.2) and the second limit in Theorem 3.4.

At this writing it is not known if, in general, the distributions of weighted Leja points taken from Σ_1 and Σ_2 converge to the equilibrium distribution (cf. Theorem 3.5 where this question is settled if there is no weight present).

The argument in the proof of Theorem 3.5 was taken from [134] (see erratum) by A. L. Levin and E. B. Saff, who considered the Zolotarjov problem for rational functions such that the ratio of the degrees of the numerator and denominator polynomials tends to some number, i.e. they considered

$$\sigma_{n,m}(E, F) = \sup_{r_{n,m}} \frac{\min\{|r_{n,m}(x)| \mid x \in F\}}{\max\{|r_{n,m}(x)| \mid x \in E\}},$$

where now the supremum is taken for all rational functions of numerator degree at most n and denominator degree at most m. They verified the convergence of the normalized Zolotarjov constant $\sigma_{n,m}(E, F)^{1/(n+m)}$, as $n/m \to \lambda$, and established a close connection with an energy problem when the two masses on the two plates of the condenser are not necessarily equal.

Considering again the case $\Sigma = \Sigma_1 \cup \Sigma_2$, $\varepsilon_1 = 1$, $\varepsilon_2 = -1$, and the rational Fekete points defined before (3.4). H. Kloke [102] proved that in the nonweighted case $w \equiv 1$ not only does

$$\mathcal{F}\left(\{z_k^{(1,n)}\}_{k=1}^n, \{z_k^{(2,n)}\}_{k=1}^n\right)^{1/n(n-1)}$$

tend to a limit, but also the $n(n-1)$-st roots of the expressions

$$\prod_{i=1}^n \prod_{k\neq i} \left|\frac{z_i^{(1,n)} - z_k^{(1,n)}}{z_i^{(1,n)} - z_k^{(2,n)}}\right| w(z_i^{(1,n)})w(z_k^{(1,n)}) \tag{8.1}$$

and

$$\prod_{i=1}^n \prod_{k\neq i} \left|\frac{z_i^{(2,n)} - z_k^{(2,n)}}{z_i^{(2,n)} - z_k^{(1,n)}}\right| \frac{1}{w(z_i^{(2,n)})w(z_k^{(2,n)})}$$

have a limit as $n \to \infty$. This easily follows even for the general case from the results in Chapters VIII, I and III. In fact, let us consider only (8.1). If we introduce the weights

$$w_n(z) = w(z) \Big/ (\prod_{k=1} |z - z_k^{(2,n)}|)^{1/n}$$

on Σ_1 and consider a Fekete set $\mathcal{F}_n$ with respect to w_n, then it easily follows from the extremality of the points $\{z_k^{(1,n)}\}$ (if we think of the points $\{z_k^{(n,2)}\}$ as being fixed) that the Vandermonde expression (III.1.1) for w_n and the expression

$$\prod_{\substack{1\leq i,k\leq n\\ i\neq k}} \frac{\left|z_i^{(1,n)} - z_k^{(1,n)}\right| w(z_i^{(1,n)})w(z_k^{(1,n)})}{\left|z_i^{(1,n)} - z_k^{(2,n)}\right| \left|z_k^{(1,n)} - z_i^{(2,n)}\right|} \tag{8.2}$$

differ only in a factor $\exp(O(n))$. But on Σ_1 the weights w_n uniformly tend to the weight

$$\tilde{w}(z) = w(z)/\exp\left(-U^{\mu_2^*}(z)\right)$$

(see Theorem 3.4) and so the above consideration gives that the $n(n-1)$-st root of the product in (8.2) tends to the weighted transfinite diameter of Σ_1 with weight $\tilde{w}$ as $n \to \infty$. But (8.1) and (8.2) differ only in the factor

$$\prod_{i=1}^n \prod_{k\neq i} \frac{1}{\left|z_k^{(1,n)} - z_i^{(2,n)}\right|},$$

the $n(n-1)$-st root of which is known to converge to

$$\int\int \log \frac{1}{|z-t|} d\mu_1^*(t) d\mu_2^*(z)$$

by Theorem 3.4 (a feature of this proof is that even if we start off with the unweighted case of [102], we need the results from Chapters I–III because in the proof a weight appears on Σ_1).

The proof also shows (use (3.7)) that the $n(n-1)$-st root of the products

$$\prod_{i=1}^{n} \prod_{k \neq i} \left| z_i^{(1,n)} - z_k^{(1,n)} \right|,$$

$$\prod_{i=1}^{n} \prod_{k \neq i} w(z_i^{(1,n)}) w(z_k^{(1,n)}),$$

and

$$\prod_{i=1}^{n} \prod_{k \neq i} \left| z_i^{(1,n)} - z_k^{(2,n)} \right|,$$

from which (8.2) is built up, separately converge, as well.

In [102] Kloke proves several estimates for the speed of convergence of the above quantities in the case $w \equiv 1$. For example, he shows that if Σ_1 and Σ_2 are compact such that the diameters of their connected components are bounded from below, then for

$$W_n = \mathcal{F}\left(\{z_k^{(1,n)}\}_{k=1}^n, \{z_k^{(2,n)}\}_{k=1}^n \right),$$

which is the (unweighted) rational Vandermonde expression, we have

$$0 \leq \frac{1}{n(n-1)} \log W_n + (F_1 + F_2) \leq \frac{C}{n} + 4 \frac{\log n}{n},$$

where F_1 and F_2 are the constants from Theorem 1.4(d) and $F_1 + F_2$ is the minimal energy. The rate $\{\log n/n\}$ seems to be common to this type of estimates.

Section VIII.5

Theorem 5.1 was proved by A. A. Gonchar [62]; see also the works of H. Widom [231, 232]. T. Ganelius [53] obtained sharper asymptotics for the best rational approximation of signum type functions for smooth sets, and he also considered the case when the sets can touch each other. His approach can be used to obtain a proof of Vjaceslavov's result [228] on the exact degree of rational approximation to $|x|^\alpha$ (see [54]).

Theorem 5.2 was taken from [218] by V. Totik.

The analogue of the approximation problem discussed in Section VI.1 for weighted rational functions $w^n r_n$, where $r_n = p_n/q_n$, $\deg p_n \leq n$, $\deg q_n \leq n$, has been investigated for some special weights w by P. Borwein, E. A. Rakhmanov, and E. B. Saff [20] and by E. A. Rakhmanov, E. B. Saff, and P. C. Simeonov [191]. In particular, they show that if $w(x) = e^{-x}$ on $[0, \infty)$, then every continuous function f on $[0, b]$ with $b < 2\pi$ is the uniform limit of weighted rationals

$\{e^{-nx} r_n(x)\}$ and that 2π cannot be replaced by any larger constant in this assertion. Furthermore, if $w(x) = x^\theta$, $\theta > 1$, on $[0, 1]$, then every continuous function on $[a, 1]$ with $a > a_\theta := \tan^4((\pi/4)((\theta - 1)/\theta))$ is the uniform limit of weighted rationals $\{x^{n\theta} r_n(x)\}$ and a cannot be replaced by any constant smaller than a_θ.

Weighted rational approximation with varying weights on compact subsets of an open set G in the complex plane has been studied by I. E. Pritsker and R. S. Varga [188]. There it is assumed that the weight is analytic in G and does not vanish.

Section VIII.6

H. Kloke [100] used Fekete points for finding the conformal mapping of ring domains onto rings. He also gave estimates for the speed of the first convergence in Theorem 6.2, which shows that for z's satisfying

$$r_1 + \frac{1}{\sqrt{n}} < \varphi(z) < r_2 - \frac{1}{\sqrt{n}},$$

the "rate" is not worse than

$$\text{const.} \frac{\log n}{\sqrt{n}}.$$

Section VIII.7

After the fundamental paper [40] by P. Erdős and P. Turán, there has been considerable interest in discrepancy estimates under various assumptions (see e.g. [41], [39]). The most relevant development regarding Section VIII.7 is due to H.-P. Blatt [15]; namely he proved Theorem 7.2 with the $\log(n/\log C_n)$ factor on the right of (7.7) replaced by $\log n$, but actually his argument can be modified so as to yield (7.7).

The content of this section was taken from [218] by V. Totik. See that paper for further references to works on discrepancy estimates that are connected with Theorem 7.1.

For further applications of potential theory in connection with discrepancy estimates, see the papers [16] by Blatt and Mhaskar and [3] and [4] by Andrievskii and Blatt.

Appendix A. The Dirichlet Problem and Harmonic Measures

In this appendix we prove Wiener's theorem on regular boundary points for Green functions and Dirichlet problems. In particular, it will follow that these two concepts are identical.

A.1 Regularity with Respect to Green Functions

Let $G \subset \overline{\mathbf{C}}$ be a domain such that ∂G is of positive capacity and, for $a \in G$, let $g_G(z, a)$ be the Green function of G with pole at a. Recall from Sections I.4 and II.4 that $g_G(z, a)$ is defined as the unique function on G satisfying the following properties:

(i) $g_G(z, a)$ is nonnegative and harmonic in $G \setminus \{a\}$ and bounded as z stays away from a,

(ii) $g_G(z, a) - \log \dfrac{1}{|z - a|}$ is bounded in a neighborhood of a,

(iii) $\lim_{z \to x,\, z \in G} g_G(z, a) = 0$ for quasi-every $x \in \partial G$

with (ii) replaced by

(ii)′ $g_G(z, a) - \log |z|$ is bounded in a neighborhood of ∞

when $a = \infty$.

We call a point $x \in \partial G$ on the boundary of G a *regular point* (with respect to the Green function $g_G(z, a)$) if

$$\lim_{z \to x,\ z \in G} g_G(z, a) = 0.$$

Soon we shall see that this notion is independent of the choice of a; therefore, we shall just speak of regular boundary points. Note also that, by definition (which, however, depends on the existence theorem for Green functions), quasi-every point on the boundary of G is a regular point.

Theorem 1.1 (Wiener's Theorem). *Let* $0 < \lambda < 1$ *and set*

$$A_n(x) := \left\{ y \,\middle|\, y \notin G,\ \ \lambda^n \le |y - x| < \lambda^{n-1} \right\}.$$

Then $x \in \partial G$, $x \neq \infty$, is a regular boundary point of G if and only if

$$\sum_{n=1}^{\infty} \frac{n}{\log(1/\mathrm{cap}(A_n(x))} = \infty. \tag{1.1}$$

In particular, regularity is a local property.

Proof. First we consider the case when G is an unbounded domain and $a = \infty$. Set $E := \overline{\mathbf{C}} \setminus G$ and

$$V(E) := \log \frac{1}{\mathrm{cap}(E)}.$$

It is immediate that the condition (1.1) does not change if $A_n(x)$ is defined as

$$A_n(x) := \left\{ y \,\middle|\, y \notin G,\ \ \lambda^n \leq |y - x| \leq \lambda^{n-1} \right\}$$

(note that equality is allowed at both places on the right); hence in what follows we can work with this definition of $A_n(x)$, which is more convenient than the original one, for then $A_n(x)$ is compact.

We may assume without loss of generality that $x = 0$. We start with the proof of the sufficiency of condition (1.1). In view of the representation

$$g_G(z, \infty) = -U^{\mu_E}(z) + V(E) \tag{1.2}$$

(see (I.4.8)), which we use to extend g_G to the whole plane, the inequality

$$U^{\mu_E}(z) \leq V(E), \tag{1.3}$$

and the lower semi-continuity of U^{μ_E}, it suffices to show that (1.1) implies

$$U^{\mu_E}(0) = V(E). \tag{1.4}$$

Assume to the contrary that (1.4) is not true, i.e.

$$\beta := V(E) - U^{\mu_E}(0) > 0.$$

Then, in view of Theorem I.4.1, we will have

$$V(E^*) - U^{\mu_{E^*}}(0) \geq \beta \tag{1.5}$$

for every compact subset E^* of E of positive capacity.

One can easily verify from the discussion at the end of the present proof that λ can be replaced by λ^k with any fixed k, so we can assume λ as small as we like. In particular, we can assume that $\lambda < 1/4$ is so small that $\log(1/(1-\lambda)) < \beta/2$ is satisfied.

If

$$\sum_{n=1}^{\infty} \frac{n}{\log(1/\mathrm{cap}(A_{2n}(0))} = \infty,$$

then we set $E_n = A_{2n}(0)$; otherwise, we choose $E_n = A_{2n-1}(0)$. In any case the sets E_n are disjoint and satisfy

$$\sum_{n=1}^{\infty} \frac{n}{V(E_n)} = \infty. \tag{1.6}$$

Finally, we choose $E^* = \{0\} \cup (\cup_{n \geq n_0} E_n)$ with an n_0 so large that we have $E^* \subset \{z \mid |z| \leq 1/2\}$. Then E^* is compact and $E^* \subset E$.

Now let $z \in E_i$ and $t \in E_j$ with $j \neq i$ and $\text{cap}(E_i) > 0$. If $j < i$ then $|z - t| \geq (1 - \lambda)|t|$, while for $j > i$ we have $|z - t| \geq |t|$. Thus, in any case

$$\log \frac{1}{|t|} + \log \frac{1}{1 - \lambda} \geq \log \frac{1}{|z - t|}.$$

Let us integrate this inequality with respect to $d\mu_{E^*}(t)$ on $E^* \setminus E_i$. Then, observing that both z and t lie in the disk $D_{1/2}(0)$, the left-hand side will be at most

$$U^{\mu_{E^*}}(0) + \log \frac{1}{1 - \lambda} < V(E^*) - \beta + \frac{\beta}{2},$$

while the right-hand side is

$$U^{\mu_{E^*}}(z) - \int_{E_i} \log \frac{1}{|z - t|} d\mu_{E^*}(t).$$

Since

$$U^{\mu_{E^*}}(z) = V(E^*)$$

for quasi-every $z \in E_i$, we obtain

$$\int_{E_i} \log \frac{1}{|z - t|} d\mu_{E^*}(t) \geq \frac{\beta}{2}, \qquad \text{for q.e. } z \in E_i.$$

Let ν_i be the restriction of μ_{E^*} to E_i. Then the preceding inequality takes the form

$$U^{\nu_i}(z) \geq \frac{\beta}{2}, \qquad \text{for q.e. } z \in E_i.$$

Thus,

$$\mu_{E^*}(E_i) = \nu_i(E_i) = \int \frac{U^{\mu_{E_i}}}{V(E_i)} d\nu_i = \frac{1}{V(E_i)} \int U^{\nu_i} d\mu_{E_i} \geq \frac{\beta}{2V(E_i)}.$$

But this implies

$$\begin{aligned} U^{\mu_{E^*}}(0) = \int \log \frac{1}{|t|} \, d\mu_{E^*}(t) \quad &\geq \quad \sum_i \left(\log \frac{1}{\lambda^{2i-2}} \right) \mu_{E^*}(E_i) \\ &\geq \quad 2\beta \log \frac{1}{\lambda} \sum_i \frac{i - 1}{V(E_i)} = \infty, \end{aligned}$$

which is a contradiction by (1.5) and (1.6). This contradiction was the result of the assumption that (1.1) holds but (1.4) does not; hence (1.1) implies regularity.

Now we turn to the necessity, and let us assume that (1.1) does not hold. Then there are arbitrary small r's such that the circle $\{z \mid |z| = r\}$ does not intersect $E := \overline{\mathbf{C}} \setminus G$ (see Lemma I.2.1). For such an r, let K_1 be the intersection of E with the disk $D_r(0) := \{z \,|\, |z| \le r\}$, and set $K_2 := E \setminus K_1$. Then both K_1 and K_2 are compact, and we can choose a bounded closed neighborhood K_2^* of K_2 such that also K_2^* is disjoint from K_1.

On applying Theorem I.4.1 as before, it is enough to prove the irregularity of $x = 0$ with respect to any domain $G_1 \subseteq G$ containing x on its boundary; thus we can assume without loss of generality that infinitely many (or all if we like) $A_n(0)$ are of positive capacity. This implies, in particular, that K_1 has positive capacity.

Let k_0 be the smallest integer with the property that for $k > k_0$ the sets $A_k := A_k(0)$ are disjoint from K_2. In what follows we shall assume that A_{k_0} is entirely contained in the disk $D_r(0)$; if this is not the case, then in the following discussion we have to split A_{k_0} into $A_{k_0} \cap D_r(0)$ and $A_{k_0} \setminus D_r(0)$ and make the necessary changes. Finally, without loss of generality we may assume $r < 1/2$ is so small that

$$\sum_{k=k_0}^{\infty} \frac{k}{V(A_k)} < \frac{1}{2\log(1/\lambda)}$$

is satisfied.

Since for $\operatorname{cap}(A_k) > 0$, $k \ge k_0$, we have with $\nu := \mu_{K_1}|_{A_k}$

$$\mu_{K_1}(A_k) = \int_{A_k} \frac{U^{\mu_{A_k}}}{V(A_k)} d\mu_{K_1} = \frac{1}{V(A_k)} \int U^{\nu} d\mu_{A_k} \le \frac{V(K_1)}{V(A_k)},$$

it follows that

$$U^{\mu_{K_1}}(0) \le \sum_{k=k_0}^{\infty} \left(\log \frac{1}{\lambda^k}\right) \mu_{K_1}(A_k) \le V(K_1) \log \frac{1}{\lambda} \sum_{k=k_0}^{\infty} \frac{k}{V(A_k)} < \frac{1}{2} V(K_1),$$

and this means that

$$g_{\overline{\mathbf{C}} \setminus K_1}(0, \infty) > 0.$$

Using the mean value inequality for the subharmonic function $g_{\overline{\mathbf{C}} \setminus K_1}(z, \infty)$ on circles $\{z \mid |z| = r\}$ that do not intersect E (as we have seen, there are such circles with arbitrary small $r > 0$) we can conclude

$$\limsup_{z \to 0,\, z \in G} g_{\overline{\mathbf{C}} \setminus K_1}(z, \infty) > 0. \tag{1.7}$$

Now let M be larger than the maximum of $g_{\overline{\mathbf{C}} \setminus K_1}$ on the boundary of K_2^* and m the minimum of g_G on the same set. Since K_2^* is a neighborhood of K_2 disjoint from K_1, we have the relation $\partial K_2^* \subset G$; therefore $m > 0$ by the minimum principle. Replacing M by a larger number if necessary, we can assume that $M/m \ge 1$. Let us now apply the generalized minimum principle (Theorem I.2.4) to the function $(M/m) g_G - g_{\overline{\mathbf{C}} \setminus K_1}$ on the set $G^* := \overline{\mathbf{C}} \setminus (K_1 \cup K_2^*)$. This is a superharmonic function in G^* which is bounded from below (around infinity it behaves like $((M/m) - 1) \log |z|$) and which has nonnegative boundary limits quasi-everywhere on $\partial G^* = \partial K_1 \cup \partial K_2^*$ (recall also that on ∂K_2^* both $g_{\overline{\mathbf{C}}\setminus K_1}$ and g_G are continuous). Therefore we can conclude that $(M/m)g_G - g_{\overline{\mathbf{C}}\setminus K_1}$ is nonnegative on G^*, and we have in view of (1.7)

$$\limsup_{z\to 0, z\in G} g_G(z,\infty) \geq \limsup_{z\to 0, z\in G} \frac{m}{M} g_{\overline{\mathbf{C}}\setminus K_1}(z,\infty) > 0,$$

which verifies that 0 is not a regular boundary point of G.

Thus, the proof of Wiener's theorem is complete in the case when $a = \infty$.

Now let us turn to the case of an arbitrary domain G with $\mathrm{cap}(\partial G) > 0$. The mapping $z \to z' := 1/(z-a)$ maps G into an unbounded domain G' and $g_G(z,a)$ is transformed to $g_{G'}(z',\infty)$. Thus, $x \in \partial G$ is a regular boundary point with respect to G if and only if $x' = 1/(x-a)$ is a regular boundary point with respect to G'. Therefore, we only have to show that the Wiener condition (1.1) is also preserved under this mapping.

Let D be a disk around x not containing a on its boundary, and let D' be its image. Then on D the mapping $z \to z'$ is a constant times a nonexpansive mapping (i.e. which can only shrink distances), and on D' the same is true for its inverse. We have mentioned after the proof of Lemma I.2.1 that a nonexpansive mapping does not increase the capacity; therefore there is a constant M such that for every subset K of D we have

$$\frac{1}{M}\mathrm{cap}(K) \leq \mathrm{cap}(K') \leq M\mathrm{cap}(K),$$

where K' denotes the image of K. In particular, for every sufficiently large n we have

$$\frac{1}{M}\mathrm{cap}(A_n(x)) \leq \mathrm{cap}(A_n(x)') \leq M\mathrm{cap}(A_n(x)). \tag{1.8}$$

The sets $A_n(x)'$ are not the sets $A_m(x')$ for the point x', but it immediately follows from what we have just said about the mapping $z \to z'$ that there is an L such that for every large n

$$A_n(x') \subseteq \bigcup_{j=n-L}^{n+L} A_n(x)',$$

and conversely

$$A_n(x)' \subseteq \bigcup_{j=n-L}^{n+L} A_n(x'). \tag{1.9}$$

The first containment immediately implies

$$\begin{aligned} \frac{1}{\log(1/\mathrm{cap}(A_n(x')))} &\leq \frac{1}{\log(1/\mathrm{cap}(\bigcup_{j=n-L}^{j=n+L} A_n(x)'))} \\ &\leq \sum_{j=n-L}^{j=n+L} \frac{1}{\log(1/\mathrm{cap}(A_n(x)'))}, \end{aligned} \tag{1.10}$$

where in the last step we also applied the inequality Theorem I.6.2(e) (for large n all of the above sets are contained in a disk of radius $1/2$, so we can choose $M = 1$ in that inequality). On multiplying (1.10) by n, summing over n and making use of (1.8) we can see that

$$\sum_n \frac{n}{\log(1/\mathrm{cap}(A_n(x')))} = \infty$$

implies

$$\sum_n \frac{n}{\log(1/\mathrm{cap}(A_n(x)))} = \infty.$$

The converse can be proved in the same way from (1.9), and so the invariance of the Wiener condition under the mapping $z \to z'$ has been verified. □

A.2 Regularity with Respect to Dirichlet Problems

Next we shall discuss regular boundary points with respect to the Dirichlet problem. First we recall the definition of the Perron–Wiener–Brelot solution of the Dirichlet problem from Section I.2.

Consider a domain $G \subset \overline{\mathbf{C}}$ such that $\overline{\mathbf{C}} \setminus G$ has positive capacity, and suppose that f is a bounded Borel measurable function defined on ∂G. The upper and lower classes of functions corresponding to f and G are defined as

$$\mathcal{H}_f^{u,G} := \{g \mid g \text{ superharmonic and bounded below on } G, \quad \liminf_{z\to x,\, z\in G} g(z) \geq f(x) \quad \text{for} \quad x \in \partial G\}$$

and

$$\mathcal{H}_f^{l,G} := \{g \mid g \text{ subharmonic and bounded above on } G, \quad \limsup_{z\to x,\, z\in G} g(z) \leq f(x) \quad \text{for} \quad x \in \partial G\},$$

and the upper and lower solutions of the Dirichlet problem for the boundary function f are given by

$$\overline{H}_f^G(z) := \inf_{g\in\mathcal{H}_f^{u,G}} g(z), \qquad z \in G,$$

and

$$\underline{H}_f^G(z) := \sup_{g\in\mathcal{H}_f^{l,G}} g(z), \qquad z \in G.$$

Always $\overline{H}_f^G \leq \underline{H}_f^G$ and if $\overline{H}_f^G \equiv \underline{H}_f^G$, then this function H_f^G is called the Perron–Wiener–Brelot solution of the Dirichlet problem on G for the boundary function f.

In what follows we shall assume that ∂G is a compact subset of $\mathbf{C}$. This can always be achieved by a fractional linear transformation.

First we show that the upper and lower solutions are harmonic functions, and if f is a continuous function, then the Perron–Wiener–Brelot solution exists, is harmonic in G, and has boundary limit $f(x)$ for quasi-every $x \in \partial G$. We shall do this in several steps.

I. Let $|f| \leq M$ on ∂G. Then in the upper and lower classes we can restrict ourselves to functions that have values in the interval $[-M, M]$ (i.e. the lower and upper limits do not change if we only take them for such functions). In fact, if, for example, g is an upper function, then $\min(M, g)$ is again an upper function, and by the minimum principle it is at least as large as $\min f \geq -M$ in G. Thus, in what follows we shall restrict our attention to functions with values in the interval $[-M, M]$.

II. Let $\overline{\Delta} \subset G$ be a closed finite subdisk of G. For every upper function $g \in \mathcal{H}_f^{u,G}$ we can construct another $g^* \in \mathcal{H}_f^{u,G}$ in such a way that $g^* \leq g$, g^* is harmonic in Δ, and the mapping $g \to g^*$ is monotone. In fact, the restriction of g to the boundary of Δ is a lower semi-continuous function, so there is an increasing sequence $\{h_m\}$ of continuous functions on $\partial\Delta$ that tends to g on $\partial\Delta$. We can extend each h_m into Δ harmonically (Corollary 0.4.4) and we denote this extension also by h_m. Now in Δ the sequence $\{h_m\}$ is a bounded and increasing sequence of harmonic functions; hence its limit h is also harmonic in Δ by Harnack's principle (Theorem 0.4.10). Now let g^* coincide with g outside Δ, and with h inside Δ. Since g is superharmonic, we have $h_m(z) \leq g(z)$ for all m and $z \in \Delta$, so the inequality $g^* \leq g$ is clear. In a similar fashion it easily follows from the minimum principle that if $g_1 \leq g_2$, then $g_1^* \leq g_2^*$. Thus it remains to prove that each g^* is in the upper class $\mathcal{H}_f^{u,G}$. Since g^* coincides with g outside Δ, only the superharmonicity of g^* has to be proved. The lower semi-continuity of g^* on $\overline{\Delta}$ follows from the fact that there $g^* = h$ is the limit of an increasing sequence of continuous functions. Since $g^* = g$ outside Δ, and g is lower semi-continuous, the lower semi-continuity of g^* on the whole domain G follows. Thus, to be able to conclude the superharmonicity of g^*, it remains to show that for every $z_0 \in G$ there is an $r_0 > 0$ such that for all $0 < r < r_0$ we have

$$\frac{1}{2\pi}\int_{-\pi}^{\pi} g^*(z_0 + re^{it})dt \leq g^*(z_0)$$

(see Remark 0.5.3). For $z_0 \in \Delta$ obviously there is such an r_0 because of the mean value property of g^* in Δ, while for $z_0 \notin \Delta$ any r_0 will be suitable for which the disk $D_{r_0}(z_0)$ lies in G because the corresponding inequality holds for g and $g^* \leq g$ but $g^*(z_0) = g(z_0)$.

III. We claim that both the upper and lower solutions to the Dirichlet problem are harmonic functions in G. Consider, for example, the upper solution. It is enough to prove the harmonicity on every finite disk $\Delta \subset G$ with $\overline{\Delta} \subset G$, for then $\overline{H}_f^G(z)$ will also be harmonic at infinity in case G is an unbounded domain (indeed, the upper solution is bounded by M and so Corollary 0.3.5 applies). Let us fix a $z_0 \in \Delta$, and let $g_n \in \mathcal{H}_f^{u,G}$ be functions such that $g_n(z_0) \to \overline{H}_f^G(z_0)$ as $n \to \infty$. Since the minimum of two upper functions is again an upper function, we can assume that $g_{n+1} \le g_n$. Consider the corresponding g_n^* constructed in the preceding paragraph. We also have $g_{n+1}^* \le g_n^*$, so these g_n^*'s converge to a g^* that is harmonic in Δ by Harnack's principle. Since $g^* \le g_n$ for all n, we clearly have $g^*(z_0) = \overline{H}_f^G(z_0)$. We claim that this equality holds at every other point of Δ, which yields the harmonicity in Δ of the upper solution $\overline{H}_f^G(z_0)$. In fact, let z_1 be another point of Δ, and choose upper functions h_n such that $\{h_n(z_1)\}$ converges to $\overline{H}_f^G(z_1)$. As before, we can assume $h_{n+1} \le h_n$, and even that $h_n \le g_n$. Then $\{h_n^*\}$ converges to a function $h^* \le g^*$ which is harmonic in Δ. But we must have $h^*(z_0) = \overline{H}_f^G(z_0) = g^*(z_0)$; therefore, the two functions h^* and g^* coincide on Δ by the minimum principle. Thus, $g^*(z_1) = h^*(z_1) = \overline{H}_f^G(z_1)$ as we claimed.

IV. Next we show that

$$\limsup_{z\to x,\, z\in G} \overline{H}_f^G(z) \le f(x) \tag{2.1}$$

and

$$\liminf_{z\to x,\, z\in G} \underline{H}_f^G(z) \ge f(x) \tag{2.2}$$

at every point $x \in \partial G$ that is a regular point for the Green functions of G and at which f is continuous. In particular, these relations hold quasi-everywhere when f is continuous.

Let us consider the first relation at a regular point $x \in \partial G$. Since G is a domain, there is a broken line in G that intersects any neighborhood of x. Thus, if $r > 0$ is sufficiently small, and $K = (\overline{\mathbf{C}} \setminus G) \cap \overline{D}_r(x)$, then x will lie on the outer boundary of the compact set K. Since regularity is a local property, we get that x is a regular boundary point with respect to Green functions in $\overline{\mathbf{C}} \setminus K$, i.e. the Green function $g_{\overline{\mathbf{C}}\setminus K}(z,\infty)$ is continuous and vanishes at x. Now let ε_{2r} be the maximum of the differences $|f(x) - f(y)|$ for all $y \in \partial G \cap \overline{D_{2r}(x)}$, γ the minimum of $g_{\overline{\mathbf{C}}\setminus K}(z,\infty)$ for $|z-x| \ge 2r$, and $M := \sup_{y\in\partial G} |f(y)|$ the maximum of $|f|$. Then $\gamma > 0$ and it immediately follows that

$$g(z) := f(x) + \varepsilon_{2r} + \frac{M}{\gamma} g_{\overline{\mathbf{C}}\setminus K}(z,\infty), \qquad z \in G,$$

is an upper function for f. Hence

$$\limsup_{z\to x,\, z\in G} \overline{H}_f^G(z) \le \limsup_{z\to x,\, z\in G} g(z) \le f(x) + \varepsilon_{2r},$$

and since $r > 0$ was arbitrary, and $\varepsilon_{2r} \to 0$ as $r \to 0$ by the assumed continuity f at x, the relation (2.1) follows.

V. Finally, we show that if f is a continuous function, then the upper and lower solutions coincide, and their boundary limits agree with f quasi-everywhere. In fact, by what we have just proved, $\underline{H}_f^G(z) - \overline{H}_f^G(z)$ is a nonpositive harmonic function in G which has boundary limit zero at quasi-every $x \in \partial G$. Thus, this function is zero by the generalized minimum principle, i.e. the upper and lower solutions coincide. This proves the existence of the solution of the Dirichlet problem with boundary function f. Inequalities (2.1) and (2.2) verify the claim concerning the boundary limits of this solution.

Note also that the preceding proof of

$$\lim_{z\to x,\, z\in G} H_f^G(z) = f(x) \tag{2.3}$$

used the continuity of f only at the point x, so if the Dirichlet problem is solvable for an f in G and $x \in \partial G$ is regular with respect to Green functions, then we have (2.3) provided f is continuous at $x \in \partial G$.

Now we are in position to prove the equivalence of the regularity of a point with respect to Dirichlet's problem and Green functions. Recall that $x \in \partial G$ is called a regular boundary point with respect to the Dirichlet problem in G if (2.3) holds for every continuous f.

Theorem 2.1. *Let G be a domain with* $\mathrm{cap}(\partial G) > 0$, *and $x \in \partial G$. Then the following properties are pairwise equivalent.*

(i) *x is regular with respect to the Dirichlet problem, i.e.* (2.3) *is true for every continuous f.*
(ii) *If the Dirichlet problem for the boundary function f is solvable in G and f is continuous at x, then* (2.3) *is true.*
(iii) *x is regular with respect to Green functions in G, i.e. if $a \in G$ and $g_G(z, a)$ is the Green function with pole at a, then*

$$\lim_{z\to x,\, z\in G} g_G(z, a) = 0. \tag{2.4}$$

(iv) *Wiener's condition holds, i.e.*

$$\sum_{n=1}^{\infty} \frac{n}{\log(1/\mathrm{cap}(A_n(x)))} = \infty,$$

where the sets $A_n(x)$ were defined in Theorem 1.1.
(v) *x is a fine limit point of $\overline{\mathbf{C}} \setminus G$.*

Proof. The equivalence of (iii) and (iv) was the content of Theorem 1.1; (ii) obviously implies (i), while it was proved above that (iii) implies (ii). That (iv) implies (v) follows from Lemma I.5.5 (the proof of this lemma used Wiener's theorem Theorem 1.1 which we have verified above, so we can use Lemma I.5.5 here). The implication (v) $\Longrightarrow$ (iii) follows from the definition of fine topology (which means via the Riesz decomposition theorem that every super or subharmonic function is continuous in that topology) as follows: the Green function is zero at every $z \notin G$ except for an F_σ-set E of zero capacity. By Lemma I.5.3, x is also the fine limit point of the set $(\overline{\mathbf{C}} \setminus G) \setminus E$; hence

$$g_G(x,a) = \lim_{z\to x,\, z\notin G\cup E} g_G(z,a) = 0,$$

and this is exactly (iii), for $g_G(z,a)$ is upper semi-continuous (in this proof we used the standard extension of $g_G(z,a)$ to a subharmonic function to $\overline{\mathbf{C}} \setminus \{a\}$).

Thus, it is left to prove that (i) implies (iii).

As before, we can assume that ∂G is a compact subset of $\mathbf{C}$. For $z \in \partial G$ let us define $f(z) := |z - x|$, and let H_f^G be the solution of the corresponding Dirichlet problem (the existence of which has already been proved). Then H_f^G is a positive function in G, so if $r > 0$ is some small fixed number, then there is a constant $c > 0$ such that $H_f^G(z) \ge c\, g_G(z,a)$ for $|z-a| = r$ (here and in what follows replace $|z-a| = r$ with $|z| = 1/r$ when $a = \infty$). But then in the domain $G^* := G \setminus \overline{D_r(a)}$ the function $H_f^G(z) - c\, g_G(z,a)$ is bounded from below, is harmonic there and has nonnegative boundary limits at quasi-every point of ∂G^*, so by the generalized minimum principle it is nonnegative on the whole G^*. Since $H_f^G(z)$ tends to $f(x) = 0$ as $z \to x$, $z \in G$, it follows that the Green function $g_G(z,a)$ also has zero boundary limit as $z \to x$, and this is exactly property (iii). □

A.3 Harmonic Measures and the Generalized Poisson Formula

Let G be a domain with $\operatorname{cap}(\partial G) > 0$, and let a be a point of G. Let us form the balayage $\widehat{\delta_a}$ of the Dirac mass δ_a onto ∂G. Then $\widehat{\delta_a}$ is called the *harmonic measure* of the point a with respect to G.

We have discussed some properties of the harmonic measures in Section II.4, where we showed that the Green function of G with pole at a coincides with

$$g_G(z,a) = \log \frac{1}{|z-a|} - U^{\widehat{\delta_a}}(z) + c_a, \tag{3.1}$$

where $c_a = 0$ if G is bounded, and $c_a = g_G(a,\infty)$ if G contains the point infinity (see formula (II.4.31)).

With the help of harmonic measures we can define the generalized Poisson integral

$$\mathcal{PI}_G(f,z) = \int_{\partial G} f(t)\, d\widehat{\delta_z}(t) \tag{3.2}$$

for functions f defined on the boundary ∂G of G. In this section we shall show that the solution of the Dirichlet problem is given by this generalized Poisson integral whenever this integral exists. We shall also show that if $a, b \in G$, then there is a positive constant $C_{a,b}$ such that $\widehat{\delta_a} \le C_{a,b}\widehat{\delta_b}$. Therefore, any two harmonic measures are comparable, and hence the integrability of f with respect to any of them is equivalent to the integrability with respect to any other one, so there will be no ambiguity in the expression "f is integrable with respect to harmonic measures".

Theorem 3.1 (Brelot's Theorem). *Let G be a domain such that ∂G is compact and of positive capacity, and let f be a finite, Borel measurable function defined on ∂G. Then the Dirichlet problem in G is solvable for f if and only if f is integrable with respect to harmonic measures, and then the solution is given by the generalized Poisson integral*

$$\mathcal{PI}_G(f, z) = \int_{\partial G} f d\widehat{\delta_z}.$$

In particular, this is true for every bounded Borel measurable function.

The proof of the theorem yields that the conclusions are true when the existence of the generalized Poisson integral is assumed in the weaker sense that it can be finite or infinite (i.e. the only noncovered case is when the integrals of both the positive and the negative parts of f are infinite).

The theorem allows us to give another meaning to harmonic measures. Let $E \subset \partial G$, and let us consider the Dirichlet problem in G with boundary function equal to 1 on E and equal to 0 on $\partial G \setminus E$. This is solvable, and the solution $\omega_{E,G}$ is (also) called the harmonic measure associated with E and G, though it is not a measure but a harmonic function. $\omega_{E,G}$ can be used to estimate harmonic functions if some information is known on them on the boundary; therefore these harmonic measures $\omega_{E,G}$ play an important role in harmonic analysis. In view of the preceding theorem $\omega_{E,G}$ is given by a generalized Poisson integral, so we deduce the formula

$$\omega_{E,G}(z) = \widehat{\delta_z}(E) \tag{3.3}$$

connecting the two notions of harmonic measures.

Proof. The proof of the theorem will be given in several steps. First we assume the integrability of f with respect to harmonic measures.

I. Here we show that *if $x \in \partial G$ is a regular boundary point, then*

$$\widehat{\delta_a} \to \delta_x \quad as \quad a \to x \tag{3.4}$$

in the weak topology.*

In fact, suppose this is not true. Then there is a $\rho < 1/2$ and an $\varepsilon > 0$ such that for a sequence of points $a = a_1, a_2, \ldots$ converging to x we have $\widehat{\delta_a}(D_\rho(x)) \le 1-\varepsilon$.

First let G be bounded. Let $K = \overline{D_{\rho/2}(x)} \cap \partial G$, and $M = \log(1/\mathrm{cap}(K))$. Then Wiener's criterion yields that x is also a regular boundary point for the domain $\overline{\mathbf{C}} \setminus K$, so

$$U^{\mu_K}(x) = M.$$

On the other hand, the equilibrium potential is strictly less than M outside $D_\rho(x)$, i.e. there is an $\varepsilon_1 > 0$ such that

$$U^{\mu_K}(z) \le M - 3\varepsilon_1, \quad z \notin D_\rho(x).$$

From Lemma I.6.10 we know that there exists an increasing sequence $\{K_n\}$ of compact subsets of K such that $\mu_K(K_n) \to 1$, and with $\mu_n = \mu_K|_{K_n}$ all the potentials U^{μ_n} are continuous. By the monotone convergence theorem it follows that for sufficiently large n we have

$$U^{\mu_n}(x) \geq M - \varepsilon\varepsilon_1,$$

while

$$U^{\mu_n}(z) \leq M - 2\varepsilon_1, \quad z \notin D_\rho(x).$$

This latter inequality also implies that $U^{\mu_n}(z) \leq M$ for every z (note that the potential of μ_n is not larger than that of μ_K in $D_\rho(x)$, which is at most M everywhere).

Now U^{μ_n} is a continuous function that is harmonic in G, so we can apply property (c) of Theorem II.4.1 to write for $a = a_j$, $j = 1, 2, \ldots$,

$$\begin{aligned} U^{\mu_n}(a) &= \int U^{\mu_n} d\widehat{\delta_a} \leq \widehat{\delta_a}(\overline{\mathbf{C}} \setminus D_\rho(x))(M - 2\varepsilon_1) + \widehat{\delta_a}(D_\rho(x))M \\ &\leq \varepsilon(M - 2\varepsilon_1) + (1 - \varepsilon)M = M - 2\varepsilon\varepsilon_1 \leq U^{\mu_n}(x) - \varepsilon\varepsilon_1 \end{aligned}$$

which, for $j \to \infty$, contradicts the continuity of U^{μ_n}. This contradiction verifies (3.4) for bounded domains.

When G is unbounded, the proof is similar if we use Riesz' formula (II.4.25). In fact, by this formula,

$$U^{\mu_n}(a) = \int U^{\mu_n} d\widehat{\delta_a} - \|\mu_n\| g_G(a, \infty),$$

and we can reason as before:

$$\begin{aligned} U^{\mu_n}(a) &= \int U^{\mu_n} d\widehat{\delta_a} - \|\mu_n\| g_G(a, \infty) \\ &\leq M - 2\varepsilon\varepsilon_1 - \|\mu_n\| g_G(a, \infty) \\ &\leq U^{\mu_n}(x) - \varepsilon\varepsilon_1 - \|\mu_n\| g_G(a, \infty), \end{aligned}$$

and since here $g_G(a_j, \infty) \to 0$ as $j \to \infty$ by the regularity of $x \in \partial G$, we arrive again at a contradiction with the assumed continuity of U^{μ_n}.

II. Suppose that f is continuous. We are going to verify below that $\mathcal{PI}_G(f, z)$ is harmonic in G. What we have just proven gives

$$\lim_{z \to x, z \in G} \mathcal{PI}_G(f, z) = f(x)$$

for every regular point $x \in \partial G$; hence this is true quasi-everywhere. Now we can apply the simple Lemma I.2.6 to conclude that $\mathcal{PI}_G(f, z)$ is indeed the solution of the Dirichlet problem, i.e. for continuous f the theorem is verified.

III. Next we show the validity of the theorem for semi-continuous boundary functions f. Let us suppose for example, that f is lower semi-continuous.

Let g be an upper function for f, and let us consider the function

$$g^*(x) := \liminf_{z \to x, z \in G} g(z), \qquad x \in \partial G. \tag{3.5}$$

Then g^* is lower semi-continuous, and $g^* \geq f$. Thus, there is a sequence $\{g_n^*\}$ of continuous functions converging monotone increasingly to g^* (on ∂G, of course). Now $\mathcal{PI}_G(g_n^*, z)$ solves the Dirichlet problem for g_n^*, and g is an upper function for the latter function, so $\mathcal{PI}_G(g_n^*, z) \leq g(z)$. For $n \to \infty$ we obtain from the monotone convergence theorem the inequality $P\mathcal{I}_G(g^*, z) \leq g(z)$. But $\mathcal{PI}_G(f, z) \leq \mathcal{PI}_G(g^*, z)$, so by taking the infinium for all upper functions g we can deduce that $\mathcal{PI}_G(f, z) \leq \overline{H}_f^G(z)$.

Note that this argument did not use the semi-continuity of f and can be repeated for lower functions, as well. Thus, we have proved that

$$\underline{H}_f^G(z) \leq \mathcal{PI}_G(f, z) \leq \overline{H}_f^G(z) \tag{3.6}$$

for every z and f for which the generalized Poisson integral converges.

Now we use the lower semi-continuity of f. It implies via (3.4) that

$$\liminf_{z \to x, z \in G} \mathcal{PI}_G(f, z) \geq f(x)$$

for every $x \in G$ that is a regular boundary point of ∂G; hence this is true quasi-everywhere. We claim that the set E where this inequality does not hold is an F_σ-set. In fact, f is lower semi-continuous, so there are continuous functions $f_n(x) < f(x)$ converging monotonically to $f(x)$ at every $x \in \partial G$. Now if

$$E := \{x \in \partial G \mid \liminf_{z \to x, z \in G} \mathcal{PI}_G(f, z) < f(z)\},$$

and

$$E_n := \{x \in \partial G \mid \liminf_{z \to x, z \in G} \mathcal{PI}_G(f, z) < f_n(z)\},$$

then $E = \cup_{n=1}^{\infty} E_n$ and each E_n is compact.

Thus, E is an F_σ set of zero capacity, so for every $z \in G$ there is a finite measure $\nu = \nu_z$ such that $U^\nu(z) < \infty$ but $U^\nu(x) = \infty$ for every $x \in E$ (see Lemma I.2.3). Now if m is the infimum of U^ν on ∂G, then for every $\varepsilon > 0$ the sum

$$\mathcal{PI}_G(f, x) + \varepsilon\,(U^\nu(x) - m)\,, \qquad \nu = \nu_z,$$

is an upper function by the choice of ν. Hence

$$\mathcal{PI}_G(f, x) + \varepsilon\,(U^\nu(x) - m) \geq \overline{H}_g^G(z), \qquad \nu = \nu_z,$$

and for $\varepsilon \to 0$ we obtain the converse of the right estimate in (3.6), by which we have verified

$$\mathcal{P}I_G(f, z) = \overline{H}_f^G(z). \tag{3.7}$$

Finally, since f is lower semi-continuous, there is an increasing sequence of continuous functions $\{f_n\}$ converging to f. By part **II** of this proof, then

$$\underline{H}_f^G(z) \geq \underline{H}_{f_n}^G(z) = \mathcal{PI}_G(f_n, z),$$

and here the right-hand side tends to $\mathcal{PI}_G(f, z)$ as $n \to \infty$, by the monotone convergence theorem. Thus, $\underline{H}_f^G(z) \geq \mathcal{PI}_G(f, z)$ holds. This and equality (3.7) prove the theorem for f.

IV. Let f be any function integrable with respect to any of the $\widehat{\delta_z}$, $z \in G$ (and then with respect to any other one, see **V** below). Then, by the Vitali-Carathéodory theorem, there is a sequence $\{g_n\}$ of lower semi-continuous, and another sequence $\{h_n\}$ of upper semi-continuous functions such that $h_n \leq f \leq g_n$, and

$$\int (g_n - h_n) d\widehat{\delta_{z_0}} \to 0, \tag{3.8}$$

where $z_0 \in G$ is some fixed point (see [195, Theorem 2.25]). Then, in view of the comparability of the harmonic measures, the same relation is true if z_0 is replaced by any $z \in G$. For h_n and g_n we can apply part **III** to conclude (see (3.6))

$$\begin{aligned}\mathcal{PI}_G(h_n, z) &= \underline{H}_{h_n}^G(z) \leq \underline{H}_f^G(z) \leq \mathcal{PI}_G(f, z) \\ &\leq \overline{H}_f^G(z) \leq \overline{H}_{g_n}^G(z) = \mathcal{PI}_G(g_n, z).\end{aligned}$$

Now the proof is completed by the observation that here the difference of the left and right-hand sides tends to zero in view of (3.8).

V. In this part of the proof we show that any two harmonic measures are comparable. More precisely, if a and b both belong to a compact subset S of G, then

$$\widehat{\delta_a} \leq C_{S,G} \widehat{\delta_b}, \tag{3.9}$$

where the positive constant $C_{S,G}$ depends only on S and G.

It is enough to prove this for domains with C^2 boundary. In fact, then for other G's we can select an increasing sequence of domains $\{G_n\}$ with C^2 boundary exhausting G. We have shown in Section II.4 that then $\widehat{\delta_a}^{\partial G_n} \to \widehat{\delta_a}^{\partial G}$ in the weak* topology, where the upper index indicates onto what set we take the balayage. It is also immediate from the proof (and from the rest of the proof below) that this convergence is uniform in $a \in S$; furthermore for large n the constants C_{S,G_n} are bounded. Then, letting $n \to \infty$ we can conclude from

$$\int h \, d\widehat{\delta_a}^{\partial G_n} \leq C_{S,G_n} \int h \, d\widehat{\delta_b}^{\partial G_n}$$

the inequality

$$\int h \, d\widehat{\delta_a} \leq C'_{S,G} \int h \, d\widehat{\delta_b}$$

for any nonnegative continuous h with compact support in $\mathbf{C}$, which is enough to conclude (3.9).

Thus, let us assume that G is of C^2 boundary. Choose a closed set $S_1 \subset G$ such that S is contained in its interior. There is a number C^* such that on ∂S_1 we have $g_G(z,a) \le C^* g_G(z,b)$ for every $z \in \partial S_1$ and $a, b \in S$. Since both of these functions vanish on the boundary ∂G, an application of the maximum modulus theorem yields that the same inequality continues to hold in all of $G \setminus S_1$. Thus, for the normal derivatives on ∂G in the direction of the inner normal, we also have

$$\frac{\partial g_G(s,a)}{\partial \mathbf{n}} \le C^* \frac{\partial g_G(s,b)}{\partial \mathbf{n}},$$

and for such domains the claim follows by Theorem II.4.10, according to which the harmonic measure is given by the normal derivative of the Green function:

$$d\widehat{\delta_a}(s) = \frac{1}{2\pi} \frac{\partial g_G(s,a)}{\partial \mathbf{n}} \, ds.$$

VI. Finally, we verify that every generalized Poisson integral is harmonic in G.

In the same way as before, we can assume that G has C^2 boundary (imagine f to be continuously extended to a neighborhood of ∂G). For $\tau > 0$, let γ_τ be the contour $\{s + \tau \mathbf{n}(s) | \in \partial G\}$ (see Fig. 3.1). It follows from the smoothness of the boundary that

$$\frac{\partial g_G(s + \tau\mathbf{n}(s), a)}{\partial \mathbf{n}(s)} \to \frac{\partial g_G(s,a)}{\partial \mathbf{n}(s)}$$

uniformly in $s \in \partial G$ as $\tau \to 0$. By the symmetry of the Green function (Theorem II.4.8), the functions $g_G(s + \tau\mathbf{n}(s), a)$ are harmonic in a, so the same is true of

$$\frac{\partial g_G(s + \tau\mathbf{n}(s), a)}{\partial \mathbf{n}(s)},$$

and then of course every integral

$$\int_{\partial G} f(s) \frac{\partial_{g_G}(s + \tau\mathbf{n}(s), a)}{\partial \mathbf{n}(s)} \, ds$$

inherits this property. Now the harmonicity of

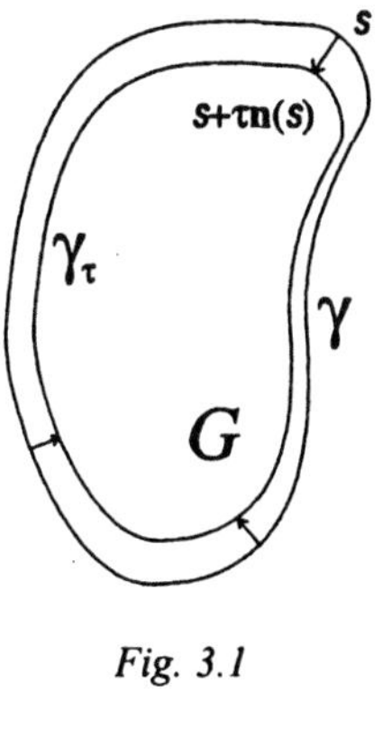

Fig. 3.1

$$\int_{\partial G} f(s) d\widehat{\delta_a}(s) = \int_{\partial G} f(s) \frac{1}{2\pi} \frac{\partial g_G(s,a)}{\partial \mathbf{n}} \, ds$$

in a follows by letting τ tend to 0.

This completes the proof of the sufficiency of the integrability condition.

We still have to prove its necessity. Thus, suppose that the Dirichlet problem is solvable for f, and for an upper function g consider the associated function g^* defined in (3.5), and analogously, for a lower function h set

$$h_*(x) := \liminf_{z \to x,\ z \in G} h(z), \qquad x \in \partial G.$$

Fix $z \in G$. We have verified in the proof above (see part **III**) that

$$\mathcal{PI}_G(g^*, z) \le g(z),$$

and similar reasoning shows that

$$h(z) \le \mathcal{PI}_G(h_*, z),$$

i.e.

$$h(z) \le \mathcal{PI}_G(h_*, z) \le \mathcal{PI}_G(g^*, z) \le g(z).$$

Observing that here, by assumption of the existence of the Dirichlet solution, the supremum of the left-hand side for all h coincides with the infimum of the right-hand side for all g, and furthermore that

$$h_*(t) \le f(t) \le g^*(t)$$

for all $t \in \partial G$, the integrability of f with respect to $\widehat{\delta_z}$ immediately follows. □

Appendix B. Weighted Approximation in $\mathbf{C}^N$

by Thomas Bloom

In this appendix we will present multidimensional versions of some of the results on weighted approximation given in Chapters I, II and III.

Those chapters rely on potential theory in one complex variable and its extension to the weighted case. For the multidimensional generalizations we will use pluripotential theory. This theory has been developed over the last 30 years or so and in particular gives the "correct" version of capacity of sets in $\mathbf{C}^N$.

We will use it to construct a weighted pluricomplex Green function $V^*_{\Sigma,Q}$, given a closed set $\Sigma \subset \mathbf{C}^N$ and an admissible weight w on Σ. The support of the associated Monge-Ampère measure is the set on which the "sup" norm of a weighted polynomial lives (see Theorems 2.6 and 2.11).

In the one variable case, this is equivalent to the following: One uses Theorem I.4.1 (restated for subharmonic functions and upper envelopes) to characterize $F_w - U^{\mu_w}$ and then applies the Riesz decomposition theorem to obtain

$$\mu_w = \frac{1}{2\pi}\Delta(F_w - U^{\mu_w}).$$

In one variable this approach loses the "electrostatic" interpretation and it is not the most straightforward way to embark on explicit computation. However in several variables it is the only option. In $\mathbf{C}^N$, the Monge-Ampère measure $(dd^c V^*_{\Sigma,Q})^N$ arises from the complex Monge-Ampère operator $(dd^c)^N$, which is the natural operator associated to pluripotential theory, and it is non-linear (if $N > 1$).

Explicit methods for computing the support of the Monge-Ampère measure are not available in the several variable case and it would be interesting to develop some.

In Section B.1 we review some basic facts from pluripotential theory.

In Section B.2 we prove the results on where the "sup" norm and sup norm of a weighted polynomial live. Our approach does not use results on Fekete points.

In Section B.3 we define Fekete points in the weighted several variable case. Theorem 3.2 gives a result on the distribution of Fekete points but a generalization of Theorem III.1.3 is not as yet known in several variables (see Problem 3.3).

B.1 Pluripotential Theory

In this section we give a brief summary of various concepts from pluripotential theory.

$\mathbf{C}^N$ denotes complex N-space. We use $z = (z_1, \ldots, z_N)$ with $z_i \in \mathbf{C}$ as coordinates for $\mathbf{C}^N$. We may identify $\mathbf{C}^N$ with $\mathbf{R}^{2N}$ (Euclidean $2N$-space) with coordinates $(x_1, y_1, \ldots, x_N, y_N)$ where $x_k = \mathrm{Re}(z_k)$ and $y_k = \mathrm{Im}(z_k)$ for $k = 1, \ldots, N$.

Under this identification of $\mathbf{C}^N$ with $\mathbf{R}^{2N}$ all the usual concepts from real analysis in Euclidean space (e.g. Lebesgue $2N$-dimensional measure) apply to $\mathbf{C}^N$. The Euclidean norm of a point $z \in \mathbf{C}^N$ is given by $|z| := (|z_1|^2 + \cdots + |z_N|^2)^{1/2}$. The open ball of center z^0 and radius r, $r > 0$, is

$$B(z^0, r) \;:=\; \{z \in \mathbf{C}^N \,|\, |z - z^0| < r\}.$$

An N multi-index $\alpha = (\alpha_1, \ldots, \alpha_N)$ is an N-tuple of non-negative integers. The monomial $(z_1^{\alpha_1}) \ldots (z_N^{\alpha_N})$ is denoted z^α. It is a monomial of degree $|\alpha| = \alpha_1 + \cdots + \alpha_N$.

A polynomial $p(z) = \sum_{|\alpha| \le n} c_\alpha z^\alpha$ is of degree n if at least one of the coefficients c_α with $|\alpha| = n$ is non-zero.

Let G be an open subset of $\mathbf{C}^N$.

Definition 1.1. A function $u : G \to [-\infty, \infty)$ is *upper semi-continuous* (u.s.c.) on G if for every $z^0 \in G$, $\limsup_{z \to z^0} u(z) \le u(z^0)$.

The function u is called *lower semi-continuous* (l.s.c.) if $-u$ is upper semi-continuous.

Given a function $f : G \to [-\infty, \infty)$, its upper semi-continuous regularization f^* is defined by

$$f^*(z^0) \;=\; \limsup_{z \to z^0} f(z) \quad \text{for all} \quad z^0 \in G. \tag{1.1}$$

Then $f^* \ge f$ and f^* is the smallest u.s.c. function with this property. That is, if u is u.s.c. on G and $u \ge f$, then $u \ge f^*$ (see also notes to Section II.2).

The concept of upper semi-continuity given here is the same as that given in Chapter 0.

In pluripotential theory, the standard convention is to work with the multivariable version of subharmonic functions (rather than the multivariable version of superharmonic functions). Thus, we have

Definition 1.2. A function $u : G :\to [-\infty, \infty)$ is plurisubharmonic (p.s.h.) if it is u.s.c. on G, $u \not\equiv -\infty$ on any component of G, and, for every $a \in G, b \in \mathbf{C}^N$ the function of the single complex variable $\lambda \to u(a + \lambda b)$ is subharmonic or identically $-\infty$ on every component of the set $\{\lambda \in \mathbf{C} \,|\, a + \lambda b \in G\}$.

We will use the notation $u \in PSH(G)$.

For example, let f be an analytic function on G with $f \not\equiv 0$. Then $\log|f(z)| \in PSH(G)$ (compare with Example 0.5.4).

The set of plurisubharmonic functions on G that are locally bounded is denoted by $PSH(G) \cap L^\infty_{\text{loc}}(G)$.

The p.s.h. functions on $\mathbf{C}^N$ of at most logarithmic growth at ∞ are

$$\mathcal{L} := \{u \in PSH(\mathbf{C}^N) \mid u \le \log^+|z| + C\}. \tag{1.2}$$

Also those p.s.h. functions on $\mathbf{C}^N$ of logarithmic growth at ∞ are

$$\mathcal{L}_+ := \{u \in PSH(\mathbf{C}^N) \mid \log^+|z| + D_1 \le u \le \log^+|z| + D_2\}. \tag{1.3}$$

Here $\log^+|z| = \max\{\log|z|, 0\}$ and in (1.2) and (1.3) the constants C, D_1, D_2 may depend on u. Clearly $\mathcal{L}_+ \subset \mathcal{L}$.

For example, let $p(z)$ be a polynomial of degree $d \ge 1$ on $\mathbf{C}^N$. Then $(1/d)\log|p(z)| \in \mathcal{L}$ but only for $N = 1$ is $(1/d)\log|p(z)| \in \mathcal{L}_+$. This is because, for $N > 1$, the zero set of a non-constant polynomial is not compact.

The starting point for potential theory in one complex variable is Laplace's equation and its solutions (harmonic functions). In pluripotential theory, the corresponding role is played by the (homogeneous) complex Monge-Ampère equation and its solutions (rather than Laplace's equation and harmonic functions on $\mathbf{R}^{2N}$). This is essentially because a "free upper envelope" of plurisubharmonic functions must satisfy the complex Monge-Ampère equation. A specific result in this direction is Theorem 1.3 below.

We will first describe the complex Monge-Ampère equation. We consider the operators defined in terms of the real coordinates of $\mathbf{R}^{2N}$ by, for $k = 1, \ldots, N$,

$$\frac{\partial}{\partial z_k} := \frac{1}{2}\left(\frac{\partial}{\partial x_k} - \sqrt{-1}\,\frac{\partial}{\partial y_k}\right), \quad \frac{\partial}{\partial \bar{z}_k} := \frac{1}{2}\left(\frac{\partial}{\partial x_k} + \sqrt{-1}\,\frac{\partial}{\partial y_k}\right) \tag{1.4}$$

and the differential forms

$$dz_k = dx_k + \sqrt{-1}\,dy_k, \quad d\bar{z}_k = dx_k - \sqrt{-1}\,dy_k. \tag{1.5}$$

Let u be real-valued and twice continuously differentiable on the open set G ($u \in C^2(G)$). Then u satisfies the complex Monge-Ampère equation if

$$\det\left(\frac{\partial^2 u}{\partial z_i \partial \bar{z}_j}\right) = 0, \tag{1.6}$$

where $i, j = 1, \ldots, N$, and det denotes determinant.

For $N = 1$,

$$\frac{\partial^2}{\partial z \partial \bar{z}} = \frac{1}{4}\left(\frac{\partial^2}{\partial x^2} + \frac{\partial^2}{\partial y^2}\right)$$

and (1.6) reduces to Laplace's equation. In contrast, however, to the one variable case, for $N > 1$ solutions of (1.6) are not necessarily real analytic. Indeed there exist solutions of (1.6) "in the sense of distributions" which are not C^2.

For $u \in C^2(G)$ we consider the $2N$-form

$$(dd^c u)^N := dd^c u \wedge \ldots \wedge dd^c u, \tag{1.7}$$

where

$$d^c u = \sqrt{-1}\sum_{j=1}^{N}\left(\frac{\partial u}{\partial \bar{z}_j}d\bar{z}_j - \frac{\partial u}{\partial z_j}dz_j\right). \tag{1.8}$$

Then,

$$dd^c u = 2\sqrt{-1}(\sum_{j,k=1}^{N}\frac{\partial^2 u}{\partial z_j \partial \bar{z}_k}dz_j \wedge d\bar{z}_k)$$

and

$$(dd^c u)^N = 4^N N! \det\left(\frac{\partial^2 u}{\partial z_j \partial \bar{z}_k}\right) dV, \tag{1.9}$$

where dV is the standard $2N$-dimensional volume form on $\mathbf{R}^{2N}$.

The operator $u \to (dd^c u)^N$ has an extension "in the sense of distributions" to locally bounded plurisubharmonic functions on G. For $u \in PSH(G) \cap L^\infty_{\text{loc}}(G)$, $(dd^c u)^N$ is a locally finite positive Borel measure on G. The proof of this fact is not a standard application of the theory of distributions as (1.9) shows that one must consider a product of distributions.

The next theorem gives an important specific result showing that solutions of (1.6) play, in the several variable case, the role that harmonic functions do in the one variable case. Given a locally bounded plurisubharmonic function u, if u does not satisfy the complex Monge-Ampère equation in a neighborhood N of a point z^0, then there exists another plurisubharmonic function $\tilde{u}$ which equals u outside N, is strictly larger than u at some points of N and $\tilde{u}$ satisfies (1.6) in a neighborhood of z^0.

The theorem implies that a "free upper envelope" of plurisubharmonic functions satisfies the complex Monge-Ampère equation.

Theorem 1.3. *Let $u \in PSH(G) \cap L^\infty_{\text{loc}}(G)$. Let z^0 be a point of G and $\overline{B(z^0, R)} \subset G$, $R > 0$. Then there exists a unique $\tilde{u} \in PSH(G) \cap L^\infty_{\text{loc}}(G)$ such that*

(i) $(dd^c \tilde{u})^N = 0$ *on* $B(z^0, R)$;
(ii) $\tilde{u} = u$ *on* $G \setminus B(z^0, R)$;
(iii) $\tilde{u} \geq u$ *on* $B(z^0, R)$.

Indeed, on $B(z^0, R)$ we can take the Perron-Bremmerman envelope

$$\tilde{u}(z) = \sup\{v(z) \mid v \in PSH(B(z^0, R)),\ \limsup_{\xi \to \xi'} v(\xi) \leq u(\xi') \quad \textit{for all } \xi' \in \partial B(z^0, R)\}.$$

This is the analogue of the Perron-Wiener-Brelot lower solution of the classical Dirichlet problem (see Section I.2).

The next definition provides the generalization to several variables of sets of capacity zero.

Definition 1.4. A set $F \subset \mathbf{C}^N$ is *pluripolar* if, for all $a \in F$, there is a neighborhood B of a and a function $u \in PSH(B)$ such that $F \cap B \subset \{u = -\infty\}$.

By a theorem of Josephson ([101], Theorem 4.7.4), given F pluripolar, there is a function $u \in PSH(\mathbf{C}^N)$ such that $F \subset \{u = -\infty\}$. In fact, we may take $u \in \mathcal{L}$ ([101], Theorem 5.2.4).

For example, if $F \subset \{z \in \mathbf{C}^N | f(z) = 0\}$ and f is analytic on $\mathbf{C}^N$, $f \not\equiv 0$, then F is pluripolar since $\log|f| \in PSH(\mathbf{C}^N)$.

As in the one variable case, we say that a property holds quasi-everywhere (q.e.) on a set S if it holds on $S \setminus F$, where F is pluripolar.

An important property of capacity of sets in one variable which holds in several variables is

Theorem 1.5. ([101], Theorem 4.7.7) *A countable union of pluripolar sets is again pluripolar.*

In one variable, sets which satisfy Definition 1.4 are known as polar sets (rather than pluripolar sets). A Borel set in the plane is of capacity zero if and only if it is polar. Given a compact set of capacity zero in the plane, Evans' Theorem (III.1.11) gives a potential whose value on the set is $+\infty$, thus exhibiting it as a polar set.

In fact, in the several variable case, there is a theory of capacity of sets in which the pluripolar sets are precisely those of outer capacity zero but we will not need this aspect of the theory (see [101]).

In pluripotential theory, specific functions are often constructed as an upper envelope of plurisubharmonic functions. The next theorem gives convenient conditions for the (regularized) upper envelope of a family of plurisubharmonic functions of at most logarithmic growth at ∞ to again be a plurisubharmonic function of at most logarithmic growth at ∞.

Theorem 1.6. ([101], Proposition 5.2.1) *Let* $\mathcal{U} = \{u_i\}_{i \in I}$ *be a family of functions in* $\mathcal{L}$. *Let* $u(z) = \sup_{i \in I} u_i(z)$. *Suppose that* $\{z \in \mathbf{C}^N \mid u(z) < +\infty\}$ *is not pluripolar. Then* $u^* \in \mathcal{L}$.

In fact u as defined in Theorem 1.6 is not, in general, u.s.c.. The function u^*, its upper semicontinuous regularization, is, and this property is needed for a function to be plurisubharmonic.

Now it is important that, in the situation of Theorem 1.6, the function u^* does not differ from u on a large set. Specifically, we have the following result of Bedford and Taylor.

Theorem 1.7. ([101], Theorem 4.7.6) *In the situation of Theorem* 1.6 *the set* $\{z \mid u^*(z) > u(z)\}$ *is pluripolar.*

We will now give the generalization to several variables of the Green function with pole at ∞. Let $\Sigma \subset \mathbf{C}^N$ be compact.

Definition 1.8. The *pluricomplex Green function* of Σ, denoted by V_Σ, is defined for $z \in \mathbf{C}^N$ by

$$V_\Sigma(z) := \sup\{u(z) \mid u \in \mathcal{L} \quad \text{and} \quad u \le 0 \quad \text{on} \quad \Sigma\}. \tag{1.10}$$

The function $V_\Sigma^*(z)$ denotes its upper semi-continuous regularization.

Then Σ is pluripolar if and only if $V_\Sigma^* \equiv +\infty$ ([101], Corollary 5.2.2).

If Σ is not pluripolar, the function $V_\Sigma^*(z)$ has the following properties (compare with those of the Green function, Section I.4 or II.4):

(i) $V_\Sigma^*(z)$ is non-negative and (by Theorem 1.3) satisfies the complex Monge-Ampère equation on $\mathbf{C}^N \setminus \Sigma$;

(ii) $V_\Sigma^* \in \mathcal{L}_+$ since $V_\Sigma \in \mathcal{L}$ (by Theorem 1.6) and for C large, $\log^+|z| - C \le 0$ on Σ so $V_\Sigma^*(z) \ge \log^+|z| - C$;

(iii) $V_\Sigma^*(z) = 0$ q.e. on Σ (by Theorem 1.7).

The Monge-Ampère measure $(dd^c V_\Sigma^*)^N$ has total mass $(2\pi)^N$ ([101], Corollary 5.5.3) and $\operatorname{supp}(dd^c V_\Sigma^*)^N \subset \Sigma$. This measure is referred to as the *equilibrium measure* of Σ.

We will give some examples of compact sets, pluricomplex Green functions and equilibrium measures without details of calculations. Further examples and a sample of the calculations involved can be found in [101].

Example 1.9.

(i) Let $\Sigma = \{z \in \mathbf{C}^N \bigm| |z| \le 1\}$. Σ is the unit Euclidean ball in $\mathbf{C}^N$.

Then $V_\Sigma = V_\Sigma^* = \log^+|z|$ and $(dd^c V_\Sigma^*)^N$ is (up to normalization) the surface area on the sphere $\{z \in \mathbf{C}^N \bigm| |z| = 1\}$.

(ii) Let $\Sigma = \{(z_1, z_2) \in \mathbf{C}^2 \bigm| |z_1| \le 1,\ |z_2| \le 1\}$. The set Σ is the unit polydisc in $\mathbf{C}^2$. Then

$$V_\Sigma = V_\Sigma^* = \max\{\log^+|z_1|,\ \log^+|z_2|\}$$

and $(dd^c V_\Sigma)^2$ is (up to normalization) the measure $d\theta_1 d\theta_2$ on

$$\{(z_1, z_2) \in \Sigma \bigm| |z_1| = 1,\ |z_2| = 1\}$$

where θ_1, θ_2 are the angular parts of polar coordinates for z_1, z_2.

$V_\Sigma(z_1, z_2)$ is not of class C^2 on $\mathbf{C}^2 \setminus \Sigma$ but nevertheless satisfies the complex Monge-Ampère equation "in the sense of distributions" on $\mathbf{C}^2 \setminus \Sigma$.

(iii) Let $\Sigma = \{(z_1, z_2) \in \mathbf{C}^2 \bigm| |z| \le 1\} \cup \{(z_1, z_2) \in \mathbf{C}^2 \bigm| z_2 = 0, |z_1| \le 2\}$.

The set Σ is the unit Euclidean ball in $\mathbf{C}^2$ union a pluripolar set. We have

$$V_\Sigma(z) = \begin{cases} \log^+|z| & \text{for } z_2 \ne 0 \\ \log^+|z_1/2| & \text{for } z_2 = 0 \end{cases}$$

and

$$V_\Sigma^*(z) = \log^+ |z|.$$

Note that

$$\{(z_1, z_2) \in \mathbf{C}^2 \mid V_\Sigma^*(z) > V_\Sigma(z)\} = \{(z_1, z_2) \in \mathbf{C}^2 \mid z_2 = 0, \quad |z_1| > 1\},$$

and this is a pluripolar set. □

Two important theorems follow:

Theorem 1.10. *Let $u \in PSH(G) \cap L^\infty_{\text{loc}}(G)$. Then the measure $(dd^c u)^N$ places zero mass on any pluripolar set.*

Theorem 1.11 (Principle of Domination). *Let $u \in \mathcal{L}$, $v \in \mathcal{L}_+$ and suppose $u \leq v$ almost everywhere with respect to the measure $(dd^c v)^N$ on* $\text{supp}(dd^c v)^N$. *Then $u \leq v$ on $\mathbf{C}^N$.*

In particular, in view of Theorem 1.10 the hypotheses of Theorem 1.11 are satisfied if $u \leq v$ q.e. on $\text{supp}(dd^c v)^N$.

In the one variable case Theorem 1.11 includes the principle of domination of Section I.3 which there is stated for (superharmonic) potentials.

B.2 Weighted Polynomials in $\mathbf{C}^N$

Let $\Sigma \subset \mathbf{C}^N$ be a closed set and w a real-valued function on Σ such that $w \geq 0$. The function w is called a *weight function.* As in the one variable case, we have

Definition 2.1. A weight function is *admissible* if it satisfies the following properties:

(i) w is upper semi-continuous;

(ii) the set

$$\{z \in \Sigma \mid w(z) > 0\} \tag{2.1}$$

is not pluripolar;

(iii) if Σ is unbounded then $|z|w(z) \to 0$ as $|z| \to \infty$, $z \in \Sigma$.

We define $Q = Q_w$ via $Q = -\log w$. Then Q is l.s.c. on Σ, the set $\{z \in \Sigma \mid Q(z) < +\infty\}$ is not pluripolar and, if Σ is unbounded,

$$\lim_{|z|\to\infty, z\in\Sigma} (Q(z) - \log|z|) = +\infty. \tag{2.2}$$

The weighted *pluricomplex Green function of Σ with respect to Q* is defined by

$$V_{\Sigma,Q}(z) := \sup\{u(z) \mid u \in \mathcal{L},\ u \leq Q \quad \text{on } \Sigma\}. \tag{2.3}$$

We let $V^*_{\Sigma,Q}$ denote its upper semi-continuous regularization. Of course, in the unweighted case, that is $w \equiv 1$, we have $Q \equiv 0$ and $V_{\Sigma,Q}$ is, for Σ compact, the pluricomplex Green function of Σ (see (1.10)).

Using condition (ii) and Theorem 1.6 we conclude that $V^*_{\Sigma,Q} \in \mathcal{L}$.

Now, for $\rho > 0$ we let $\Sigma_\rho := \{z \in \Sigma \mid |z| \le \rho\}$. We will now show that even if Σ is unbounded there is a compact subset of Σ that determines $V_{\Sigma,Q}$.

Lemma 2.2. *For ρ sufficiently large, $V_{\Sigma,Q} = V_{\Sigma_\rho,Q}$.*

Proof. By condition (ii) Σ is not pluripolar. Thus, by Theorem 1.5, for some $\sigma > 0$, Σ_σ is not pluripolar. Then, since $V^*_{\Sigma_\sigma,Q} \in \mathcal{L}$ we have, for some constant $c > 0$,

$$V^*_{\Sigma_\sigma,Q} \le \log^+|z| + c. \tag{2.4}$$

Using condition (2.2), we may choose $\rho > \sigma$ so large that

$$Q(z) - \log|z| \ge c + 1 \quad \text{for} \quad z \in \Sigma \setminus \Sigma_\rho. \tag{2.5}$$

Now suppose $u \in \mathcal{L}$ and $u \le Q$ on Σ_ρ. Then since $u \le V^*_{\Sigma_\sigma,Q}$ we obtain from (2.4) and (2.5) that $u(z) \le Q(z)$ for $z \in \Sigma \setminus \Sigma_\rho$. We conclude that $u \le Q$ on Σ if $u \le Q$ on Σ_ρ. This means that $V_{\Sigma_\rho,Q} \le V_{\Sigma,Q}$. The reverse inequality being obvious, the proof is concluded. □

The Borel measure $(dd^c V^*_{\Sigma,Q})^N$ has compact support since, by Lemma 2.2, it is equal to $(dd^c V^*_{\Sigma_\rho,Q})^N$ whose support is contained in Σ_ρ.

We will use the notation

$$\mu_w := (dd^c V^*_{\Sigma,Q})^N, \tag{2.6}$$

$$\mathcal{S}_w := \operatorname{supp}(\mu_w), \tag{2.7}$$

$$\mathcal{S}^*_w := \{z \in \Sigma \mid V^*_{\Sigma,Q}(z) \ge Q(z)\}. \tag{2.8}$$

This is the same notation as used in the one variable case ($\mathcal{S}_w$ and μ_w are introduced in Theorem I.1.3 and $\mathcal{S}^*_w$ in Theorem III.1.2). We will justify the use of the same notation as the one-dimensional case in Lemma 2.4. That is, for $\Sigma \subset \mathbf{C}$ and w an admissible weight on Σ we may use Theorem I.1.3, minimizing an energy integral to construct μ_w or using the procedure of this section we construct the measure $\frac{1}{2\pi} dd^c V^*_{\Sigma,Q}$. Then, in fact, both methods give the same Borel measure.

First, however, we prove (compare with Theorem I.1.3(e)) the following lemma.

Lemma 2.3. *Let $\Sigma \subset \mathbf{C}^N$ be a closed set and w an admissible weight function. Then $\mathcal{S}_w \subset \mathcal{S}^*_w$.*

Proof. Suppose $z^0 \in \Sigma \setminus \mathcal{S}^*_w$. That is $V^*_{\Sigma,Q}(z^0) < Q(z^0)$. We will show that $(dd^c V^*_{\Sigma,Q})^N$ has zero mass on a ball centered at z^0 i.e. $V^*_{\Sigma,Q}$ satisfies the complex Monge-Ampère equation in a neighborhood of z^0 so $z^0 \notin \mathcal{S}_w$.

Now, since $V^*_{\Sigma,Q}$ is u.s.c. and Q is l.s.c., there is a ball $B(z^0,r)$, $r>0$, centered at z^0 such that

$$\sup_{z\in B(z^0,r)} V^*_{\Sigma,Q}(z) < \inf_{z\in B(z^0,r)\cap\Sigma} Q(z).$$

Applying Theorem 1.3 to $u=V^*_{\Sigma,Q}$ we obtain $\tilde{u}\geq u$ on $B(z^0,r)$ and $\tilde{u}=u$ on $\partial B(z^0,r)$, $\tilde{u}\in\mathcal{L}$ and $(dd^c\tilde{u})^N\equiv 0$ on $B(z^0,r)$. This implies, by the maximum principle, that for $z\in B(z^0,r)$,

$$\tilde{u}(z) \leq \sup_{z\in B(z^0,r)} V^*_{\Sigma,Q}(z) < \inf_{z\in B(z^0,r)\cap\Sigma} Q(z).$$

(Although plurisubharmonic functions satisfy a maximum principle, in the above case it suffices to use the maximum principle for the subharmonic functions of one complex variable $\lambda\to\tilde{u}(z^0+\lambda v)$ for all unit vectors v in $\mathbf{C}^N$.) Hence $\tilde{u}(z)\leq Q(z)$ for all $z\in\Sigma$ and so $\tilde{u}=V^*_{\Sigma,Q}$. Thus $V^*_{\Sigma,Q}$ satisfies the Monge-Ampère equation in $B(z^0,r)$. □

Lemma 2.4. *Let $\Sigma\subset\mathbf{C}$ be closed and w an admissible weight function on Σ. Then*

$$V^*_{\Sigma,Q} = F_w - U^{\mu_w} \tag{2.9}$$

and

$$\mu_w = (1/2\pi)dd^cV^*_{\Sigma,Q}, \tag{2.10}$$

where F_w and U^{μ_w} are as in Theorem I.1.3.

Proof. Rephrasing Theorem I.4.1 in terms of subharmonic functions and upper envelopes (rather than superharmonic functions and lower envelopes) we have

$$F_w - U^{\mu_w}(z) =$$
$$\sup\{u(z)\,\big|\,u\in\mathcal{L},\ u \text{ is harmonic for } |z| \text{ large and } u\leq Q \text{ q.e. on } \Sigma\}. \tag{2.11}$$

Now $V^*_{\Sigma,Q}$ is a member of the family of functions on the right side of (2.11) since $V_{\Sigma,Q}\leq Q$ on Σ so, by Theorem 1.7, $V^*_{\Sigma,Q}\leq Q$ q.e. on Σ. Thus $F_w-U^{\mu_w}\geq V^*_{\Sigma,Q}$ and we must now prove the reverse inequality.

The function $F_w-U^{\mu_w}$ is in $\mathcal{L}$ and by Theorem I.1.3(d), $F_w-U^{\mu_w}\leq Q$ q.e. on Σ. Since $\operatorname{supp}(dd^cV^*_{\Sigma,Q})$ is contained in Σ, using the principle of domination (Theorem 1.11), we have

$$F_w - U^{\mu_w} \leq V^*_{\Sigma,Q} \quad \text{for all} \quad z\in\mathbf{C}.$$

This proves (2.11).

By the Riesz decomposition theorem

$$\mu_w = (1/2\pi)\Delta(F_w - U^{\mu_w}).$$

But for u subharmonic, $dd^cu = \Delta u\, dm$, where dm is Lebesgue measure on $\mathbf{R}^2$ and this proves (2.10). □

Now, we have in $\mathbf{C}^N$ (compare with Theorem I.1.3) the following result.

Theorem 2.5. *Let $\Sigma \subset \mathbf{C}^N$ be a closed set and w an admissible weight function. Then the following properties hold:*

(i) $\mathcal{S}_w$ *is not pluripolar;*

(ii) $\mathcal{S}_w \subset \mathcal{S}_w^*$;

(iii) $V_{\Sigma,Q}^* \le Q$ *q.e. on* Σ;

(iv) $V_{\Sigma,Q} = Q$ *for q.e.* $z \in \mathcal{S}_w$ *(or* $\mathcal{S}_w^*$*).*

Proof. Property (i) follows from Theorem 1.10. Property (ii) is the same as Lemma 2.3. Property (iii) follows from Theorem 1.7, since $V_{\Sigma,Q} \le Q$, and property (iv) follows from (ii) and (iii). □

A weighted polynomial is, as in the one variable case, a function of the form $w^n P_n$ where P_n is a polynomial of degree $\le n$. We have the following estimates.

Theorem 2.6. *Suppose P_n is a polynomial of degree at most n and $|w^n P_n(z)| \le M$ for q.e. $z \in \mathcal{S}_w$. Then*

(i) $|P_n(z)| \le M \exp(n V_{\Sigma,Q}^*(z))$ *for all* $z \in \mathbf{C}^N$;

(ii) $|w^n P_n(z)| \le M \exp[n(V_{\Sigma,Q}^*(z) - Q(z))]$ *for all* $z \in \Sigma$;

(iii) $|w^n P_n(z)| \le M$ *for q.e.* $z \in \Sigma$.

Proof. The hypothesis implies that

$$\frac{1}{n}\log\left|\frac{P_n(z)}{M}\right| \le Q(z) \quad \text{for q.e.} \quad z \in \mathcal{S}_w.$$

Hence, by Theorem 2.5(iv)

$$\frac{1}{n}\log\left|\frac{P_n(z)}{M}\right| \le V_{\Sigma,Q}^*(z) \quad \text{for q.e.} \quad z \in \mathcal{S}_w.$$

Then, by the principle of domination,

$$\frac{1}{n}\log\left|\frac{P_n(z)}{M}\right| \le V_{\Sigma,Q}^*(z) \quad \text{for all} \quad z \in \mathbf{C}^N,$$

which proves (i) and (ii). Using Theorem 2.5(iii) and (ii) above we obtain (iii). □

In particular, for $z \in \Sigma \setminus \mathcal{S}_w^*$, we have

$$|w^n P_n(z)| < \|w^n P_n\|_\Sigma. \tag{2.12}$$

We will use the notation $\|f\|_K^*$ for the "sup" of a function on a set K. That is,

$$\|f\|_K^* = \inf\{\|f\|_{K\setminus F} \mid F \text{ is pluripolar, } F \subset K\}. \tag{2.13}$$

Remark 2.7. If K is not pluripolar in a neighborhood of any of its points (i.e. $\{z \in K \mid |z - z^0| < \delta\}$ is not pluripolar for any $z^0 \in K, \delta > 0$) and f is continuous on K, then $\|f\|_K = \|f\|_K^*$.

Theorem 2.6(iii) may be reformulated as follows:

$$\|w^n P_n\|_{\mathcal{S}_w}^* = \|w^n P_n\|_\Sigma^* \tag{2.14}$$

for all polynomials P_n of degree at most n ($n = 1, 2, \ldots$).

We will show that $\mathcal{S}_w$ is the smallest compact set such that (2.14) is satisfied for all weighted polynomials. That is, the "sup" norm of a weighted polynomial lives on $\mathcal{S}_w$ (the precise statement is Theorem 2.11). Furthermore, if Σ is not pluripolar at any of its points and w is continuous, then the actual sup norm of any weighted polynomial lives on $\mathcal{S}_w$ (the precise statement is Theorem 2.12). These results generalize to $\mathbf{C}^N$ Theorem III.2.3 and Corollary III.2.6.

First, we will need to represent the weighted pluricomplex Green function as an upper envelope of functions of the form $[\deg(P)]^{-1} \log |P|$, rather than, as it is defined, an upper envelope of functions in $\mathcal{L}$.

Let Σ be closed in $\mathbf{C}^N$ and w an admissible weight function on Σ. For $n = 1, 2, \ldots$ consider

$$\Phi_{\Sigma,n}(z) = \Phi_n(z) := \sup\{|P_n(z)| \mid \deg P_n \le n \quad \text{and} \quad \|w^n P_n\|_\Sigma \le 1\}. \tag{2.15}$$

Then, for each $z \in \mathbf{C}^N$ we have

$$\Phi_n(z)\Phi_m(z) \le \Phi_{n+m}(z). \tag{2.16}$$

We let

$$\Phi_\Sigma(z) = \Phi(z) := \lim_{n\to\infty} (\Phi_n(z))^{1/n} = \sup_{n\ge1} (\Phi_n(z))^{1/n}, \tag{2.17}$$

since from (2.16) it follows that the above limit exists and equals the supremum (see notes to Section III.3). The function Φ_Σ is called the *Siciak extremal function*.

Similarly, we define

$$\psi_{\Sigma,n}(z) = \psi_n(z) := \sup\{|P_n(z)| \mid \deg P_n \le n \quad \text{and} \quad \|w^n P_n\|_\Sigma^* \le 1\} \tag{2.18}$$

and

$$\psi_\Sigma(z) = \psi(z) := \lim_{n\to\infty} (\psi_n(z))^{1/n} = \sup_{n\ge1} (\psi_n(z))^{1/n}. \tag{2.19}$$

Theorem 2.8. *We have*

(i) $\log \Phi(z) = V_{\Sigma,Q}(z)$;

(ii) $(\log \psi(z))^* = V_{\Sigma,Q}^*(z)$.

Proof. We may assume Σ is compact, since replacing Σ by Σ_ρ all the quantities in the statement of Theorem 2.8 remain unchanged.

By definition, for $n = 1, 2, \ldots$,

$$\frac{1}{n} \log |\Phi_n(z)| \le Q(z) \quad \text{for all} \quad z \in \Sigma,$$

so by definition of the weighted pluricomplex Green function,

$$\frac{1}{n} \log |\Phi_n(z)| \le V_{\Sigma,Q}(z) \quad \text{for all} \quad z \in \mathbf{C}^N.$$

Using (2.17), we have

$$\log \Phi(z) \le V_{\Sigma,Q}(z) \quad \text{for all} \quad z \in \mathbf{C}^N.$$

Similarly, if P_n is a polynomial in the family defined by the right side of (2.18)

$$\frac{1}{n} \log |P_n(z)| \le Q(z) \quad \text{for q.e.} \quad z \in \Sigma.$$

Thus, by Theorem 2.5(iv),

$$\frac{1}{n} \log |P_n(z)| \le V_{\Sigma,Q}(z) \quad \text{for q.e.} \quad z \in \mathcal{S}_w$$

and by the principle of domination,

$$\frac{1}{n} \log |P_n(z)| \le V^*_{\Sigma,Q}(z) \quad \text{for all} \quad z \in \mathbf{C}^N,$$

and so

$$\log \psi(z) \le V^*_{\Sigma,Q}(z) \quad \text{for all} \quad z \in \mathbf{C}^N.$$

Now to prove (i) it suffices to show that

$$V_{\Sigma,Q} \le \log \Phi. \tag{2.20}$$

But this will also prove (ii) since if (i) holds, $(\log \Phi)^* = V^*_{\Sigma,Q}$ and since $\log \psi \ge \log \Phi$ we get $(\log \psi)^* \ge V^*_{\Sigma,Q}$.

We will use the following theorem of Siciak which shows that a general function $u \in \mathcal{L}$ may be approximated by functions of the form $[\deg(P)]^{-1} \log |P(z)|$.

Theorem 2.9. *Let $u \in \mathcal{L}$. Then there exists a sequence of functions $\{u_k\}$, $k = 1, 2, \ldots$, satisfying, for all $z \in \mathbf{C}^N$,*

(i) $u_{k+1}(z) \le u_k(z), \quad k = 1, 2, \ldots;$

(ii) $\lim_{k \to \infty} u_k(z) = u(z);$

(iii) *for each k, there exist finitely many polynomials $\{P_{j,k}\}_{1 \le j \le t_k}$ each of degree $\le n_{j,k}$ such that*

$$u_k(z) = \sup_{1 \le j \le t_k} \frac{1}{n_{j,k}} \log |P_{j,k}(z)|.$$

Now, we continue the proof of Theorem 2.8. Suppose $u \in \mathcal{L}$ and $u \leq Q$ on Σ. To prove (2.20) we must show that $u \leq \log \Phi$. Consider a sequence $\{u_k\}$ as in Theorem 2.9. We may suppose u is bounded below on Σ (since Q is l.s.c. it is bounded below by, say, C on Σ and we may replace u by $\mathrm{Max}(u, C)$ if necessary). The sequence u_k decreases monotonically to u which is less than or equal to Q on Σ. By Dini's theorem, given $\varepsilon > 0$ there exists an integer k_0 such that

$$u(z) \leq u_k(z) \leq Q(z) + \varepsilon \quad \text{for all} \quad z \in \Sigma \quad \text{and all} \quad k \geq k_0.$$

Now, possibly adding a term $cz_1^{n_{j,k}}$ to $P_{j,k}$, with c small, we may assume $n_{j,k} = \deg(P_{j,k})$ and

$$u(z) - \varepsilon \;\leq\; u_k(z) \;\leq\; Q(z) + 2\varepsilon \quad \text{for all} \quad z \in \Sigma, \quad \text{all} \quad k \geq k_0.$$

Thus, for $d = d(k) = \sup_{1 \leq j \leq t_k} \deg(P_{j,k})$,

$$u_k(z) - 2\varepsilon \;=\; \sup_{1 \leq j \leq t_k} \frac{\log |e^{-2\varepsilon n_{j,k}} P_{j,k}(z)|}{n_{j,k}} \leq \frac{1}{d} \log \Phi_d(z).$$

Hence $u - 3\varepsilon \;\leq\; \log \Phi$. Since ε is arbitrary, (2.20) follows. □

Now we have (compare with Theorem III.2.9) the following corollary.

Corollary 2.10. *The function $V^*_{\Sigma,Q} = \log \psi$ q.e. on Σ and if $V^*_{\Sigma,Q}$ is continuous at z^0, then*

$$V^*_{\Sigma,Q}(z^0) \;=\; V_{\Sigma,Q}(z^0) \;=\; \log \Phi(z^0) \;=\; \log \psi(z^0).$$

Proof. The first statement follows from Theorem 1.7 and the second statement from Theorem 2.8. □

Theorem 2.11. *Let S be any closed subset of Σ such that $\|w^n P_n\|^*_S = \|w^n P_n\|^*_\Sigma$ for all polynomials P_n of degree at most n ($n = 1, 2, \ldots$). Then $S \supset \mathcal{S}_w$.*

Proof. By Theorem 2.8(ii), we have $V^*_{S,Q} \;=\; V^*_{\Sigma,Q}$. Hence

$$\mathcal{S}_w \;=\; \mathrm{supp}(dd^c V^*_{\Sigma,Q})^N \;=\; \mathrm{supp}(dd^c V^*_{S,Q})^N \subset S.$$

□

Theorem 2.12. *Let Σ be a closed subset of $\mathbf{C}^N$ that is not pluripolar in a neighborhood of any of its points. Let w be a continuous admissible weight function on Σ. Then $\|w^n P_n\|_\Sigma \;=\; \|w^n P_n\|_{\mathcal{S}_w}$ for all polynomials P_n of degree at most n ($n \;=\; 1, 2, \ldots$). Furthermore if S is any closed subset of Σ such that $\|w^n P_n\|_\Sigma = \|w^n P_n\|_S$ for all polynomials P_n of degree at most n ($n = 1, 2, \ldots$), then $S \supset \mathcal{S}_w$.*

Proof. Use Remark 2.7 and Theorem 2.11. □

In the case that the hypotheses of Theorem 2.12 are not satisfied, it is convenient to introduce a new weight function $\bar{w}$ defined by

$$\bar{w}(z) = \lim_{\delta\to 0} \|w\|^*_{B(z,\delta)\cap\Sigma}. \tag{2.21}$$

The proofs of Theorems 2.13 and 2.14 are (replacing "capacity zero" by "pluripolar") identical to the proofs in the one variable case (see proof of Theorem III.2.3) and so we merely state the results.

Theorem 2.13. *The function $\bar{w}$ is admissible, $\bar{w} \le w$, and $\bar{w} = w$ q.e. on Σ.*

Theorem 2.14.

$$\|w^n P_n\|^*_{S_w} = \|\bar{w}^n P_n\|_{S_w} = \|\bar{w}^n P_n\|_\Sigma = \|w^n P_n\|^*_\Sigma.$$

We also have the following relation between w and $\bar{w}$ (here $\overline{Q} = -\log\bar{w}$).

Theorem 2.15. $V^*_{\Sigma,Q} = V^*_{\Sigma,\overline{Q}}$.

Proof. Since $\overline{Q} \le Q$, $V^*_{\Sigma,\overline{Q}} \le V^*_{\Sigma,Q}$ and we need only prove the reverse inequality.

Suppose $u \in \mathcal{L}$ and $u \le Q$ on Σ. Then $u \le \overline{Q}$ q.e. on Σ so $u \le \overline{Q}$ q.e. on $S_{\bar{w}}$. It follows that $u \le V^*_{\Sigma,\overline{Q}}$ q.e. on $S_{\bar{w}}$ and by the principle of domination $u \le V^*_{\Sigma,\overline{Q}}$ on $\mathbf{C}^N$. This proves Theorem 2.15. □

Remark 2.16. We note that the immediate consequences of Theorem 2.15 (see also Theorem III.2.3) are

$$\mu_w = \mu_{\bar{w}} \quad\text{and}\quad S_w = S_{\bar{w}}.$$

B.3 Fekete Points

We will consider Fekete points for subsets of $\mathbf{C}^N$.

First we consider the monomials to be ordered lexicographically. That is $z^\alpha > z^\beta$ if $|\alpha| > |\beta|$ or if $|\alpha| = |\beta|$ and $\alpha_i = \beta_i$ for $i = 1, \dots, j$ but $\alpha_{j+1} > \beta_{j+1}$ (with an obvious meaning if $j = 0$). We use the notation $e_k(z)$ for the k-th monomial under this ordering. For $e_k(z) = z^\alpha$ we write $\alpha = \alpha(k)$. For example, in $\mathbf{C}^2$, the first six monomials under this ordering are

$$e_1 = 1,\ e_2 = z_1,\ e_3 = z_2,\ e_4 = z_1^2,\ e_5 = z_1 z_2 \quad\text{and}\quad e_6 = z_2^2.$$

The Vandermonde determinant in several variables is defined as follows. Let T be a positive integer, $T \ge 2$, and $\zeta_1, \dots, \zeta_T$ points in $\mathbf{C}^N$. The Vandermonde determinant of order T is the $T \times T$ determinant

$$V(\zeta_1, \dots, \zeta_T) = \det \begin{pmatrix} 1 & \dots & \dots & 1 \\ e_1(\zeta_1) & \dots & \dots & e_1(\zeta_T) \\ \vdots & & & \vdots \\ \vdots & & & \vdots \\ e_{T-1}(\zeta_1) & \dots & \dots & e_{T-1}(\zeta_T) \end{pmatrix}. \tag{3.1}$$

It may be considered as a polynomial in NT variables (the coordinates of $\zeta_1, \dots, \zeta_T$).

Let Σ be a closed subset of $\mathbf{C}^N$ and w an admissible weight function on Σ. We also consider the functions

$$W(\zeta_1, \dots, \zeta_T) := V(\zeta_1, \dots, \zeta_T) w(\zeta_1)^{|\alpha(T)|} \dots w(\zeta_T)^{|\alpha(T)|}. \tag{3.2}$$

Note that fixing $T-1$ of the entries on W we have a weighted polynomial of degree $|\alpha(T)|$ in the remaining entry.

Definition 3.1. A T-th Fekete set for Σ (associated to w) consists of points $\xi_1, \dots, \xi_T \in \Sigma$ such that $|W(\xi_1, \dots, \xi_T)| = |\sup_{\zeta_i \in \Sigma} W(\zeta_1, \dots, \zeta_T)|$.

For $N = 1, \alpha(T) = T - 1$ and (3.1) gives

$$W(\zeta_1, \dots, \zeta_T) = \prod_{1 \le i < j \le T} |\zeta_i - \zeta_j| w(\zeta_i) w(\zeta_j)$$

which is equation (III.1.1).

The number of monomials of degree $\le n$ in N variables is denoted by $m_n := \binom{n+N}{n}$. Let $\mathcal{F}_n := (\xi_1^{(n)}, \dots, \xi_{m_n}^{(n)})$ be an m_n-th Fekete set for Σ (associated to w). We let

$$L_w^i(z, \mathcal{F}_n) := \frac{W(\xi_1^{(n)}, \dots, \xi_{i-1}^{(n)}, z, \xi_{i+1}^{(n)}, \dots, \xi_{m_n}^{(n)})}{W(\xi_1^{(n)}, \dots, \xi_{m_n}^{(n)})}. \tag{3.3}$$

Then $L_w^i(z, \mathcal{F}_n)$ is a weighted polynomial of degree at most n for $i = 1, \dots, m_n$, which satisfies

$$L_w^i(\xi_j^{(n)}, \mathcal{F}_n) = \begin{cases} 1 & i = j \\ 0 & i \neq j \end{cases} \tag{3.4}$$

and

$$\|L_w^i\|_\Sigma = 1. \tag{3.5}$$

By Lagrange's formula we have

$$w^n P_n(z) = \sum_{j=1}^{m_n} w^n P_n(\xi_j) L_w^j(z, \mathcal{F}_n) \tag{3.6}$$

so that (see (III.1.12))

$$\|w^n P_n\|_\Sigma \le m_n \|w^n P_n\|_{\mathcal{F}_n}. \tag{3.7}$$

From (3.4), (3.5), and (2.12) we may conclude that $\xi_j^{(n)} \in \mathcal{S}_w^*$ (see (III.1.2)) and $\mathcal{F}_n \subset \mathcal{S}_w^*$ for $n = 1, 2, \dots$.

We will derive further information on the distribution of Fekete points. Namely with $F = \overline{\bigcup_{n=1}^\infty \mathcal{F}_n}$ we have

Theorem 3.2. $\mathcal{S}_w \subset F$.

Proof. We will show $V_{F,Q} = V_{\Sigma,Q}$; hence

$$\mathcal{S}_w = \operatorname{supp}(dd^c V_{\Sigma,Q}^*)^N = \operatorname{supp}(dd^c V_{F,Q}^*)^N \subset F.$$

Now, suppose $\|w^n P_n\|_F \leq 1$. Then, by (3.7), $\|w^n P_n\|_\Sigma \leq m_n$. Thus (see (2.15)) $\Phi_{F,n} \leq m_n \Phi_{\Sigma,n}$. Since $\lim_{n\to\infty} m_n^{1/n} = 1$, we conclude from Theorem 2.8(i) that

$$V_{F,Q} \leq V_{\Sigma,Q}.$$

The reverse inequality being obvious, we are done. □

Problem 3.3. Consider the sequence of normalized counting measures

$$\nu_n = \frac{1}{m_n} \sum_{j=1}^{m_n} \langle \xi_j^{(n)} \rangle$$

where $\langle\ \rangle$ denotes the Dirac δ-measure at the indicated point.

For $N = 1$, Theorem III.1.3 shows that the weak* limit of the sequence $\{\nu_n\}$ is $\mu_w = (1/2\pi) dd^c V_{\Sigma,Q}^*$. For $N > 1$ it is not known what measures are weak* limits of the sequence ν_n although a reasonable conjecture is that the sequence $\{\nu_n\}$ converges weakly to $(1/2\pi)^N (dd^c V_{\Sigma,Q}^*)^N$. Even in the unweighted case ($w \equiv 1$, $N > 1$) the problem is unsolved.

Problem 3.4. Define Fekete polynomials, for each integer $T \geq 1$ by

$$p_T(z) = \frac{V(\xi_1, \ldots, \xi_T, z)}{V(\xi_1, \ldots, \xi_T)}.$$

$p_T(z)$ is a polynomial of degree $|\alpha(T+1)|$. Is

$$\overline{\lim_{T\to\infty}} \frac{1}{|\alpha(T+1)|} \log \frac{|p_T(z)|}{\|w^{|\alpha(T+1)|} p_T\|_\Sigma} = V_{\Sigma,Q}^*(z) \quad \text{on} \quad \mathbf{C}^N \setminus \Sigma\ ?$$

For $N = 1$, using Lemma 2.4 this is Corollary III.1.10 (slightly modified).

B.4 Notes and Historical References

Section B.1

The definition of $(dd^c u)^N$ for u, a locally bounded plurisubharmonic function, is due to Bedford and Taylor ([8] also [101], Section 3.4). It is based on an earlier estimate due to Chern, Levine, and Nirenberg.

For $N > 1$ a "natural" domain for the operator $u \to (dd^c u)^N$ is not known, see [101], Section 3.8. Theorem 1.3 is due to Bedford and Taylor ([8], Theorem 9.1]). Theorem 1.6 is due to Siciak [206]. Theorem 1.11 is due to Bedford and Taylor ([10], Theorem 6.5]).

Section B.2

Lemma 2.3 is essentially the same as Proposition 9.3 in [10].

Theorem 2.8 (part (i) and w continuous) appears in [206].

The fact that the pluricomplex Green function can be represented as in Theorem 2.8(i), due, in this general case, to Siciak, is a crucial step in using pluripotential theory to obtain results on approximation by polynomials in several variables (see e.g. [18]).

Theorem 2.9 is in [208]. Similar (though not identical) results are in [206] and [207]. Theorem 2.12, in the unweighted case, is a consequence of Theorem 7.1 in [9].

Section B.3

Problem 3.3 in the unweighted case is stated in [207]. Fekete points, in the multivariable weighted setting, were defined in [206].

Interesting results relating Fekete points and Chebyshev constants in the multivariable case had previously been proven by Zaharjuta [237].

Leja points in the (unweighted) multivariable setting were defined in [85] and [19].

Basic Results of Potential Theory

Harnack's Principle	Theorem 0.4.10, p. 17
Helly's theorem	Section 0.1.3, p. 3
Isolated singularities of harmonic functions	Theorem 0.3.4, p. 11
Kellogg's theorem	Section IV.2, p. 210
Leja-Siciak extremal function	Section III.5, p. 177
Lower envelope theorem	Theorem I.6.9, p. 73
Maximum principle for harmonic functions	Theorem 0.2.6, p. 8
Maximum principle for logarithmic potentials	Corollary II.3.3, p. 104
Maximum principle for Green potentials	Corollary II.5.9, p. 131
Mean-value of potential	Theorem II.1.2, p. 84
Mean-value property for harmonic functions	Theorem 0.2.4, p. 7
Poisson's formula	Theorems 0.4.1, A.3.1, pp. 12, 459
Principle of descent	Theorem I.6.8, p. 70
Principle of domination	Theorem II.3.2, p. 104
Principle of domination for Green potentials	Theorem II.5.8, p. 130
Privaloff's theorem	Section IV.2, p. 210
Recovering a measure from its potential	Section II.1, p. 83
Regular points	Section I.4, Appendix A, pp. 54, 449
Riesz decomposition theorem	Theorem II.3.1, p. 100
Schwarz's theorem	Theorem 0.4.2, p. 13
Transfinite diameter	Section III.1, p. 142
Unicity theorem	Theorem II.2.1, p. 97
Unicity for Green potentials	Theorem II.5.3, p. 126
Uniqueness of measures	Lemma I.1.8, p. 29
Uniqueness for Green potentials	Theorem II.5.6, p. 129
Wiener's theorem	Theorem I.4.6, Appendix A, pp. 54, 449

Bibliography

[1] N. I. AKHIEZER: *Theory of Approximation*. Ungar, New York, 1956

[2] N. I. AKHIEZER: On the weighted approximation of continuous functions by polynomials on the entire real axis. *AMS transl., Ser. 2*, 22:95–137, 1962

[3] V. V. ANDRIEVSKII and H.-P. BLATT: A discrepancy theorem on quasiconformal curves. *Constr. Approx.*, 13:363–379

[4] V. V. ANDRIEVSKII and H.-P. BLATT: Erdős-Turán theorems on piecewise smooth curves and arcs. *J. Approx. Theory* (to appear)

[5] T. H. BAGBY: The modulus of a plane condenser. *J. Math. Mech.*, 17:315–329, 1967

[6] T. H. BAGBY: On interpolation by rational functions. *Duke Math. J.*, 36:95–104, 1969

[7] W. C. BAULDRY: Estimates of Christoffel functions of generalized Freud type weights. *J. Approx. Theory*, 46:217–229, 1986

[8] E. BEDFORD and B. A. TAYLOR: A new capacity for plurisubharmonic functions. *Acta Math.*, 149:1–40, 1982

[9] E. BEDFORD and B. A. TAYLOR: Fine topology, Silov boundary and $(dd^c)^n$. *J. Func. Anal.*, 72:225–251, 1987

[10] E. BEDFORD and B. A. TAYLOR: Plurisubharmonic functions with logarithmic singularities. *Ann. de l'Inst. Fourier (Grenoble)*, 38:133–171, 1988

[11] D. BENKŐ: Fast decreasing polynomials. Master's thesis, József Attila University, Szeged, Hungary, 1995

[12] A. S. BESICOVIC: A general theorem of the covering principle and relative differentiation of additive functions. *Proc. Cambridge Phil. Soc.*, 41:103–110, 1945

[13] D. BESSIS, C. ITZYKSON, and J. ZUBER: Quantum field theory techniques in graphical enumeration. *Adv. Appl. Math.*, 1:109–157, 1980

[14] E. BISHOP: A minimal boundary for function algebras. *Pacific J. Math.*, 9:629–642, 1959

[15] H.-P. BLATT: On the distribution of simple zeros of polynomials. *J. Approx. Theory*, 69:250–268, 1992

[16] H.-P. BLATT and H. N. MHASKAR: A general discrepancy theorem. *Ark. Math.*, 31:219–246, 1993

[17] H.-P. BLATT, E. B. SAFF, and M. SIMKANI: Jentzsch-Szegő type theorems for the zeros of best approximants. *J. London Math. Soc.*, 38:307–316, 1988

[18] T. BLOOM: On the convergence of multivariable Lagrange interpolants. *Constr. Approx.*, 5:415–435, 1989

[19] T. BLOOM, L. BOS, C. CHRISTENSEN, and N. LEVENBERG: Polynomial interpolation of holomorphic functions in **C** and $\mathbf{C}^n$. *Rocky Mountain J. Math.*, 22(2):441–470, 1992

[20] P. BORWEIN, E. A. RAKHMANOV, and E. B. SAFF: Rational approximation with varying weights I. *Constr. Approx.*, 12:223–240, 1996

[21] P. BORWEIN and E. B. SAFF: On the denseness of weighted incomplete approximations. In A. A. Gonchar and E. B. Saff, editors, *Progress in Approximation Theory*, pp. 419–429. Springer-Verlag, 1992

[22] G. BOULIGAND: Sur les fonctions bornées et harmoniques dans un domaine infini, nulles sur sa frontière. *C. R. Acad. Sci. Paris*, 169:763–766, 1919
[23] G. BOULIGAND: Domaine infinis et cas d'exception du problème de Dirichlet. *C. R. Acad. Sci. Paris*, 178:1054–1057, 1924
[24] A. BOUTET DE MONVEL, L. PASTUR, and M. SHCHERBINA: On the statistical mechanics approach in the random matrix theory: Integrated density of states. *J. Stat. Physics*, 79:585–611, 1995
[25] M. BRELOT: *Eléments de la Théorie Classique du Potentiel.* Les cours Sorbonne, Paris, 1959
[26] M. BRELOT: *Lectures on Potential Theory.* Tata Institute, Bombay, 1960
[27] E. BRÉZIN and A. ZEE: Universality of the correlations between eigenvalues of large random matrices. *Nucl. Phys.*, B402:613–627, 1993
[28] L. CARLESON: Mergelyan's theorem on uniform polynomial approximation. *Math. Scand.*, 15:167–175, 1964
[29] H. CARTAN: Théorie générale du balayage en potentiel newtonien. *Ann. Univ. Grenoble*, 22:221–280, 1946
[30] A. CORNEA: An identity theorem for logarithmic potentials. *Osaka J. Math.*, 28:829–836, 1991
[31] R. COURANT and D. HILBERT: *Methods of Mathematical Physics, Vol. I.* Interscience Publishers, Inc., New York, 1953
[32] CH. J. DE LA VALLÉE-POUSSIN: *Le Potentiel Logarithmique.* Gauthier-Villars, Paris, 1949
[33] P. DEIFT, T. KRIECHERBAUER, and K. T-R. MCLAUGHLIN: New results for the asymptotics of orthogonal polynomials and related problems via the inverse spectral method. *J. Approx. Theory* (to appear)
[34] J. DENY: Sur les infinis d'un potentiel. *C. R. Acad. Sci. Paris Sér. I. Math.*, 224:524–525, 1947
[35] R. A. DEVORE: *The Approximation of Continuous Functions by Positive Linear Operators*, volume 293 of *Lecture Notes in Math.*, Springer-Verlag, Berlin-Heidelberg, New York, 1972
[36] P. DRAGNEV and E. B. SAFF: Constrained energy problems with applications to orthogonal polynomials of a discrete variable. *J. d'Analyse Math.* (to appear)
[37] F. DYSON: Statistical theory of energy levels of complex systems I–III. *J. Math. Phys.*, 3:140–156, 157–165, 166–175, 1962
[38] F. DYSON: A class of matrix ensembles. *J. Math. Phys.*, 13:90–97, 1972
[39] P. ERDŐS: On the uniform distribution of the roots of certain polynomials. *Ann. Math.*, 43:59–64, 1942
[40] P. ERDŐS and P. TURÁN: On the uniformly dense distribution of certain sequences of points. *Ann. Math.*, 41:162–173, 1940
[41] P. ERDŐS and P. TURÁN: On the distribution of roots of polynomials. *Ann. Math.*, 43:59–64, 1942
[42] G. C. EVANS: Potentials and positively infinite singularities of harmonic functions. *Monatsh. Math.*, 43:419–424, 1936
[43] L. FEJÉR: Über die Lage der Nullstellen von Polynomen, die aus Minimumforderungen gewisser Art entspringen. *Math. Ann.*, 85:41–48, 1922
[44] M. FEKETE: Über den transfiniten Durchmesser ebener Punktmengen. *Math. Z.*, 32:108–114, 1930
[45] G. FREUD: On an inequality of Markov type. *Soviet Math. Dokl.*, 12:570–573, 1971
[46] G. FREUD: On two polynomial inequalities, I. *Acta Math. Acad. Sci. Hungar.*, 22:109–116, 1971
[47] G. FREUD: On the converse theorems of weighted polynomial approximation. *Acta Math. Acad. Sci. Hungar.*, 24:363–371, 1973

[48] G. FREUD: On weighted L_1-approximation by polynomials. *Studia Math.*, 46:125–133, 1973
[49] G. FREUD: On the coefficients in the recursion formulae of orthogonal polynomials. *Proc. Roy. Irish Acad. Sect. A(1)*, 76:1–6, 1976
[50] G. FREUD: On the zeros of orthogonal polynomials with respect to measures with noncompact support. *Anal. Numér. Théor. Approx.*, 6:125–131, 1977
[51] O. FROSTMAN: Potentiel d'équilibre et capacité des ensembles avec quelques applications à la théorie des fonctions, Thesis. *Meddel. Lunds Univ. Mat. Sem.*, 3:1–118, 1935
[52] O. FROSTMAN: La méthode de variation de Gauss et les fonctions sousharmoniques. *Acta Sci. Math.*, 8:149–159, 1936–37
[53] T. GANELIUS: Rational approximation in the complex plane and on the line. *Ann. Acad. Sci. Fenn.*, 2:129–145, 1976
[54] T. GANELIUS: Rational approximation to x^α on [0, 1]. *Anal. Math.*, 5:19–33, 1979
[55] J. B. GARNETT: *Analytic Capacity and Measure.* Springer-Verlag, Berlin, 1970
[56] J. B. GARNETT: *Bounded Analytic Functions.* Academic Press, New York, 1981
[57] C. F. GAUSS: Allgemeine Lehrsätze. Werke, 5. p. 232
[58] G. M. L. GLADWELL: *Contact Problems in the Classical Theory of Elasticity.* Sijthoff & Noordhoff, Alphen aan den Rijn, The Netherlands, 1980
[59] M. v. GOLITSCHEK: Approximation by incomplete polynomials. *J. Approx. Theory*, 28:155–160, 1980
[60] M. v. GOLITSCHEK, G. G. LORENTZ, and Y. MAKOVOZ: Asymptotics of weighted polynomials. In A. A. Gonchar and E. B. Saff, editors, *Progress in Approximation Theory*, pages 431–451. Springer-Verlag, New York, 1992
[61] M. GOLUZIN: *Geometric Theory of Functions of a Complex Variable*, volume 26 of *Translations of Mathematical Monographs.* Amer. Math. Soc., Providence, R. I., 1969
[62] A. A. GONCHAR: On the speed of rational approximation of some analytic functions. English transl.: *Math. USSR-Sb.*, 34:131–145, 1978
[63] A. A. GONCHAR and E. A. RAKHMANOV: Equilibrium measure and the distribution of zeros of extremal polynomials. *Mat. Sb.*, 125(167):117–127, 1984
[64] A. A. GONCHAR and E. A. RAKHMANOV: On the equilibrium problem for vector potentials. *Uspekhi Mat. Nauk*, 40(4(244)):155–156, 1985. English transl.: *Russian Math. Surveys* 40(1985)
[65] J. GÓRSKI: Méthode des points extrémaux de résolution du problème de Dirichlet dans l'espace. *Ann. Math. Polon.*, 1:418–429, 1950
[66] J. GÓRSKI: Remarque sur le diamètre transfini des ensembles plans. *Ann. Soc. Math. Polon.*, 23:90–94, 1950
[67] J. GÓRSKI: Sur certaines fonctions harmoniques jouissant des propriétés extrémales par rapport à un ensemble. *Ann. Soc. Math. Polon.*, 23:259–271, 1950
[68] J. GÓRSKI: Sur un problème de F. Leja. *Ann. Soc. Math. Polon.*, 25:273–278, 1952
[69] J. GÓRSKI: Sur certaines propriétés de points extrémaux liés à un domaine plan. *Ann. Polon. Math.*, 3:32–36, 1957
[70] J. GÓRSKI: Sur la représentation conforme d'un domaine multiplement connexe. *Ann. Polon. Math.*, 3:218–224, 1957
[71] J. GÓRSKI: Distributions restreintes des points extrémaux liés aux ensembles dans l'espace. *Ann. Math. Polon.*, 4:325–339, 1957-1958
[72] J. GÓRSKI: Les suites de points extrémaux liés aux ensembles dans l'espace à 3 dimensions. *Ann. Math. Polon.*, 4:14–20, 1957-1958
[73] J. GÓRSKI: Solution of some boundary-value problems by the method of F. Leja. *Ann. Polon. Math.*, 8:249–257, 1960
[74] J. GÓRSKI: Application of the extremal points method to some variational problems in the theory of schlicht functions. *Ann. Math. Polon.*, 17:141–145, 1965

[75] W. K. HAYMAN and P. B. KENNEDY: *Subharmonic Functions 1*, volume 9 of *London Math. Soc. Monographs*. Academic Press, London, 1976

[76] X. HE and X. LI: Uniform convergence of polynomials associated with varying weights. *Rocky Mountain J. Math.*, 21:281–300, 1991

[77] L. L. HELMS: *Introduction to Potential Theory*. Wiley–Interscience, New York, 1969

[78] P. HENRICI: *Applied and Computational Complex Analysis, Vol. 2*. John Wiley & Sons, New York, 1977

[79] P. HENRICI: *Applied and Computational Complex Analysis, Vol. 3*. John Wiley & Sons, New York, 1986

[80] E. HILLE: *Analytic Function Theory II*. Ginn and Company, Boston, 1962

[81] I. I. HIRSCHMAN: *The Decomposition of Walsh and Fourier series*, volume 15 of *Memoires of the Amer. Math. Soc.,*. American Mathematical Society, Providence, R. I., 1955

[82] K. HOFFMAN: *Banach Spaces of Analytic Functions*. Prentice-Hall, London, 1962

[83] K. G. IVANOV, E. B. SAFF, and V. TOTIK: Approximation by polynomials with locally geometric rate. *Proc. Amer. Math. Soc.*, 106:153–161, 1989

[84] K. G. IVANOV and V. TOTIK: Fast decreasing polynomials. *Constr. Approx.*, 6:1–20, 1990

[85] M. JEDRZEJOWSKI: Transfinite diameter and extremal points for a compact subset of $\mathbf{C}^n$. *Univ. Iagel. Acta Math.*, 29:65–70, 1992

[86] R. JENTZSCH: *Untersuchungen zur Theorie Analytischer Funktionen*. Inaugural-dissertation, Berlin, 1914

[87] K. JOHANSSON: On fluctuations of eigenvalues of random hermitian matrices. (to appear)

[88] S. KAMETANI: Positive definite integral quadratic forms and generalized potentials. *Proc. Imp. Acad. Japan*, 20:7–14, 1944

[89] O. D. KELLOGG: Recent progress with Dirichlet problem. *Bull. Amer. Math. Soc.*, 32:601–625, 1926

[90] O. D. KELLOGG: *Foundations of Potential Theory*, volume 31 of *Grundlehren der mathematischen Wissenschaften*. Springer-Verlag, Berlin, 1967

[91] W. KLEINER: Démonstration du théorème de Osgood-Carathéodory par la méthode des points extrémaux. *Ann. Polon. Math.*, 2:67–72, 1955

[92] W. KLEINER: Démonstration du théorème de Carathéodory par la méthode des points extrémaux. *Ann. Polon. Math.*, 11:217–224, 1962

[93] W. KLEINER: Sur l'approximation du diamètre transfini. *Ann. Polon. Math.*, 12:171–173, 1962

[94] W. KLEINER: Sur la condensation de masses. *Ann. Polon. Math.*, 15:85–90, 1964

[95] W. KLEINER: Sur la détermination numérique des points extrémaux de Fekete-Leja. *Ann. Polon. Math.*, 15:91–96, 1964

[96] W. KLEINER: Sur l'approximation de la représentation conforme par la méthode des points extrémaux de F. Leja. *Ann. Polon. Math.*, 14:131–140, 1964

[97] W. KLEINER: Sur les approximations de F. Leja dans le problème plan de Dirichlet. *Ann. Polon. Math.*, 15:203–209, 1964

[98] W. KLEINER: Une condition de Dini-Lipschitz dans la théorie du potentiel. *Ann. Polon. Math.*, 14:117–130, 1964

[99] W. KLEINER: Une variante de la méthode de F. Leja pour l'approximation de la représentation conforme. *Ann. Polon. Math.*, 15, 1964

[100] W. KLEINER: A variant of Leja's approximations in Dirichlet's plane problem. *Ann. Polon. Math.*, 16:201–211, 1965

[101] M. KLIMEK: *Pluripotential Theory*. Oxford University Press, 1991

[102] H. KLOKE: On the capacity of a plane condenser and conformal mapping. *J. Reine Angew. Math.*, 358:179–201, 1985

[103] A. KNOPFMACHER, D. S. LUBINSKY, and P. NEVAI: Freud's conjecture and approximation of reciprocals of weights by polynomials. *Constr. Approx.*, 4:9–20, 1988
[104] A. B. J. KUIJLAARS: A note on weighted polynomial approximation with varying weights. *J. Approx. Theory*, 87:112–115, 1996
[105] A. B. J. KUIJLAARS: The role of the endpoint in weighted polynomial approximation with varying weights. *Constr. Approx.*, 12:287–301, 1996
[106] A. B. J. KUIJLAARS: Weighted approximation with varying weights: the case of a power-type singularity. *J. Math. Anal. & Appl.*, 204:409–418, 1996
[107] A. B. J. KUIJLAARS and W. VAN ASSCHE: A contact problem in elasticity related to weighted polynomials on the real line. In Proceedings of Conference on Functional Analysis and Approximation Theory, Maratea, Italy, 1996 (to appear)
[108] A. B. J. KUIJLAARS and W. VAN ASSCHE: A problem of Totik on fast decreasing polynomials. *J. Approx. Theory* (to appear)
[109] A. B. J. KUIJLAARS and P. D. DRAGNEV: Equilibrium problems associated with fast decreasing polynomials. (to appear)
[110] M. LACHANCE, E. B. SAFF, and R. S. VARGA: Bounds for incomplete polynomials vanishing at both endpoints of an interval. In C. V. Coffman and G. J. Fix, editors, *Constructive Approaches to Mathematical Models*, pages 421–437. Academic Press, New York, 1979
[111] N. S. LANDKOF: *Foundations of Modern Potential Theory*. Grundlehren der mathematischen Wissenschaften. Springer-Verlag, Berlin, 1972
[112] F. LEJA: Sur les suites de polynômes bornés presque partout sur la frontière d'un domaine. *Math. Ann.*, 108:517–524, 1933
[113] F. LEJA: Sur les suites de polynômes, les ensembles fermés et la fonction de Green. *Ann. Soc. Math. Polon.*, 12:57–71, 1934
[114] F. LEJA: Sur une famille de fonctions harmoniques dans le plan liées à une fonction donnée sur la frontière d'un domaine. *Bull. Internat. Acad. Polon. Sci. Lettres, S''ere A*, pages 79–92, 1936
[115] F. LEJA: Sur les polynômes de Tchebycheff et la fonction de Green. *Ann. Soc. Math. Polon.*, 19:1–6, 1946
[116] F. LEJA: Une condition de régularité et d'irrégularité des frontières dans le problème de Dirichlet. *Ann. Soc. Math. Polon.*, 20:223–228, 1947
[117] F. LEJA: Une méthode d'approximation des fonctions réelles d'une variable complexe par des fonctions harmoniques. *Rendi. Acad. Nazionale Lincei*, 8:292–302, 1950
[118] F. LEJA: Une méthode élémentaire de résolution du problème de Dirichlet dans le plan. *Ann. Soc. Math. Polon.*, 23:230–245, 1950
[119] F. LEJA: Sur une famille de fonctions analytiques extrémales. *Ann. Soc. Math. Polon.*, 25:1–16, 1952
[120] F. LEJA: Polynômes extrémaux et la représentation conforme des domaines doublement connexes. *Ann. Polon. Math.*, 1:13–28, 1955
[121] F. LEJA: Distributions libres et restreintes de points extrémaux dans les ensembles plans. *Ann. Polon. Math.*, 3:147–156, 1957
[122] F. LEJA: Propriétes des points extrémaux des ensembles plans et leur application à la représentation conforme. *Ann. Polon. Math.*, 3:319–342, 1957
[123] F. LEJA: *Teoria Funkeji Analityeznych*. Akad., Warszawa, 1957. (In Polish)
[124] F. LEJA: Sur certaines suites liées aux ensembles plans et leur application à la représentation conforme. *Ann. Polon. Math.*, 4:8–13, 1957-1958
[125] F. LEJA: Sur les moyennes arithmétiques, géométriques et harmoniques des distances mutuelles des points d'un ensemble. *Ann. Polon. Math.*, 9:211–218, 1961
[126] A. LEVENBERG and L. REICHEL: A generalized ADI iterative method . *Numer. Math.*, 66:215–233, 1993
[127] A. L. LEVIN and D. S. LUBINSKY: Canonical products and the weights $\exp(-|x|^{\alpha})$, $\alpha > 1$, with applications. *J. Approx. Theory*, 49:149–169, 1987

[128] A. L. LEVIN and D. S. LUBINSKY: Weights on the real line that admit good relative polynomial approximation, with applications. *J. Approx. Theory*, 49:170–195, 1987
[129] A. L. LEVIN and D. S. LUBINSKY: Christoffel Functions and Orthogonal Polynomials and Nevai's conjecture for Freud Weights. *Constr. Approx.*, 24:463–535, 1993
[130] A. L. LEVIN and D. S. LUBINSKY: *Christoffel Functions and Orthogonal Polynomials for Exponential Weights*, volume 111/535 of *Memoirs Amer. Math. Soc.* Amer. Math. Soc., Providence, R. I., 1994
[131] A. L. LEVIN and D. S. LUBINSKY: L_p Markov-Bernstein inequalities for Freud weights. *J. Approx. Theory*, 77:229–248, 1994
[132] A. L. LEVIN and D. S. LUBINSKY: Orthogonal polynomials and Christoffel functions for $\exp(-|x|^\alpha)$, $\alpha \leq 1$. *J. Approx. Theory*, 80:219–252, 1995
[133] A. L. LEVIN, D. S. LUBINSKY, and T. Z. MTHEMBU: Christoffel functions and orthogonal polynomials for Erdős weights . *Rendiconti Math.*, 14:199–289, 1994
[134] A. L. LEVIN and E. B. SAFF: Optimal ray sequences of rational functions connected with the Zolotarev problem. *Constr. Approx.*, 10:235–273, 1994. Erratum: 12:437–438, 1996
[135] A. L. LEVIN and E. B. SAFF: Fast decreasing rational functions. (to appear)
[136] G. G. LORENTZ: Approximation by incomplete polynomials (problems and results). In E. B. Saff and R. S. Varga, editors, *Padé and Rational Approximations: Theory and Applications*, pages 289–302. Academic Press, New York, 1977
[137] D. S. LUBINSKY: Gaussian quadrature, weights on the whole real line, and even entire functions with nonnegative order derivatives. *J. Approx. Theory*, 46:297–313, 1986
[138] D. S. LUBINSKY: A survey of general orthogonal polynomials for weights on finite and infinite intervals. *Acta Appl. Math.*, 10:237–296, 1987
[139] D. S. LUBINSKY: Even entire functions absolutely monotone in $[0, \infty)$ and weights on the whole real line. In C. Brezinski et al., editor, *Orthogonal Polynomials and Their Applications*. Springer-Verlag, Berlin, 1988
[140] D. S. LUBINSKY: *Strong Asymptotics for Extremal Errors and Polynomials Associated with Erdős–type Weights*, volume 202 of *Pitman Research Notes*. Longman House, Harlow, UK, 1988
[141] D. S. LUBINSKY, H. N. MHASKAR, and E. B. SAFF: Freud's conjecture for exponential weights. *Bull. Amer. Math. Soc.*, 15:217–221, 1986
[142] D. S. LUBINSKY and E. B. SAFF: *Strong Asymptotics for Extremal Polynomials Associated with Weights on* **R**, volume 1305 of *Lecture Notes in Math.* Springer-Verlag, New York, 1988
[143] D. S. LUBINSKY and E. B. SAFF: Uniform and mean approximation by certain weighted polynomials, with applications. *Constr. Approx.*, 4:21–64, 1988
[144] D. S. LUBINSKY and E. B. SAFF: Markov-Bernstein and Nikolskii inequalities, and Christoffel functions for exponential weights on $[-1, 1]$. *SIAM J. Math. Anal.*, 24:528–556, 1993
[145] D. S. LUBINSKY and V. TOTIK: How to discretize a logarithmic potential? *Acta. Sci. Math.(Szeged)*, 57:419–428, 1994
[146] D. S. LUBINSKY and V. TOTIK: Weighted polynomial approximation with Freud weights. *Constr. Approx.*, 10:301–315, 1994
[147] AL. MAGNUS: A proof of Freud's conjecture about orthogonal polynomials related to $|x|^p \exp(-x^{2m})$. In C. Brezinski et al., editor, *Orthogonal Polynomials and Their Applications*, Lecture Notes in Math., pages 362–372. Springer-Verlag, Heidelberg, 1985
[148] AL. MAGNUS: On Freud's equations for exponential weights. *J. Approx. Theory*, 46:65–99, 1986
[149] A. J. MARIA: The potential of positive mass and the weight function of Wiener. *Proc. Nat. Acad. Sci. USA*, 20:485–489, 1934

[150] C. MARKETT: Nikolskiĭ-type inequalities for Laguerre and Hermite expansion. In J. Szabados, editor, *Functions, Series, Operators*, pages 811–834. Colloquia Math. Soc. J. Bolyai, North-Holland, Amsterdam, 1984

[151] A. I. MARKUSHEVICH: *Theory of Functions of a Complex Variable, Vol. III*. Prentice-Hall, Englewood Cliffs, NJ, 1967

[152] MADAN LAL MEHTA: *Random Matrices*. Academic Press, Inc., Boston, second edition, 1991

[153] K. MENKE: Über die Verteilung von gewissen Punktsystemen mit Extremaleigenschaften. *J. Reine Angew. Math.*, 283/284:421–435, 1976

[154] K. MENKE: On Tsuji points in a continuum. *Complex Variables*, 2:165–175, 1983

[155] H. N. MHASKAR: Weighted polynomial approximation. *J. Approx. Theory*, 46:100–110, 1986

[156] H. N. MHASKAR and C. MICCHELLI: On the n-width for weighted approximation of entire functions. In L. L. Schumaker C. K. Chui and J. D. Ward, editors, *Approximation Theory VI*, pages 429–432. Academic Press, San Diego, 1989

[157] H. N. MHASKAR and E. B. SAFF: Extremal problems for polynomials with exponential weights. *Trans. Amer. Math. Soc.*, 285:204–234, 1984

[158] H. N. MHASKAR and E. B. SAFF: Polynomials with Laguerre weights in L^p. In P. R. Graves-Morris, E. B. Saff, and R. S. Varga, editors, *Rational Approximation and Interpolation*, volume 1105, pages 511–523. Springer-Verlag, Berlin, 1984

[159] H. N. MHASKAR and E. B. SAFF: Weighted polynomials on finite and infinite intervals: A unified approach. *Bull. Amer. Math. Soc.*, 11:351–354, 1984

[160] H. N. MHASKAR and E. B. SAFF: A Weierstrass-type theorem for certain weighted polynomials. In S. P. Singh, editor, *Approximation Theory and Applications*, pages 115–123. Pitman Publishing Ltd., Harlow, 1985

[161] H. N. MHASKAR and E. B. SAFF: Where does the sup norm of a weighted polynomial live? (A generalization of incomplete polynomials). *Constr. Approx.*, 1:71–91, 1985

[162] H. N. MHASKAR and E. B. SAFF: The distribution of zeros of asymptotically extremal polynomials. *J. Approx. Theory*, 65:279–300, 1991

[163] H. N. MHASKAR and E. B. SAFF: Weighted analogues of capacity, transfinite diameter and Chebyshev constant. *Constr. Approx.*, 8:105–124, 1992

[164] D. S. MOAK, E. B. SAFF, and R. S. VARGA: On the zeros of Jacobi polynomials $P_n^{(\alpha_n,\beta_n)}(x)$. *Trans. Amer. Math. Soc.*, 249:159–162, 1979

[165] N. I. MUSKHELISHVILI: *Singular Integral Equations*. P. Noordhoff N. V., Groningen, 1953

[166] Z. NEHARI: *Conformal Mapping*. McGraw-Hill, New York, 1952

[167] P. NEVAI: Orthogonal polynomials on the real line associated with the weight $|x|^\alpha \exp(-|x|^\beta)$ I. *Acta Math. Acad. Sci. Hungar.*, 24:335–342, 1973. (In Russian)

[168] P. NEVAI: *Orthogonal Polynomials*, volume 213 of *Memoirs Amer. Math. Soc.* Amer. Math. Soc., Providence, RI, 1979

[169] P. NEVAI: Géza Freud, orthogonal polynomials and Christoffel functions. A case study. *J. Approx. Theory*, 48:3–167, 1986

[170] P. NEVAI and J. S. DEHESA: On asymptotic average properties of zeros of orthogonal polynomials. *SIAM J. Math. Anal*, 10:1184–1192, 1979

[171] P. NEVAI and V. TOTIK: Weighted polynomial inequalities. *Constr. Approx.*, 2:113–127, 1986

[172] P. NEVAI and V. TOTIK: Sharp Nikolskii inequalities with exponential weights. *Anal. Math.*, 13:261–267, 1987

[173] N. NINOMIYA: Méthode de variation du minimum dans la théorie du potentiel. *Sem. Théory du potentiel*, 5:9 pp., 1958–59

[174] N. NINOMIYA: Sur le principe du maximum et le balayage. *Jap. J. Math.*, 29:68–77, 1959

[175] M. Ohtsuka: On potentials in locally compact spaces. *J. Sci. Hiroshima Univ.i, Ser. A-I Math.*, 25:135–352, 1961
[176] L. Pastur and A. Figotin: *Spectra of Random and Almost-Periodic Operators*, volume 297 of *Grundlehren der mathematischen Wissenschaften*. Springer-Verlag, Berlin, 1992
[177] L.A. Pastur: Spectral and probabalistic aspects of matrix models. (to appear)
[178] L.A. Pastur: On the universality of the level spacing distribution for some ensembles of random matrices. *Letters in Math. Physics*, 25:259–265, 1992
[179] F. Peherstorfer: Minimal polynomials for compact sets of the complex plane. *Constr. Approx.*, 12:481–488, 1996
[180] A. Pinkus: *n-widths in Approximation Theory*. Springer-Verlag, New York, 1985
[181] Ch. Pommerenke: On the hyperbolic capacity and conformal mapping. *Proc. Amer. Math. Soc.*, 14:941–947, 1963
[182] Ch. Pommerenke: Polynome und konforme Abbildung. *Monatsh. Math.*, 69:58–61, 1965
[183] Ch. Pommerenke: On the logarithmic capacity and conformal mapping. *Duke Math J.*, 35:321–326, 1968
[184] Ch. Pommerenke: *Univalent Functions*. Vandenhoeck & Ruprecht, Göttingen, 1975
[185] I. E. Pritsker: Polynomial approximation with varying weights on compact sets of the complex plane. *Proc. Amer. Math. Soc.* (to appear)
[186] I. E. Pritsker and R. S. Varga: Weighted polynomial approximation in the complex plane. *Constr. Approx.* (to appear)
[187] I. E. Pritsker and R. S. Varga: The Szegő curve, zero distribution and weighted approximation. *Trans. Amer. Math. Soc.* (to appear)
[188] I. E. Pritsker and R. S. Varga: Weighted rational approximation in the complex plane. (manuscript)
[189] E. A. Rakhmanov: On asymptotic properties of polynomials orthogonal on the real axis. *Mat. Sb.*, 119(161):163–203, 1982. English transl.: *Math. USSR-Sb.*, 47:155-193, 1984
[190] E. A. Rakhmanov: Equilibrium measure and zero distribution of extremal polynomials of a discrete variable. *Mat. Sb.*, 187(8):109–124, 1996. (in Russian)
[191] E. A. Rakhmanov, E. B. Saff, and P. C. Simeonov: Rational approximation with varying weights II. *J. Approx. Theory* (to appear)
[192] T. Ransford: *Potential Theory in the Complex Plane*. Cambridge University Press, Cambridge, 1995
[193] F. Riesz: Sur les fonctions subharmoniques et leur rapport à la théorie du potentiel. I. *Acta Math.*, 48:329–343, 1930. II. *Acta Math.*, 59:321-360, 1930
[194] W. Rudin: *Principles of Mathematical Analysis*. McGraw-Hill, New York, 3rd edition, 1964
[195] W. Rudin: *Real and Complex Analysis*. McGraw-Hill, New York, 2nd edition, 1974
[196] E. B. Saff: Incomplete and orthogonal polynomials. In C. K. Chui, L. L. Schumaker, and J. D. Ward, editors, *Approximation Theory IV*, pages 219–255. Academic Press, New York, 1983
[197] E. B. Saff, J. L. Ullman, and R. S. Varga: Incomplete polynomials: An electrostatics approach. In E. W. Cheney, editor, *Approximation Theory II*, pages 769–782. Academic Press, San Diego, 1980
[198] E. B. Saff and R. S. Varga: On incomplete polynomials. In L. Collatz, G. Meinardus, and H. Werner, editors, *Numerische Methoden der Approximationstheorie, ISNM*, volume 42, pages 281–298. Birkhäuser-Verlag, Basel, 1978
[199] E. B. Saff and R. S. Varga: Uniform approximation by incomplete polynomials. *Internat. J. Math. Math. Sci.*, 1:407–420, 1978
[200] E. B. Saff and R. S. Varga: The sharpness of Lorentz's theorem on incomplete polynomials. *Trans. Amer. Math. Soc.*, 249:163–186, 1979

[201] J. SICIAK: Sur la distribution des points extrémaux dans les ensembles plans. *Ann. Math. Polon*, 4:214–219, 1957-1958

[202] J. SICIAK: On some extremal functions and their applications in the theory of functions of several complex variables. *Trans. Amer. Math. Soc.*, 105:322–357, 1962

[203] J. SICIAK: Some applications of the method of extremal points. *Colloq. Math.*, XI:209–249, 1964

[204] J. SICIAK: Asymptotic behaviour of harmonic polynomials bounded on a compact set. *Ann. Math. Polon.*, 20:267–278, 1968

[205] J. SICIAK: Degree of convergence of some sequences in the conformal mapping theory. *Colloq. Math.*, 16:49–59, 1976

[206] J. SICIAK: Extremal plurisubharmonic functions in $\mathbf{C}^n$. *Ann. Pol. Math.*, 39:175–211, 1981

[207] J. SICIAK: Extremal plurisubharmonic functions and capacities in $\mathbf{C}^n$. *Sophia Kokyusoku in Mathematics*, 14, 1982

[208] J. SICIAK: A remark on Tchebysheff polynomials in $\mathbf{C}^n$. Preprint, Jagellionian University, 1996

[209] P. SIMEONOV: *Weighted Polynomial and Rational Approximation with Varying Weights*. Ph.D. thesis, University of South Florida, Tampa, FL, 1997

[210] H. STAHL: Beiträge zum Problem der Konvergenz von Padé approximierenden. Technical report, Dissertation, TU–Berlin, 1976

[211] H. STAHL: A note on a theorem of H. N. Mhaskar and E. B. Saff. In E.B. Saff, editor, *Approximation Theory, Tampa*, volume 1287 of *Lecture Notes in Math.*, pages 176–179. Springer-Verlag, Berlin, 1987

[212] H. STAHL and V. TOTIK: *General Orthogonal Polynomials*, volume 43 of *Encyclopedia of Mathematics*. Cambridge University Press, New York, 1992

[213] M. H. STONE: The generalized Weierstrass approximation theorem. *Math. Magazine*, 21:167–184, 237–254, 1948

[214] G. SZEGŐ: Über die Nullstellen von Polynomen, die in einem Kreise gleichmässig konvergieren. *Sitzungsber. Ber. Math. Ges.*, 21:59–64, 1922. (see also Gábor Szegő, *Collected Papers I*, R. Askey, editor, Birkhäuser Verlag, Basel 1982, 535–543)

[215] G. SZEGŐ: Bemerkungen zu einer Arbeit von Herrn M. Fekete: Über die Verteilung der Wurzeln bei gewissen algebraischen Gleichungen mit ganzzahligen Koeffizienten". *Math. Zeitscher*, 21:203–208, 1924. (see also Gábor Szegő, *Collected Papers I*, R. Askey, editor, Birkhäuser Verlag, Basel 1982, 637–643)

[216] G. SZEGŐ: *Orthogonal Polynomials*, volume 23 of *Colloquium Publications*. Amer. Math. Soc., Providence, R. I., 1975

[217] V. TOTIK: Approximation by algebraic polynomials. In E. W. Cheney, C. K. Chui, and L. L. Schumaker, editors, *Approximation Theory VII*, pages 227–249. Academic Press, San Diego, 1992

[218] V. TOTIK: Distribution of simple zeros of polynomials. *Acta Math.*, 170:1–28, 1993

[219] V. TOTIK: Fast decreasing polynomials via potentials. *J. D'Analyse Math.*, 62:131–154, 1994

[220] V. TOTIK: *Weighted Approximation with Varying Weights*, volume 1300 of *Lecture Notes in Math.* Springer-Verlag, Berlin-Heidelberg-New York, 1994

[221] V. TOTIK and J. L. ULLMAN: Local asymptotic distribution of zeros of orthogonal polynomials. *Trans. Amer. Math. Soc.*, 62:131–154, 1994

[222] M. TSUJI: *Potential Theory in Modern Function Theory*. Maruzen, Tokyo, 1959

[223] J. L. ULLMAN: On the regular behaviour of orthogonal polynomials. *Proc. London Math. Soc.*, 24:119–148, 1972

[224] J. L. ULLMAN: Orthogonal polynomials associated with an infinite interval. *Michigan Math. J.*, 27:353–363, 1980

[225] J. L. ULLMAN: Orthogonal polynomials for general measures II. In C. Brezinski et al., editor, *Polynômes Orthogonaux et Applications; Proceedings, Bar-le-Duc 1984*, volume 1171 of *Lecture Notes in Math.*, pages 247–254. Springer-Verlag, New York, 1986

[226] J. L. ULLMAN and M. F. WYNEKEN: Weak limits of zeros of orthogonal polynomials. *Constr. Approx.*, 2:339–347, 1986

[227] J. L. ULLMAN, M. F. WYNEKEN, and L. ZIEGLER: Norm oscillatory weight measures. *J. Approx. Theory*, 46:204–212, 1986

[228] N. S. VJACESLAVOV: On the uniform approximation of $|x|$ by rational functions. *Dokl. Akad. Nauk SSSR*, 220:512–515, 1975. (In Russian). English transl.: *Soviet Math. Dokl.*, 16 (1975), 100-104

[229] J. L. WALSH: *Interpolation and Approximation by Rational Functions in the Complex Domain*, volume 20 of *Colloquium Publications*. Amer. Math. Soc., Providence, 1960

[230] H. WIDOM: Polynomials associated with measures in the complex plane. *J. Math. Mech.*, 16:997–1013, 1967

[231] H. WIDOM: An inequality for rational functions. *Proc. Amer. Math. Soc.*, 24:415–416, 1970

[232] H. WIDOM: Rational approximation and n-dimensional diameter. *J. Approx. Theory*, 5:343–361, 1972

[233] N. WIENER: Discountinuous boundary conditions and the Dirichlet problem. *Trans. Amer. Math. Soc.*, 25:307–314, 1923

[234] N. WIENER: Certain notions in potential theory. *J. Math. Phys.*, 3:24–51, 1924

[235] N. WIENER: The Dirichlet problem. *J. Math. Phys.*, 3:127–146, 1924

[236] E. P. WIGNER: On the statistical distribution of the widths and spacings of nuclear resonance levels. *Proc. Cambridge Phil. Soc.*, 47:790–795, 1951

[237] V. P. ZAHARJUTA: Transfinite diameter, Čebyšev constants, and capacity for compacta in $\mathbf{C}^n$. *Math. USSR-Sb.*, 25:350–364, 1975

[238] R. A. ZALIK: Inequalities for weighted polynomials. *J. Approx. Theory*, 37:137–146, 1983

[239] E. I. ZOLOTARJOV: *Collected Works, II.* USSR Acad. Sci, 1932. (In Russian)

[240] A. ZYGMUND: *Trigonometric Series*: Cambridge University Press, New York, 1959

List of Symbols

a_n	Mhaskar-Rakhmanov-Saff number, p. 204
a_n	weighted Leja points, p. 258
$B(x; \beta, \alpha)$	beta function, p. 344
$\mathbf{C}$	complex plane, p. 1
$\overline{\mathbf{C}}$	Riemann sphere, p. 1
$C(E, F)$	condenser capacity, pp. 132, 393
$C_0(\mathcal{D})$	continuous functions with compact support in $\mathcal{D}$, p. 101
$C_0^2(\mathcal{D})$	two-times continuously differentiable functions in $C_0(\mathcal{D})$, p. 101
$C_0(O)$	continuous functions vanishing outside O, p. 281
$\mathrm{cap}(\Sigma)$	logarithmic capacity, p. 25
$\mathrm{cap}(w, E)$	weighted capacity for w restricted to E, p. 64
χ_E	characteristic function of E, p. 64
$\chi_\varepsilon(z)$	characteristic function of disk $\|z\| \leq \varepsilon$, p. 98
c_μ	minimal carrier capacity, p. 373
$\mathbf{C}^N$	complex N-space, p. 466
$\mathrm{Con}(A)$	convex hull of A, p. 165
$C_r(z_0)$	circle $\|z - z_0\| = r$, p. 35
c_w	weighted capacity, p. 63
$c(w, \Sigma_1, \Sigma_2)$	weighted "signed" capacity for condenser (Σ_1, Σ_2), p. 396
Δf	Laplacian, p. 20
δ_n	$n(n-1)/2$-th root of maximum Vandermonde, p. 142
$\delta(\Sigma)$	transfinite diameter of Σ, p. 142
δ_t	unit mass at t, p. 112

δ_n^w	$n(n-1)/2$-th root of maximum weighted Vandermonde, p. 143
δ_w	weighted transfinite diameter, p. 144
$\mathrm{dist}\,(A, B)$	Hausdorff distance, p. 196
$d_n(Y, X)$	Kolmogorov n-width, p. 349
$D(r_1, r_2)$	ring $\{z \mid r_1 \le \|z\| \le r_2\}$, p. 421
$D'_R(z_0)$	punctured disk $0 < \|z - z_0\| < R$, p. 11
$D_r(z_1)$	open disk $\|z - z_1\| < r$, p. 8
$\mathcal{E}$	signed measures with finite Green energy, p. 128
$\mathcal{E}^+$	positive measures with finite Green energy, p. 128
$e_k(z)$	k-th monomial in several variables, p. 478
$E_{n,p}(w)$	$\inf\left\{\|\|wP_n\|\|_{L^p} \,\Big\|\, P_n \in \Pi_n\right\}$, p. 372
ε_j	sign of charge on Σ_j, p. 382
e_n^W	$\inf_{r_n} \|\|W^n(\chi - r_n)\|\|_\Sigma$, p. 418
$F(K)$	F-functional, p. 194
$\|\|f\|\|_K^*$	supremum on K ignoring zero capacity sets, p. 154
$\mathcal{F}_n$	n point Fekete set, p. 143
$\mathcal{F}_j^{(n)}$	(rational) Fekete set on Σ_j, p. 395
F_w	modified Robin constant ($= V_w - \int Q d\mu_w$), p. 27
F_w^c	Robin constant ($= F_1 + F_2$) for condenser, p. 418
$\|\|f\|\|_{X_E(w)}$	$\sup_{z\in E} \|f(z)w(z)\|$, p. 349
γ_λ	constant $\Gamma(\lambda/2)\Gamma(1/2)/2\Gamma((\lambda+1)/2)$, p. 239
$\gamma_n(\mu)$	leading coefficient of orthogonal polynomials, p. 360
$g_G(z, a)$	Green function of G with pole at a, pp. 53, 109
$H(\alpha, \beta)$	continuous weights on $\mathbf{R}$ satisfying $\lim_{\|x\|\to\infty} \dfrac{\log(1/w(x))}{x^\beta} = \alpha$, p. 344
$h * g(z)$	convolution, p. 98
$\mathcal{H}_f^{l,R}$	set of lower functions on R, p. 41
$\mathcal{H}_f^{u,R}$	set of upper functions on R, p. 41
H_f^R	Perron-Wiener-Brelot solution of Dirichlet problem, p. 41

$\overline{H}_f^R$, $\underline{H}_f^R$,	upper and lower solutions of Dirichlet problem on R, p. 41
$h_n(\alpha_1, \beta_1, \alpha_2, \beta_2)$	$\sup\left\{\|P_n w_1\|/\|P_n w_2\| \;\middle|\; \deg P_n \le n\right\}$, p. 343
$H_n(x)$	Hermite polynomials, p. 361
$I(\mu)$	energy integral, p. 24
$I_w(\mu)$	weighted energy integral, p. 26
$\text{ind}_\gamma(z)$	winding number of z with respect to γ, p. 422
$\text{“inf”}_{z\in H}$	infimum ignoring zero capacity sets, p. 43
$\log_\varepsilon(z)$	$\min\left(\log\frac{1}{\|z\|}, \log\frac{1}{\varepsilon}\right)$, p. 98
$L(U^\mu; z_0, r)$	mean value of U^μ over circle $C_r(z_0)$, p. 84
m	two-dimensional Lebesgue measure, p. 83
$\mathcal{M}$	signed measures of the form $\sum_{j=1}^N \varepsilon_j \sigma_j$, $\sigma_j \ge 0$, p. 382
$\mathcal{M}(\Sigma)$	unit Borel measures on Σ, p. 24
MAX U^ν	set of maximum points of U^ν, p. 374
$\mathcal{M}_\mu$	set of weak* limit measures of $\nu(p_n(\mu))$, p. 374
μ_+	$\sum_{\varepsilon_j=1} \mu_j$, p. 391
μ_-	$\sum_{\varepsilon_j=-1} \mu_j$, p. 391
$\mu\|_D$	measure μ restricted to D, p. 97
$\mu_\Sigma, \omega_\Sigma$	equilibrium distribution for Σ, p. 24
μ_E^G	Green equilibrium distribution for E, p. 132
μ_w	equilibrium distribution for weight w, p. 27
μ_w^G	Green equilibrium distribution for w, p. 132
$\|\|\mu\|\|$	total mass of measure μ, p. 38
$\|\|\mu\|\|_e^2$	Green energy $\iint g_G(z,\zeta)\,d\mu(z)\,d\mu(\zeta)$, p. 127
$(\mu, \nu)_e$	mutual Green energy $\iint g_G(z,\zeta)\,d\mu(z)\,d\nu(\zeta)$, p. 127
$\mu(w, E)$	equilibrium distribution for w restricted to E, p. 64

(ν, λ)	mutual logarithmic energy $\iint \log \frac{1}{\lvert z-t\rvert} d\nu(z) d\lambda(t)$, p. 387
$\hat{\nu}$	balayage of measure ν, p. 110
ν_H	normalized counting measure on H, p. 145
$\nu(P_n(\mu))$	normalized zero counting measure, p. 373
ω_K, μ_K	equilibrium distribution for K, p. 194
∂D	boundary of D, p. 17
$\partial_\infty E$	outer boundary of E, p. 68
$\partial/\partial \mathbf{n}$	normal derivative, p. 83
$\mathrm{Pc}(\Sigma)$	polynomial convex hull of Σ, p. 53
Φ_n^w, Φ_n	Fekete polynomial associated with w, p. 150
$\Phi(z)$	Leja-Siciak function, p. 177
$\Phi_\Sigma(z)$	Siciak extremal function for $\Sigma \subset \mathbf{C}^N$, p. 475
Π_n	set of monic polynomials of degree n, p. 364
$p_n(\mu; x)$	orthonormal polynomials with respect to $d\mu$, p. 360
$P(t, z)$	Poisson kernel, p. 13
q.e.	quasi-everywhere, p. 25
Q_w, Q	external field $\log(1/w)$ for weight w, p. 26
$\overline{R}$	closure of R, p. 83
R_w	$\{z \in \Sigma \mid (U^{\mu_w} + Q)(z) < F_w\}$, p. 157
Σ_+	$\bigcup_{\varepsilon_j=1} \Sigma_j$, p. 391
Σ_-	$\bigcup_{\varepsilon_j=-1} \Sigma_j$, p. 391
Σ^*	$\{z \in \Sigma \mid \Sigma$ has positive capacity in every neighborhood of $z\}$, p. 269
Σ_0	$\{z \in \Sigma \mid w(z) > 0\}$, p. 26
$\overline{\sigma}_j$	$\sum_{i \neq j} \varepsilon_i \sigma_i$, p. 388
$s_\lambda(t)$	Ullman distribution, p. 238
"$\subset$"	inclusion except for a set of zero capacity, p. 196
"$\sup_{z \in H}$"	supremum ignoring zero capacity sets, p. 43
$\mathrm{supp}(\mu)$	support of μ, p. 3

$\mathcal{S}_w$	support of μ_w, p. 27
$\mathcal{S}_w^*$	$\{z \in \Sigma \mid (U^{\mu_w} + Q)(z) \leq F_w\}$, p. 144
S^w	restricted support of μ_w, p. 281
τ_n	weighted Zolotarjov numbers, p. 398
t_w	Chebyshev constants, p. 163
t_n^w	Chebyshev numbers, p. 163
$\tilde{t}_w$	restricted Chebyshev constant, p. 163
$\tilde{t}_n^w$	restricted Chebyshev numbers, p. 163
T_n^w, T_n	Chebyshev polynomials, p. 163
U^μ	logarithmic potential, p. 21
U_G^ν	Green potential, p. 124
V	minimal logarithmic energy, p. 24
V_w	minimal weighted energy, p. 27
$V(x_1, x_2, \ldots, x_n)$	Vandermonde determinant, p. 142
$V(\zeta_1, \ldots, \zeta_T)$	Vandermonde determinant in several variables, p. 479
w	weight function, p. 26
$\overline{w}(z)$	$\lim_{\delta \to 0^+} \|w\|^*_{\overline{D_\delta(z)}}$, p. 154
$X_E(w)$	function space, p. 349

Index

Grundlehren der mathematischen Wissenschaften

A Series of Comprehensive Studies in Mathematics

A Selection

210. Gihman/Skorohod: The Theory of Stochastic Processes I
211. Comfort/Negrepontis: The Theory of Ultrafilters
212. Switzer: Algebraic Topology – Homotopy and Homology
215. Schaefer: Banach Lattices and Positive Operators
217. Stenström: Rings of Quotients
218. Gihman/Skorohod: The Theory of Stochastic Processes II
219. Duvaut/Lions: Inequalities in Mechanics and Physics
220. Kirillov: Elements of the Theory of Representations
221. Mumford: Algebraic Geometry I: Complex Projective Varieties
222. Lang: Introduction to Modular Forms
223. Bergh/Löfström: Interpolation Spaces. An Introduction
224. Gilbarg/Trudinger: Elliptic Partial Differential Equations of Second Order
225. Schütte: Proof Theory
226. Karoubi: K-Theory. An Introduction
227. Grauert/Remmert: Theorie der Steinschen Räume
228. Segal/Kunze: Integrals and Operators
229. Hasse: Number Theory
230. Klingenberg: Lectures on Closed Geodesics
231. Lang: Elliptic Curves. Diophantine Analysis
232. Gihman/Skorohod: The Theory of Stochastic Processes III
233. Stroock/Varadhan: Multidimensional Diffusion Processes
234. Aigner: Combinatorial Theory
235. Dynkin/Yushkevich: Controlled Markov Processes
236. Grauert/Remmert: Theory of Stein Spaces
237. Köthe: Topological Vector Spaces II
238. Graham/McGehee: Essays in Commutative Harmonic Analysis
239. Elliott: Probabilistic Number Theory I
240. Elliott: Probabilistic Number Theory II
241. Rudin: Function Theory in the Unit Ball of C^n
242. Huppert/Blackburn: Finite Groups II
243. Huppert/Blackburn: Finite Groups III
244. Kubert/Lang: Modular Units
245. Cornfeld/Fomin/Sinai: Ergodic Theory
246. Naimark/Stern: Theory of Group Representations
247. Suzuki: Group Theory I
248. Suzuki: Group Theory II
249. Chung: Lectures from Markov Processes to Brownian Motion
250. Arnold: Geometrical Methods in the Theory of Ordinary Differential Equations
251. Chow/Hale: Methods of Bifurcation Theory
252. Aubin: Nonlinear Analysis on Manifolds. Monge-Ampère Equations
253. Dwork: Lectures on ρ-adic Differential Equations
254. Freitag: Siegelsche Modulfunktionen
255. Lang: Complex Multiplication
256. Hörmander: The Analysis of Linear Partial Differential Operators I
257. Hörmander: The Analysis of Linear Partial Differential Operators II
258. Smoller: Shock Waves and Reaction-Diffusion Equations
259. Duren: Univalent Functions
260. Freidlin/Wentzell: Random Perturbations of Dynamical Systems
261. Bosch/Güntzer/Remmert: Non Archimedian Analysis – A System Approach to Rigid Analytic Geometry
262. Doob: Classical Potential Theory and Its Probabilistic Counterpart

263. Krasnosel'skiĭ/Zabreĭko: Geometrical Methods of Nonlinear Analysis
264. Aubin/Cellina: Differential Inclusions
265. Grauert/Remmert: Coherent Analytic Sheaves
266. de Rham: Differentiable Manifolds
267. Arbarello/Cornalba/Griffiths/Harris: Geometry of Algebraic Curves, Vol. I
268. Arbarello/Cornalba/Griffiths/Harris: Geometry of Algebraic Curves, Vol. II
269. Schapira: Microdifferential Systems in the Complex Domain
270. Scharlau: Quadratic and Hermitian Forms
271. Ellis: Entropy, Large Deviations, and Statistical Mechanics
272. Elliott: Arithmetic Functions and Integer Products
273. Nikol'skiĭ: Treatise on the Shift Operator
274. Hörmander: The Analysis of Linear Partial Differential Operators III
275. Hörmander: The Analysis of Linear Partial Differential Operators IV
276. Ligget: Interacting Particle Systems
277. Fulton/Lang: Riemann-Roch Algebra
278. Barr/Wells: Toposes, Triples and Theories
279. Bishop/Bridges: Constructive Analysis
280. Neukirch: Class Field Theory
281. Chandrasekharan: Elliptic Functions
282. Lelong/Gruman: Entire Functions of Several Complex Variables
283. Kodaira: Complex Manifolds and Deformation of Complex Structures
284. Finn: Equilibrium Capillary Surfaces
285. Burago/Zalgaller: Geometric Inequalities
286. Andrianaov: Quadratic Forms and Hecke Operators
287. Maskit: Kleinian Groups
288. Jacod/Shiryaev: Limit Theorems for Stochastic Processes
289. Manin: Gauge Field Theory and Complex Geometry
290. Conway/Sloane: Sphere Packings, Lattices and Groups
291. Hahn/O'Meara: The Classical Groups and K-Theory
292. Kashiwara/Schapira: Sheaves on Manifolds
293. Revuz/Yor: Continuous Martingales and Brownian Motion
294. Knus: Quadratic and Hermitian Forms over Rings
295. Dierkes/Hildebrandt/Küster/Wohlrab: Minimal Surfaces I
296. Dierkes/Hildebrandt/Küster/Wohlrab: Minimal Surfaces II
297. Pastur/Figotin: Spectra of Random and Almost-Periodic Operators
298. Berline/Getzler/Vergne: Heat Kernels and Dirac Operators
299. Pommerenke: Boundary Behaviour of Conformal Maps
300. Orlik/Terao: Arrangements of Hyperplanes
301. Loday: Cyclic Homology
302. Lange/Birkenhake: Complex Abelian Varieties
303. DeVore/Lorentz: Constructive Approximation
304. Lorentz/v. Golitschek/Makovoz: Construcitve Approximation. Advanced Problems
305. Hiriart-Urruty/Lemaréchal: Convex Analysis and Minimization Algorithms I. Fundamentals
306. Hiriart-Urruty/Lemaréchal: Convex Analysis and Minimization Algorithms II. Advanced Theory and Bundle Methods
307. Schwarz: Quantum Field Theory and Topology
308. Schwarz: Topology for Physicists
309. Adem/Milgram: Cohomology of Finite Groups
310. Giaquinta/Hildebrandt: Calculus of Variations I: The Lagrangian Formalism
311. Giaquinta/Hildebrandt: Calculus of Variations II: The Hamiltonian Formalism
312. Chung/Zhao: From Brownian Motion to Schrödinger's Equation
313. Malliavin: Stochastic Analysis
314. Adams/Hedberg: Function Spaces and Potential Theory
315. Bürgisser/Clausen/Shokrollahi: Algebraic Complexity Theory
316. Saff/Totik: Logarithmic Potentials with External Fields
317. Rockafellar/Wets: Variational Analysis
318. Kobayashi: Hyperbolic Complex Spaces

Springer and the environment

At Springer we firmly believe that an international science publisher has a special obligation to the environment, and our corporate policies consistently reflect this conviction.

We also expect our business partners – paper mills, printers, packaging manufacturers, etc. – to commit themselves to using materials and production processes that do not harm the environment. The paper in this book is made from low- or no-chlorine pulp and is acid free, in conformance with international standards for paper permanency.

Springer

Printing: Mercedesdruck, Berlin
Binding: Buchbinderei Lüderitz & Bauer, Berlin

Printed in the United Kingdom
by Lightning Source UK Ltd.
119601UK00006BB/27